ADVANCES IN MATERIALS SCIENCE AND TECHNOLOGY

The themes of the conference are related to applied materials like, structural materials, energy materials, electronic materials, biomaterials, computational materials science and sustainable materials. Globally the emphasis is on materials which can cater to reduction in weight of the system for structural applications, miniaturization of electronic devices, materials which improves the efficiency and performance for energy industry at the same time being environmentally sustainable, can be recycled/ reused with emphasis on cost reduction either through material or manufacturing/synthesis routes. Nano materials are continuously being explored to meet these needs. The conference theme deals with the application of modelling techniques to explore synthesis new materials, study their behaviour and predict failures and manufacturing. The proceedings hence covers application of materials in very important areas where new technology is developed for a sustainable development. The proceedings gives the glimpse of new materials synthesis, nanomaterials, property evaluation for various fields, usage of computational methods and AI/ML for materials science and engineering field. The conference is very relevant to the young engineers and science researchers, who are focusing their work in the materials sciences, with emphasis on AI & ML predictive models for synthesis and analysis

ADVANCES IN MATERIALS SCIENCE AND TECHNOLOGY

Proceedings of the 2ⁿᵈ International Conference on Advances in Materials Science & Technology (ICAMST-2024), July 29-31, 2024, Bangalore, India

Editor

Dr. Srikari Srinivasan
Dr. Suresh R
Dr. Rahul M Cadambi

CRC Press
Taylor & Francis Group
Boca Raton London New York

CRC Press is an imprint of the
Taylor & Francis Group, an **informa** business

First edition published 2026
by CRC Press
4 Park Square, Milton Park, Abingdon, Oxon, OX14 4RN

and by CRC Press
2385 NW Executive Center Drive, Suite 320, Boca Raton FL 33431

CRC Press is an imprint of Informa UK Limited

British Library Cataloguing-in-Publication Data
A catalogue record for this book is available from the British Library

ISBN: 978-1-041-12341-5 (hbk)
ISBN: 978-1-041-12342-2 (pbk)
ISBN: 978-1-003-66427-7 (ebk)

DOI: 10.1201/9781003664277

Typeset in Times LT Std
by Aditiinfosystems

Contents

List of Figures

List of Tables

Advances in Materials Science and Technology – Dr. Srikari Srinivasan et al. (eds)
© 2025 Taylor & Francis Group, London, ISBN 978-1-041-12342-2

About the RUAS

Ramaiah University of Applied Sciences (RUAS) a top private University in Bangalore, Karnataka, was founded in December 2013, under the Karnataka University Act. The creation of RUAS brought together several well-established educational institutions of the Ramaiah Group, reorienting them to a changing present and an unpredictable future. The academic programs of RUAS focus on student centric higher education. Students experience an integrated approach to academia, research, training, reallife problem solving and entrepreneurship. RUAS is a comprehensive University with 13 Faculties / Schools / Constituent Colleges. The University offers undergraduate, postgraduate and doctoral programs in Engineering and Technology, Mathematical and Physical Sciences, Art and Design, Management and Commerce, Hospitality Management and Catering Technology, Pharmacy, Dental Sciences, Life and Allied Health Sciences and Medical and Nursing Sciences. RUAS aspires to be the premier university of choice in Asia for student-centric professional education and service with a strong focus on applied research whilst maintaining the highest academic and ethical standards in a creative and innovative environment.

Advances in Materials Science and Technology – Dr. Srikari Srinivasan et al. (eds)
© 2025 Taylor & Francis Group, London, ISBN 978-1-041-12342-2

About the Conference

The themes of the conference are related to applied materials like, structural materials, energy materials, electronic materials, biomaterials, computational materials science and sustainable materials. Globally the emphasis is on materials which can cater to reduction in weight of the system for structural applications, miniaturization of electronic devices, materials which improves the efficiency and performance for energy industry at the same time being environmentally sustainable, can be recycled/ reused with emphasis on cost reduction either through material or manufacturing/synthesis routes. Nano materials are continuously being explored to meet these needs. The conference theme deals with the application of modelling techniques to explore synthesis new materials, study their behavior and predict failures and manufacturing. The proceedings hence covers application of materials in very important areas where new technology is developed for a sustainable development. The proceedings gives the glimpse of new materials synthesis, nanomaterials, property evaluation for various fields, usage of computational methods and AI/ML for materials science and engineering field. The conference is very relevant to the young engineers and science researchers, who are focusing their work in the materials sciences, with emphasis on AI & ML predictive models for synthesis and analysis.

Patorns

Dr. M. R. Jayaram, Honorable Chancellor, RUAS

Prof. Kuldeep Kumar Raina, Vice-Chancellor, RUAS

Advisory Committee

Prof. Kuldeep Kumar Raina, Vice Chancellor, RUAS

Dr. G. Venkatesh, Registrar, RUAS

Dr. P. S. Anil Kumar, IISc, Bangalore, India

Prof. Achintya Bezbaruah, North Dakota State University, USA

Prof. Suhasini Gururaja, Auburn University, USA

Prof. C. V. Ramana, Texas University, El Paso, USA

Dr. Mallikarjuna Nadagouda, USA

Dr. M. V. Reddy, New Graphite World, Quebec, Canada

Prof. B. Srinivasa Murthy, Director, IIT-Hyderabad, India

Dr. B. L. V. Prasad, Directo, CSIR-NCL, India

Prof. Giridhar U. Kulakarni, Director, JNCASR

Dr. Shakti Singh Chauhan, Director, IPIRTI

Dr. Manmeet Kaur, BARC India

Dr. Debadatta Ratna, NMRL, DRDO

Prof. Ramesh Chandra Mallik, IISc

Dr. Ramesh Kumar, NPOL, DRDO

Dr. Uday Kumar, CSIR-NAL

Dr. Manjunatha C. M., CSIR-NAL, India

Dr. Shankar M. Venugopal, Vice President, Mahindra & Mahindra, India

Prof. Indradev Samajdar, IIT-Bombay

Prof. James Raju, University of Hyderabad

Prof. S. B. Krupanidhi, IISc, Bangalore

Prof. Govind R. Kadambi, Pro-VC, RUAS, Bangalore

Key Note Speaker

Prof. Bikramjit Basu, Materials Research Centre, IISc, Bengaluru

Prof. Pikee Priya, Materials Engineering Department, IISc, Bengaluru

Dr. Harish C. Barshilia, Chief Scientist & Head, Surface Engineering Division, CSIR-NAL, Bengaluru

Dr. Andi Udaykumar, Chief Scientist, Materials Science Division, CSIR-NAL, Bengaluru

Dr. Manmeet Kaur, Head, Gas Sensing Devices Section, Technical Physics Division, BARC, Trombay

Dr. Debadatta Ratna, Scientist-G and Head, Directorate of Polymer Science and Technology, NMRL-DRDO, Ambernath

Dr. Kishora Shetty, Boeing, Bengaluru

Prof. Suhasini Guruaj, Auburn University, Alabama, USA

Prof. Katsuya Inoue, Hiroshima University, Japan

Members of the Programme Committee

Prof. J. V. Desai, RUAS

Dr. Srikari S., RUASDr. Rahul M. Cadambi., RUAS

Ÿ Dr. B. S. Dayananda, RUAS

Dr. Jayahar Sivasubramanian, RUAS

Dr. T. Niranjana Prabhu, RUAS

Dr. Premakumar H. B., RUAS

Dr. Nayana N. Patil, RUAS

Dr. B. S. Nandakumar, RUAS

Dr. Sheetal R. Batakurki, RUAS

Dr. Vibha A. Shetty, RUAS

Dr. Sivaranjani Gali, RUAS

Dr. Ashmita Shetty, RUAS

Dr. P. N. Anantharamaiah, RUAS

Dr. B. S. Ravikumar, RUAS

Dr. Suresh R., RUAS

Dr. Deepak Kumar, RUAS

Dr. K. J. Mallikarjunaiah, RUAS

Dr. Shivanand Madolappa, RUAS

Mr. Divakar L., RUAS

Mr. Eshwar Reddy H. N., RUAS

Mr. Parikshith N., RUAS

Subject Area

1. Structural Materials: Aerospace, Automotive, Mechanical, Civil Engineering

2. Bio-materials and dental materials related to Medical/ Dental/ Pharmacy

3. Energy Materials

4. Materials in Electronics and Sensors

5. Sustainable Materials and Technology

6. Computational Science for all the Materials

Keywords

Structural Materials; Energy materials; Materials in Electronics and Sensors; Biomaterials; Dental Materials; Sustainable Materials; Computational science in Materials

Editor's Biographical Information

1. Dr. Srikari S, working as Professor in Automotive and Aeronautical Engineering Department. Specialization in Ceramics and composites, coatings, biomaterials, selection of materials and manufacturing processes for aerospace, automotive applications. Working has Head of Bio-materials Research Center, previously worked has Associated Dean Research and Head of the Department. Had worked as PI and Co-PI for more than 8 sponsored projects from DST, AR&DB, GTRE, ICMR, DBT, BRNS and Faurecia Clean Mobility. Sponsored projects were aimed at development and studies on coatings for various applications like biomaterials, energy and sensors. Have conducted several corporate trainings on Selection of Materials and Manufacturing for Automotive and Aerospace Industries. Have supervised more than 30 PG students on material and manufacturing solutions for automotive and aerospace applications and 2 Ph.D. Scholars.

2. Dr Suresh R, working as Professor in the Department of Mechanical and Manufacturing Engineering, MS Ramaiah University of Applied Sciences, Bangalore, India. His passion for research runs deep into areas such as the processing of hard and refractory materials, traditional and non-traditional machining, surface Engineering, metal matrix composites, Joining Process, Additive Manufacturing and robotics and automation system. He has organized several national and international conferences and workshops in the field of material science and manufacturing. He has authored more than 150 peer reviewed international journals papers and received research grants from AICTE, DST and KCTU, GoK. He received "Outstanding Scientist Award" from the Centre for advanced Research and Design Chennai in 2015 and ''Annual Exemplary Faculty-Research 2020'' Award from RUAS in 2020. He is an official reviewer of original research articles in the field of materials and manufacturing. He is also a member of MISTE, SAE India and other Professional Societies, and is an active consultant to many reputed manufacturing industries and R&D organizations.

3. Dr. Rahul M Cadambi, working as Professor at RUAS. He now heads the Composites Materials and Technologies Research Centre (CMTRC) which he was instrumental in establishing in 2017 in collaboration with Kazan National Technical and Research University (KNRTU-KAI), Russia. He is the Centre Coordinator for Centre of Excellence in 'Computational Mechanics' for the Ramaiah Group which is mentored by Prof. J. N. Reddy, Texas A&M University. He was involved in incubating a private company, Valdel Advanced Technologies Pvt. Ltd. in Peenya campus. As an investigator, he is actively involved in various funded and consultancy projects of around Rs. 450 lakhs for various Defence organizations,

Armed Forces and other R&D laboratories. He coordinated to enlist RUAS as a nodal centre for BRDs (Indian Air Force). His 16-year-long career is marked by industrial, research and academic accomplishments. His research interests include polymer composites, synthesis of nanomaterials, elastomeric and polymer nanocomposites, damage tolerant design of composite materials and computational modeling of defects, fracture and fatigue processes in composites.

External Reviewers

Dr. Akshat Jain
Process Engineer III, Applied Materials, Santa Clara, CA, USA,
akshatj@alumni.upenn.edu

Dr. Trivikram R. Moluguool
Asst. Research Scientist, Texas A&M University, College Station, TX, USA
Email: mtvrao@gmail.com

Dr. Sajal Kumar Ghosh
Associate Professor, School of Natural Sciences, Shiv Nadar University,Delhi NCR
Email: sajal.ghosh@snu.edu.in

Dr. R. Y. RAVI KUMAR
Head, Production Engineering Section
National Institute of Technology Warangal (NITW), Warangal-506 004, Telangana State, INDIA. Email:yrk@nitw.ac.in

Dr. Manoj V Mane
Assistant Professor
Center for Nano and Material sciences (CNMS), Jain University, Bengaluru, Karnataka.
Email: manoj.mane@jainuniversity.ac.in

Dr. Srukari Shinivasan
Professor in Automotive and Aeronautical Engineering Department, MSRUAS, Bangalore
srikari.aae.et@msruas.ac.in

Prof. BM Nagabhushana
Dept of Chemistry, MS Ramaih Institute of Technology,
bmnshan@yahoo.com

Prof Gurumurthy Hegde
Dept of Chemistry, Christ University, Bengaluru
murthyhegde@gmail.com

Dr. Suresh R
Professor, Department of Mechanical Engineering, MSRUAS, Bangalore, India
Suresh.me.et@msruas.ac.in

Dr. Dibya Prakash Rai
Associate Professor
Department of Physics, Pachhunga University College, Mizoram University, Aizawl-796001
Email: dibya@pucollege.edu.in   & dibyaprakashrai@gmail.com

Dr. Niranjana Prabhu
Professor and Dean
Research & Innovation, Technology Campus, MSRUAS, Bangalore
Email: assodeanresearch.rtc@msruas.ac.in

Foreword

It gives me great pleasure to present the proceedings of this conference, which showcase cutting-edge research across diverse domains of applied materials—including structural, energy, electronic, biomaterials, computational, and sustainable materials. As global challenges demand lighter, more efficient, and environmentally responsible solutions, materials science plays a pivotal role in advancing cost-effective and sustainable innovations. The focus on recyclable materials, novel synthesis and manufacturing methods, and the rising importance of nanomaterials highlights the drive toward a more sustainable future.

This conference also emphasizes the growing use of computational tools, including Artificial Intelligence (AI) and Machine Learning (ML), in materials modeling, property prediction, and performance analysis. These technologies are reshaping how new materials are developed and optimized. The papers in this volume reflect these advancements and serve as a valuable resource for young scientists and engineers working at the intersection of materials science and intelligent design. I applaud the contributors for their impactful research and hope this collection inspires continued innovation and collaboration in the field.

Prof. Kuldeep Kumar Raina,
Vice Chancellor, RUAS, Bangalore, India

Preface

The field of materials science and engineering continues to evolve rapidly, driven by the need for innovative solutions across structural, energy, biomedical, and electronic applications. This book, comprising the proceedings of the **International Conference on Advances in Materials Science and Technology (ICAMST-2024),** brings together a curated selection of high-quality research papers that address current challenges and emerging trends in applied materials. It serves as a comprehensive resource for academicians, researchers, industry professionals, and students working at the forefront of materials development and application.

The contributions compiled in this volume represent a rich and diverse spectrum of research in the field of materials science and technology. These studies explore the synthesis and in-depth characterization of a variety of advanced materials, including metal and polymer matrix composites, nanomaterials, functional coatings, and bio-compatible materials. Additionally, several chapters delve into computational materials science, where modeling and simulation techniques are used to predict material behavior, optimize design parameters, and explore the impact of defects and processing conditions. A significant focus is placed on the development of sustainable materials those that can be recycled, reused, or manufactured through eco-friendly processes with the goal of minimizing environmental impact without compromising performance. The integration of Artificial Intelligence (AI) and Machine Learning (ML) into materials research is another prominent theme, demonstrating how predictive algorithms are revolutionizing the way new materials are designed, tested, and deployed, particularly in high-demand sectors such as aerospace, electronics, and renewable energy.

The book, titled *Advances in Materials Science and Technology*, is meticulously organized into 43 chapters, each authored by researchers and experts from academia, research institutions, and industry. Every chapter offers an in-depth exploration of a specialized topic, ranging from experimental techniques for material synthesis and property evaluation to the use of advanced computational tools for simulating material behavior under various conditions. Discussions include case studies and application-driven approaches that highlight the role of materials in addressing current and emerging technological challenges. Readers will find comprehensive coverage on energy-efficient materials, lightweight structural materials, biodegradable polymers, smart sensors, and next-generation electronic components. The book serves as a valuable reference for students, researchers, and professionals aiming to deepen their understanding of the evolving landscape of materials science and contribute meaningfully to the development of innovative and sustainable material solutions.

We extend our sincere thanks to all authors, reviewers, and organizing committee members who contributed to the success of this conference and the compilation of this book.

Dr. Srikari S
Dr. Suresh R
Dr. Rahul M Cadambi

Acknowledgements

The editors acknowledge their gratitude to CRC press/ Taylor and Francis Group for this opportunity and their professional support. Finally, we would like to thank all chapter authors for their interest and availability to work on this project.

Advances in Materials Science and Technology – Dr. Srikari Srinivasan et al. (eds)
© *2025 Taylor & Francis Group, London, ISBN 978-1-041-12342-2*

1

Development of a Nanocrystalline Cellulose-Based Film to Combat the Staining Potential of Silver Diamine Fluoride

Lalitha S. Jairam,
Aishwarya Arun Koparde*

Department of Pediatric and Preventive Dentistry,
Faculty of Dental Sciences, M S Ramaiah University of Applied Sciences,
New Bel Road, Bangalore, India

Akshay Arjun,
Likhit Balse, H. B. Premkumar

Department of Physics,
Faculty of Mathematical and Physical Sciences,
M S Ramaiah University of Applied Sciences, Peenya Industrial Area,
Bangalore, India

Garima Goenka, Hafsa Khalid,
Harshavardhana R, Javeriya Zain, K. Shwetha Shree

Faculty of Dental Sciences, M S Ramaiah University of Applied Sciences,
New Bel Road, Bangalore, Karnataka, India

ABSTRACT: Silver Diamine Fluoride (SDF) is an effective, inexpensive and easy-to-use topical fluoride however treatment acceptance rate is very low as it causes staining of the arrested carious lesion. Multiple efforts to reduce staining using potassium iodide (KI) and glutathione (GSH) have had limited success. So, aim was to assess the effectiveness of nanocrystalline cellulose-based film incorporated with SDF in the prevention of tooth staining. Nanocrystalline cellulose (NCN) was synthesized from Darbha Grass and characterized using by FTIR, SEM, XRD. To these films 2%,4%,6% and 8%SDF was incorporated and characterized by FTIR. Film thickness and folding endurance testing was performed. Staining potential of the SDF- NCN film was assessed *in vitro* by assigning 75 human extracted teeth to 5 groups of 15 teeth each. Color change determined using CIELAB system. XRD and FTIR peak values were suggestive of nanocellulose. Average thickness of films was 27 μm - 84 μm. Mean Folding Endurance value was 250 times. ΔL values in CLELAB system were significantly lower when compared to SDF group. Within limitations of this study, it can be concluded that nanocellulose can be used as an effective medium for delivery of SDF for oral applications.

KEYWORDS: Film, Nanocrystalline cellulose, Silver diamine fluoride, Tooth staining

1. Introduction

Dental Caries is one of the most prevalent chronic infectious diseases affecting a large number of people worldwide (Selwitz et al., 2007). Several strategies have been adopted to prevent and treat this disease, topical fluoride application being one among them (Horst et al., 2018). Silver diamine fluoride (38%) is a form of professionally applied fluoride

*Corresponding author: aishukoparde@gmail.com

DOI: 10.1201/9781003664277-1

(Baik et al., 2021). It is a colorless liquid containing silver particles and 38% (44,800 ppm) fluoride ions (which is the highest fluoride concentration available for dental use) at pH 10 it contains 25% silver, 8% ammonia, 5% fluoride, and 62% water (Mohamed et al., 2023; Patel et al., 2019). SDF is an effective, inexpensive, and easy-to-use cariostatic agent that provides a non-invasive treatment modality for managing dental caries, especially in high-risk populations (Zhao et al., 2018). Randomized clinical trials have found it to be more effective than other topical fluorides such as 5% Sodium Fluoride varnish thus attracting the attention of several researchers (Mohamed et al., 2023) Even though the material has high clinical efficacy, the treatment acceptance rate is very low because of the staining of the arrested carious lesion after SDF application (Cappiello et al., 2024). Various research studies are being carried out to control the staining due to SDF and to increase the patient and parent's acceptance of the treatment (Magno et al., 2019). Multiple efforts have been made to reduce the staining with the use of potassium iodide (KI) and glutathione (GSH) but with limited success (Kamble et al., 2021; Nguyen et al., 2017; Sayed et al., 2018). Jackson et al. conducted a study on the use of silver diamine fluoride and silver salt crosslinked nanocrystalline cellulose (CNC) films as antibacterial agents where they analyzed the controlled release of silver to possibly treat oral bacteria (Jackson et al., 2021). Nanocrystalline cellulose can be synthesized by darbha grass. *Desmostachya bipinnata* also commonly known as Darbha or Kusha grass in India (P et al., 2020). The literature on Ayurveda emphasizes its therapeutic benefits. Figure 1.1 depicts conventional treatments to obtain cellulose nanoparticles adapted from (Rojas et al, 2015).

With this impetus, the study aims to develop an SDF-nanocrystalline cellulose-based film and assess its effectiveness in the prevention of tooth staining.

2. Methodology

Synthesize of nanocrystalline cellulose - Materials that were required in this study include – Well-dried and finely powdered Darbha (*Desmostachya bipinnata*) leaves sourced locally, Sodium Hydroxide (NaOH), Sodium Hypochlorite, Acetate buffer (pH 4.5), Sulphuric acid (99%). The procedure followed was from previous published literature (Wulandari et al., 2016; Jonoobi et al., 2015). Darbha grass was sourced from local farmers, dried, and finely ground to a powder. The following steps were employed: delignification, bleaching, and acid hydrolysis. The obtained nanocrystalline cellulose crystals were characterized by FTIR, SEM, and XRD.

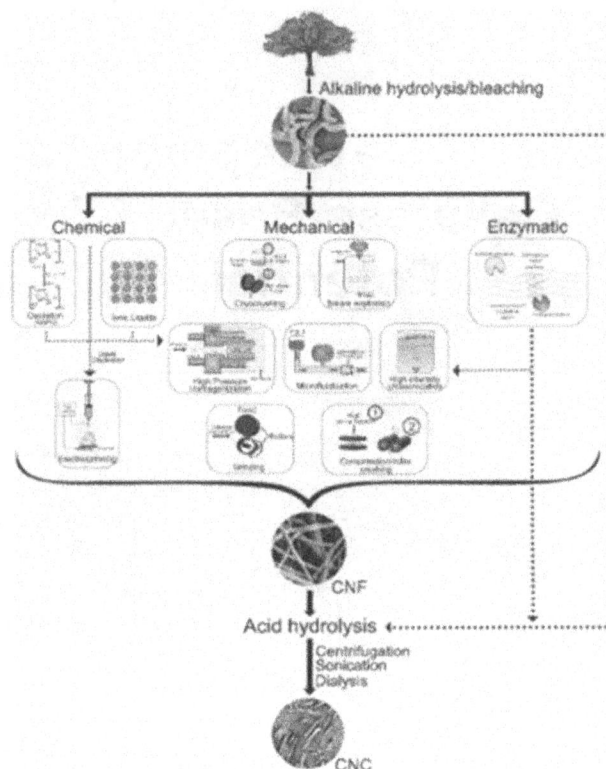

Fig. 1.1 Conventional treatments to obtain cellulose nanoparticles

Source: Rojas et al, 2015

Preparation of Nanocrystalline cellulose-based film - 20 g of prepared nanocrystalline cellulose was added into 20 ml distilled water slowly using a spatula. A magnetic stirrer was used to obtain a 20% (w/w) concentration of CNC by adjusting the rotations to 450 rpm at 37°C for 20 mins. Prepared nanocrystalline cellulose was poured into petri dishes and placed in an oven at 37°C as depicted in Fig. 1.2.

After drying the contents of the petri dish for 24-36 hrs in the oven at 37°C, the film was obtained by gentle teasing with forceps to form nanocrystalline cellulose-based film.

Preparation of samples - To incorporate 2% of SDF into the film 20 ml of solution with 2 % SDF was poured into petri dishes and placed in an oven at 37°C.Similarly, 4%,6%, and 8% (w/w) concentrations were prepared as shown in figure 3. After 24 -36 hrs, the films were removed from the petri dish by gentle teasing with forceps. The obtained films were characterized by FTIR. The following other tests were done for the films:

Film thickness testing: Thickness of film specimens (2 × 2 cm^2) was measured using a calibrated digital caliper at

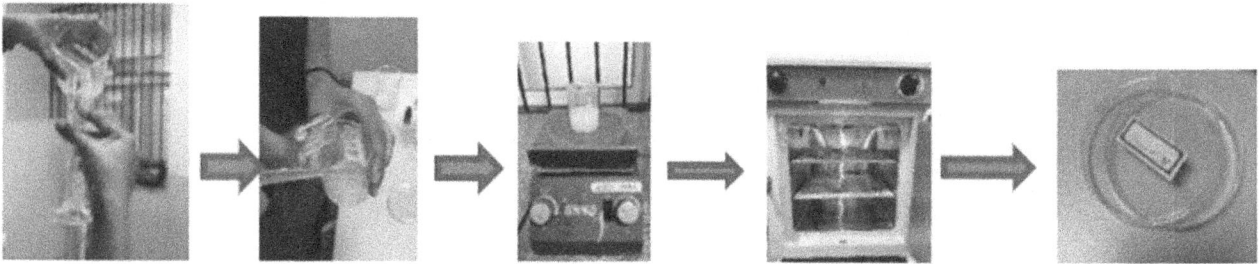

Fig. 1.2   Preparation of nanocrystalline cellulose based film

Source: Author's compilation

three random positions on film and the mean thickness calculated (Bala et al., 2013).

Folding endurance: 3 film specimens $2 \times 2cm^2$ from each formulation were repeatedly folded at same place, number of times the film was folded before breaking or any visible crack was detected was taken as the folding endurance value and the mean was calculated (Chandrashekar et al., 2022).

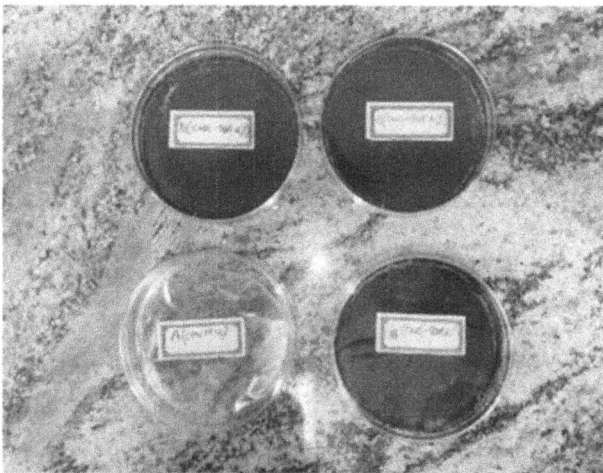

Fig. 1.3   2% 4%, 6%, and 8% (w/w) concentrations of SDF into the film

Source: Author's compilation

In vitro **assessment of the staining potential of the SDF- nanocrystalline cellulose film** - A total of 75 human extracted teeth were assigned to 5 groups of 15 teeth each. The samples were randomly allocated to five groups: Control group: SDF, Group 1: Nanocrystalline cellulose-SDF 2% film; 2: Nanocrystalline cellulose-SDF 4% film;3: Nanocrystalline cellulose-SDF 6% film; Group 4: Nanocrystalline cellulose-SDF 8% film. After the application of the film to the samples, the samples were hermetically sealed in artificial saliva and changed every 7 days. An *in-vitro* assessment of color was done at baseline (before the application of the film), after 2

weeks, and 4 weeks from the day of application. Dentin staining was determined based on the parameters of the CIELAB system through the formula (DE=[(L*2-L*1) 2b(a*2-a*1) 2b (b*2- b*1) 2] 1/2) and DL (DL*=L*2-L*1) (Chandrashekar et al., 2022)

3. Results and Discussion

3.1 FTIR Analysis

The absorption peaks at 3335 cm^{-1} and around 2904 cm^{-1} were attributed to the O-H and C-H stretching vibrations, respectively. The peak at 1408 cm^{-1} was attributed to C-H stretching in the polysaccharide ring of cellulose as depicted in Fig. 1.4.

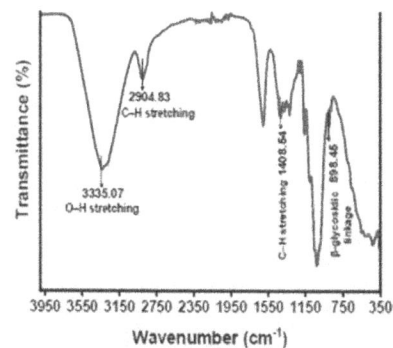

Fig. 1.4   Nanocellulose obtained from darbha grass – FTIR spectrum

Source: Author's compilation

3.2 Scanning Electron Microscopy (SEM) Analysis

SEM image of cellulose nanofibers prepared from darbha grass showed well-developed nanocellulose as shown in Fig. 1.5.

3.3 XRD Analysis

Figure 1.6 shows XRD pattern intensity peaks at 15.6o,19.2 o,23o and 32 o which is suggestive of nanocellulose.

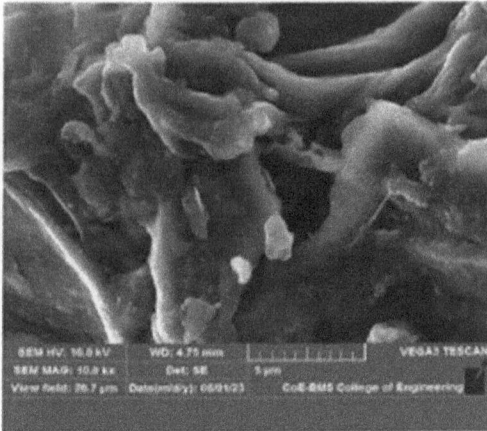

Fig. 1.5   SEM images of nanocellulose

Source: Author's compilation

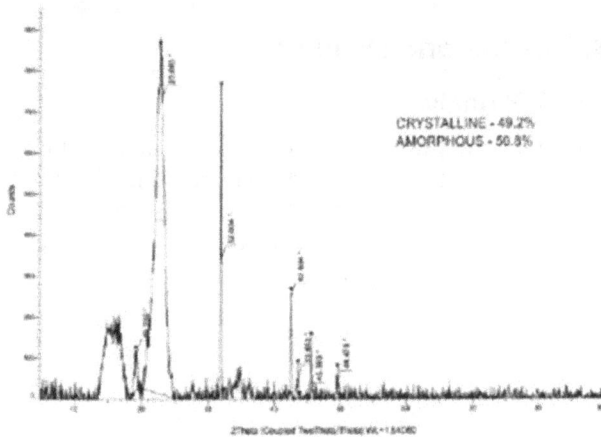

CRYSTALLINE - 49.2%
AMORPHOUS - 50.8%

Fig. 1.6   X-ray diffraction pattern of nanocellulose

Source: Author's compilation

3.4 FTIR Analysis – Nanocellulose and SDF

FTIR spectrum of the nanocellulose with SDF showed characteristic peaks of sodium fluoride at 3200 cm^{-1} which corresponds to the O-H stretching of hydroxyl groups. The peak at 1650 cm^{-1} is of amide group. The characteristic peaks for alkanes functional groups can be observed at 1400cm^{-1} and SDF at 3400 cm^{-1} and 1380 cm^{-1}. The peak at 1160cm^{-1} and 1064cm^{-1} are suggestive of nanocellulose as shown in Fig. 1.7.

Film Thickness

The average thickness of the films ranged from 27 μm to 84 μm as shown in Table 1.1.

Folding Endurance

The films were folded manually till breakage or visible cracking and it was found that the films did not break even after folding over 250 times as shown in Table 1.2.

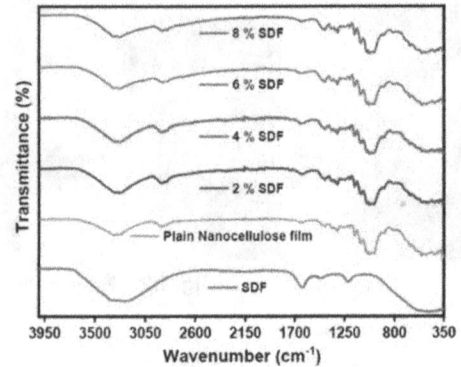

Fig. 1.7   Nanocellulose/SDF FTIR spectrum

Source: Author's compilation

Table 1.1   Mean thickness of films

Film	Mean thickness value (µm)
Control film - Without SDF	84.1857
2% SDF	27.8935
4% SDF	27.5006
6% SDF	38.1641
8% SDF	63.1061

Source: Author's compilation

Table 1.2   Folding endurance of the films

Films	Mean Folding Endurance Value
Plain Nanocrystalline cellulose Film	350 times
Nanocrystalline cellulose-SDF 2% film	300 times
Nanocrystalline cellulose-SDF 4% film	250 times
Nanocrystalline cellulose-SDF 6% film	300 times
Nanocrystalline cellulose-SDF 8% film	300 times

Source: Author's compilation

3.5 Staining Property: CIELAB Values

The CIELAB values for plain SDF application on the tooth show that the ΔL value is high which indicates that the surface of the tooth has darkened. Films containing nanocellulose with SDF, it can be observed from Table 1.3 that the ΔL values are significantly lower when compared to the SDF group. The group with 4% SDF showed the lowest values indicating that it would be the preferred concentration to be added to the films. Further, as the concentration of SDF was increased, it was seen that the ΔL values also increased which is undesirable. The management of dental caries has been heavily reliant on fluorides (Jullien et al., 2021). In recent times, SDF has been extensively used in certain countries for the prevention

Table 1.3 CIELAB values of the films

Sample Tooth	Before treatment				After treatment				ΔE	ΔL*
	L*	a*	b*	C	L*	a*	b*	C		
Plain SDF	53.89	2.83	14.58	14.85	24.33	1.67	2.42	2.94	31.98	29.56
Plain Nanocellulose Film	77.34	4.73	11.36	12.31	70.36	4.67	9.91	10.96	7.13	6.98
2% SDF	65.12	5.38	16.21	17.08	60.12	0.31	17.06	17.06	7.17	5.00
4% SDF	72.37	1.64	5.74	5.97	70.43	4.96	16.09	16.84	11.04	1.94
6% SDF	72.95	3.79	19.17	19.54	67.99	2.56	18.16	18.34	5.21	4.96
8% SDF	61.57	4.56	19.57	20.09	44.87	0.41	16.38	16.39	17.71	16.70

Source: Author's compilation

of dental caries but drawbacks of staining impede its use (Seifo et al., 2020). In a study by *Jackson et al*, they found that the combination of cellulose nanocrystals was suitable for incorporating SDF. However, they used lab-grade crystalline nanocellulose which is expensive. Cellulose, which is found in a variety of materials including hardwood, soft wood, agricultural debris, bacteria, algae, tunicates, etc., is the source of nanocellulose (Thomas et al., 2018). In the present study, we have synthesized nanocellulose from darbha grass which makes it more cost effective and also yields better properties as it is based on plant source. The synthesis of nanocellulose from darbha grass was successful with the acid hydrolysis method. The nanocellulose obtained was confirmed by FTIR and XRD which were similar to the spectrum seen in previous literature (Ververis et al., 2004; Kaur et al., 2021). The preparation of nanocellulose films with incorporation of SDF was achieved by incorporating 2%, 4%, 6%, and 8% of SDF to assess which concentration yielded better results in reduction of staining. The film thickness of dental bonding agents usually is in the range of 50-100 μm (Peter et al., 1997) and the films obtained in the present study were well within this range. The characterization of the films showed that SDF was incorporated into the nanocellulose matrix which was confirmed by the FTIR spectrum. Assessment of the staining properties using the CIELAB showed that the direct application of SDF caused the maximum darkening of the tooth structure. The incorporation of SDF into nanocellulose films greatly decreased the staining. It was noted that the addition of 4% SDF to nanocellulose films yielded the least ΔL scores. Literature supports the claim that nanocellulose can be used as an effective matrix for the delivery of SDF (Errokh et al., 2019). Clearly, there is a need to explore the potential application of SDF in a cost-effective matrix such as nanocellulose so as to combat the staining caused by it (Toro et al., 2021).

4. Conclusion

This study demonstrates the successful preparation of nanocellulose from a plant source namely darbha grass. The obtained nanocellulose was used as a medium to disperse SDF. Different weight percentages (2%,4%,6% and 8%) of SDF were incorporated into the films and a control film was also prepared. The measurement of staining revealed SDF incorporated in nanocellulose demonstrated a marked decrease in the staining potential. Within the limitations of this study, it can be concluded that nanocellulose can be used as an effective medium for delivery of SDF for oral applications.

5. Acknowledgment

The authors acknowledge M.S. Ramaiah University of Applied Sciences, Faculty of Pharmacy and Faculty of Dental Sciences. The authors are thankful to the faculty of BMS College of Engineering, Bangalore for providing the necessary facilities for material characterization.

6. Funding

We acknowledge the financial support provided by the Indian Council of Medical Research under the ICMR-STS funding 2023 (**Reference Id: 2023-05061**)

References

1. Bala, R., Pawar, P., Khanna, S. & Arora, S. (2013) 'Orally dissolving strips: A new approach to oral drug delivery system', *International Journal of Pharmaceutical Investigation*, 3(2), p. 67.
2. Bhandari, M.P. (2019) 'Bashudaiva Kutumbakkam. The entire world is our home and all living beings are our relatives. Why we need to worry about climate change, with reference to pollution problems in the major cities of India, Nepal, Bangladesh and Pakistan', *Advances in Agricultural and Environmental Sciences*, 2(1), pp. 8–35.

3. Bhokardankar, P. S., and B. Rathi. "Indigenous wisdom of Ayurvedic drugs to treat urinary tract infections." *International Journal of Ayurvedic Medicine* 11, no. 3 (2020): 370–377.

4. Chandrashekar, A., Vargheese, S., Vijayan, J.G., Gopi, J.A. & Prabhu, T.N. (2022) 'Highly efficient removal of Rhodamine B dye using nanocomposites made from cotton seed oil-based polyurethane and silylated nanocellulose', *Journal of Polymers and the Environment*, 30(12), pp. 4999–5011.

5. Chibinski, A.C., Wambier, L.M., Feltrin, J., Loguercio, A.D., Wambier, D.S. & Reis, A. (2017) 'Silver diamine fluoride has efficacy in controlling caries progression in primary teeth: a systematic review and meta-analysis', *Caries Research*, 51(5), pp. 527–541.

6. Chieng, B.W., Lee, S.H., Ibrahim, N.A., Then, Y.Y. & Loo, Y.Y. (2017) 'Isolation and characterization of cellulose nanocrystals from oil palm mesocarp fiber', *Polymers*, 9(8), p. 355.

7. Errokh, A., Magnin, A., Putaux, J.L. & Boufi, S. (2019) 'Hybrid nanocellulose decorated with silver nanoparticles as reinforcing filler with antibacterial properties', *Materials Science and Engineering: C*, 105, p. 110044.

8. Horst, J.A. (2018) 'Silver fluoride as a treatment for dental caries', *Advances in Dental Research*, 29(1), pp. 135–140.

9. Jackson, J., Dietrich, C., Shademani, A. & Manso, A. (2021) 'The manufacture and characterization of silver diammine fluoride and silver salt cross-linked nanocrystalline cellulose films as novel antibacterial materials', *Gels*, 7(3), p. 104.

10. Jonoobi, M., Oladi, R., Davoudpour, Y., Oksman, K., Dufresne, A., Hamzeh, Y. & Davoodi, R. (2015) 'Different preparation methods and properties of nanostructured cellulose from various natural resources and residues: A review', *Cellulose*, 22, pp. 935–969.

11. Jullien, S., Huss, G. & Weigel, R. (2021) 'Supporting recommendations for childhood preventive interventions for primary health care: Elaboration of evidence synthesis and lessons learnt', *BMC Pediatrics*, 21, pp. 1–2.

12. Kamble, A.N., Chimata, V.K., Katge, F.A., Nanavati, K.K. & Shetty, S.K. (2021) 'Comparative evaluation of effect of potassium iodide and glutathione on tooth discoloration after application of 38% silver diamine fluoride in primary molars: An in vitro study', *International Journal of Clinical Pediatric Dentistry*, 14(6), p. 752.

13. Magno, M.B., Silva, L.P., Ferreira, D.M., Barja-Fidalgo, F. & Fonseca-Gonçalves, A. (2019) 'Aesthetic perception, acceptability and satisfaction in the treatment of caries lesions with silver diamine fluoride: a scoping review', *International Journal of Paediatric Dentistry*, 29(3), pp. 257–266.

14. Mohamed, Y. & Ashraf, R. (2023) 'Remineralization potential of phosphorylated chitosan and silver diamine fluoride in comparison to sodium fluoride varnish: an in vitro study', *European Archives of Paediatric Dentistry*, 24(3), pp. 327–334.

15. Murthy, A.R. & Mahajan, B. (2016) 'Medicinal importance of Darbha: A review', *Journal of Ayurvedic and Herbal Medicine*, 2(3), pp. 89–95.

16. Nguyen, V., Neill, C., Felsenfeld, J. & Primus, C. (2017) 'Potassium iodide: The solution to silver diamine fluoride discoloration', *Health*, 5(1), p. 555655.

17. Patel, J., Anthonappa, R.P. & King, N.M. (2019) 'Silver diamine fluoride: A critical review and treatment recommendations', *Dental Update*, 46(7), pp. 626–632.

18. Peter, A., Paul, S.J., Lüthy, H. & Schärer, P. (1997) 'Film thickness of various dentine bonding agents', *Journal of Oral Rehabilitation*, 24(8), pp. 568–573.

19. P.S. & Rathi, B. (2020) 'Indigenous wisdom of Ayurvedic drugs to treat urinary tract infections', *International Journal of Ayurvedic Medicine*, 11(3), pp. 370–377.

20. Qin, Z.Y., Tong, G., Chin, Y.F. & Zhou, J.C. (2011) 'Preparation of ultrasonic-assisted high carboxylate content cellulose nanocrystals by TEMPO oxidation', *BioResources*, 6(2), pp. 1136–1146.

21. Rojas, J., Bedoya, M. & Ciro, Y. (2015) 'Current trends in the production of cellulose nanoparticles and nanocomposites for biomedical applications', *Cellulose: Fundamental Aspects and Current Trends*, pp. 193–228.

22. Sayed, M., N. Matsui, N. Hiraishi, T. Nikaido, M. F. Burrow, and J. Tagami. "Effect of glutathione biomolecule on tooth discoloration associated with silver diammine fluoride." *International Journal of Molecular Sciences* 19, no. 5 (2018): 1322.

23. Seifo, N., Robertson, M., MacLean, J., Blain, K., Grosse, S., Milne, R., Seeballuck, C. & Innes, N. (2020) 'The use of silver diamine fluoride (SDF) in dental practice', *British Dental Journal*, 228(2), pp. 75–81.

24. Toro, R. G., A. M. Adel, T. de Caro, F. Federici, L. Cerri, E. Bolli, et al. "Evaluation of Long-Lasting Antibacterial Properties and Cytotoxic Behavior of Functionalized Silver-Nanocellulose Composite." *Materials* 14, no. 15 (2021): 4198.

25. Ververis, C., K. Georghiou, N. Christodoulakis, P. Santas, and R. Santas. "Fiber Dimensions, Lignin and Cellulose Content of Various Plant Materials and Their Suitability for Paper Production." *Industrial Crops and Products* 19, no. 3 (2004): 245–54.

26. Wulandari, W.T., Rochliadi, A. & Arcana, I.M. (2016) 'Nanocellulose prepared by acid hydrolysis of isolated cellulose from sugarcane bagasse', *IOP Conference Series: Materials Science and Engineering*, 107(1), p. 012045.

27. Zhao, I. S., S. S. Gao, N. Hiraishi, M. F. Burrow, D. Duangthip, M. L. Mei, E. C. Lo, and C. H. Chu. "Mechanisms of silver diamine fluoride on arresting caries: a literature review." *International Dental Journal* 68, no. 2 (2018): 67–76.

Advances in Materials Science and Technology – Dr. Srikari Srinivasan et al. (eds)
© 2025 Taylor & Francis Group, London, ISBN 978-1-041-12342-2

2

Papilla Reconstruction with Titanium-Prepared Platelet-Rich Fibrin: A Case Report and Literature Review

Amrutha Rao*,
Bhavya Shetty, Rohit Prasad, and Akshatha Raj
Department of Periodontics,
Faculty of Dental Sciences, M S Ramaiah University of Applied Sciences,
Bangalore, Karnataka, India

ABSTRACT: Interdental papilla loss which leads to appearance of black triangles between teeth can cause aesthetic disorders, phonetic problems, and food accumulation. Titanium has good corrosion resistance and passivates itself in vivo by forming an adhesive oxide layer due to which it is more useful in activating platelets as compared to glass tubes. Due to its great hemocompatibility, it forms profuse, highly polymerized and organized fibrin meshwork thus holding numerous growth factors for longer time period. Application of this highly biocompatible titanium-prepared platelet-rich fibrin (T-PRF) involves minimally invasive technique with low risk and satisfactory clinical outcomes for papilla regeneration. The aim of the study was to assess effectiveness of autologous T-PRF in interdental papilla reconstruction. A 45-year-old woman presented with interdental papilla loss in maxillary anterior region. Periodontal variables such as keratinized tissue width, pocket depth, and papilla height loss were recorded before and after the treatment. Papilla reconstruction was done with T-PRF using tunneling technique. Follow-up was done after 45 days. A comprehensive computer-based search for literature review combined the following databases into one search: Medline, google scholar, and crossref. This search also used key words. In this case report, the use of T-PRF resulted in regeneration of the interdental papilla thereby eliminating black triangle.

KEYWORDS: Interdental papilla loss, Papilla height loss, Papilla reconstruction, Titanium prepared platelet-rich fibrin

1. Introduction

Interdental papilla morphology is determined by factors such as age, shape of crown, contact position and angle between the roots of adjacent teeth, alveolar crest-interdental contact point distance, interdental embrasure volume, crestal bone height, soft tissue appearance, periodontal disease and history of periodontal surgery. Interdental papilla loss which leads to appearance of black triangles between teeth can cause aesthetic disorders, phonetic problems, and food accumulation (Chow et al, 2010 and Sharma et al, 2010).

Though various non-surgical approaches through prosthetic, restorative, and orthodontic procedures have been used for papilla reconstruction, they have not always yielded satisfactory results. Several full or partial thickness, free or pedicle tissue management surgical techniques including roll, rotated, or advanced flaps have been proposed for papilla reconstruction as stated by (Deepalakshmi et al, 2007 and Pinto et al, 2010). Several

*Corresponding author: amrutharao9588@gmail.com

DOI: 10.1201/9781003664277-2

studies state that Connective tissue grafts provide better results than conventional Platelet rich fibrin despite their disadvantages such as presence of a second wound site and variable papilla reconstruction outcomes (Oszcan et al, 2022 and Singh D et al, 2019).

Tunali et al., in 2013 harnessed titanium biomedical tubes for the construction of Titanium coated titanium platelet-rich fibrin (T-PRF). Titanium has been more useful in stimulating platelets compared to glass tubes and forms a thicker fibrin clot exhibiting longer resorption span and profuse fibrin meshwork provoking increase cellularity at the prescribed loci thereby inducing periodontal genesis.

With Titanium having one of the greatest bio-compatibility, the application of titanium-prepared platelet-rich fibrin involves minimally invasive technique with low risk and satisfactory clinical outcomes, but limited research is done on titanium-prepared platelet-rich fibrin. To the best of our knowledge, no studies are done to evaluate the success of titanium-prepared platelet-rich fibrin in papilla reconstruction.

2. Methodology

The participant provided oral informed consent, and the study was conducted in accordance with the Helsinki Declaration of 1975, as revised in 2000. The study protocol was approved by the Institutional Committee of Ethics in Dental Research of the Faculty of Dentistry at Ramaiah University of Applied Sciences.

2.1 Inclusion Criteria

Teeth having contact points, at least 2 mm keratinized tissue width (KTW), Probing depth (PD) ≤3 mm adjacent to the open embrasure

2.2 Exclusion Criteria

Patients with cardiovascular disease, type 2 diabetes mellitus, or osteoporosis, having anti-epileptics, immunosuppressants, corticosteroids, non-steroidal anti-inflammatory drugs or hormones, using tobacco/alcohol, Pregnant and lactating women.

A 45-year-old systemically healthy female patient reported to Department of Periodontology, with the chief complaint of black space between the upper front teeth region, causing displeasing smile and food entrapment. Intraoral examination done with periodontal probe revealed probing depth of 2 mm, keratinised tissue width of 3 mm and loss of 4 mm of interdental papilla- Nordland and Tarnow (1998) class I (The tip of interdental papilla lies between the interdental contact point and the coronal most extent of Cemento-enamel junction) between teeth 12 and 13

(Fig. 2.1a and 2.1b). The treatment was planned as papilla reconstruction between teeth 12 and 13 by tunneling technique with T-PRF membrane.

2.3 Presurgical Procedures

Patient underwent scaling and root planing in phase I therapy. An acrylic stent was prepared from 13-23 teeth region for assessing the treatment outcome.

2.4 Preparation of T PRF

The titanium tubes were produced from grade IV titanium. The blood sample of 9 mL was drawn from the antecubital vein of the subject's arm in one attempt was transferred to the titanium tube. The blood was quickly collected, and the tubes were immediately centrifuged at 2,800 rpm for 12 minutes with a specific table centrifuge, at room temperature as stated by Tunali [7]. After centrifugation, the T-PRF clots were removed from the tubes using sterile tweezers, separated from the RBC base using scissors (Fig. 2.1f), and placed on sterile woven gauze. T-PRF membrane was procured by compressing the fibrin clot in the PRF box.

2.5 Surgical Technique

The surgical procedure was performed using local infiltration anaesthesia (2% lignocaine with 1:80,000 concentration of adrenaline) in relation to 12, 13 interdental papillary regions. A split thickness semilunar incision was performed 3 mm apically from the mucogingival junction facial to the interdental area between 12 and 13(Fig. 2.1c). Intrasulcular incisions were made around the necks of the adjacent teeth extending from the buccal to the palatal surface (Fig. 2.1d), to free the connective tissue attachment from the root surface and to allow coronal displacement of the gingival-papillary unit as stated by Jenabian[8]. Following this, tunnel preparation was done connecting the semilunar and intrasulcular incisions using tunnelling knives (Fig. 2.1e). T-PRF membrane was then tucked in and pushed coronally to support and provide bulk to the coronally positioned interdental tissue. Then the flap was stabilised using 4-0 vicryl sutures (Fig. 2.1g).

2.6 Post Surgical Instructions

Coe-pak periodontal dressing was placed in the surgical site and prescribed a course of antibiotics (Capsule Amoxicillin 500 mg, thrice a day for 5 days) and analgesic (Tablet Aceclofenac 100 mg + Paracetamol 325 mg, twice a day for 3 days). Post-surgical instructions were given to the patient. Patient was instructed not to brush over the pack and to avoid hot foods during the first 3 hours after surgery to permit the pack to harden. Patient was advised not to smoke.

2.7 Post-operative Follow-up

The periodontal dressing and the sutures were removed after 2 weeks. Healing was found to be uneventful.

2.8 Literature Review

Table 2.1 Literature review on treatment of interdental papilla loss

S No.	Methodology	Inference	Reference
1.	In a randomized control trial study, patients underwent Papilla Reconstruction (PR) with Platelet Rich Fibrin at 28 sites and Connective tissue graft (CTG) at 27 sites placed in the maxillary anterior region with Semilunar incision technique.	CTG provided better PR outcomes. Periodontal parameters such as Gingival Index, Periodontal Index, and Probing Depth showed a slight increase at 1st month and then, turned to their BL levels. The periodontal parameters showed significant improvement after both treatment modalities. No inter-group difference was found except for Keratinized Tissue Width, which was in favor of CTG	(Bulut et al, 2022)
2.	For the preparation of T-PRF blood samples were collected from 10 healthy male volunteers. The blood samples were drawn using a syringe. Nine milliliters was transferred to a dry glass tube, and 9 mL was transferred to a titanium tube. Half of each clot (i.e., the blood that was clotted using T-PRF or L-PRF) was processed with a scanning electron microscope (SEM). The other half of each clot was processed for fluorescence microscopy analysis and light microscopy analysis	The T-PRF samples seemed to have a highly organized network with continuous integrity compared to the other L-PRF samples. Histomorphometric analysis showed that T-PRF fibrin network covers larger area than L-PRF fibrin network; also fibrin seemed thicker in the T-PRF samples.	(Tunali et al, 2013)
3.	This case report deals with a variant method of Beagle's technique on 37-year-old systemically healthy male patient with the chief complaint of black space between the upper front teeth region, causing displeasing smile and food entrapment. Intraoral examination done with periodontal probe revealed the loss of 2 mm of interdental papilla. Patient was treated surgically with Advanced Platelet Rich Fibrin (A-PRF) membrane for papilla reconstruction in the upper anterior aesthetic zone.	This case report with A-PRF could contribute to evidence in the subject of papilla reconstruction in future.	(Vijayalakshmi et al, 2020)
4.	Four databases MEDLINE (by PubMed), Cochrane database, EBSCO, and Google Scholar were explored to identify the studies in English from 2015 up to March 2021. An additional hand search of relevant journals was also done. All the reviewers screened the retrieved articles using the particular inclusion criteria. Randomized control trials evaluating the use of T-PRF to enhance the treatment outcome of intrabony defect in chronic periodontitis patients were included in the study.	Use of T-PRF induced a reduction in Pocket probing depth, gain in Relative attachment level along with open flap debridement in the management of intrabony defects	(Mahale et al, 2022)

3. Results

There was increase in papillary fill by 1mm when the patient was reviewed after 45 days follow-up [Fig. 2.1h and 2.1i]. The distance from the contact point to the papilla tip was 3 mm as compared to the baseline value of 4 mm which was before the surgery, although probing depth and keratinized tissue width remained same.

4. Conclusion

Papilla reconstruction techniques are an exciting area of interest in periodontal therapy. This case report with T-PRF resulted in papilla regeneration thereby eliminating black triangle which often has unaesthetic appearance and associated problems such as tissue discomfort due to food lodgement and phonetics. Use of T PRF thus is relatively less invasive as compared to gold standard, connective tissue graft. The study also could contribute to evidence in the subject of papilla reconstruction in future. Controlled clinical trials in the future could add up to the current evidence.

Acknowledgments

The authors acknowledge M S Ramaiah University of Applied Sciences and Department of Periodontology for providing support to the study.

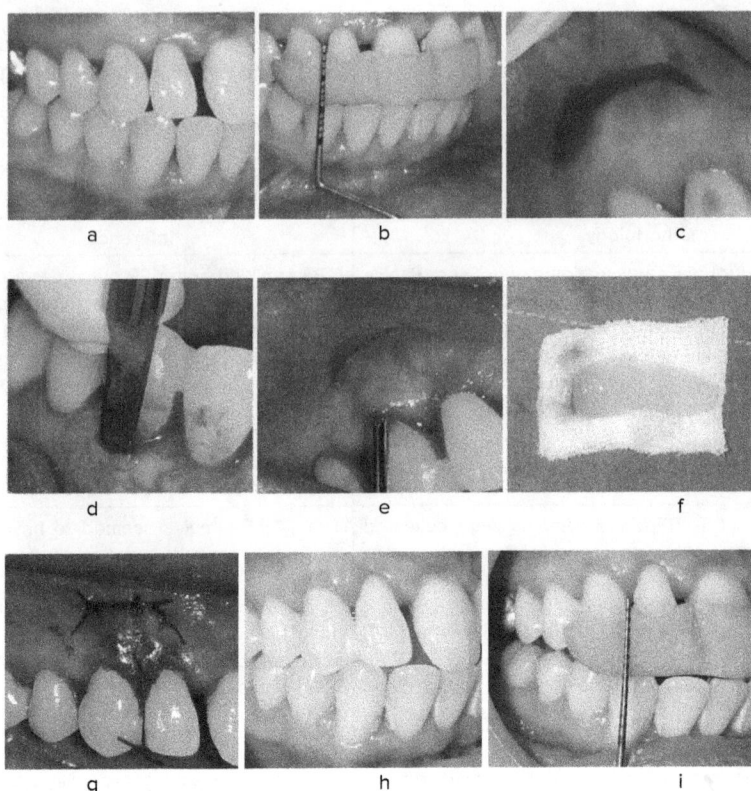

Fig. 2.1 a) Preoperative view showing interdental papilla loss between 12 and 13 teeth. b) Papilla Height loss of 3mm measured using UNC-15 probe. c) Semilunar incision placement. d) Intrasulcular incision placement. e) Tunnel preparation using tunneling knife. f) T-PRF procurement. g) Suture placement aided by composite stop. h) Follow up showed healing after 30 days. i) Papilla Height gain of 1mm

Source: Author

References

1. Chow, Y.C., Eber, R.M., Tsao, Y.P., Shotwell, J.L. and Wang, H.L., 2010. Factors associated with the appearance of gingival papillae. *Journal of clinical periodontology*, *37*(8), pp.719–727.

2. De Castro Pinto, R.C.N., Colombini, B.L., Ishikiriama, S.K., Chambrone, L., Pustiglioni, F.E. and Romito, G.A., 2010. The subepithelial connective tissue pedicle graft combined with the coronally advanced flap for restoring missing papilla: A report of two cases. *Quintessence International*, *41*(3).

3. Deepalakshmi, D., Ahathya, R.S., Raja, S. and Kumar, A., 2007. Surgical reconstruction of lost interdental papilla: a case report. *Periodontal Practice Today*, *4*(3).

4. Jenabian, N., Rad, M.R., Bijani, A. and Ghahari, P., 2018. The comparison of papilla preservation technique and semilunar incision with sub-epithelial connective tissue graft in dark triangle treatment. *Caspian Journal of Dental Research*, *7*(1), pp. 21–26.

5. Mahale, S.A., Katkurwar, A., Bhandare, J.V. and Mahale, A., 2022. The use of titanium platelet-rich fibrin to enhance the treatment outcome of intrabony defect in the chronic periodontitis patients–A systematic review and meta-analysis. *Journal of Dental Research and Review*, *9*(1), pp.1–8.

6. Ozcan Bulut, S., Ilhan, D., Karabulut, E., Caglayan, F. and Keceli, H.G., 2022. Efficacy of platelet-rich fibrin and connective tissue graft in papilla reconstruction. *Journal of Esthetic and Restorative Dentistry*, *34*(7), pp.1096–1104.

7. Sharma, A.A. and Park, J.H., 2010. Esthetic considerations in interdental papilla: remediation and regeneration. *Journal of esthetic and restorative dentistry*, *22*(1), pp.18–28.

8. Singh, D., Jhingran, R., Bains, V.K., Madan, R. and Srivastava, R., 2019. Efficacy of platelet-rich fibrin in interdental papilla reconstruction as compared to connective tissue using microsurgical approach. *Contemporary Clinical Dentistry*, *10*(4), pp.643–651.

9. Tunalı, M., Özdemir, H., Küçükodacı, Z., Akman, S., Yaprak, E., Toker, H. and Fıratlı, E., 2014. A novel platelet concentrate: titanium-prepared platelet-rich fibrin. *BioMed research international*, *2014*(1), p.209548.

10. Vijayalakshmi, R., Sathyapriya, R., Prashanthi, P. And Burnice, C., 2020. Advanced-Platelet Rich Fibrin Assisted Papilla Reconstruction by Modified Beagle's Technique-A Novel Approach. *Journal of Clinical & Diagnostic Research*, *14*(4).

Advances in Materials Science and Technology – Dr. Srikari Srinivasan et al. (eds)
© *2025 Taylor & Francis Group, London, ISBN 978-1-041-12342-2*

3

Evaluation of Platelet Rich Fibrin with Biodentine in Direct Pulp Therapy of Permanent Teeth— An in Vivo Study

Arbiya Anjum,
Swaroop Hegde, Rhea S. Mathew[1],
Shruthika Mahajan[2], Shruthi Nagaraja
Department of Conservative Dentistry and Endodontics,
Faculty of Dental Sciences, M S Ramaiah University of Applied Sciences,
Bangalore, Karnataka, India

ABSTRACT: Vital pulp therapy aims to preserve and maintain pulp tissue that is in a healthy or reversibly inflamed state. Biomaterials used for this procedure should focus on minimally invasive strategies for maintenance of the pulp vitality and possess regenerative potential. The need arises for developing a pulp capping material with these properties that will contribute to favorable treatment outcomes. To evaluate and compare the outcome of direct pulp therapy with Biodentine alone versus Biodentine with PRF in permanent teeth with reversible pulpitis. A total number of 50 patients requiring direct pulp capping were randomly allocated equally into two treatment groups (n=25 each). Direct pulp capping was performed in Group I using Biodentine (control group), and in Group II using Platelet Rich Fibrin (PRF) with Biodentine. Clinical and radiographic evaluations were performed at baseline, 6 and 12 months. The data obtained was blindly analyzed using the McNemar Chi-square test. The short term clinical and radiological success rates of PRF with Biodentine (92%) and Biodentine alone (84%) were comparable. However, at 12 months, there was favorable long-term clinical and radiological outcomes with success rate of 76% in Group II as compared to Group I (64%). PRF could be used as a suitable autologous adjunct to Biodentine in direct pulp capping procedures of permanent teeth with reversible pulpitis. Use of PRF with Biodentine can be a novel strategy to treat reversible pulpitis in human permanent teeth.

KEYWORDS: Regeneration, Biomaterials, Platelet rich fibrin, Biodentine, Pulp capping

1. Introduction

Minimally invasive treatment approach focusses upon maintenance of tooth in healthy state and preservation of pulpal vitality. Vital pulp therapy (VPT) is one such treatment which aims at preservation and maintenance of the pulp tissue that has been compromised (Hanna et al., 2020). This therapy involves primarily the use of newer synthetic materials that provide better seal than the traditional calcium hydroxide such as Mineral Trioxide Aggregate (MTA) and Biodentine (Keswani et al., 2014). However, MTA has numerous disadvantages, including its tendency to discolor tooth, the presence of toxic elements such as arsenic, difficult handling properties, and high cost (Torabinajed et al., 2018).

Recently, a new calcium silicate-based material, Biodentine (Septodont, France), has been introduced which was designed as a "dentin replacement" material with properties similar to MTA without its disadvantages.

[1]shruthi.cd.ds@msruas.ac.in, [2]rhea.mathew.93.rm@gmail.com

DOI: 10.1201/9781003664277-3

When used as direct pulp capping material, it has shown favorable response for up to two years (Gupta et al., 2016). It also promotes formation of reparative dentin in direct contact with vital pulp tissue. The second treatment modality involves the use of biologically-based procedures such as Regenerative Endodontics to physiologically replace damaged tooth structures as well as cells of the pulp-dentin complex (Murray et al., 2006) Regeneration using Protein Rich Fibrin (PRF) has shown to be capable of Inducing re-vascularization of dam-aged/necrotic exposed pulp tissue, fighting off bacterial infection and secreting a wide-range of growth factors capable of inducing pulp cell proliferation and differentiation, which is an ideal requisite of a pulp capping agent(Choukroun et al., 2006) It is an autologous fibrin-based biomaterial, derived from human blood which is rich in platelets, cytokines and growth factors. It is more efficient clinically compared with Platelet Rich Protein (Huang et al., 2010). The combination of host cells, a three-dimensional fibrin matrix and growth factors contained within PRF act to synergistically enhance faster and more potent tissue wound healing and regeneration.

The present study evaluated the combined effect of PRF at the exposure site to provide growth factors for better wound healing and Biodentine, owing to its superior sealing ability to reduce the risk of marginal micro-leakage when in contact with moisture content in PRF to improve the long term treatment outcomes of vital pulp therapy.

Hence the aim of this study was to evaluate and compare the outcome of direct pulp therapy with Biodentine alone and Biodentine with PRF in permanent teeth with reversible pulpitis.

2. Methodology

The study was conducted in the department of Conservative Dentistry and Endodontics, Ramaiah University. Healthy adult patients (18-60 years) with clinical diagnosis indicating reversible pulpitis (normal response to thermal and electric pulp testing along with radiological features of carious lesion approximating pulp with no periradicular changes), with presence of controllable bleeding at site of exposure, and diameter of exposed area between 0.5-1.5mm after caries excavation were chosen for the study. Pregnant women, immature teeth, patients on antibiotics, un-restorable teeth or periodontally compromised teeth were all excluded from the study. The study was evaluated and approved by the Institutional Review Board and Ethics committee, MSRUAS. A total of the 90 patients visiting the Outpatient Department of Conservative Dentistry and Endodontics, FDS-RUAS, were assessed for eligibility, wherein 50 patients met the inclusion criteria and were enrolled.

Following the diagnosis, local anesthetic (Xylocaine, Astra Zeneca, NSW) was ad-ministered. The teeth were isolated using Rubber Dam Kit (Hygienic, Coltene/Whaledent,Germany) and disinfected with 3 % NaOCl (VIP, Vensons India, Bengaluru). The cavity preparation and removal of undermined enamel was done using a sterile diamond/carbide bur (SS WHITE) with high speed airotor hand piece (NSK, Japan). On nearing the pulp, the dentin was air dried followed by application of Caries detector dye (Kurary, Osaka, Japan) for 10 seconds. Slow speed micromotor hand piece with a sterile round carbide bur was used to remove the remaining soft carious tissue and finally with spoon excavator (GDC Marketing, Hoshiarpur) was used in areas where the dye preferentially stained dentin. Removal of caries was done until no/minimal light pink color of dye was noticeable. During the removal of caries in case of pulp exposure, hemostasis was achieved using 3% NaOCl for 20-60 seconds. The bleeding was assessed from site of exposure. If the bleeding was not or if hemostasis was not achieved within 1-10 minutes, non-surgical root canal treatment was planned (Matsuo et al., 1996). Indirect pulp therapy was performed in those cases with no pulpal exposure, and remaining dentine thickness less than or equal to 0.5mm.

2.1 Restoration Procedure – On the Exposed Pulp

Group 1: The Biodentine (Septodont, 2015) was mixed and 2mm of Biodentine was placed followed by an interim restoration with GIC (GC Fuji IX GP).

Group 2: Bleeding was induced with a sterile probe over the exposure site. Platelet Rich fibrin was obtained from the patient's own blood drawn and centrifuged using a centrifugal machine (REMI R-8C) at 3000 revolutions per minute (rpm) for 10 minutes (Fig. 3.1) (Saluja et al., 2022).

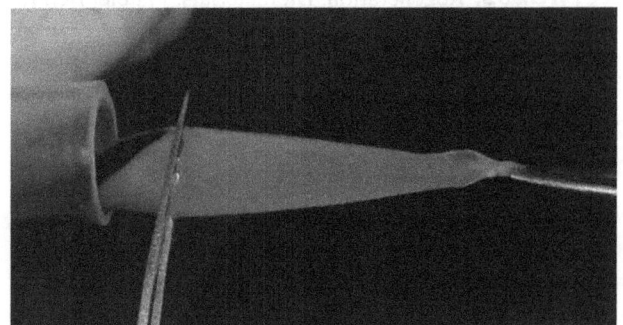

Fig. 3.1   Platelet rich protein obtained from patients own blood

Source: Author's compilation

After 10 minutes, the centrifuged PRF was cut in 2mm thickness and placed at the exposure site with gentle

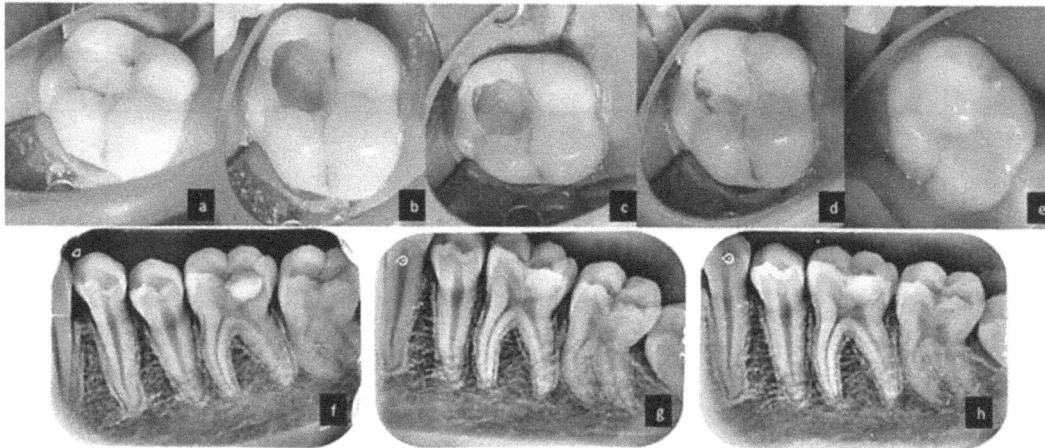

Fig. 3.2 a) Preoperative clinical picture of 36. b) Pinpoint exposure found. c) PRF placed on the exposure. d) Biodentine base placed. e) Final restoration with Type IX GIC. f) Preoperative radiograph. g) Radiograph after 24 hours. h) 4 month follow up radiograph

Source: Author's compilation

pressure applied using a ball burnisher to make it in a thin membrane. Biodentine was placed as 2mm thickness over the PRF and restored with interim restoration using Type 9 GIC (GC Fuji IX GP). Occlusal adjustment was checked and corrected if required. (Fig. 3.2).

Baseline clinical and radiological evaluation was done. Patient was asked to immediately report if any pain or other symptoms appeared between follow-up visits.

2.2 Recall Protocol

The Patient was recalled after 6 months and was informed about subsequent follow up at 12 months interval. In the subsequent clinical recall, subjective symptomology was evaluated. The patient was interrogated about sensitivity, mastication discomfort, and presence of pain. Electric pulp testing [Dental Pulp Tester, Minen Industries, Henan] was performed following which cold test (Roeko Endo Frost) and heat test (Heated gutta percha stick) was performed along with radiograph evaluation. The integrity of the interim restoration, presence/absence of swelling, periapical pathology, reparative dentin formation, pulpal calcifications, and canal obliteration were evaluated. GIC was reduced to a cavity base (dentine substitute level) and the teeth was restored using Resin Composite material. Treatment was considered successful based on the following features: absence of signs and symptoms of pulpal pathosis; lack of pain and tenderness to percussion; no soft tissue swelling, fistula, or abnormal mobility; absence of periapical rarefaction, internal or external resorption, and root canal obliteration; and normal pulp viability.

3. Results and Discussion

The data was entered in the excel spread sheet. Statistical analysis was done using SPSS (statistical Package for Social Sciences) version 20 (IBM SPSS statistics [IBM corp. released 2011]. The data obtained were blindly analyzed using the McNemar chi-square test.

Following the recall protocol, at 6 months follow-up, clinical and radiological findings were recorded. Vitality tests and follow-up radiographs revealed no adverse changes. At the 12 months follow-up, there were 3 (12%) subjects lost in both the groups, for various reasons, however on reaching out to the subjects via telephone conversation, they gave no adverse subjective symptoms. These subjects were considered as successful outcomes taking the lack of any adverse symptomology as a positive outcome [Table 3.1]. Clinical and radiological follow-up at 12 months revealed 3 (12%) subjects to have symptoms in the PRF with Biodentine group and the difference from baseline was not statistically significant (p=0.25). In the Biodentine group, 6 (24%) subjects had symptoms and the difference from baseline was statistically significant (p=0.031). Two patients developed pulp stones at 12 months follow- up which was identified radiographically but were asymptomatic.

This study focused on the biological approach to management of vital pulp exposures that inadvertently occur during management of deep caries lesions. A very specific criteria is used while selecting cases for vital pulp therapy which determine the outcome including nature of exposure (iatrogenic, caries, traumatic), size, aseptic condition at site, pulpal status, and materials to be used

Table 3.1 Distribution of clinical findings and radiological findings between baselines to 12 months amongst subjects

Groups	Baseline clinical findings	12 months - clinical findings		Lost to follow up	P Value (McNemar Test)
		Present	Absent		
PRF with Biodentine (N=25)	0	3 (12%)	19 (76%)	3 (12%)	0.25
Biodentine (N=25)	0	6 (24%)	16 (64%)	3 (12%)	0.031*
		12 months- radiological findings		Lost to follow up	
PRF with Biodentine (N=25)	0	3 (12%)	19 (76%)	3 (12%)	0.25
Biodentine (N=25)	0	6 (24%)	16 (64%)	3 (12%)	0.031*

*Significant

Source: Author

(Duncan et al., 2022,2019). Management of deep caries may be performed as selective, stepwise or non-selective removal. This study followed a selective removal approach using caries indicator dye, and sealed site of exposure in a single stage. Hemostasis is considered a key factor in determining a successful outcome (Barros et al., 2020). An inability to achieve hemostasis initially could be indicative of an inflammatory pulpal stasis. Uncontrolled bleeding also leads to poor moisture control and contamination of dentin at exposure site (Kalharo et al 2008). This would impair the overall seal of restorative material, and increase the likelihood of microbial contamination (Garcia-Godoy et al., 2005, Costa et al., 2001). In this study, hemostasis was achieved by placing cotton pellet soaked in sodium hypochlorite 3%, for 20-60 seconds.

Advances in biotechnology has paved the way for obtaining different platelet concentrates by using specific centrifugation methods (El Bagdadi et al., 2019). Vital pulp therapies have seen a paradigm shift with an increased focus on using bioactive agents (PRF, CGF), enzymes, biomimetic materials (MTA and calcium silicate derivatives). Pulp capping materials should possess ideal characteristics including induction of reparative dentin formation, sealing the exposure site by binding to dentin and the overlying restoration, and contributing to continual vitality of pulp (Cohen et al 1994).

3.1 Platelet Rich Fibrin as Pulp Capping Agent

PRF is a second-generation platelet concentrate that is obtained through a centrifugation process using patient's own blood.It is an autologous fibrin matrix that acts as a resorbable membrane rich in platelet cytokines, growth factors and cells. A sustained release of growth factors include Interleukin 1,4 and 6 (IL-1,4,6), Transforming growth factor BETA 1, Insulin like growth factor 1 and 2, Platelet derived growth factor and Cytokine

vascular endothelial growth factor.PRF when applied for regeneration provides a sustained release of cytokines and growth factors (Kang et al., 2011). This is in contrast to PRP which releases most of its growth factors in the first few hours and dissolves completely within 3 days. It has a major role in self-regulation of inflammatory and infectious phenomenon [6], promoting cell proliferation and differentiation (Choukroun et al., 2017)

Antibacterial activity has been demonstrated by PRF, however PRP has shown to be superior in this regard. In terms of tissue regeneration, root elongation, increased dentinal wall thickness and apical closure, PRF has been found to have more potential compared to PRP and blood clot derivatives (Narang et al., 2015, Murray et al., 2018). Better handling characteristics without the need for an external anticoagulant or biochemical agents, simple preparation process and high biocompatibility are its advantages over first generation PRP (Goswami et al., 2024). This study results are comparable to study done (Keswani et al., 2014) to evaluate MTA and PRF as pulpotomy agent in permanent teeth with incomplete root development in 62 patients. It was concluded that PRF could be a better alternative to MTA as pulpotomy agent.

A recent study evaluating PRF and MTA as pulpotomy agent concluded that at 6 months 90% success rate was observed in PRF and 92 % success rate in MTA group. There was statistically significant difference between the groups (Patidar et al., 2017).

3.2 Biodentine as Pulp Capping Agent

Biodentine, a calcium silicate-based cement bioactive material which is used as a dentin substitute owing to similar mechanical properties. It possess advantages of better marginal sealing ability and good mechanical properties when compound to Calcium hydroxide, and shorter setting time as well as easier handling properties as compared to MTA. It has positive effect on progenitor pulp stem cells, forming a reparative dentin-like matrix (Kunert et al., 2020) Rate of reparative dentin formation in biodentine was at faster rate compared to other materials such as MTA(Laurent et al., 2008, Bachoo et al., 2013)

Placement of a well-sealed restoration is crucial to the overall success. Owing to the definite and shorter setting time of biodentine, it was appropriately selected for this study. The use of glass ionomer cement was performed as the immediate restorative material over biodentine that was applied to the exposure site.

In this study the radiological success rates of PRF with biodentine (92%) and biodentine (84%) were comparable. However, at 12 months interval, alternate hypothesis stating that there is significant difference in the clinical and radiographic out-come of direct pulp capping performed with biodentine and PRF with biodentine in direct pulp therapy of permanent teeth was accepted. Thus, there was favorable long-term clinical and radiological outcome with success rate of 76% in PRF with biodentine group as compared to biodentine with success rate of 64%. This observation indicates that pulp capping may be a reliable treatment for pulp exposure in adult teeth treated with PRF with biodentine. This study had a total of 9 cases were categorized as failures (3 in PRF with biodentine and 6 in biodentine group). Clinically, the 9 cases demonstrated negative response to EPT, cold test and heat test, and radiological widening of periodontal space (n=4) and periapical radiolucency (n=5) was evident. At 12 months, 3 (12%) patients in the PRF with biodentine group and 6 (24%) in the biodentine group had clinical and radiological symptoms. Which implied those initial pre-operative diagnoses was correctly made.

3.3 Other Variables

Out of 50 cases selected, two cases one each in Biodentine and PRF with Biodentine showed presence of pulp stones. No change in the pulp stone size was observed in both the groups and absence of clinical response and radiological findings were observed. Thus, calcification did not essentially indicate a case of failure which is in accordance with literature available. However, presence of calcifications could increase the difficulty of subsequent endodontic procedure, if indicated. Favorable outcomes of vital pulp therapy have been reported in young permanent teeth. This study found no significant difference in terms of age and gender (Cushley et al., 2021). Histological evidence is considered gold standard for assessing success of direct pulp capping, however requires destroying the tooth. CBCT has been used to assess dentin bridge

formation, 31/40 and this may be considered a limitation to this study's design (Ludlow et al., 2015).

In this study a standardized technique for obtaining treatment radiographs employed the use of a stent designed to stabilize the position of the guide ring. Radiological assessment performed subsequently was to assess the outcome in terms of occurrence of any periapical changes (Nowicka et al., 2015). Notwithstanding the promising results of this study, there were certain limitations. Firstly, there was no quantitative analysis of reparative dentin formation was done in this study. Secondly, it is a technique sensitive and tedious procedure, requiring specialized equipment and training. Thirdly, the limited sample size may have contributed to the absence of statistically significant differences in the success rates of the different groups. Correlation of age, gender and demographics could be considered confounding factors.

4. Funding

No conflicts of interest.

5. Conclusion

In conclusion, the continuously and rapidly advancing field of Regenerative dentistry demands development of an ideal pulp capping agent with both hermetic seal and effective dentin regeneration. This study attempted to evaluate the effectiveness of using pulp capping agent Platelet Rich Fibrin with biodentine in permanent teeth with reversible pulpitis and found it to be superior to when performed with biodentine alone. Further research on this promising potential pulp capping agent is required in terms of longer follow-up period and specific biomechanical mechanisms involved.

References

1. Bachoo, I.K., Seymour, D. and Brunton, P., (2013). A biocompatible and bioactive replacement for dentine: is this a reality? The properties and uses of a novel calcium-based cement. *British dental journal, 214*(2),.E5–E5
2. Barros, M.M.A.F., De Queiroz Rodrigues, M.I., Muniz, F.W.M.G. and Rodrigues, L.K.A., (2020). Selective, stepwise, or nonselective removal of carious tissue: which technique offers lower risk for the treatment of dental caries in permanent teeth? A systematic review and meta-analysis. *Clinical oral investigations, 24*:521–532.
3. Choukroun, J. and Miron, R.J., (2017). *Platelet rich fibrin in regenerative dentistry* (pp. 1-14). Oxford: Wiley-Blackwell..
4. Choukroun, J., Diss, A., Simonpieri, A., Girard, M.O., Schoeffler, C., Dohan, S.L., Dohan, A.J., Mouhyi, J. and Dohan, D.M., (2006). Platelet-rich fibrin (PRF): a

second-generation platelet concentrate. Part V: histologic evaluations of PRF effects on bone allograft maturation in sinus lift. *Oral Surgery, Oral Medicine, Oral Pathology, Oral Radiology, and Endodontology, 101*(3):299–303.

5. Cohen, B.D. and Combe, E.C., (1994). Development of new adhesive pulp capping materials. *Dental update, 21*(2): 57–62

6. Costa, C.A., Edwards, C.A. and Hanks, C.T., (2001). Cytotoxic effects of cleansing solutions recommended for chemical lavage of pulp exposures. *American Journal of Dentistry, 14*(1):25–30.

7. Cushley, S., Duncan, H.F., Lappin, M.J., Chua, P., Elamin, A.D., Clarke, M. and El-Karim, I.A., (2021). Efficacy of direct pulp capping for management of cariously exposed pulps in permanent teeth: a systematic review and meta-analysis. *International endodontic journal, 54*(4):556–571.

8. Duncan, H.F., (2022). Present status and future directions—Vital pulp treatment and pulp preservation strategies. *International Endodontic Journal, 55:*497–511.

9. El Bagdadi, K., Kubesch, A., Yu, X., Al-Maawi, S., Orlowska, A., Dias, A., Booms, P., Dohle, E., Sader, R., Kirkpatrick, C.J. and Choukroun, J., 2019. Reduction of relative centrifugal forces increases growth factor release within solid platelet-rich-fibrin (PRF)-based matrices: a proof of concept of LSCC (low speed centrifugation concept). *European Journal of Trauma and Emergency Surgery, 45:*467–479.

10. European Society of Endodontology (ESE) developed by:, Duncan, H.F., Galler, K.M., Tomson, P.L., Simon, S., El-Karim, I., Kundzina, R., Krastl, G., Dammaschke, T., Fransson, H. and Markvart, M., (2019). European Society of Endodontology position statement: Management of deep caries and the exposed pulp. *International Endodontic Journal, 52*(7):923–934.

11. Garcia-Godoy, F. and Murray, P.E., (2005). Systemic evaluation of various haemostatic agents following local application prior to direct pulp capping. *Braz J Oral Sci, 4*(14):791–797

12. Goswami, P., Chaudhary, V., Arya, A., Verma, R., Vijayakumar, G. and Bhavani, M., (2024). Platelet-Rich Fibrin (PRF) and its Application in Dentistry: A Literature Review. *Journal of Pharmacy and Bioallied Sciences, 16*(Suppl 1):S5–S7

13. Gupta, A., Makani, S., Vachhani, K., Sonigra, H., Attur, K. and Nayak, R., (2016). Biodentine: an effective pulp capping material. *Sch J Dent Sci, 3*(1):15–19..

14. Hanna, S.N., Alfayate, R.P. and Prichard, J., (2020). Vital pulp therapy an insight over the available literature and future expectations. *European endodontic journal, 5*(1):46

15. Huang, F.M., Yang, S.F., Zhao, J.H. and Chang, Y.C., (2010). Platelet-rich fibrin increases proliferation and differentiation of human dental pulp cells. *Journal of endodontics, 36*(10):1628–1632

16. Kalhoro, F.A. And Katpar, S., (2008). Direct Pulp Capping: Degree of Bleeding is a Strong Candidate as Prognosis Index, In Vivo Study.

17. Kang, Y.H., Jeon, S.H., Park, J.Y., Chung, J.H., Choung, Y.H., Choung, H.W., Kim, E.S. and Choung, P.H., (2011). Platelet-rich fibrin is a Bioscaffold and reservoir of growth factors for tissue regeneration. *Tissue Engineering Part A, 17*(3-4): 349–359.

18. Keswani, D., Pandey, R.K., Ansari, A. and Gupta, S., (2014). Comparative evaluation of platelet-rich fibrin and mineral trioxide aggregate as pulpotomy agents in permanent teeth with incomplete root development: a randomized controlled trial. *Journal of endodontics, 40*(5): 599–605

19. Kunert, M. and Lukomska-Szymanska, M., 2020. Bio-inductive materials in direct and indirect pulp capping—a review article. *Materials, 13*(5):1204

20. Laurent, P., Camps, J., De Méo, M., Déjou, J. and About, I., (2008). Induction of specific cell responses to a Ca3SiO5-based posterior restorative material. *Dental materials, 24*(11):1486–1494..

21. Ludlow, J.B., Timothy, R., Walker, C., Hunter, R., Benavides, E., Samuelson, D.B. and Scheske, M.J., 2015. Effective dose of dental CBCT—A Meta analysis of published data and additional data for nine CBCT units. *Dentomaxillofacial Radiology, 44*(1):20140197.

22. Murray, P. E., Garcia-Godoy, F., & Hargreaves, K. M. (2007). Regenerative Endodontics: A Review of Current Status and a Call for Action. Journal of Endodontics, 33(4), 377–390. https://doi.org/10.1016/j.joen.2006.09.013

23. Murray, P.E., 2018. Platelet-rich plasma and platelet-rich fibrin can induce apical closure more frequently than blood-clot revascularization for the regeneration of immature permanent teeth: a meta-analysis of clinical efficacy. *Frontiers in bioengineering and biotechnology, 6:*139..

24. Narang, I., Mittal, N. and Mishra, N., (2015). A comparative evaluation of the blood clot, platelet-rich plasma, and platelet-rich fibrin in regeneration of necrotic immature permanent teeth: a clinical study. *Contemporary clinical dentistry, 6*(1):63–68

25. Nowicka, A., Wilk, G., Lipski, M., Kołecki, J. and Buczkowska-Radlińska, J., (2015). Tomographic evaluation of reparative dentin formation after direct pulp capping with Ca (OH) 2, MTA, Biodentine, and dentin bonding system in human teeth. *Journal of endodontics, 41*(8):1234–1240.

26. Patidar, S., Kalra, N., Khatri, A. and Tyagi, R., (2017). Clinical and radiographic comparison of platelet-rich fibrin and mineral trioxide aggregate as pulpotomy agents in primary molars. *Journal of Indian Society of Pedodontics and Preventive Dentistry, 35*(4):367–373.

27. Saluja, H., Dehane, V. and Mahindra, U., (2011). Platelet-Rich fibrin: A second generation platelet concentrate and a new friend of oral and maxillofacial surgeons. *Annals of maxillofacial surgery, 1*(1):53–57..

28. Torabinejad, M., Parirokh, M. and Dummer, P.M., (2018). Mineral trioxide aggregate and other bioactive endodontic cements: an updated overview–part II: other clinical applications and complications. *International endodontic journal, 51*(3):284–317.

Advances in Materials Science and Technology – Dr. Srikari Srinivasan et al. (eds)
© 2025 Taylor & Francis Group, London, ISBN 978-1-041-12342-2

4

Formulation and Characterization of a Novel Drug Loaded Nano-Sponge for Vital Pulp Therapy

Ayesha Najam, Shruthi Nagaraja*

Department of Conservative Dentistry and Endodontics,
Faculty of Dental Sciences, M S Ramaiah University of Applied Sciences,
Bengaluru

Deveswaran Rajamanickam

Professor and Head,
Department of Pharmaceutics, Drug Design and Development Centre,
Faculty of Pharmacy, M S Ramaiah University of Applied Sciences,
Bengaluru, Bangalore, Karnataka, India

Sylvia Mathew,
Poornima Ramesh, Nivaskumar

Department of Conservative Dentistry and Endodontics,
Faculty of Dental Sciences, M S Ramaiah University of Applied Sciences,
Bengaluru

ABSTRACT: To formulate and evaluate the pH, drug release, calcium release and cytotoxicity of the novel drug formulation. Nano-sponge, Ly-PRF, nano-calcium were prepared separately, along with Diclofenac sodium mixed with water in the ratio of 1:2:2:1 by weight, and centrifuged. The supernatant obtained was lyophilized to obtain the novel Drug Loaded Nano-sponge (DLNS) and evaluated for pH, drug release, calcium release and cytotoxicity, with MTA as a control group. The DLNS displayed an increase in pH from 8.1 to 9.6 after 28 days while MTA had a rise in pH from 11.7 to 12.24 after 28 days. The drug and calcium release were sustained and increased with time in DLNS. Both the groups exhibited excellent biocompatibility and no cytotoxicity at 24 and 120 hours. The formulated DLNS exhibited favourable properties that closely resembles the ideal properties of a biomaterial beneficial for pulp therapy.

KEYWORDS: Local drug delivery, Lyophilized platelet rich fibrin (Ly PRF), Nano-sponge, Platelet concentrates, Pulp capping materials, Vital pulp therapy

1. Introduction

Vital Pulp Therapy (VPT) is gaining popularity over conventional root canal therapy owing to predictable outcomes and long-term tooth retention 9 (Yu et al., 2007, Langeland., 1987). There is also a broadened horizon over the indications of VPT wherein, even teeth with irreversible pulpitis and mild inflammation are also being managed without conventional root canal therapy. However, the success of VPT depends on the ability of pulp capping

---

*Corresponding author: shruthi.cd.ds@msruas.ac.in

DOI: 10.1201/9781003664277-4

materials to reduce pulpal inflammation, promote continued formation of dentin-pulp complex, and restore vitality by stimulating re-growth of pulpal tissue (Andelin et al, 2003). Although Calcium Hydroxide [Ca(OH)$_2$] has been the gold standard amongst pulp capping materials, it has inherent drawbacks such as tunnel defects in dentin bridge formation, ineffective long-term seal against microleakage and degradation under permanent restorations. Mineral Trioxide Aggregate (MTA), Biodentine, modifications of Calcium-silicate based materials, along with scaffolds, enzymes, growth factors and platelet concentrates have been used as pulp capping materials (Kunert et al., 2020). However, MTA has also exhibited numerous drawbacks such as difficulty in handling, long setting time, and potential discoloration to the tooth (Parirokh, M. et al, 2010). There is thus a need to develop a bioengineered pulp capping material that will have predictable outcomes in terms of pulp tissue regeneration, maintenance and restoration of pulp vitality and improved longevity.

Local drug delivery systems have the potential to carry and release a therapeutic agent at a desired concentration at an optimal time with a preferred effect on specific target sites. In VPT, biomaterials may be transferred to the area of pulp exposure on a specially designed vehicle which may prepare the environment conducive to healing of pulp tissue. The direct transfer and subsequent release of loaded biomaterials on wounded pulp in the required dose, in a controlled manner should lead to reduction in inflammation, promoting cell differentiation (Parhizkar et al., 2021). One such system includes nano-sponges that are primarily nanosized drug carriers with a three-dimensional structure created by crosslinking polymers, that can provide a controlled drug release with targeted drug delivery. The potential of nano-sponges to act as a carrier for pulp capping agents needs to be explored.

Platelet concentrates such as Platelet-rich fibrin (PRF) have gained popularity in VPT due to its rich reservoir of autologous growth factors including platelet-derived growth factor (PDGF), transforming growth factor beta (TGFβ), and vascular endothelial growth factors (VEGF). To maximize therapeutic applicability of PRF, a viable strategy to improve storage, maintain growth factor release and microarchitecture, lyophilization or a freeze-drying method was proposed Ngah, (N.A., et al, 2021). Literature reports improved clinical success rates for lyophilized platelet derived preparation with Ca(OH)$_2$ as pulpotomy agents in primary molars (Kalaskar et al, 2004).).

The ability of pulp capping materials to mediate reparative dentin synthesis and the quality of dentin bridge formation is essential for successful VPT (Bahammam et al, 2024). Synthetic calcium phosphate has excellent biocompatibility

due to its similarity in chemical properties with the inorganic component of calcified tissues (Cai et al, 2008). Nanostructured calcium phosphates (NCP) are highly soluble with small particle size and large surface area that facilitates rapid release of Ca^{2+} and PO4$^-$ ions, enhancing potential for tooth remineralization (Ridi et al, 2017).

With this background, it is proposed that a nano-sponge based delivery system containing lyophilized PRF (Ly-PRF), nano-calcium along with an anti-inflammatory drug such as diclofenac sodium could potentially help reduce inflammation of the pulp aiding in its regeneration and repair, especially in VPT procedures. Such a formulation has not been assessed previously in literature and thus the aim of the present study was to formulate a novel pulp capping agent and evaluate its pH, drug release, calcium release and cytotoxicity.

2. Methodology

Ethics clearance was obtained from the University Ethics Committee for Human Trials (Ref no. EC-2021/PG/064) prior to commencement of study. The novel formulation consisting of a nano-sponge, Ly-PRF and nano-calcium were fabricated separately following specifications by Ngah et al 2021 and Aidaros et al 2018, along with addition of diclofenac sodium. The details of fabrication of DLNS and characterization are described in the following sections.

2.1 Developing A Novel Drug Loaded Nano-Sponge

Preparation of β-Cyclodextrin Nanosponges

8.6 gm of β-cyclodextrin was dissolved in 50 ml of dimethylformamide in a round bottom flask and 6 gm of glutaraldehyde solution was added dropwise, to react for 4 hours at 1000C until completely dry. 30 ml of deionized water was added to remove any unreacted dimethylformamide. Residual by-products were dried in a water bath at 1000C for 2 days and subsequently ground to a fine powder. This was dispersed in water and lyophilized at a temperature of -500C with 13.33 mbar pressure for 8 hours to recover nano-sponges (Gangadharappa et al, 2017).

Preparation of Lyophilized PRF

Following obtaining informed consent, venous blood was collected from 6 healthy volunteers and centrifuged at 3000 rpm for 12 minutes for obtaining PRF. This was frozen overnight in a refrigerator and lyophilized at -54°C for 8 hours. Ly-PRF was stored in air-tight bags at 200C until usage (Ngah et al., 2021).

Formulation of Nano-Calcium

7 g of calcium phosphate was dissolved in 100ml of distilled water and sonicated by probe sonicator for 15 minutes. The mixture was spread on a petridish and dried in a tray dryer at 400C until a dry residue was obtained, which was collected and weighed (Aidaros et al 2018).

Preparation of Drug Loaded Nanosponges

Nano-sponge, Ly-PRF, nano calcium and diclofenac sodium in the ratio of 1:2:2:1 by weight were suspended in 20 ml of distilled water using a magnetic stirrer and stirred continuously for 24 hours. This suspension was then centrifuged for 2000 rpm for 10 mins to separate uncomplexed drug as a residue below the colloidal supernatant. This supernatant was lyophilized at a temperature of -50°C and 13.33 mbar pressure for 6 hours. The resulting DLNS were recovered and stored in a dessicator until further use, designated as Group 1.

2.2 Preparation of MTA (Control)

The novel formulation was compared with MTA as a control. MTA was mixed according to manufacturer instructions using a glass slab and metal spatula. The mixed cement was placed into stainless steel molds to obtain samples designated as Group 2.

2.3 pH Analysis

A digital pH meter (Model No. Mk –Vi Systronics, India) was first calibrated by a buffer solution of pH 4 and 7. pH of all samples of Group 1 & 2 was measured by placing the electrode inside a falcon tube containing each sample and deionized water at 3 hours, 24 hours, 14 days and 28 days. The electrode was washed with distilled water and wiped dry between readings.

2.4 Calcium Ion Release

EDTA Titration method was used for calcium ion release analysis for both groups. The same solution used to test pH was analyzed for Ca2+ release (Tanomaru-Filho et al 2009). The calcium ion concentration was measured at 3 hours, 2 hours, 14 days and 28 days for both groups (Itoh et al 1970).

2.5 Drug Release of Novel Formulation (DLNS)

The release profile was obtained by soaking triplicate samples of DLNS in 10 mL of simulated body fluid (SBF) at pH = 7.4 and placed in a water bath at 37 °C. The concentration of the drug release (diclofenac sodium) into SBF was evaluated by UV spectroscopy (Uv-1900i Uv-Vis Spectrophotometer, Europe) at 250 nm and 383 nm and assessed from 15 to 90 minutes.

2.6 Cytotoxicity

The cytotoxicity of both the groups on the Human dental pulp tissue-derived stem cells (hDPSCs) were assessed using MTT assay at 24 hrs and 120 hrs on samples with 5 different drug concentrations of 31.25, 62.5, 125, 250, 500 µg/ml along with blank, untreated and standard concentration samples of 12.5 µg/ml. The absorbance was read on a spectrophotometer at 570nm and 630nm.

3. Statistical Analysis

The Statistical software SPSS 19.0 was used for the analysis of the data and Microsoft word and Excel have been used to generate graphs, tables etc. Unpaired t test was applied to compare mean of different variables across groups. A $p<0.05$ was considered as statistically significant and $p<0.01$ as highly significant.

4. Results

4.1 pH Analysis

Both groups revealed an alkaline pH at all tested intervals (Table 4.1). DLNS (Group 1) revealed a pH of 8.1 at the end of 3 hours, which significantly increased to 9.34 at the end of 28 days. MTA control (Group 2) revealed a pH of 11.74 at the end of 3 hours, which significantly increased to 12.24 at the end of 28 days.

Table 4.1 Multiple pairwise comparison of pH at different time intervals for DLNS (Group 1) and MTA control (Group 2) pH Analysis

Variable	Group	Mean	SD	T	p Value
DLNS (Group 1)	3 Hours	8.11	0.02	572.2	0.0001*
	24 Hours	9.14	0.02	177.3	
	14 Days	9.23	0.02	479.5	
	28 Days	9.34	0.02	411.07	
MTA control (Group 2)	3 Hours	11.74	0.02	572.2	0.0001*
	24 Hours	12.02	0.06	177.3	
	14 Days	12.12	0.02	479.5	
	28 Days	12.24	0.02	411.07	

4.2 Calcium Ion Release

Group 1 revealed a Ca2+ release of 17.7mg/g at 3 hours which significantly increased to 76.83 mg/g at the end of 28 days. Group 2 revealed a release of 11.3 mg/g at 3 hours which significantly increased to 79.23 mg/g at the end of 28 days (Table 4.2).

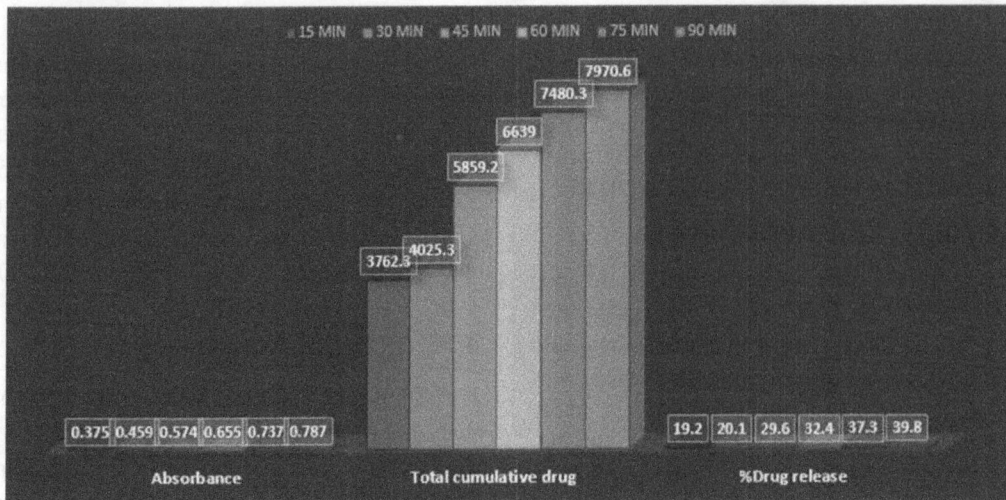

Fig. 4.1 Mean drug absorbance, total cumulative drug and percentage of drug release (diclofenac sodium) during different time intervals by the drug loaded nano-sponge

Table 4.2 Multiple pairwise comparison of calcium release at different time intervals for DLNS (Group 1) and MTA control (Group 2) pH analysis

Variable	Group	Mean	SD	T	p Value
DLNS (Group 1)	3 Hours	17.7	0.31	32.2	0.0001*
	24 Hours	34.7	0.94	70.4	
	14 Days	69.13	0.71	64.4	
	28 Days	76.83	0.65	5.2	
MTA control (Group 2)	3 Hours	11.3	0.37	32.2	0.0001*
	24 Hours	6.84	0.24	70.4	
	14 Days	90.49	0.39	64.4	
	28 Days	79.23	0.93	5.2	

4.3 Drug Release of DLNS

The release of diclofenac sodium from DLNS was assessed at time intervals of 15 mins, 30 mins, 45 mins, 60 mins, 75 mins and 90 mins (Fig. 4.1). The absorbance rate of diclofenac sodium increased with time. There was a sustained increase in drug release (diclofenac sodium) with time which peaked at 90mins.

4.4 Cytotoxicity

The MTT assay suggested that the hDPSCs treated with DLNS did not reveal any toxicity at both 24 hours and 120 hours. The cell viability was highest at a drug concentration of 250 µg/ml with 98% and 95 % cell viability at 24 hours and 120 hours respectively (Fig. 4.2). MTA group also showed no toxicity, after the treatment periods of 24hrs and

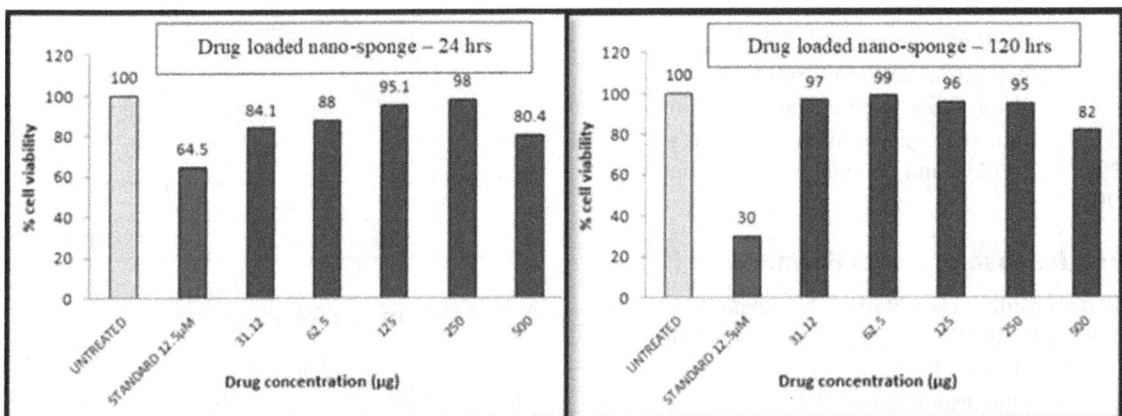

Fig. 4.2 Drug concentration and cell viability assessed by MTT assay for drug loaded nano sponge (Group 1) at 24 hours and 120 hours

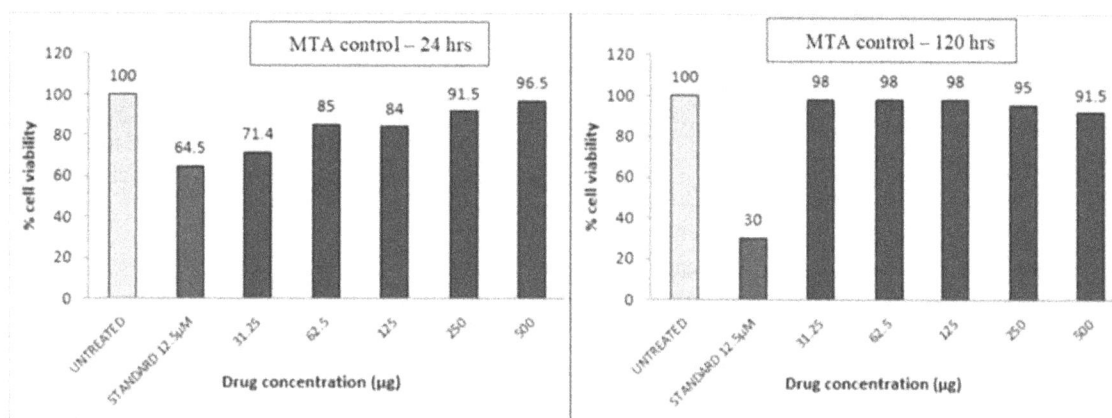

Fig. 4.3 Drug concentration and cell viability assessed by MTT assay for MTA control group (Group 2) at 24 hours and 120 hours

120 hrs(Fig. 4.3). This suggests that the cell viability of the DLNS was similar to MTA.

5. Discussion

The main role of biomaterials used for VPT is to seal the pulpal wound, stimulate formation of tertiary dentine, promote dentinogenic potential of pulpal cells to ultimately retain the tooth as a functional unit (Asgary et al, 2018, Tziafas et al, 2017). It should also have the potential to lower inflammation and induce pulpal healing. None of the biomaterials available today have been able to satisfy all the requirements of an ideal agent for VPT. Ca(OH)2, although considered as a gold standard for pulp capping, has poor adhesion to dentin, porosity in the dentinal bridge with tunnel defects, and an inability to provide a long-term seal against microleakage (Cohenca et al, 2013). Long setting times of MTA is an obstacle for its usage as a pulp capping material since MTA needs to be layered with other materials while it is still fresh (Bogen et al, 2008). Advances in biomaterial research are focused on the development of material that can generate the most beneficial tissue response to improve clinical outcomes in VPT. In this study, an attempt was made to formulate a novel DLNS that can potentially combine therapeutic benefits of Ly-PRF, nano-calcium, nano-sponge, and diclofenac sodium.

Nano-sponges are a type of LDD system that can encapsulate both hydrophilic and lipophilic moieties, to be transferred to a target site. In this study, we used solvent method for preparation of Beta-cyclodextrine nano-sponge. Cyclodextrine based nano-sponges offer a unique advantage of controlled release, are biologically safe and can incorporate a wide range of compounds (Sherje et al, 2017). The scale-up process is easy and hence can be commercialized. They exhibit higher degree of entrapment

efficiency due to their specialized porous structure and are well-tolerated in cell culture conditions (Wagle et al, 2020). In this study as well, the DLNS had high cell viability of 98% and 95% at a concentration of 250 µg/ml at 24 hours and 120 hours respectively, similar to that of MTA. This demonstrates that the DLNS has good potential and holds promise for clinical usage. However further animal studies are necessary to assess odontogenic and osteogenic potential of this novel pulp capping material on the hDPSCs, prior to clinical usage. Ly-PRF was added to the novel formulation since it is a rich reservoir of growth factors that can facilitate reparative dentin formation. The main advantage of Ly-PRF is the ability to store the same and improved shelf life as opposed to conventional PRF (Ngah et al., 2021).

In the present study, both groups recorded an alkaline pH at all time intervals. It is known that MTA releases hydroxyl ions that create an alkaline environment which adversely affects bacterial survival and facilitates mineralization through release of alkaline phosphatase and bone morphogenetic protein 2 (Parirokh et al, 2010). The alkaline pH of DLNS is possibly due to high calcium and hydroxyl ion release owing to the presence of nano-calcium in its formulation. Moreover, diclofenac sodium also has an alkaline pH of 8.15 which adds to the alkalinity of this formulation.

The DLNS had a higher initial release of calcium at 3 hours and 24 hours which was sustained and increased over a period of 28 days as the calcium in nano form improves its bioavailability and surface action (Cai et al, 2008). Ca2+ release is essential for successful regeneration as it increases the proliferation of hDPSCs in a dose-dependent manner and enhances the activity of pyro-phosphatase that help in dentin bridge formation (Gandolfi et al, 2014). In our study, MTA also displayed a sustained high release of Ca2+ up to the end of 28 days which is in accordance with existing literature (Natale et al, 2015). Thus, the novel DLNS presents a viable option for use as a pulp capping agent as it can provide a favorable environment for enhanced mineralization.

The ability of a biomaterial to resolve pulpal inflammation to achieve predictable regeneration of the dentin-pulp complex has remained elusive. Although the dentin-pulp complex can react naturally to injury by forming a bridge of reparative dentin, this process is significantly impaired if inflammation persists. Since secretion of inflammatory cytokines by injured pulpal cells causes significant pain and discomfort to patients, it is critical to resolve pulpal inflammation to create a micro-environment conducive for pulpal healing (Arora et al, 2021). An anti-inflammatory medication mixed with a drug delivery vehicle for pulp-capping can potentially curb inflammatory processes in a caries exposed pulp tissue (Louwakul et al, 2012). A previous study observed that the association of NSAIDs did not interfere in the pH of Ca (OH)2 pastes and significantly increased antimicrobial activity with the best results being found with diclofenac sodium (de Freitas et al, 2017). In this study, diclofenac sodium was added to the novel formulation to provide a mild anti-inflammatory effect that is required to stimulate dentin bridge formation and healing while diminishing post-operative pain. Here, it was found that the release of diclofenac sodium from the drug loaded nano-sponge increased over time and peaked at 90 mins.

This study attempted to formulate a novel drug loaded nano-sponge for potential use as a pulp capping agent. The preliminary findings found in this study show favorable results to further explore its potential usage as a pulp capping agent. However, it is essential to perform further physico-chemical and biological characterization of the novel formulation before validating this experimental DLNS for clinical use.

6. Conclusion

The DLNS exhibited favorable properties such as alkaline pH, calcium ion release, drug release and no cytotoxicity which was comparable to MTA. Within the limitations of this study, we conclude that the novel formulation of DLNS containing Ly-PRF, nano-calcium and diclofenac sodium has potential to be explored for usage as a pulp capping agent. Further studies are needed to evaluate other physical, chemical and biological parameters of the novel formulation along with animal studies to determine potential clinical usage.

Acknowledgments

The authors acknowledge M.S. Ramaiah University of Applied Sciences, Faculty of Pharmacy and Faculty of Dental Sciences. The authors would like to thank Lakshmi Maternity, Bengaluru and Averin Biotech Laboratory for supporting the study.

References

1. Aidaros, N.H., Niazy, M.A. and El-yassaky, M.A. (2018). The Effect of Incorporating Nanocalcium Phosphate Particles into Biodentine on Pulpal Tissue Response (In Vitro Study). Al-Azhar Dental Journal for Girls, 5(2):181–186.
2. Andelin, W.E., Shabahang, S., Wright, K. and Torabinejad, M. (2003). Identification of hard tissue after experimental pulp capping using dentin sialoprotein (DSP) as a marker. Journal of endodontics, 29(10):646–650.
3. Arora, S., Cooper, P.R., Friedlander, L.T., Rizwan, S., Seo, B., Rich, A.M. and Hussaini, H.M. (2021). Potential application of immunotherapy for modulation of pulp inflammation: opportunities for vital pulp treatment. International Endodontic Journal, 54(8):1263–1274.
4. Asgary, S., Hassanizadeh, R., Torabzadeh, H. and Eghbal, M.J. (2018). Treatment outcomes of 4 vital pulp therapies in mature molars. Journal of endodontics, 44(4):529–535.
5. Bahammam, L.A., Alsharqawi, W., Bahammam, H.A., Mounir, M. and Mounir Sr, M.M. (2024). Histological evaluation of pulpal response and dentin bridge formation after direct pulp capping using recombinant Amelogenin and mineral trioxide aggregate (MTA). Cureus, 16(2).
6. Bogen, G., Kim, J.S. and Bakland, L.K. (2008). Direct pulp capping with mineral trioxide aggregate. The Journal of the American Dental Association, 139 (3):305–15.
7. Cai, Y. and Tang, R. (2008). Calcium phosphate nanoparticles in biomineralization and biomaterials. Journal of Materials Chemistry, 18(32):3775–3787.
8. Cohenca, N., Paranjpe, A. and Berg, J. (2013). Vital pulp therapy. Dental Clinics, 57(1):59–73.
9. de Freitas, R.P., Greatti, V.R., Alcalde, M.P., Cavenago, B.C., Vivan, R.R., Duarte, M.A.H., Weckwerth, A.C.V.B. and Weckwerth, P.H. (2017). Effect of the association of nonsteroidal anti-inflammatory and antibiotic drugs on antibiofilm activity and pH of calcium hydroxide pastes. Journal of endodontics, 43(1):131–134.
10. Gandolfi, M.G., Siboni, F., Primus, C.M. and Prati, C. (2014). Ion release, porosity, solubility, and bioactivity of MTA Plus tricalcium silicate. Journal of endodontics, 40(10):1632–1637.
11. Gangadharappa, H.V., Prasad, S.M.C. and Singh, R.P. (2017). Formulation, in vitro and in vivo evaluation of celecoxib nanosponge hydrogels for topical application. Journal of Drug Delivery Science and Technology, 41:488–501.

12. Itoh, A. and Ueno, K. (1970). Evaluation of 2-hydroxy-1-(2-hydroxy-4-sulpho-1-naphthylazo)-3-naphthoic acid and hydroxynaphthol blue as metallochromic indicators in the EDTA titration of calcium. Analyst, 95(1131):583–589.

13. Kalaskar, R.R. and Damle, S.G. (2004). Comparative evaluation of lyophilized freeze dried platelet derived preparation with calcium hydroxide as pulpotomy agents in primary molars. Journal of Indian Society of Pedodontics and Preventive Dentistry, 22(1):24–29.

14. Kunert, M. and Lukomska-Szymanska, M. (2020). Bio-inductive materials in direct and indirect pulp capping—a review article. Materials, 13(5):1204.

15. Langeland, K. (1987). Tissue response to dental caries. Dental Traumatology, 3(4):149–171.

16. Louwakul, P. and Lertchirakarn, V. (2012). Incorporation of anti-inflammatory agent into calcium hydroxide pulp capping material: an in vitro study of physical and mechanical properties. Dental materials journal, 31(1):32–39.

17. Natale, L.C., Rodrigues, M.C., Xavier, T.A., Simões, A., De Souza, D.N. and Braga, R.R. (2015). Ion release and mechanical properties of calcium silicate and calcium hydroxide materials used for pulp capping. International endodontic journal, 48(1):89–94.

18. Ngah, N.A., Dias, G.J., Tong, D.C., Mohd Noor, S.N.F., Ratnayake, J., Cooper, P.R. and Hussaini, H.M. (2021). Lyophilised platelet-rich fibrin: physical and biological characterisation. Molecules, 26(23):7131.

19. Parhizkar, A. and Asgary, S. (2021). Local drug delivery systems for vital pulp therapy: a new hope. International Journal of Biomaterials, 2021(1):5584268.

20. Parirokh, M. and Torabinejad, M. (2010). Mineral trioxide aggregate: a comprehensive literature review—part I: chemical, physical, and antibacterial properties. Journal of endodontics, 36(1):16–27.

21. Parirokh, M. and Torabinejad, M. (2010). Mineral trioxide aggregate: a comprehensive literature review—part III: clinical applications, drawbacks, and mechanism of action. Journal of endodontics, 36(3):400–413.

22. Ridi, F., Meazzini, I., Castroflorio, B., Bonini, M., Berti, D. and Baglioni, P. (2017). Functional calcium phosphate composites in nanomedicine. Advances in Colloid and Interface Science, 244:281–295.

23. Sherje, A.P., Dravyakar, B.R., Kadam, D. and Jadhav, M. (2017). Cyclodextrin-based nanosponges: A critical review. Carbohydrate polymers, 173:37–49.

24. Tanomaru-Filho, M., Faleiros, F.B.C., Saçaki, J.N., Duarte, M.A.H. and Guerreiro-Tanomaru, J.M. (2009). Evaluation of pH and calcium ion release of root-end filling materials containing calcium hydroxide or mineral trioxide aggregate. Journal of endodontics, 35(10):1418–1421.

25. Tziafas, D., Smith, A.J. and Lesot, H. (2000). Designing new treatment strategies in vital pulp therapy. Journal of dentistry, 28(2):77–92.

26. Wagle, S.R., Kovacevic, B., Walker, D., Ionescu, C.M., Shah, U., Stojanovic, G., Kojic, S., Mooranian, A. and Al-Salami, H. (2020). Alginate-based drug oral targeting using bio-micro/nano encapsulation technologies. Expert Opinion on Drug Delivery, 17(10):1361–1376.

27. Yu, C. and Abbott, P.V. (2007). An overview of the dental pulp: its functions and responses to injury. Australian dental journal, 52:S4–S6.

Note: All the figures and tables in this chapter were made by the author.

Advances in Materials Science and Technology – Dr. Srikari Srinivasan et al. (eds)
© *2025 Taylor & Francis Group, London, ISBN 978-1-041-12342-2*

5

Comparison of the Retentive Properties of a Removable Partial Denture Clasp Fabricated by 3-D Printing and Laser Sintering — An In-Vitro Study

Akshitha H. M.
Department of Prosthodontics,
Krishnadevaraya College of Dental Sciences,
Bangalore

Gayathridevi S. K.*, Ravishnakar
Department of Prosthodontics,
Faculty of Dental Sciences, M S Ramaiah University of Applied Sciences,
Bangalore

Afreen Shadan
Healthcare Management Student,
Cape Breton University (Nova Scotia, Canada) and Clinical trainee at
Park Lane Dental Specialists (Halifax, Canada)
Bangalore, Karnataka, India

Babashankar Alva
Department of Prosthodontics,
Faculty of Dental Sciences, M S Ramaiah University of Applied Sciences,
Bangalore

ABSTRACT: In patients who are partially edentulous, and the practicality of a fixed prosthesis is unknown, removable partial dentures remain a successful treatment option. The most commonly utilized retention component is Acker's clasp, and retention is one of the most important aspects for RPD to function well. The lost wax process has historically been used to make removable partial dentures. Since CAD/CAM's introduction, the 3-D printed wax pattern method and Direct Metal Laser Sintering (DMLS) technologies have simplified, if not replaced, the traditional lost wax procedure in dentistry. The retentive qualities of the Ackers clasp manufactured using CAD/CAM techniques have not been thoroughly researched in the literature. As a result, the current study examined the retentive force of ackers clasps manufactured using 3-D printing of wax patterns and DMLS technology. Retentive forces were tested using a Universal Testing Machine (68TM-30). The Independent Student t-test showed that DMLS clasps had considerably higher mean retention values than cast clasps. The current study's findings demonstrate how laser sintering might improve the quality and price of partial dentures, increasing accessibility. However, this technique may alter Co- Cr alloys' mechanical, physical, and biocompatibility properties, impacting clinical performance. Addressing study limitations and optimizing fabrication parameters in future research will further improve outcomes.

KEYWORDS: Circumferential clasp, Cobalt-chromium, Direct metal laser sintering, Retention

*Corresponding author: gayathridevi.pr.ds@msruas.ac.in

DOI: 10.1201/9781003664277-5

1. Introduction

Millions of partly edentulous people throughout the world can live better lives with removable partial dentures. These dental prostheses, being cost-effective, are used for replacing lost teeth in those who are partly edentulous. (Campbell., 2017). When it comes to mastication and functional muscle movements, a removable partial denture should offer enough retention to withstand dislodging pressures. One of the key elements of the RPD that accounts for most of the retention is retainers. Retention is provided by occlusal rests, proximal plates, dental clasps, and implant attachments that make contact with the abutment teeth. Since clasps are flexible and have the possibility of contacting the abutment's undercut in a removable partial denture, they can be used as direct retainers. Among the most popular direct retainers are Ackers or circumferential clasps. (Alageel.,2019). Alloys of nickel-chrome and cobalt-chrome are commonly used in clasp construction for their simplicity and excellent retention. Historically, the lost wax process has been utilized to create patient-specific removable partial dentures. The fact that it requires additional stages and is technique-sensitive are its primary disadvantages. (Phoenix., 2008, Carr., 2015). CAD/CAM technology is now often used in dentistry to produce metal frames for RPD. By enabling 3D models from working cast images, this method streamlines manufacturing, lowers expenses, and saves time. Enhancing repeatability and reducing variations among dental workers are other benefits. (Abdel-Rahim., 2016)

In clinical practice, CAD data is applied in two types for creating frameworks: additive manufacturing with the help of a 3D printer for creation of a resin design that is then injected and cast (Alageel.,2018; Brockhurst., 1996; Örtorp., 2011) direct metal laser sintering (Van Noort., 2012; Williams., 2006; Almufleh., 2018). This study investigates laser-sintering for processing metal frameworks for dental prostheses and offers suggestions based on a specific stress and undercut analysis for circumferential clasps on molar and premolar teeth. Pull-out tests and finite element analysis were used to confirm the findings, which gave insight into how to make circumferential clasps for both premolar and molar teeth. The findings suggest that, while molars can firmly accommodate these clasps, premolars may not provide adequate stability, making their usage less dependable (Mansour., 2016). Studies have indicated that while frameworks made by casting 3-D printed resin models tested well on the working cast, they needed minor clasp changes when tested in the patient's mouth. (Williams., 2006). On the other hand, a prosthodontist who evaluated SLS-fabricated frames found that they suited patients well (Anes., 2023). The retentive pressures of two varieties of digitally produced removable partial denture clasps—back-action and reverse back-action clasps for a premolar tooth—were investigated in another study. In order to measure the retentive force and determine the retention change over time, a tooth was inserted and removed 20,000 times using the Instron 5544 universal testing equipment. The future scope of the study stated that comparing conventional and digital methods is crucial due to limited literature (Charan., 2014) To simplify the process of using laser sintering technology to produce removable partial dentures, there is a need to compare the retentive properties of clasps made with direct metal laser sintering procedures against those made using conventional casting techniques of 3-D printed resin templates. Although there is little literature on the retention forces of RPD frameworks made with laser sintering technology and no comparable study comparing the retentive potential of these clasps, the current study compares the retentive qualities of clasps created with traditional 3-D printed pattern casting versus the DMLS technology.

2. Materials and Methodology

In the present study, the clasp's retentive qualities were examined utilizing two alternative production procedures. The extracted maxillary second premolar and maxillary second molar teeth used in this study to replicate the clinical scenario of Kennedy's class III were provided by the Ramaiah University Department of Oral and Maxillofacial Surgery. Cobalt chromium alloy was supplied from Confident Dental Lab, Bengaluru. For traditional casting, as well as the Direct Metal Laser Sintering manufacture of clasps from cobalt chromium alloy in powdered form, these materials were employed.

2.1 Sample Size

In the present study, a total sample size of 20 was carefully considered to ensure reliable results. This sample size was divided equally into two groups, with 10 samples in each group, for a balanced evaluation of retentive forces and performance, enhancing the validity and reliability of the result findings. In Group 1: The fabrication of the clasp was done by the Direct Metal Laser Sintering technique and in Group 2: The fabrication of the clasp was done by conventional casting technique from a 3D-printed resin pattern.

2.2 Preparation of the Samples

A partially edentulous condition simulating Kennedy's Class 3 design, where Ackers clasp the most recommended extra coronal direct retainer, was considered for the study. The extracted maxillary second premolar and maxillary second molar were mounted at the gap of 10.5mm of an acrylic block to simulate clinical conditions. The mesiodistal width of the missing maxillary first molar was duplicated by this gap. Rest seats were made on the molar and premolar of standard size to accommodate an Akers clasp structure and guiding plane preparation after the tooth was mounted on an acrylic block (Zhang, M. 2022), Thus, a customized assembly of an acrylic block with molar and premolar teeth embedded was designed, as shown in Fig. 5.1.

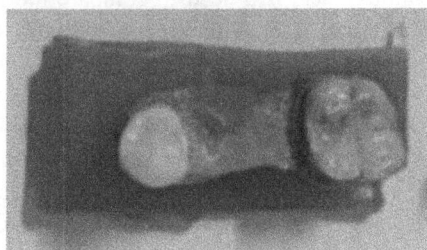

Fig. 5.1 Assembly of acrylic block

2.3 Designing and Fabricating the Samples

Followed by mounting and rest seat preparation the tooth models/ customized assembly were scanned using the Lab Scanner (3Shape E1 Dental Lab Scanner) to acquire digital data. With the aid of CAD Software, an Akers clasp was designed on the prepared tooth. Both clasps were connected to a horizontal rod; from the middle of the horizontal rod, a vertical projection was designed to measure the clasp's retentive pull during function, as shown in Fig. 5.2. The resulting CAD file was saved in open STL format, which was then used for fabrication. As shown in Fig. 5.3, the samples from both groups were equally split into two groups: Group 1: Direct Metal Laser Sintering:

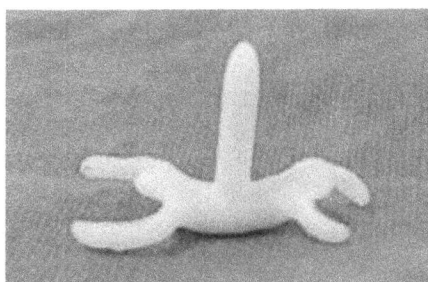

Fig. 5.2 Printed resin wax pattern

The cobalt-chromium alloy clasp was printed using the digitally designed framework saved in STL format (EOS M 290 system). Group 2: Milled wax patterns were casted by conventional lost wax technique. The digitally designed framework saved in STL format was used for 3D resin pattern printing. (Next Dent 5100). The printed resin pattern was checked for fit on customized assembly and later invested and casted using the lost wax technique.

2.4 Sample Testing of the Retentive Force

The retentive forces were checked using a Universal testing Machine - an electromechanical testing system. This universal testing machine is a Nano-hydraulic testing machine (BISS) that checks load capacity up to 25 KN. The testing of polymers and metals is commonly used by this Nano hydraulic testing machine. The electromagnetic testing system has the following components. This equipment consists of a load frame with an integrated controller, a crosshead-mounted load cell, grips for tension testing, and table-mounted anvils on a platen for compression testing. Bluehill software is installed on a PC owned by BISS. Twenty samples from each group were individually put on the customized assembly, which was secured to the lowest half of the testing machine. The testing sample's vertical rod was fastened to the upper membrane and pull test was then conducted. Tables 5.1 and 5.2 list the materials and equipment utilized in the present study.

Table 5.1 Materials used

Material	Application	Source
Extracted teeth (mandibular premolar and molar)	To simulate a Kennedy's class III clinical situation	Dept of OMFS, Ramaiah Dental College
Cobalt-Chromium alloy	Conventional casting of the clasps	Confident Dental Lab, Bengaluru
Cobalt chromium alloy powder	Clasps fabricated by DMLS technique	Confident Dental Lab, Bengaluru

3. Statistical Analysis

The retentive force values from the test samples were subjected to multiple linear regression analysis using SPSS statistical software (SPSS Statistics 20, IBM, USA). Paired t-tests were used to assess the clasp's fit correctness and

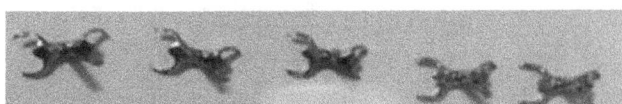

(a)

(b)

Fig. 5.3 (a) Direct metal laser sintering (b) Milled wax patterns conventionally casted

Table 5.2 Equipment's used

Equipment	Application	Source
Universal testing Machine (68TM-30)	To test the retentive forces of the clasps	Reva Institute of Technology, Bengaluru
Cobalt-Chromium alloy casting machine (Ultra Cast-D)	Conventional casting of the clasps	Confident Dental Lab, Bengaluru
DMLS 3-D printing machine EOS M 290 system	Clasps fabricated by DMLS technique	Confident Dental Lab, Bengaluru
Resin 3-D printing machine Next Dent 5100	Fabrication of resin pattern that was used for investment for conventional casting technique	Confident Dental Lab, Bengaluru
Lab Scanner 3Shape E1 Dental Lab Scanner		Confident Dental Lab, Bengaluru

retentive force prior to and after cyclic load. The retentive force measurements at the clasp tip, clasp shoulders, and rest areas were evaluated using one-way ANOVA, followed by Tukey's multiple comparison test, with a significance level of $\alpha = 0.001$.

4. Results and Discussion

Figure 5.4 shows the tensile test report produced for each tested material. The above formula was used to generate these tensile test results. After removing the clasp, the change in retention was determined using the formula $\Delta F = F0 - Fx$, where ΔF is the change in retention, $F0$ is the initial force needed to pull the clasp out at 0 cycles, and Fx is the force needed to pull the clasp out after x cycles. Retentive forces were evaluated at 1,000-cycle intervals over a total of 20,000 cycles. (Charan., 2013). The retention force of a removable partial denture (RPD) is calculated by subtracting the cumulative dislodging force applied on a replacement tooth from the sum of the clasp retention forces. Each clasp framework's retentive qualities were then evaluated by tabulating the findings, as seen in Tables 5.3 and 5.4. These provide the average values and range of pressures needed to remove each structure. All force measurements were expressed in MPa, and the results were graphically shown in Fig. 5.5. A precise comparison of the retentive forces across the various manufacturing techniques was made possible by this

Table 5.3 Converted values of retentive forces between two groups

Clasps	DMLS Clasps	Cast Clasps
Sample 1	17	12.8
Sample 2	15.8	12.2
Sample 3	18	11
Sample 4	13	13.2
Sample 5	22	13.4
Sample 6	16.1	15.2
Sample 7	17.5	11.5
Sample 8	19	13.1
Sample 9	21.3	15
Sample 10	18.4	14.4
Mean	**17.81**	**13.18**

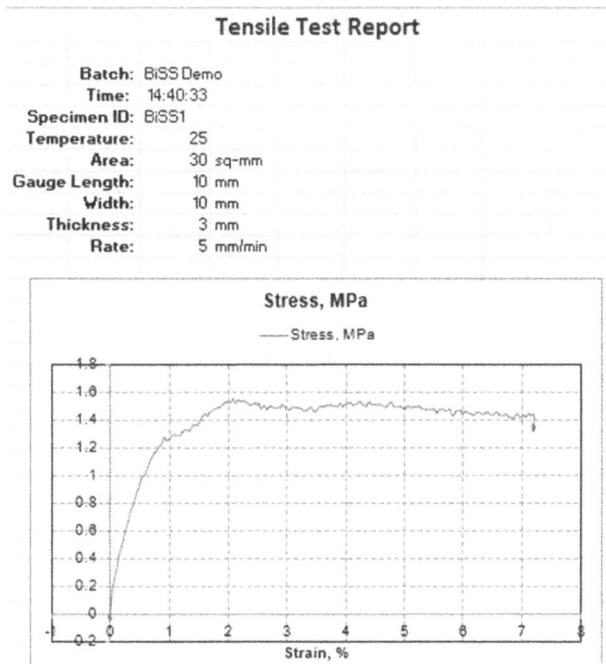

Tensile Test Report

Batch: BiSS Demo
Time: 14:40:33
Specimen ID: BiSS1
Temperature: 25
Area: 30 sq-mm
Gauge Length: 10 mm
Width: 10 mm
Thickness: 3 mm
Rate: 5 mm/min

Fig. 5.4 Tensile test report

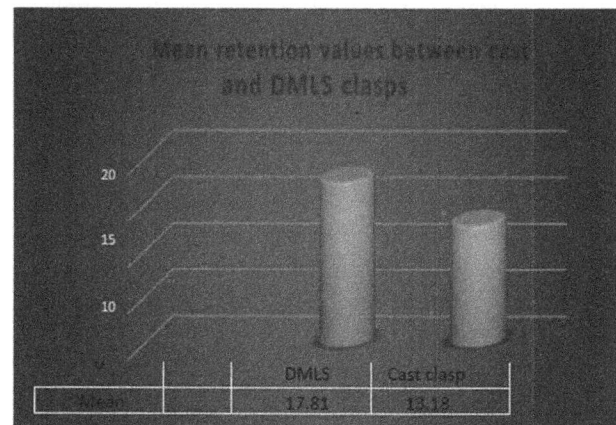

Fig. 5.5 Retentive forces required to dislodge each framework between two groups

Table 5.4 Comparison of mean retention values between cast & DMLS clasps using independent St udentTest

Group	N	Mean	SD	Mean Diff	95% CI of the diff.		t	P-Value
					Lower	Upper		
DMLS Clasps	10	17.81	2.63	4.63	2.65	6.61	4.911	<0.001*
Cast Clasps	10	13.18	1.40					

* Statistically Significant

thorough investigation. The independent Student's t-test findings showed that the DMLS clasps' mean retention values (17.81 ± 2.63 MPa) were considerably greater than the cast clasps' (13.18 ± 1.40 MPa). Between the two groups, the mean difference was 4.63 MPa, with a 95% CI between 2.65 and 6.61 MPa. With a p-value of less than 0.001, this difference in retention values was statistically significant, suggesting that DMLS clasps significantly improved retention over cast clasps. Computer-aided design (CAD) and computer-aided manufacturing (CAM) are commonly utilized by dentists for a number of applications, such as milling metal alloys and wax patterns, zirconia, hybrid prostheses, and removable partial denture (RPD) framework patterns. The study's goal was to compare the retentive forces of cobalt-chromium (Co-Cr) clasps manufactured with DMLS to those made with traditional casting techniques and 3D-printed wax models (Zhang., 2022). The null hypothesis holds that there will be no obvious difference in mean retention values between DMLS and conventionally cast clasps. Retention is a critical component of the effectiveness of cast partial dentures because it directly impacts patient comfort, longevity, and the stability and performance of the prosthesis. Some factors that may affect clasp retention include the design of direct retainers, the masticatory stresses used, and how well the clasps fit on the abutment teeth.

In order to improve patient comfort and prosthesis lifetime, effective retention makes sure the device stays firmly in place while in use. The retention of the clasp is not an independent criterion for its selection and use. Other aspects to be considered while designing the clasp include undercuts, tooth form, survey lines, and the guiding plane (Charan., 2013). In clinical scenarios, saliva can also act as a buffer, reducing the force necessary to remove a direct retainer. In addition, depending on the clinical state, an indirect retainer may be utilized to increase the retention of a removable partial denture

(Rodrigues., 2002)

The methods for producing dental restorations and prostheses have been completely transformed by the application of CAD/CAM technology in dentistry. This technology allows for precise digital designing and manufacturing, leading to improved accuracy, fit, and overall quality of dental components. The digital workflow, from scanning to final production, ensures consistency and reduces the likelihood of errors that may occur in traditional manual methods.

In contrast to traditional casting procedures, frameworks fabricated using DMLS benefit from a complete digital workflow, ensuring optimal clasp retention and minimizing manual intervention. The DMLS technique offers high precision in producing metal frameworks, which can lead to better mechanical properties and enhanced retention compared to conventional methods. On the other hand, 3D-printed wax patterns used for traditional casting provide a high level of detail but may introduce variability in the final product due to the casting process itself. Using high-speed milling, repeated laser sintering, and direct casting, Nakata et al. created the Akers clasp and measured the average retentive force (NakataT., 2017). When additive and subtractive manufacturing are used together, high accuracy and surface properties can be obtained without a lot of human post-processing. The initial retentive forces of laser-sintered specimens and direct casting clasps were 12.3±2.6 N and 12.9±3.5 N, respectively. Comparing CAD-CAM clasps to baseline retentive force values of 16.1±0.8 N, 12.3±2.6 N, and 21.5±5.3 N, Torii et al. developed the Akers clasp. The validated initial retentive force of cast clasps was 12.9±3.5 N (Torii, M., 2018). In this study, the casts exhibited a loss in retentive values by 41.1% while that of CAM clasps were about 14.3–30.8% after aging through 10,000 cycles. Software parameters such as offset have an effect and act as variables for variable changes. Alageel et al. investigated the pressures that food imparted to acrylic resin teeth during simulated mastication, as well as the retention forces given by various clasp types, including wrought wire, circumferential, and I-bar clasps. This created an algorithm that could predict the retention of an RPD and estimate the ideal number of clasps needed. Co-Cr clasps made with the DMLS technology had a substantially larger retentive force than cast clasps in the current investigation, confirming the theory that DMLS offers a more desirable function. These better results are probably a result of DMLS's increased accuracy and decreased requirement for human post-processing. Clinically, patients may benefit from more stable and pleasant prostheses as a result of improved DMLS clasp retention. The absence of aging modeling and the study's concentration on a single clasp design are two of its drawbacks, though. To confirm these results, future studies should investigate different clasp designs and production methods. Indirect retainers and the impact of saliva on retention in clinical settings are further elements that should be considered.

5. Conclusion

In conclusion, RPDs may be made via laser sintering rather than casting. By reducing the cost and improving the quality of RPDs, laser sintering can make the treatment more widely available. However, Co-Cr alloys' mechanical, physical, and biocompatibility properties as well as their clinical performance may be impacted by laser sintering. Variations in manufacturing techniques, such as laser beam power, scanning speed, metal particle size, and layer thickness, can all have an influence on the characteristics of laser sintered alloy.

Acknowledgments

The authors thank Reva Institute of Technology, Bengaluru, for the Universal Testing Machine (68TM-30) and to Confident Dental Lab, Bengaluru, for their support with the Cobalt-Chromium casting machine, EOS M 290 DMLS, Next Dent 5100 resin printer, and 3Shape E1 Lab Scanner.

References

1. Campbell, S. D., Cooper, L., Craddock, H., Hyde, T. P., Nattress, B., Pavitt, S. H., & Seymour, D. W. (2017). Removable partial dentures: The clinical need for innovation. The Journal of prosthetic dentistry, 118(3), 273–280.
2. Alageel, O., Alsheghri, A. A., Algezani, S., Caron, E., & Tamimi, F. (2019). Determining the retention of removable partial dentures. The Journal of Prosthetic Dentistry, 122(1), 55–62.
3. Phoenix RD, Cagna DR, DeFreest CF, Stewart KL. Stewart's Clinical Removable Partial Prosthodontics. Quintessence Publishing (IL); 2008.
4. Carr AB, Brown DT. McCracken's Removable Partial Prosthodontics - E-Book. Elsevier Health Sciences; 2015.
5. Abdel-Rahim, N. Y., Abd El-Fattah, F. E., & El-Sheikh, M. M. (2016). Laboratory comparative study of three different types of clasp materials. Tanta Dental Journal, 13(1), 41–49.
6. Alageel, O., Abdallah, M. N., Alsheghri, A., Song, J., Caron, E., & Tamimi, F. (2018). Removable partial denture alloys processed by laser-sintering technique.

Journal of Biomedical Materials Research Part B: Applied Biomaterials, 106(3), 1174–1185.
7. Brockhurst, P. J. (1996). A new design for partial denture circumferential clasp arms. Australian dental journal, 41(5), 317–323.
8. Örtorp, A., Jönsson, D., Mouhsen, A., & von Steyern, P. V. (2011). The fit of cobalt–chromium three-unit fixed dental prostheses fabricated with four different techniques: A comparative in vitro study. Dental Materials, 27(4), 356–363.
9. Van Noort, R. (2012). The future of dental devices is digital. Dental materials, 28(1), 3–12.
10. Williams, R. J., Bibb, R., Eggbeer, D., & Collis, J. (2006). Use of CAD/CAM technology to fabricate a removable partial denture framework. The Journal of prosthetic dentistry, 96(2), 96–99.
11. Almufleh, B., Emami, E., Alageel, O., de Melo, F., Seng, F., Caron, E., & Tamimi, F. (2018). Patient satisfaction with laser-sintered removable partial dentures: a crossover pilot clinical trial. The Journal of prosthetic dentistry, 119(4), 560–567.
12. Van Noort, R. 2012., Williams, R. J., 2006 and Almufleh, B 2018
13. Mansour, M., Sanchez, E., & Machado, C. (2016). The use of digital impressions to fabricate tooth-supported partial removable dental prostheses: a clinical report. Journal of Prosthodontics, 25(6), 495–497.
14. Anes, V., Neves, C. B., Bostan, V., Gonçalves, S. B., & Reis, L. (2023). Evaluation of the retentive forces from removable partial denture clasps manufactured by the digital method. Applied Sciences, 13(14), 8072.
15. Charan, J., & Biswas, T. (2013). How to calculate sample size for different study designs in medical research?. Indian journal of psychological medicine, 35(2), 121–126.
16. Zhang, M., Gan, N., Qian, H., & Jiao, T. (2022). Retentive force and fitness accuracy of cobalt-chrome alloy clasps for removable partial denture fabricated with SLM technique. Journal of Prosthodontic Research, 66(3), 459–465.
17. Alsheghri, A. A., Alageel, O., Caron, E., Ciobanu, O., Tamimi, F., & Song, J. (2018). An analytical model to design circumferential clasps for laser-sintered removable partial dentures. Dental Materials, 34(10), 1474–1482.
18. Rodrigues, R. C. S., Ribeiro, R. F., de Mattos, M. D. G. C., & Bezzon, O. L. (2002). Comparative study of circumferential clasp retention force for titanium and cobalt-chromium removable partial dentures. The Journal of prosthetic dentistry, 88(3), 290–296.
19. Nakata, T., Shimpo, H., & Ohkubo, C. (2017). Clasp fabrication using one-process molding by repeated laser sintering and high-speed milling. Journal of prosthodontic research, 61(3), 276–282.
20. Torii, M., Nakata, T., Takahashi, K., Kawamura, N., Shimpo, H., & Ohkubo, C. (2018). Fitness and retentive force of cobalt-chromium alloy clasps fabricated with repeated laser sintering and milling. journal of prosthodontic research, 62(3), 342–346.

Advances in Materials Science and Technology – Dr. Srikari Srinivasan et al. (eds)
© 2025 Taylor & Francis Group, London, ISBN 978-1-041-12342-2

6

Characterization of a Novel Drug Delivery System to Enhance Regeneration in Endodontics

Harshini Prakash,
Anu Elsa Swaroop, Sylvia Mathew*, Shruthi Nagaraja
Department of Conservative Dentistry and Endodontics,
Faculty of Dental Sciences, M S Ramaiah University of Applied Science,
Bengaluru

Deveswaran Rajamanickam
Department of Pharmaceutics,
Drug Design and Development Centre, Faculty of Pharmacy,
M S Ramaiah University of Applied Sciences,
Bengaluru

Abstract: To develop and characterize a nanosponge based novel drug delivery system infused with anti-microbials and lyophilised PRF for disinfection and regeneration in endodontics.Cyclodextrin based Nanosponges were fabricated and infused with the selected antibiotics and lyophilized PRF. Its antimicrobial activity was assessed against selected endodontic flora (Enterococcus faecalis, Streptococcus mutans and Fusobacterium nucleatum) by agar disc diffusion Assay. Biocompatibility and wound healing capacity of the Drug Loaded Nanosponge (DLNS) were assessed using MTT Assay and Scratch Assay respectively.The DLNS exhibited favourable antibacterial activity against the selected microorganisms. It was also biocompatible with human Dental Pulp Stem Cells (hDPSCs) and showed effective wound healing with good cell migration rates.Within the limitations of this invitro study, it can be concluded that the novel formulation had good anti-microbial effect against the selected microorganisms, was biocompatible and promoted wound healing.

Keywords: Biomaterials, Local drug delivery, Regeneration, Lyophilised PRF, Cyclodextrin

1. Introduction

Regenerative Endodontics is an emerging field in dentistry focused on preserving the pulp, dental structures and maintaining their functions within the oral cavity (Lee et al., 2015). One innovative approach within this field is the "Regeneration Protocol in Endodontics," which addresses the treatment of necrotic teeth with open apex. A critical component for achieving successful outcomes is ensuring a clean environment through the disinfection of the root canal system (Dahlkemper et al., 2016).

Intracanal medication plays a vital role in promoting regeneration by disinfecting the radicular area (Murray et al., 2007). Calcium hydroxide (CH) has been the intracanal medicament of choice for a long time (Huang et al., 2009). Other medicaments such as triple antibiotic paste (TAP), a combination of ciprofloxacin, metronidazole, and minocycline

---

*Corresponding author: sylviamathew.cd.ds@msruas.ac.in

DOI: 10.1201/9781003664277-6

and its variations has been explored with unfavourable outcomes, such as crown discoloration, changes in dentin properties, and cytotoxic effects at high concentrations (Wigler et al., 2013; Parhizkar et al., 2018; Nosrat et al., 2011). Various drug delivery systems and combinations have been investigated to overcome these drawbacks which shows promising potential as a disinfection strategy (Staffoli et al., 2019; Alasqah et al., 2020; Bottinoet al., 2019).

Platelet concentrates such as platelet rich fibrin (PRF) finds numerous applications in regenerative endodontics serving as a scaffold for tissue ingrowth with evidence of progressive thickening of dentinal walls, root lengthening, reduction in periapical lesions, and closure of the apex Lyophilization of Platelet Rich Fibrins (L-PRF) helps preserve its bioactivity, enhances storage stability, extends half-life, and preserves growth factors effectively to facilitate regeneration and wound healing (Ngah et al., 2021).

Local Drug delivery systems (LDDS) aim to deliver maximum therapeutic concentrations of drugs directly to the target tissue while minimizing adverse effects and ensuring controlled drug release (Puri et al., 2013). Various micro-vehicles such as microparticles, nanoparticles, micelles, and liposomes have been investigated for this purpose (Rashid et al., 2021). Polymers like cyclodextrin, polylactic co-glycolic acid, hydroxyapatite, ethyl cellulose, methyl cellulose, and gelatin are also considered effective drug delivery vehicles due to their degradation capabilities and controlled drug release properties (Liang et al., 2021; Shahri et al., 2020). Cyclodextrin-based nanosponges (CDNS) have been explored in the pharmaceutical industry over the past decade due to their spongy structure, which allows them to form inclusion and non-inclusion complexes with various drugs. This makes them effective therapeutic vehicles for delivering drugs with low bioavailability (Krabicová et al., 2020; Sherje et al., 2017). The aim of this study was to develop a biocompatible LDDS combining antimicrobials with L-PRF to disinfect the root canal and promote tissue repair and regeneration for predictable treatment outcome and to transform the practice of Endodontics.

2. Materials and Methods

The study design was approved by the University Ethics Committee for Human trials of M S Ramaiah University of Applied Sciences (Reference no: EC-2022/PG/176).

2.1 Developing A Polymer Based Local Drug Delivery Carrier Infused with Anti-Microbials and Lyophilized Platelet Concentrate

Preparation of β-Cyclodextrin Nanosponges

50 ml of dimethylformamide (500 ml, S D Fine Chem Limited, Mumbai) was mixed with 8.6 g of β-cyclodextrin (Grm1605-100g, Himedia Laboratory Pvt Ltd, Thane West). Then, 5.66 ml of glutaraldehyde solution (25% Solution, Spectrochem Pvt Ltd, Mumbai) was added dropwise (Santanu et al., 2012). The solution was heated until dry, ground into a powder, and washed with 30 ml deionized water. The residue was dried for 2 days, ground again, dispersed in water, and lyophilized (Christ Alpha 1-2 LDplus, Germany) to recover the Nanosponge (Singh et al., 2019).

Preparation of Lyophilized PRF

Venous blood was drawn from healthy volunteers after obtaining informed consent. Blood was drawn and transferred into vacutainers, and centrifuged (Get Eltek Tc 650 D - Multispin Centrifuge, Elektrocraft India Pvt Ltd) at 3000 rpm for 10 minutes (Dohan et al., 2010). The blood separated into three layers: an upper acellular layer, a middle PRF layer, and a lower RBC layer. The PRF layer was removed, compressed, and stored overnight in a refrigerator. For lyophilization, PRF was freeze-dried in a lyophilizer for eight hours (Christ Alpha 1-2 LDplus, Germany), ground into granules, and stored at 4°C (Ngah et al., 2021).

Preparation of Drug Loaded Nanosponges (DLNS)

Nanosponge (100 mg), Lyophilized PRF (100 mg), Ciprofloxacin (200 mg), Metronidazole (200 mg), and Clindamycin (200 mg) were mixed in a 1:1:2:2:2 ratio, suspended in 20 ml of distilled water, and stirred in a magnetic stirrer (1 Mlh 1ltr, Remi Sales and Engineering Ltd. India) for 24 hours. The suspension was centrifuged at 3000 rpm for 10 minutes to separate uncomplexed drugs. The supernatant was lyophilized at -51°C and 13.33 mbar pressure for 8 hours to recover the DLNS, which was then stored for further use.

Assessing the Anti-microbial Efficacy of the DLNS using Disc Diffusion Method

The microorganisms used for antimicrobial analysis, Enterococcus faecalis (MTCC 439), Streptococcus mutans (MTCC 497), and Fusobacterium nucleatum (MTCC 430), were sourced from the Microbial Type Culture Collection and Gene Bank (MTGC), Chandigarh. These cultures were inoculated on MTGE Anaerobic Agar plates and incubated anaerobically at 37°C for 48 hours using a GasPak system (SPW Industrial, CA), which creates an oxygen-free environment. The cell density was adjusted to the 0.5 McFarland standard, equivalent to 1.5×10^8 CFU/ml (Bonnet et al., 2020). The procedure involved swabbing the plates with bacteria and incubating them anaerobically for 48 hours (Kohner et al., 1995). Three 6 mm wells were then bored into the inoculated media. Sample-loaded discs (1 mg/ml) were placed on the plates, which were subsequently incubated at 37°C for 24 hours. Inhibition zones were measured in millimeters to determine antibacterial activity:

resistant (ZOI <7 mm), intermediate (ZOI 8-10 mm), and sensitive (ZOI >11 mm) (Assam et al., 2010). Streptomycin (10 µg) served as the standard control, and distilled water was used as the negative control.

Cell Viability Assay

To determine the effect of DLNS on hDPSCs viability, 3-(4,5-dimethylthiazol-2-yl)-2,5-diphenyltetrazolium bromide (MTT) assay was employed. Cells (200 µl) were seeded into 96-well plates at 20,000 cells per well and grown for 24 hours. Test agents at 6.25, 12.5, 25, 50, and 100 µg/mL concentrations were added, and plates were incubated for another 24 hours at 37°C in a 5% CO_2 atmosphere. After incubation, media was removed, MTT reagent (0.5 mg/mL) was added, and plates were incubated for 3 hours in darkness. Formazan crystals were dissolved in 100 µl of Dimethyl Sulfoxide, and absorbance was measured at 570 nm. Cell viability was calculated as:[(Mean absorbance of treated cells / Mean absorbance of untreated cells) × 100]. This procedure ensured consistent evaluation of cytotoxic effects across concentrations (Ghasemi et al., 2021).

Wound Healing Assay

Cells were cultured in 10% Fetal Bovine Serum until 70-80% confluent in T25 cm² flasks. For the assay, cells were seeded at 0.25 million cells/well in 12-well plates, allowed to reach 80-100% confluence in 48 hours (Ma et al., 2023). A perpendicular scratch was made across each well using a 200 µl pipette tip, followed by a second scratch to create a cross pattern. After washing, cells were treated with 100ug concentration of the test agent and incubated at 37°C with 5% CO_2 for 48 hours. Images (10x magnification) were taken at 0, 24, and 48 hours using consistent settings.

Gap distance was quantified using Image J, and % wound healing area was calculated as:

[% Wound healing area = (Initial area – Final area) / Initial area × 100], assessing cell migration and wound closure.

3. Statistical Analysis

Statistical analysis was done with SPSS software (version 22). Data was analyzed statistically using ANOVA and Tukey's post hoc test. P < 0.05 was regarded as statistically significant.

4. Results and Discussion

The data was analysed using SPSS version 26.0 software. Intergroup comparisons were analysed using One way ANOVA and post hoc Tukey test. Intragroup comparisons were made using Repeated Measures ANOVA and Bonferroni post hoc test. A p<0.05 was considered as statistically significant and p<0.01 as highly significant.

4.1 Antibacterial Activity of the DLNS

The DLNS showed maximum zone of inhibition against all the three tested organisms as compared to the standard control. Negative control did not show any zone of inhibition.

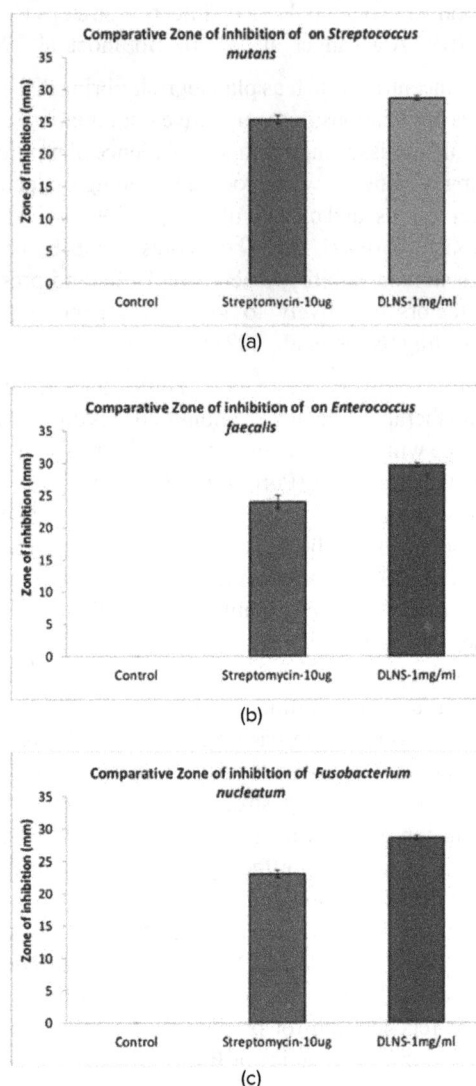

(a)

(b)

(c)

Fig. 6.1 (a) Zone of inhibition (mm) of DLNS against streptococcus mutans at 1mg/ml concentration after the incubation period of 24hrs. The presented values were the average of 3 independent individual experiments (n=3). (b) Zone of inhibition (mm) of DLNS against enterococcus faecalis at 1mg/ml concentration after the incubation period of 24hrs. The presented values were the average of 3 independent individual experiments (n=3). (c) Zone of inhibition (mm) of DLNS against fusobacterium nucleatum at 1mg/ml concentration after the incubation period of 24hrs. The presented values were the average of 3 independent individual experiments (n=3)

4.2 MTT Assay

The viability of hDPSCs post exposure to the DLNS was concentration dependent. Absorbance readings showed that untreated cells had 100% viability. As concentrations increased, viability decreased by approximately 1% per concentration level (99.71%, 99.11%, 97.96%, 96.67%, and 95.28% for 6.25 µg, 12.5 µg, 25 µg, 50 µg, and 100 µg, respectively) but were not statistically significant, with lower concentrations demonstrating higher cell viability.

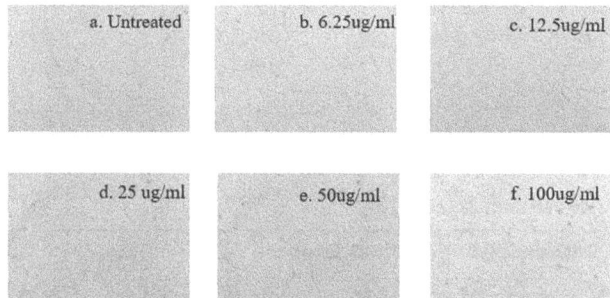

Fig. 6.2 MTT assay - Morphological appearance of hDPSCs after addition of different concentrations of the DLNS

4.3 Wound Healing

At zero hours, there was no significant difference in wound area among the untreated group (1404491.04 ± 197307.13), IFN- γ (1370328 ± 148789.55), and DLNS (1382051.135 ± 149910.21) (p < 0.01). Cell migration was significantly higher in the control group than in the untreated and DLNS group at all time intervals. DLNS demonstrated comparable efficacy to IFN- γ in promoting wound closure showing 59% and 92% closure at 24 and 48 hours respectively similar to IFN- γ 61% and 99.7% closure. These results highlight DLNS's potential to promote cell proliferation and migration and wound healing similar to IFN-γ (Table 6.1, 6.2).

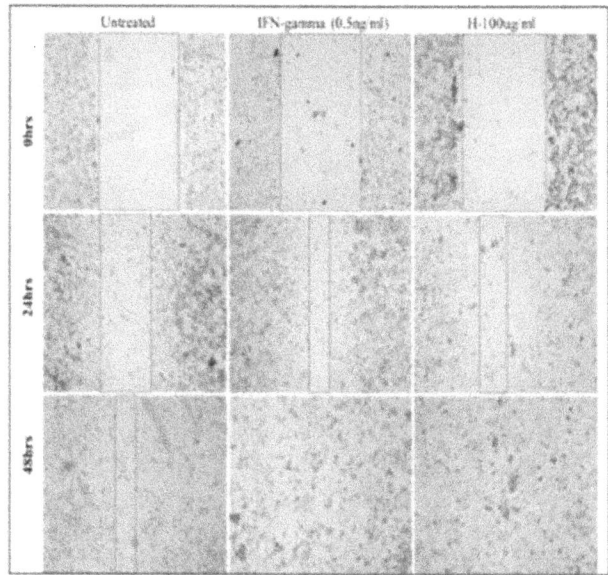

Fig. 6.4 Invitro wound healing activity of DLNS with 100ug/ml at different time intervals of 0, 24, 48 hours on hDPSCs

Table 6.1 % of wound closure − DPSCs in untreated, IFN gamma and DLNS group

Incubation	Untreated	IFN gamma	DLNS
0 hour	0	0	0
24 hours	33.41 ± 8	61.93 ± 10	51.16 ± 9
48 hours	67.57 ± 4.8	99.73 ± 0.15	92.35 ± 7

The anti-microbial efficacy and biocompatibility of a Novel Cyclodextrin-based nanosponge infused with lyophilised PRF and antimicrobials (DLNS) was assessed for its potential use as an intracanal medicament for disinfection and to promote regeneration in endodontics. The agar disc diffusion tests indicate that DLNS exhibited superior

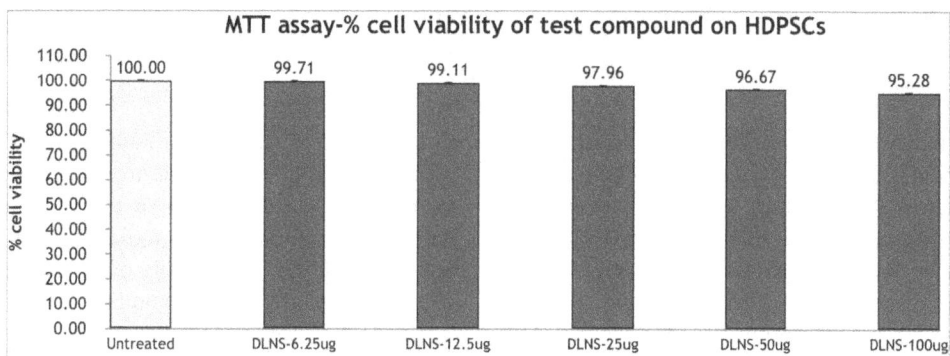

Fig. 6.3 Graph depicts the % of cell viability of DLNS on hDPSCs

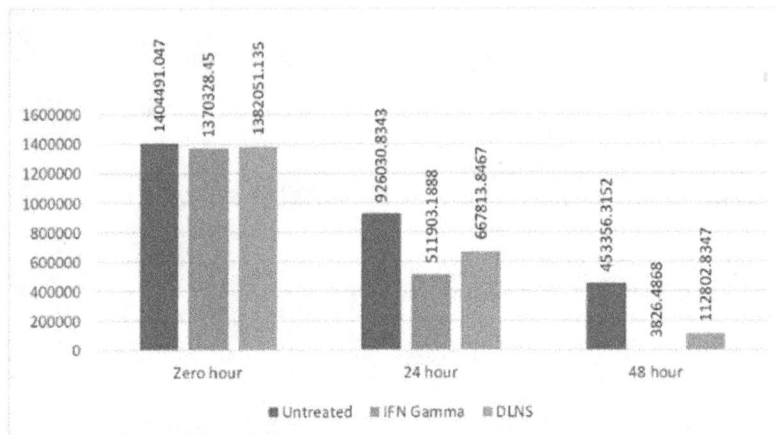

Fig. 6.5 Wound healing assay - Distribution between untreated, IFN gamma and DLNS

Table 6.2 Pair wise comparison (Post-hoc) at 0,24,48hrs

Dependent Variable	(I) Group	(J) Group	Mean Difference (I-J)	Std. Error	Sig.
0 hour	Untreated	IFN Gamma	34162.59700	136252.63264	.966
		DLNS	22439.91167	136252.63264	.985
	IFN Gamma	Untreated	-34162.59700	136252.63264	.966
		DLNS	-11722.68533	136252.63264	.996
	DLNS	Untreated	-22439.91167	136252.63264	.985
		IFN Gamma	11722.68533	136252.63264	.996
24 hours	Untreated	IFN Gamma	414127.6455*	59692.33745	.001 HS
		DLNS	258216.9876*	59692.33745	.012 S
	IFN Gamma	Untreated	-414127.6450*	59692.33745	.001 HS
		DLNS	-155910.65783	59692.33745	.089
	DLNS	Untreated	-258216.98767*	59692.33745	.012 S
		IFN Gamma	155910.65783	59692.33745	.089
48 hours	Untreated	IFN Gamma	449529.82833*	65527.04732	.001 HS
		DLNS	340553.48050*	65527.04732	.005 HS
	IFN Gamma	Untreated	-449529.82833*	65527.04732	.001 HS
		DLNS	-108976.34783	65527.04732	.293
	DLNS	Untreated	-340553.48050*	65527.04732	.005 HS
		IFN Gamma	108976.34783	65527.04732	.293

antimicrobial efficacy against three selected oral pathogens: E. faecalis, S. mutans, and F. nucleatum as compared to the standard control Streptomycin, suggesting its efficacy as an effective intracanal medicament (Karczewski et al., 2018). The present study used two vital assays to assess the biocompatibility of the DLNS. The quantitative analysis of cell viability was assessed through MTT Assay. The wound healing assay gave insights of cell proliferation and migration.

Cytotoxicity evaluation through the MTT assay relies on the ability of living cells to metabolically reduce the yellow tetrazole MTT into a purple formazan product within their mitochondria, indicating active metabolism and cell viability. The intensity of the purple colour formed correlates directly with the number of viable cells, as only live cells with functional mitochondria can carry out this reduction process (Stockert et al., 2018; Zhou et al., 2013; Mosmann et al., 1983).The results of the assay indicate that the viability of hDPSCs post exposure to the DLNS was concentration dependent with lower concentration of the drug showing good viability at the end of 24 hours.

Biomaterials should not only be biocompatible but also promote cellular activities such as proliferation, migration, and tissue regeneration (Jonkman et al., 2014; He et al., 2017). Scratch Assay is an economical laboratory technique used to study 2D cell migration and cell–cell interaction, it can mimic to some extent the cell migration in vivo. The results of the cell migration assay revealed good cell migration and wound closure for the DLNS at the end of 48 hours which was comparable to the standard control IFN-γ.

Cyclodextrin-based nanosponges has been shown to have improved pharmaceutical and physicochemical properties like stability, solubility, and bioavailability of active moieties. Loading antibiotics into nanosponges ensures sustained and targeted delivery within the intricate and often inaccessible root canal system (Caldera et al., 2017; Tejashri et al., 2013; Parhizkar et al., 2020; Swaroop et al., 2025).

PRF is derived from the patient's own blood and contains a concentrated mixture of growth factors essential for tissue repair and regeneration (Choukroun et al., 2017; Dohan et al., 2010). These growth factors, including PDGF, VEGF, and TGFβ, promote wound healing by stimulating cell proliferation, angiogenesis, and extracellular matrix formation. Lyophilization, or freeze-drying, of PRF preserves its biological activity and extends its shelf-life, overcoming limitations associated with fresh PRF, such as rapid degradation and limited storage options (Andia et al., 2010; Nireesha et al., 2013; Vocetkova et al., 2017).

The combined approach of integrating L-PRF with nanosponge-encapsulated antibiotics aimed to synergistically enhance antimicrobial efficacy and promote tissue regeneration within the root canal environment. By harnessing the benefits of nanotechnology and bioactive substances, this innovative system aims to overcome challenges associated with conventional endodontic therapies, including incomplete disinfection, and limited regenerative potential. Moreover, the study highlighted the potential of this integrated system to mitigate systemic side effects typically associated with high-dose antibiotic treatment delivering antibiotics locally in a controlled manner to maximize therapeutic outcomes while minimizing adverse effects.

Success in regenerative endodontics is also based on the interplay of many factors such as age of the patient, systemic health, canal morphology, stage of root development. Besides these, effective infection control protocol plays a significant role in the success of any endodontic treatment especially regenerative procedures. The present study was conducted in a controlled in-vitro environment, which may not fully replicate the dynamic conditions of the oral cavity. Given that endodontic infections are biofilm-mediated, investigating DLNS's efficacy in biofilm elimination is crucial. Further characterization of the biomaterial and followed by its impact on root development, periapical healing, and tissue regeneration needs to be explored.

5. Conclusion

Within the limitations of the present in vitro study, it can be concluded that the novel DLNS formulation has effective antibacterial activity against the tested oral micro-organisms. It is biocompatible and promotes wound healing as evinced through enhanced cell migration. It therefore offers a system for targeted, sustained delivery of anti-microbials for root canal disinfection and regeneration.

Acknowledgments

We acknowledge the technical and laboratory support provided by Faculty of Dental Sciences, Faculty of Pharmacy, RUAS and Averin Biotech Bangalore.

Funding/Conflicts of interests if any

The research leading to these results received funding from RUAS Seed Grant under Grant Agreement No ORI-SG/FDS/002/2023. No conflicts of interest.

References

1. Alasqah, M., Khan, S. I. R., Alfouzan, K., & Jamleh, A. (2020). Regenerative Endodontic Management of an Immature Molar Using Calcium Hydroxide and Triple Antibiotic Paste: a Two-Year Follow-Up. Case reports in dentistry, 2020(1):9025847.

2. Andia, I., Perez-Valle, A., Del Amo, C. and Maffulli, N. (2020). Freeze-drying of platelet-rich plasma: the quest for standardization. International journal of molecular sciences.21 (18):6904.

3. Assam JP, A., Dzoyem, J.P., Pieme, C.A. and Penlap, V.B. (2010). In vitro antibacterial activity and acute toxicity studies of aqueous-methanol extract of Sida rhombifolia Linn.(Malvaceae). BMC complementary and alternative medicine. 10:1–7.

4. Bonnet, M., Lagier, J.C., Raoult, D. and Khelaifia, S. (2020). Bacterial culture through selective and non-selective conditions: the evolution of culture media in clinical microbiology. New microbes and new infections, 34:100622.

5. Bottino, M.C., Albuquerque, M.T., Azabi, A., Münchow, E.A., Spolnik, K.J., Nör, J.E. and Edwards, P.C. (2019). A novel patient-specific three-dimensional drug delivery construct for regenerative endodontics. Journal

of Biomedical Materials Research Part B: Applied Biomaterials. 107(5):1576–1586.

6. Caldera, F., Tannous, M., Cavalli, R., Zanetti, M., & Trotta, F. (2017). Evolution of cyclodextrin nanosponges. International journal of pharmaceutics, 531(2):470–479.

7. Calixto, G., Bernegossi, J., Fonseca-Santos, B. and Chorilli, M., (2014). Nanotechnology-based drug delivery systems for treatment of oral cancer: a review. International journal of nanomedicine:3719–3735.

8. Choukroun, J. and Miron, R.J., (2017). Platelet rich fibrin in regenerative dentistry:1-14. Oxford: Wiley-Blackwell.

9. Dahlkemper PE., (2016). American Association of Endodontists: Guide to clinical endodontics.

10. Dohan Ehrenfest, D.M., Del Corso, M., Diss, A., Mouhyi, J. and Charrier, J.B., (2010). Three-dimensional architecture and cell composition of a Choukroun's platelet-rich fibrin clot and membrane. Journal of periodontology, 81(4): 546–555.

11. Ghasemi, M., Turnbull, T., Sebastian, S. and Kempson, I., (2021). The MTT assay: utility, limitations, pitfalls, and interpretation in bulk and single-cell analysis. International journal of molecular sciences, 22(23):12827.

12. He, X., Jiang, W., Luo, Z., Qu, T., Wang, Z., Liu, N., Zhang, Y., Cooper, P.R. and He, W., (2017). IFN-γ regulates human dental pulp stem cells behavior via NF-κB and MAPK signaling. Scientific reports, 7(1):40681.

13. Huang, G.J., (2009). Apexification: the beginning of its end. International endodontic journal, 42(10):855–866.

14. Jonkman, J.E., Cathcart, J.A., Xu, F., Bartolini, M.E., Amon, J.E., Stevens, K.M. and Colarusso, P., (2014). An introduction to the wound healing assay using live-cell microscopy. Cell adhesion & migration, 8(5):440–451.

15. Karczewski, A., Feitosa, S.A., Hamer, E.I., Pankajakshan, D., Gregory, R.L., Spolnik, K.J. and Bottino, M.C., (2018). Clindamycin-modified triple antibiotic nanofibers: a stain-free antimicrobial intracanal drug delivery system. Journal of endodontics, 44(1):155–162.

16. Kohner, P.C., Rosenblatt, J.E. and Cockerill 3rd, F.R., (1994). Comparison of agar dilution, broth dilution, and disk diffusion testing of ampicillin against Haemophilus species by using in-house and commercially prepared media. Journal of clinical microbiology, 32(6):1594–1596.

17. Krabicová, I., Appleton, S.L., Tannous, M., Hoti, G., Caldera, F., Rubin Pedrazzo, A., Cecone, C., Cavalli, R. and Trotta, F., (2020). History of cyclodextrin nanosponges. Polymers, 12(5):1122.

18. Lee, B.N., Moon, J.W., Chang, H.S., Hwang, I.N., Oh, W.M. and Hwang, Y.C., (2015). A review of the regenerative endodontic treatment procedure. Restorative dentistry & endodontics, 40(3):179–187.

19. Liang, J., Peng, X., Zhou, X., Zou, J. and Cheng, L., (2020). Emerging applications of drug delivery systems in oral infectious diseases prevention and treatment. Molecules, 25(3):516.

20. Ma, Y., Liu, Z., Miao, L., Jiang, X., Ruan, H., Xuan, R. and Xu, S., (2023). Mechanisms underlying pathological

scarring by fibroblasts during wound healing. International Wound Journal, 20(6):2190–2206.

21. Mosmann, T., (1983). Rapid colorimetric assay for cellular growth and survival: application to proliferation and cytotoxicity assays. Journal of immunological methods, 65(1-2):55–63.

22. Murray, P.E., Garcia-Godoy, F. and Hargreaves, K.M., (2007). Regenerative endodontics: a review of status and a call for action. Journal of endodontics, 33(4):377–390.

23. Ngah, N.A., Dias, G.J., Tong, D.C., Mohd Noor, S.N.F., Ratnayake, J., Cooper, P.R. and Hussaini, H.M., (2021). Lyophilised platelet-rich fibrin: physical and biological characterisation. Molecules, 26(23):7131.

24. Nireesha, G.R., Divya, L., Sowmya, C., Venkateshan, N.N.B.M. and Lavakumar, V., (2013). Lyophilization/freeze drying-an review. International journal of novel trends in pharmaceutical sciences, 3(4):87–98.

25. Nosrat, A., Seifi, A. and Asgary, S., (2011). Regenerative endodontic treatment (revascularization) for necrotic immature permanent molars: a review and report of two cases with a new biomaterial. Journal of endodontics, 37(4):562–567.

26. Parhizkar, A., Nojehdehian, H. and Asgary, S., (2018). Triple antibiotic paste: momentous roles and applications in endodontics: a review. Restorative dentistry & endodontics, 43(3).

27. Parhizkar, A., Nojehdehian, H., Tabatabaei, F. and Asgary, S., (2020). An innovative drug delivery system loaded with a modified combination of triple antibiotics for use in endodontic applications. International journal of dentistry, 2020(1):8859566.

28. Puri, K. and Puri, N., (2013). Local drug delivery agents as adjuncts to endodontic and periodontal therapy. Journal of medicine and life, 6(4):414.

29. Rashid, M., Kaur, V., Hallan, S.S., Sharma, S. and Mishra, N., (2016). Microparticles as controlled drug delivery carrier for the treatment of ulcerative colitis: A brief review. Saudi Pharmaceutical Journal, 24(4):458–472.

30. Santanu, R., Hussan, S.D., Rajesh, G. and Daljit, M., (2012). A review on pharmaceutical gel. The International Journal of Pharmaceutical Research and Bio-Science:1(5).

31. Shahri, F. and Parhizkar, A., (2020). Pivotal local drug delivery systems in endodontics; a review of literature. Iranian Endodontic Journal, 15(2):65.

32. Sherje, A.P., Dravyakar, B.R., Kadam, D. and Jadhav, M., (2017). Cyclodextrin-based nanosponges: A critical review. Carbohydrate polymers, 173:37–49.

33. Singh, R.K., Lund, F.W., Haka, A.S. and Maxfield, F.R., (2019). High-density lipoprotein or cyclodextrin extraction of cholesterol from aggregated LDL reduces foam cell formation. Journal of Cell Science, 132(23):237271.

34. Staffoli, S., Plotino, G., Nunez Torrijos, B.G., Grande, N.M., Bossù, M., Gambarini, G. and Polimeni, A., (2019). Regenerative endodontic procedures using contemporary endodontic materials. Materials, 12(6):908.

35. Stockert, J.C., Horobin, R.W., Colombo, L.L. and Blázquez-Castro, A., (2018). Tetrazolium salts and formazan products in Cell Biology: Viability assessment, fluorescence imaging, and labeling perspectives. Acta histochemica, 120(3): 159–167.

36. Swaroop, A.E., Mathew, S., Harshini, P. and Nagaraja, S., (2025). Local drug delivery for regeneration and disinfection in endodontics: A narrative review. Journal of Conservative Dentistry and Endodontics, 28(2):119–125.

37. Tejashri, G., Amrita, B. and Darshana, J., (2013). Cyclodextrin based nanosponges for pharmaceutical use: A review. Acta pharmaceutica, 63(3):335–358.

38. Vocetkova, K., Buzgo, M., Sovkova, V., Rampichova, M., Staffa, A., Filova, E., Lukasova, V., Doupnik, M., Fiori, F. and Amler, E., (2017). A comparison of high throughput core–shell 2D electrospinning and 3D centrifugal spinning techniques to produce platelet lyophilisate-loaded fibrous scaffolds and their effects on skin cells. Rsc Advances, 7(85):53706–53719.

39. Wigler, R., Kaufman, A.Y., Lin, S., Steinbock, N., Hazan-Molina, H. and Torneck, C.D., (2013). Revascularization: a treatment for permanent teeth with necrotic pulp and incomplete root development. Journal of endodontics, 39(3):319–326.

40. Zhou, H.M., Shen, Y., Wang, Z.J., Li, L., Zheng, Y.F., Häkkinen, L. and Haapasalo, M., (2013). In vitro cytotoxicity evaluation of a novel root repair material. Journal of endodontics, 39(4):478–483.

Note: All the figures and tables in this chapter were made by the author.

Advances in Materials Science and Technology – Dr. Srikari Srinivasan et al. (eds)
© 2025 Taylor & Francis Group, London, ISBN 978-1-041-12342-2

7

Effect of Surface Characteristics on Streptococcus Mutans Adhesion on Nickel Titanium (Niti) and Coaxial Stainless Steel Archwires—A Comparative Prospective Clinical Study

Nirajita Bhaduri[1]

Dept. of Orthodontics and Dentofacial Orthopaedics,
Faculty of Dental Sciences,
Ramaiah University of Applied Sciences,
Bengaluru, Karnataka, India

Rabindra S. Nayak[2]

Dept. of Orthodontics and Dentofacial Orthopaedics,
M. R. Ambedkar Dental College,
Bengaluru, Karnataka, India

ABSTRACT: The insertion of orthodontic wires tends to create new surfaces for plaque formation and adhesion of microorganisms like Streptococcus mutans which contribute to enamel demineralization. Therefore, the purpose of this in-vivo study was to compare the adhesion of *Streptococcus mutans* to the surface of Nickel Titanium (NiTi) and coaxial Stainless steel archwires and to correlate the adhesion to surface free energy and surface roughness of these wires.10 patients were screened as per the inclusion and exclusion criteria, and 0.016" NiTi and 0.0155" coaxial Stainless Steel were placed on random allocation. The surface roughness and surface free energy of the wires were analysed using a profilometer and a goniometer respectively. Comparisons of *S.mutans* adhesion was done by Mann Whitney test and the surface roughness and surface free energy were compared by the Student t test. Spearman correlation test was done to compare the *S.mutans* adhesion to the surface characteristics of the wires. The *S.mutans* adhesion was higher in Coaxial SS wire than NiTi and the difference was statistically significant. Surface roughness and surface free energy of the wires increased after 28 days of intraoral exposure. NiTi displayed greater surface free energy whereas Coaxial SS displayed greater surface roughness. Surface roughness showed a positive correlation and surface free energy showed a negative correlation to *S.mutans* adhesion.The present study concluded that there is significant difference in the *S.mutans* adhesion on the different archwires and surface roughness of the wires showed a positive correlation to *S.mutans* adhesion.

KEYWORDS: Coaxial stainless steel, Colony forming units, NiTi, S. mutans adhesion, Surface free energy, Surface roughness

[1]nirajitab@gmail.com, [2]nirajitab@gmail.com

DOI: 10.1201/9781003664277-7

1. Introduction

The insertion of orthodontic wire tends to create new surfaces available for plaque formation and therefore increases the level of microorganisms in the oral cavity. It has long been suggested that orthodontic bands and wires lead to an increased plaque accumulation and elevated levels of *streptococci* and *lactobacilli*. An increase in plaque accumulation and retention areas and consequent elevated levels of streptococci and lactobacilli places the patient at a higher risk for enamel demineralization and white spot lesions. Organic acids (with a pH as low as 2.5) produced by *Streptococcus mutans* cause enamel demineralization. It has been suggested that a *Streptococcus mutans* count higher than 105 cfu/ml of saliva is related to higher caries risk (Leal and Mickenautsch., 2010). Therefore, prevention of bacterial attachment to orthodontic wires is a critical concern for orthodontists.

Stainless steel orthodontic wires have the advantage of good strength and springiness, corrosion resistance and low cost. One method of increasing the flexibility of stainless-steel arch wires was the development of a multistranded wire, which are generated by twisting two or more strands of a small diameter wire (≤ 0.01 inch). Nickel Titanium (NiTi) wires possess superelasticity, thermal shape memory, corrosion resistance, and biocompatibility and a low elastic modulus. However, since it contains approximately 50% of nickel, which is a known allergen, the issue of nickel release is important. An increase in the corrosion of metal dental materials occurs when in contact with *Streptococcus mutans*, giving rise to possible allergic and toxic reactions as well as deterioration of the mechanical properties of wires (Kim et al., 2014).

In vitro bacterial adhesion is influenced by surface characteristics of biomaterials, particularly surface roughness and surface free energy (SFE). It has been suggested that rougher surfaces promote bacterial adhesion by increasing the adhesion areas and preventing dislodgement of bacterial colonies (Eliades and Bourauel et al.,2005). A material with high SFE attracts more bacteria to its surface than one with low SFE. Although the physical and mechanical properties of orthodontic wires have been extensively studied, there is a paucity of studies on bacterial adhesion based on their surface characteristics.

Hence, the purpose of this study is to compare the adhesion of *Streptococcus mutans* to the surface of 0.016" Nickel Titanium (NiTi) and 0.0155" coaxial Stainless steel archwires and to correlate the adhesion to surface characteristics (surface free energy and surface roughness) of these wires.

2. Methodology

10 patients who were to undergo Orthodontic treatment were screened as per the inclusion and exclusion criteria, and the diagnosis and treatment plan was established for each patient, after evaluation of the diagnostic record. Written informed consent was obtained from all patients and the study was approved by the ethical clearance committee. Oral prophylaxis was performed on each patient. The teeth of the patients were then cleaned with a non-fluoridated prophylaxis paste and rubber prophylactic cups for 10 seconds, rinsed and dried. The teeth were then etched with 37% ortho-phosphoric acid (Scotchbond, 3M Unitek Orthodontic Products, Monrovia, California, USA) for 15 seconds following which a thin layer of primer (Transbond XT, 3M Unitek Orthodontic Products, Monrovia, California, USA) was applied. The patients were then bonded with MBT brackets of 0.022" x 0.028" slot (Gemini series, 3M Unitek Orthodontic Products, Monrovia, California, USA) using Transbond XT light cure adhesive (3M Unitek Orthodontic Products, Monrovia, California, USA) and light cured with an LED curing light (3M ESPE Elipar) for 20 seconds.

2.1 Archwire Insertion and Retrieval

Following this, 0.016" Nickel Titanium wires and 0.0155" Coaxial Stainless steel wires were placed on random allocation: in 5 patients, 0.016" NiTi was placed in the upper arch and 0.0155" Coaxial SS was placed in the lower arch. In the other 5 patients, 0.0155" Coaxial SS was placed in the upper arch and 0.016" NiTi was placed in the lower arch.(Fig. 7.1) This was done to rule out inter arch oral hygiene maintenance bias. Patients were instructed on oral hygiene measures and the usage of identical dentifrices twice daily. The patients were then recalled 28 days post insertion of the archwires (T_1), and the archwires were removed cautiously to avoid any contact with the oral mucosa. (Fig. 7.2). A small portion (2mm) of the retrieved archwires were cut and stored in a phosphate buffer solution (PBS) in a 50 ml sterile falcon tube.

2.2 Streptococcus Mutans Adhesion

The samples were diluted by 100-fold in 1x sterile phosphate-buffered saline and plated on *Mitis Salivarius Bacitracin* (MSB) agar. The MSB agar was composed of *Mitis salivarius* agar (HiMedia, India) and supplemented with 15% of sucrose, 1% of agar, 0.0001% potassium tellurite solution and 0.2 units/ml of bacitracin (HiMedia, India). The plates were incubated anaerobically at 37°C for 48 hours. After the incubation period, the colonies were identified on the basis of colony morphology (Fig. 7.3).

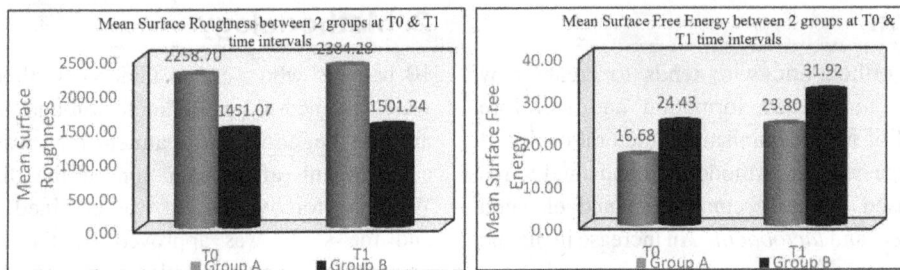

Fig. 7.1 Comparison of mean surface roughness and mean surface free energy between two archwires at both time intervals

Note: Group A – 0.0155" Coaxial Stainless Steel, Group B – 0.016" NiTi

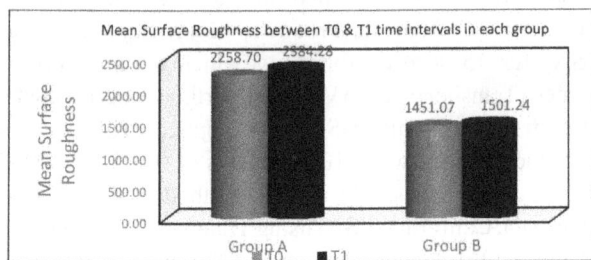

Fig. 7.2 Association between the mean surface roughness between two different time points of the two different archwires.

Note: Group A – 0.0155" Coaxial Stainless Steel, Group B – 0.016" NiTi

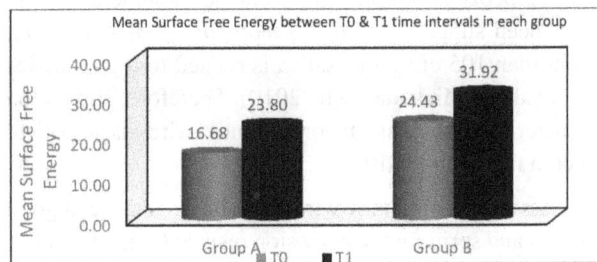

Fig. 7.3 Association between the mean surface free energy at two different time points of the two different archwires.

Note: Group A – 0.0155" Coaxial Stainless Steel, Group B – 0.016" NiTi

The overnight bacterial cultures were stored in 80% glycerol stock at −20°C.

2.3 Surface Roughness and Surface Free Energy

Surface Roughness of both groups of archwires was evaluated at T_0 (as received condition) and T_1 (after 4 weeks of intraoral exposure) using 3D Surface Profilometry (Dektak XT- Bruker, Mass). Contact angle measurements were performed to calculate the SFE of both archwire materials. To calculate the surface free energy (SFE), advancing and receding contact angles were measured for the 0.016" NiTi and 0.0155" Coaxial Stainless steel archwires in as-received condition (T_0) and after 4 weeks of intraoral exposure (T_1) using a goniometer with a CCD camera.

3. Statistical Analysis

Statistical Package for Social Sciences [SPSS] for Windows Version 22.0 Released 2013. Armonk, NY: IBM Corp., was used to perform statistical analyses. Descriptive analysis of all the explanatory and outcome parameters will be done using frequency and proportions for categorical variables, whereas in Mean & SD for continuous variables. Mann Whitney Test was used to compare the OD values & CFUs of S. mutans between 2 groups. Independent Studentt Test was used to compare the Mean Surface Roughness & Free Energy between 2 groups at T0 & T1 time intervals.

4. Results and Discussion

The Optical Density (OD) values and Colony Forming Units (CFU) of S. mutans were assessed by Mann Whitney test as given in Table 7.1. Mann Whitney test revealed that 0.0155" Coaxial Stainless Steel exhibited more S. mutans adhesion when compared to 0.016" NiTi. The comparison of mean surface roughness and surface energy between coaxial stainless steel and NiTi at T_0 and T_1 is given in Graph 1. On comparing the mean surface roughness between the two archwires at two time points, it was found that 0.0155" Coaxial Stainless Steel exhibited greater surface roughness at both time points when compared to 0.016" NiTi and the values were statistically significant. However, on comparing the mean surface free energy it was found that 0.016" NiTi exhibited greater surface free energy at both time points when compared to 0.0155" Coaxial Stainless Steel and the values were statistically significant.

Table 7.1 Mann whitney test

Comparison of OD values & CFUs of S. mutans between 2 groups using Mann Whitney Test						
Variable	Groups	N	Mean	SD	Mean Diff	P-Value
OD	Group A	10	**0.54**	0.07	0.21	<0.001*
	Group B	10	0.33	0.04		
CFU	Group A	10	71.09	20.81	36.93	<0.001*
	Group B	10	**34.16**	7.33		

* - Statistically Significant
Note: Group A – 0.0155" Coaxial Stainless Steel, Group B – 0.016" NiTi

The comparison of mean surface roughness and surface energy of both coaxial stainless steel and NiTi between T_0 and T_1 are given in Fig. 7.2 and 7.3. On comparing the mean surface roughness and surface free energy of the two archwires between the two time points it was found that the surface roughness and surface free energy of both archwires increased from T_0 to T_1 and the value was statistically significant.

The relationship between the S. mutans adhesion, surface roughness and surface free energy in each group was assessed using the Spearman's correlation test as given in Table 7.2.

On assessing the relationship between S.mutans adhesion, surface roughness and surface free energy of the two archwires, a strong correlation between S. mutans adhesion and surface roughness was seen in the 0.0155" Coaxial Stainless steel group and the value was statistically significant.

In the 0.016" NiTi group, very weak correlation between S. mutans adhesion and surface roughness was seen and the value was statistically insignificant (Fig. 7.4). In contrast, on comparing the surface free energy and S. mutans adhesion in both types of archwires, a negative correlation

Table 7.2 Spearman's correlation test

Spearman's correlation to assess the relationship b/w CFUs, Surface Roughness & Free Energy in each group					
Variable	Values	Group A		Group B	
		SR	SFE	SR	SFE
CFU	Rho	0.76	-0.06	0.14	-0.59
	P-Value	0.01*	0.87	0.70	0.04*

* - Statistically Significant
SR - Surface Roughness; SFE - Free Energy, Group A – 0.0155" Coaxial Stainless Steel, Group B – 0.016" NiTi. The correlation coefficients are denoted by 'rho'.Minus sign denotes negative correlation.

Correlation coefficient range	
0.0	No Correlation
0.01 - 0.20	Very Weak Correlation
0.21 - 0.40	Weak Correlation
0.41 - 0.60	Moderate Correlation
0.61 - 0.80	Strong Correlation

was seen and the value was statistically significant in the 0.016" NiTi group (Fig. 7.4).

Dental plaque is a highly complex organization in a biofilm form and is considered the main causative factor in dental caries and periodontal disease. In dental biofilm, which is correlated with dental caries, *Streptococcus mutans*, *Streptococcus sobrinus*, and *Lactobacillus* are the most frequently isolated microorganisms. These species grow 6 or 12 weeks after orthodontic appliance bonding (Bergamo et al., 2019).

The idea of dividing comprehensive orthodontic treatment into stages was emphasized by Dr. Raymond Begg, who proposed three major stages which are now used universally. In nearly every patient with maligned teeth, the root apices are closer to the normal position than the

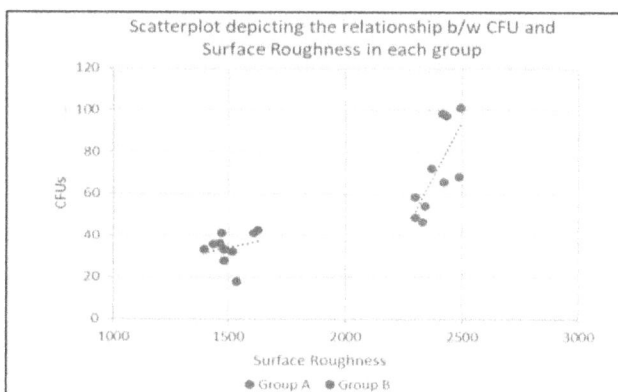

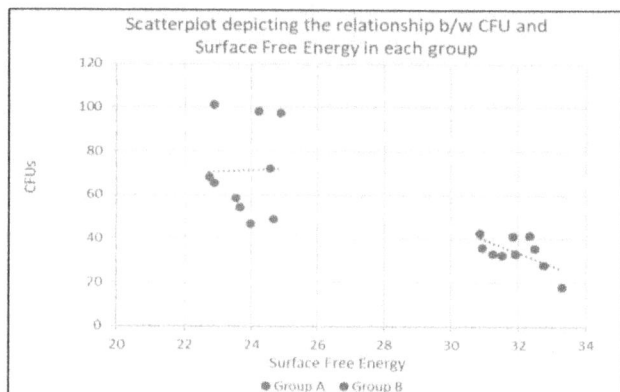

Fig. 7.4 Spearman correlation test

crowns, because malalignment almost always develops as the eruption paths of the teeth are deflected. Thus, the goals of the first stage of orthodontic treatment, alignment and leveling, are to bring the teeth into alignment and correct vertical discrepancies by leveling out the arches. The wires for initial alignment require a combination of excellent strength, excellent springiness, and a long range of action. Thus, 0.0155" Coaxial Stainless Steel and 0.016" NiTi wires were selected for the study.

The results revealed that 0.0155" Coaxial Stainless Steel exhibited greater *Streptococcus mutans* adhesion (71.09 CFUs/1000 mL) when compared to 0.016" NiTi (34.16 CFUs/1000mL) with the difference being statistically significant according to Mann Whitney Test. This can be attributed to increased surfaces for plaque accumulation and subsequent S. mutans colonization on Coaxial Stainless-steel wire, a multistranded wire when compared to a round NiTi archwire.

On comparing the mean surface roughness between the two archwires at two time points, it was found that 0.0155" Coaxial Stainless Steel exhibited greater surface roughness at both time points when compared to 0.016" NiTi and the values were statistically significant. However, on comparing the mean surface free energy it was found that 0.016" NiTi exhibited greater surface free energy at both time points when compared to 0.0155" Coaxial Stainless Steel and the values were statistically significant. Similar studies have revealed that NiTi displays greater surface roughness and surface free energy when compared to Stainless steel (Leal and Mickenautsch.,2010; Lim et al.,2008). Titanium exhibits the highest surface reactivity among orthodontic biomaterials, and hence NiTi has greater surface free energy compared to stainless steel. However, since coaxial Stainless steel is a multibraided wire, it exhibited greater surface roughness when compared to NiTi by surface profilometry in the present study.

On comparing the mean surface roughness of the two archwires between the two time points it was found that the surface roughness and surface free energy of both archwires increased from T_0 to T_1 and the value was statistically significant. Orthodontic alloys immerse in saliva and biological fluids in the oral cavity, which enables ionic conduction and are thus subject to corrosive degradation. In several in vitro and in vivo studies, investigators have tried to assess the effects of exposure of orthodontic archwires to the oral cavity (Vaid et al., (2015). Generally, it has been shown that the intraoral exposure of wires alters the topography and morphology of the material surface by pitting or crevice corrosion. Generally, in vivo-aged materials are exposed to the synergistic action of a multifactorial oral environment such as the application of complex masticatory forces and subsequent flexion of the orthodontic wire, frictional resistance, as well as the action of oral microbiota such as the acidogenic *Streptococcus mutans*.

Literature findings also confirm that the frictional properties of archwires were improved if a surface treatment was applied, e.g. Teflon, polyethylene or ion implantation. Further investigations are required to analyse the bacterial adhesion to orthodontic appliances, the effect of various intraoral parameters as well as surface characteristics of the appliances. A larger sample size may be helpful in correlation of surface characteristics to S. mutans adhesion of archwires and different orthodontic materials.

5. Conclusion

It would be prudent to select orthodontic appliances appropriately, especially in patients who have a predilection towards plaque accumulation and caries. Similarly, in older patients displaying greater chances of periodontal breakdown and disease, appliances should be selected with utmost care.

Acknowledgments

The authors would like to express my appreciation for Chromogene Biotech Pvt Ltd, Bengaluru for facilitating testing of the samples of the study. I would also like to thank Dr. Santhosh, Biostatistician, Bengaluru, for carrying out the statistical analysis.

References

1. Bergamo AZ, Matsumoto MA, Nascimento C, Andrucioli MC, Romano FL, Silva RA, Silva LA, Nelson-Filho P. (2019). Microbial species associated with dental caries found in saliva and in situ after use of self-ligating and conventional brackets. J. Appl. Oral Sci.;27: e20180426.
2. Eliades T, Bourauel C (2005). Intraoral aging of orthodontic materials: the picture we miss and its clinical relevance. Am J Orthod Dentofacial Orthop;127:403–412.
3. Kim IH, Park HS, Kim YK, Kim KH, Kwon TY (2014). Comparative short-term in vitro analysis of mutans streptococci adhesion on esthetic, nickel-titanium, and stainless-steel arch wires. Angle Orthod.;84: 680–686.
4. Leal SC, Mickenautsch S. (2010) Salivary Streptococcus mutans count and caries outcome—a systematic review. J Minim Interv Dent.;3:137–147.
5. Lim BS, Lee SJ, Lee JW, Ahn SJ.(2008) Quantitative analysis of adhesion of cariogenic streptococci to orthodontic raw materials. Am J Orthod Dentofacial Orthop;133(6): 882–8.
6. Vaid N, Vandekar M, Doshi V, Fadia D. (2015) Plaque accumulation and Streptococcus mutans levels around self-ligating bracket clips and elastomeric modules: A randomized clinical trial. APOS Trends Orthod ;5(3): 97–102.

Note: All the figures and tables in this chapter were made by the author.

Advances in Materials Science and Technology – Dr. Srikari Srinivasan et al. (eds)
© 2025 Taylor & Francis Group, London, ISBN 978-1-041-12342-2

8

Evaluation of *Cynodon Dactylon* Gel as an Avulsion Medium: An In Vitro Study

Kulkarni Radhika Ashok*, Shwetha G

Department of Pediatric and Preventive Dentistry,
Faculty of Dental Sciences, Ramaiah University of Applied Sciences,
Bangalore, Karnataka, India

Deveswaran Rajamanickam

Drug Design and Development Centre Department of Pharmaceutics,
Faculty of Pharmacy, Ramaiah University of Applied Sciences,
Bangalore, Karnataka, India

ABSTRACT: *Cynodon dactylon* (CD) also known as Bermuda grass. It has many medicinal activities. However, its use as an avulsion media to maintain cell viability is not explored. The aim of study was to assess the efficacy of CD gel as a new storage medium for avulsed tooth in maintaining the viability of periodontal ligament cells. Bermuda Grass Leaf Powder was mixed with 70% ethanol and the compounds in the extracted powder were identified using Gas Chromatography–Mass Spectrometry (GC–MS) analysis. The extracted powder was then converted to gel form. The HPDLFs (Human Periodontal Ligament fibroblasts) was isolated by Enzyme digestion method from HPDLFs. Plates containing confluent HDLFs was soaked in the various media for 3, 6, 24, 48 and 72 h at 37C and 20 °C. After incubation, viability of the cells was determined using the tetrazolium salt-based colorimetric (MTT) assay after 6, 24, 48 and 72h of incubation at 20 °C. The GC-MS analysis identified that the leaves of CD contain Neophytadiene, Phytol, Undecanoic acid and Pentadecanal. MTT assay results showed that 108.26% HDLFs cells were viable in 1000 ug CD gel. CD gel exhibits antimicrobial, anti-inflammatory, antifungal, and anticancer properties. It is non-cytotoxic and promotes cell proliferation in HPDLFs, showing higher cell viability than untreated cells for up to 24 hours.

KEYWORDS: Avulsion media, Bermuda grass, *Cynodon dactylon*, isolation, MTT assay

1. Introduction

Tooth avulsion, characterized by the complete displacement of a tooth from its socket, accounts for 0.5–3% of traumatic injuries to permanent teeth (Andreasen et al., 2010). This injury severely damages the supporting tissues, blood vessels, and nerves, necessitating prompt and correct emergency management for favorable outcomes (Khinda et al.,2017). Immediate replantation is the recommended treatment to avoid further damage to the periodontal ligament (PDL), thereby reducing the risk of post-replantation resorption, whether inflammatory or replacement in nature (Sanghavi al.,2103).

The prognosis of a replanted tooth is dependent on several factors such as the viability of PDL cells remaining on the root surface, the integrity of the root cementum, and minimal bacterial contamination. These factors are influenced by the extra-alveolar time, storage type after avulsion, and root surface alterations (Marwah et al.,2018). Immediate

*Corresponding author: radhikakulkarni1196@gmail.com

DOI: 10.1201/9781003664277-8

replantation positively affects PDL cell viability, leading to PDL healing in up to 85% of mature teeth. Replanting a tooth within 5 minutes ensures the prompt return of PDL cells to their normal function. However, after 15 minutes of dry storage, the precursor cells, progenitor cells, or stem cells lose their ability to differentiate into fibroblasts. After 30 minutes, almost all PDL cells on the tooth root are likely to have become necrotic. Factors hindering immediate replantation include the victim's and witnesses' emotional states, awareness of appropriate actions, proximity to a dentist or clinic, and consent issues (Udoye et al., 2012).

When immediate replantation is not feasible, storing the tooth in an appropriate medium is essential to preserve PDL cell viability. A suitable storage medium is a physiological solution mimicking oral conditions, aiding in conserving PDL cells post-avulsion (Ingle et al., 2009). However, none of the currently used media are perfect, and the search for a medium that can address the shortcomings of these options is ongoing (Rajendran et al., 2011). Many different storage media for avulsed teeth have been evaluated to discover the ideal solution. The most extensively examined and approved ideal storage media were Hank balanced salt solution, which has favorable biological characteristics, and readily available milk. Other storage media, such as saliva, culture medium, ViaSpan, propolis, and green tea extract, have been tested.

Cynodon dactylon, known as Bermuda grass or Doob grass, belongs to the *Poaceae* family and has several medicinal properties, including antidiabetic, diuretic, antioxidant, and anti-allergic effects. It is also used for wound healing. This plant contains numerous chemicals such as alkaloids, flavonoids, glycosides, ß-sitosterol, carotene, stigmasterol, phytol, and fryer phenols. *Cynodon dactylon* is recognized in Ayurveda for its wound healing properties and has been used for this purpose for a long time (Sharma et al., 2023). The aim of the present study is to assess the efficacy of *Cynodon dactylon* gel as a new storage medium for avulsed tooth in maintaining the viability of periodontal ligament cells.

2. Materials and Methods

The study protocol was reviewed and approved by the University Ethics Committee for Human trials of M S Ramaiah University of Applied Sciences, Bengaluru.

Preparation of *Cynodon Dactylon* Extract and Formulate Gel - 50 g of commercially available 'Pure Bermuda Grass Leaf Powder' powder was mixed with 20 ml 70% ethanol and microwave extraction was done using Microwave synthesizer and then it was incubated on water bath to get residue of *Cynodon dactylon* powder. Then the

Cynodon dactylon residue was converted to gel by adding xanthum gum, propylene glycol (5ml), methyl paraben (100mg) and distilled water (see Fig. 8.1).

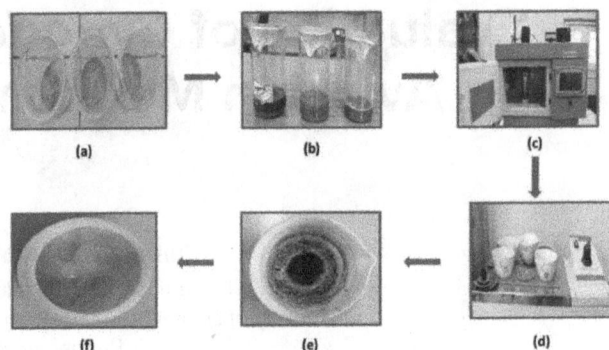

Fig. 8.1 (a) Pure Bermuda Grass Leaf Powder (b) solution containing Pure Bermuda Grass Leaf Powder and ethanol (c) Microwave synthesizer (d) Water bath (e) Residue of the powder (f) *Cynodon dactylon* gel

Source: Author's compilation

Isolation and Purification of Compounds from *Cynodon Dactylon* Gel- The compounds in the extracted *Cynodon dactylon* powder will be identified using Gas Chromatography–Mass Spectrometry (GC–MS) analysis.

GC-MS- The Clarus 680 GC was used in the analysis employed a fused silica column, packed with Elite-5MS (5% biphenyl 95% dimethylpolysiloxane, 30 m × 0.25mm ID × 250µm df) and the components were separated using Helium as carrier gas at a constant flow of 2 ml/min. The injector temperature was set at 220°C during the chromatographic run. The 1µL of extract sample injected into the instrument the oven temperature was as follows: 50 °C (2 min); followed by 150 °C at the rate of 15 °C min−1; and 150 °C, where it was held for

2min and then followed by 250°C at the rate of 30°C min−1; it was held for 8.00 min. The mass detector conditions were: Inlet line temperature 250 °C; ion source temperature 230 °C; and ionization mode electron impact at 70 eV, a scan time 0.2 sec and scan interval of 0.1 sec. The fragments from 40 to 600 Da. The spectrums of the components were compared with the database of spectrum of known components stored in the GC-MS NIST (2014) library.

Determination of pH and Osmolality of *Cynodon Dactylon* Gel- 10ml of 1mg/ml of *Cynodon dactylon* gel dissolved in Double distilled water was subjected to measure the Osmolality by Osmometer and pH by digital pH meter respectively.

Determination of Viability and Proliferation of HPDLFs (Human Periodontal Ligament Fibroblasts) in *Cynodon*

Dactylon **gel using MTT Assay-** Cell lines: HPDLFs – Human Periodontal Ligament fibroblasts isolated from Human Periodontal ligament tissue by Enzyme digestion method. The cells were maintained in DMEM with high glucose media supplemented with 10% FBS along with the 1% antibiotic-antimycotic solution in the atmosphere of 5% CO_2, 18-20% O_2 at 37^0C temperature in the CO_2 incubator and sub-cultured for every 2 days.

10000 cells will be seeded in each well of 96 well plate which will be allowed to grow for about 48 hours. Then cells will be treated with desired concentrations of *Cynodon dactylon* gel and incubated for 1, 3 6, 24 and 48 hours at 37°C in 5% CO2 atmosphere. 0.5 mg/ml MTT reagent of total volume of each well of 96 well plate will be added. The availability of cells will be determined by tetrazolium salt-based colorimetric MTT assay and cultures will then be incubated for 4 hours and then will be replaced with 150 ul of dimethyl sulfoxide afterwards (Misurya et al., 2022).

Formula: Cell viability (%) = [Mean abs of treated cells/ Mean abs of Untreated cells] x 100.

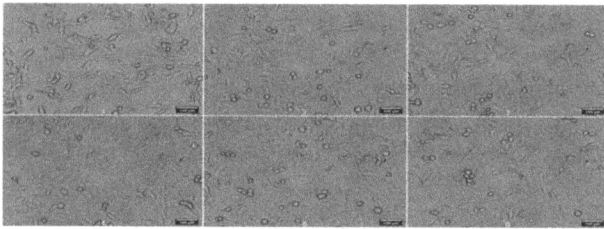

Fig. 8.2 Morphology of HPDLFs in non-treated and *Cynodon dactylon* gel with different concentrations treated conditions after the 24hours of incubation

Source: Author's compilation

3. Results and Discussion

GCMS- The GC-MS analysis identified that the leaves of *Cynodon dactylon* contain Neophytadiene (17.2%), Phytol (11.1%), undecanoic acid (10.9%) and Pentadecanal (4.28%).

pH and Osmolality of gel: pH of *Cynodon dactylon* gel was 7.41 and osmolality around 280 mOsmol/kg in all 5 concentrations (62.5, 125, 250, 500, 1000ug/ml).

MTT Assay- The results of cell viability study performed by MTT assay (see Table 8.1). suggest that the given test compound was non-toxic as well as cell proliferative in nature on time dependent fashion till the 24hours respectively. The MTT cell viability results suggest us that, given test compound, *Cynodon dactylon* Gel was non-cytotoxic as well as cell proliferative in nature on Human Periodontal Ligament fibroblasts (HPDLFs) with

Table 8.1 Human periodontal ligament fibroblasts with increased cell viability values on time dependent fashion till the 24hours respectively

Condition	% cell viability	Maximum non-toxic dose (ug/ml)
Untreated	100.00	
VEGF-10ng	106.14	
CD Gel-62.5ug	100.08	
CD Gel-125ug	100.54	1000ug/ml
CD Gel-250ug	101.13	
CD Gel-500ug	104.26	
CD Gel-1000ug	108.26	

Source: Author's compilation

increased cell viability values than the untreated cells on concentration dependent fashion and confirmed the cell proliferative effect till the 24hours. Highest dose of 1mg/ml maintained best proliferative effect on HPDLFs without losing rate of viability values till 24 hours and clearly double concluded that it has better angiogenic effect on HPDLFs (see Fig. 8.3).

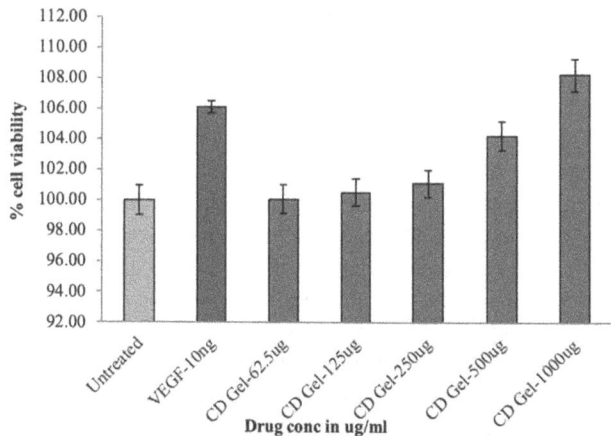

Fig. 8.3 Percentage of cell viability values of *Cynodon dactylon* Gel treated HPDLFs

Source: Author's compilation

Replanting an avulsed tooth requires the presence of functional PDL cells in the root for the tooth to be healed successfully. PDL cell viability is affected by both extraoral time and storage media (Martins et al., 2004). Avulsed teeth that have been exposed to dry conditions for an extended period of time have a poorer prognosis. An appropriate storage media for an avulsed tooth should closely simulate the oral environment (Gjertsen et al., 2011). Physiological pH, osmolality, and nutrient requirements for cell metabolism are considered when analyzing storage media. Furthermore, the ideal storage medium should have antibacterial properties, maintain the viability of PDL cells, and be easy to use and readily accessible (Poi et al., 2013).

GC-MS analysis in the current study demonstrated that *Cynodon dactylon* gel contain Neophytadiene (17.2%), Phytol (11.1%), undecanoic acid (10.9%) and Pentadecanal (4.28%). Neophytadiene is a diterpene that is known for its antimicrobial, anti-inflammatory, and analgesic properties (Prakash, D et al., 2007). Phytol is an acyclic diterpene alcohol that plays a crucial role as a precursor for the synthesis of vitamins E and K1. It has antioxidant properties and is also known for its anti-inflammatory and antimicrobial activities (Sudhakar, M et al., 2006). Undecanoic acid, also known as undecylenic acid, is a fatty acid that exhibits antifungal and antibacterial properties (Borgers et al., 2005). Pentadecanal is an aldehyde shown to possess antimicrobial properties. These constituents collectively contribute to the beneficial properties of the sample, including antimicrobial, anti-inflammatory, and antioxidant effects, which are important for maintaining the viability and health of periodontal ligament cells.

For the successful healing of an avulsed tooth after replantation, it is essential to ensure the viability of the periodontal ligament (PDL) cells on the root. The survival of these PDL cells is influenced by the duration the tooth remains outside the oral environment and the type of storage medium used (Khademi et al., 2008). Extended exposure to dry conditions negatively impacts the prognosis. Ideally, a storage medium should closely mimic the oral environment, considering factors such as physiological pH, osmolality, and the availability of nutrients required for cell metabolism (Gjertsen et al., 2011). Hammer, in 1955, was the first to highlight the importance of PDL cell viability before replanting an avulsed tooth, demonstrating that the amount of viable periodontal membrane is critical for the survival of the replanted tooth (Bijlani et al., 2022). Research indicates that to maintain PDL cell viability, the storage medium should have a pH between 6.6 and 7.8 and an osmolality between 230 and 400 mOsmol/kg. The study found that the pH of *Cynodon dactylon* gel was 7.41 and its osmolality approximately 280 mOsmol/kg.

This study compared the percentage of viable PDL cells in a novel storage medium, *Cynodon dactylon* gel, with Vascular Endothelial Growth Factor (VEGF). The MTT cell viability results showed that untreated HPDLFs were 100% viable before treatment with *Cynodon dactylon* gel and VEGF. Upon treatment with VEGF, HPDLFs exhibited 106.14% cell viability, indicating they were viable and proliferative. When HPDLFs were treated with concentrations 62.5 µg, 125 µg, 250 µg, 500 µg, and 1000 µg of *Cynodon dactylon* gel, cell viability was 100.08%,

100.54%, 101.13%, 104.26%, and 108.26%, respectively, suggesting that HPDLFs remained viable and proliferative in *Cynodon dactylon* gel for up to 24 hours. *Cynodon dactylon* Gel is non-cytotoxic, with a maximum non-toxic dose of 1000 µg/ml.

The results of this study align with those of Bijlani et al., who used ice apple as a storage medium to assess cell viability. After 24 hours of exposure of PDL cells to the ice apple fruit extract (IAFPE) storage medium, it was evident that 10% IAFPE, with 97.33% viable cells (mean OD = 0.458), was the most effective in maintaining PDL cell viability among the tested media. Coconut water has also been proven as an effective storage medium for avulsed teeth, and the findings of the current study concur with those from studies conducted on coconut water (Gopikrishna et al., 2008).

4. Conclusion

Cynodon dactylon gel exhibits antimicrobial, anti-inflammatory, antifungal, and anticancer properties. It is non-cytotoxic and promotes cell proliferation in Human Periodontal Ligament Fibroblasts (HPDLFs), showing higher cell viability than untreated cells for up to 24 hours. However, further in-vivo as well as in-vitro studies have to be conducted to determine the molecular mechanism behind cell proliferative properties of the *Cynodon dactylon* Gel.

Acknowledgment

The authors thank Dr Pushpalatha C, Head of Department of Pediatric and Preventive Dentistry for the support and motivation. The authors are thankful to Deans of Faculty of Dentistry and Faculty of Pharmacy for providing the necessary facilities.

References

1. Andreasen, J.O. and Andreasen, F.M. (2010) *Essentials of traumatic injuries to the teeth: A step-by-step treatment guide.* Oxford: John Wiley & Sons.
2. Biswas, T.K., Pandit, S., Chakrabarti, S., Banerjee, S., Poyra, N. and Seal, T. (2017) 'Evaluation of *Cynodon dactylon* for wound healing activity', *Journal of Ethnopharmacology*, 197, pp. 128–137.
3. Bijlani, S. and Shanbhog, R.S. (2022) 'An in vitro evaluation of ice apple as a novel storage medium to preserve the viability of human periodontal ligament fibroblasts', *International Journal of Clinical Pediatric Dentistry*, 15(6), p. 699.
4. Borgers, M., Degreef, H. and Cauwenbergh, G. (2005) 'Fungal infections of the skin: Infection process and antimycotic therapy', *Current Drug Targets*, 6(8), pp. 849–862.

5. Gopikrishna, V., Baweja, P.S., Venkateshbabu, N. *et al.* (2008) 'Comparison of coconut water, propolis, HBSS, and milk on PDL cell survival', *Journal of Endodontics*, 34(5), pp. 587–589. Available at: https://doi.org/10.1016/j.joen.2008.01.018

6. Gjertsen, A.W., Stothz, K.A., Neiva, K.G. and Pileggi, R. (2011) 'Effect of propolis on proliferation and apoptosis of periodontal ligament fibroblasts', *Oral Surgery, Oral Medicine, Oral Pathology, Oral Radiology, and Endodontology*, 112(6), pp. 843–848.

7. Ingle, J.I. (ed.) (2009) *PDQ Endodontics*. Shelton, CT: PMPH USA.

8. Khademi, A.A., Saei, S., Mohajeri, M.R., Mirkheshti, N. and Ghassami, F. (2008) 'A new storage medium for an avulsed tooth', *Journal of Contemporary Dental Practice*, 9(6), pp. 25–32.

9. Khinda, V.I., Kaur, G., Brar, G.S., Kallar, S. and Khurana, H. (2017) 'Clinical and practical implications of storage media used for tooth avulsion', *International Journal of Clinical Pediatric Dentistry*, 10(2), p. 158.

10. Martins, W.D., Westphalen, V.P.D. and Westphalen, F.H. (2004) 'Tooth replantation after traumatic avulsion: A 27-year follow-up', *Dental Traumatology*, 20(2), pp. 101–105.

11. Marwah, N. (2018) *Textbook of pediatric dentistry*. New Delhi: JP Medical Ltd.

12. Misurya, R., Sharma, S., Ismail, P.M.S., Gupta, N., Rajan, R., Kaur, R. and Babaji, P. (2022) 'An in vitro evaluation of efficacy of ViaSpan, Aloe Vera, Gatorade solution, and propolis storage media for maintaining the periodontal ligament cell viability', *Annals of African Medicine*, 21(1), pp. 34–38.

13. Poi, W.R., Sonoda, C.K., Martins, C.M., Melo, M.E., Pellizzer, E.P., Mendonça, M.R.D. and Panzarini, S.R. (2013) 'Storage media for avulsed teeth: A literature review', *Brazilian Dental Journal*, 24(5), pp. 437–445.

14. Prakash, D., Singh, B.N. and Upadhyay, G. (2007) 'Antioxidant and free radical scavenging activities of phenols from onion (*Allium cepa*)', *Food Chemistry*, 102(4), pp. 1389–1393.

15. Rajendran, P., Varghese, N.O., Varughese, J.M. and Murugaian, E. (2011) 'Evaluation, using extracted human teeth, of Ricetral as a storage medium for avulsions—An in vitro study', *Dental Traumatology*, 27(3), pp. 217–220.

16. Sanghavi, T., Shah, N., Parekh, V. and Singbal, K. (2013) 'Evaluation and comparison of efficacy of three different storage media, coconut water, propolis, and oral rehydration solution, in maintaining the viability of periodontal ligament cells', *Journal of Conservative Dentistry*, 16(1), pp. 71–74.

17. Sharma, S., Ismail, P.M.S., Agwan, M.A.S., Chandra, A., Jain, A., Alessa, N. and Srivastava, S. (2023) 'Evaluation of viability of periodontal ligament cells using propolis, coconut water, aloe vera, and soy milk: An in vitro comparative study', *Journal of Pharmacy and Bioallied Sciences*, 15(Suppl. 2), pp. S901–S903.

18. Sudhakar, M., Rao, C.V., Rao, P.M. and Raju, D.B. (2006) 'Evaluation of phytol in inflammatory disorders: Inhibition of TNF-α production and suppression of TNF-α receptor expression', *Biological and Pharmaceutical Bulletin*, 29(5), pp. 948–953.

19. Udoye, C.I., Jafarzadeh, H. and Abbott, P.V. (2012) 'Transport media for avulsed teeth: A review', *Australian Endodontic Journal*, 38(3), pp. 129–136.

Advances in Materials Science and Technology – Dr. Srikari Srinivasan et al. (eds)
© *2025 Taylor & Francis Group, London, ISBN 978-1-041-12342-2*

9

Evaluation of Antimicrobial Effect of Osteopontin against Caries Causing Bacteria

Roopa Narayanan* and Dhananjaya Gaviappa

Department of Pediatric and Preventive Dentistry,
Faculty of Dental Sciences, M S Ramaiah University of Applied Sciences,
Bangalore, Karnataka, India

ABSTRACT: Dental caries is a dynamic process involving enamel demineralization and remineralization, with initial stages marked by white-spot lesions. Remineralization products help manage early demineralization (Nurrohman et al.,2022) (Ishizuka et al.,2023). *Streptococcus mutans* and *Streptococcus salivarius* contribute to teeth decay (Gross et al.,2012). Probiotics, beneficial microorganisms, produce antimicrobial substances that inhibit harmful bacteria and disrupt plaque biofilms (Zaky et al.,2021). Osteopontin, involved in tooth remineralization, may enhance the effects of probiotics in dental care. Lyophilized Osteopontin was mixed with probiotic paste at different concentrations by the Minimum Inhibitory Concentration (MIC) and Minimum Bactericidal Concentration (MBC) which is determined using dilution methods. Three samples at different concentrations were evaluated: 1.5ml OPN, 3ml OPN, and Probiotic paste. Results prove that both MIC and MBC values of OPN showed a dose-dependent effect, confirming its antibacterial activity against S. mutans and S. salivarius. The 3ml OPN sample exhibited better efficacy than the 1.5ml OPN in inhibiting and killing the bacteria. This study suggests that Osteopontin-incorporated probiotic paste can effectively enhance dental caries management by using its antimicrobial activity as well.

KEYWORDS: Antimicrobial activity, Demineralization, Dental caries, Osteopontin, Probiotics, Remineralization

1. Introduction

Dental caries, a multifactorial disease affecting all ages, remains a global challenge. It results from enamel demineralization caused by bacterial acid production from dietary carbohydrates in dental plaque (Featherstone et al.,2000). Despite fluoride use and improved oral hygiene, its high prevalence necessitates innovative strategies. Research has increasingly explored bioactive molecules and microbial interventions for caries prevention. Osteopontin (OPN), an extracellular matrix protein, plays a key role in biomineralization, immune modulation, and tissue repair (Bosshardt et al.,1998) (Bellahcènet

et al.,2008). Expressed in dental tissues, OPN regulates bacterial adhesion and influences oral inflammation, suggesting therapeutic potential in caries management (Gajjeraman et al.,2007) (Bonaventura et al.,2016). Probiotics, live microorganisms that confer health benefits, have also emerged as promising oral health agents (Cagetti et al.,2013). Strains like lactobacilli and bifidobacteria exhibit antimicrobial activity against Streptococcus mutans through competitive inhibition, antimicrobial production, and immune modulation (Soderling et al.,2000). Studies highlight their ability to reduce caries incidence, offering a non-invasive, cost-effective complement to conventional prevention (Twetman et al.,2012) (Keller et al.,2012).

*Corresponding author: roopanair24@gmail.com

DOI: 10.1201/9781003664277-9

2. Methodology

The study was conducted following the recommendations for good clinical practice approved by the University Ethics Committee for Human trials of M S Ramaiah University of Applied Sciences.

2.1 Antimicrobial Analysis

Lyophilized Osteopontin will be obtained from R&D Systems (Bovine Osteopontin protein 30029052) and mixed with probiotic paste at different concentrations The concentration of OPN was established based on Schlafer et al.'s reference. The study aimed to determine the Minimum Inhibitory Concentration (MIC) and Minimum Bactericidal concentration (MBC) The samples details as below:

Table 9.1 Details of sample and microbial species used for the study

Sample Name/ Code	Concentrations used	Bacterial Species
1.5ml OPN with probiotic paste	Dilutions: 6 (100%, 50%, 25%, 12.5%, 6.25%, 3.12%)	*Streptococcus mutans*
3ml OPN with probiotic paste		*Streptococcus salivarius*
Purexa		

Source: Author

Minimum Inhibitory Concentration (MIC), which is the lowest concentration of the agent that stops microbial growth is defined as the lowest concentration that reduces initial bacterial inoculum viability by ≥99.9% (Andrews et al.,2001).

2.2 Test Organisms

- *S. mutans (MTCC No-890)*
- *S. salivarius (ATCC No-7073)*

Microbial species were tested for Minimum Inhibitory Concentration (MIC) using the microtiter broth method by Mogana (Mogana et al.,2020). Serial dilutions of a compound were prepared from 4 mg/ml to 0.1 mg/ml in sterile nutrient broth, with a bacterial inoculum of approximately 5×105 colony-forming units/ml added to each dilution. Controls included Ciprofloxacin (5 g/ml), nutrient broth only, and microbe only. After incubation at 37°C for 24-48 hours, absorbance at 600 nm was measured. The MIC was the lowest concentration that completely inhibited visible growth. For Minimum Bactericidal Concentration (MBC), the method by Ozturk & Ercisli was used to determine the lowest concentration that killed 99.9% of bacteria. The MBC was recorded as the concentration yielding fewer than 10 colonies after incubation at 37°C for 18-24 hours, with each experiment repeated at least three times (Ozturk et al.,2006).

3. Results and Discussion

3.1 5ml OPN with Probiotic Paste

As shown in Table 9.2 depicts that for both S. mutans and S. salivarius, the MIC and MBC are 12.5% and 6.25%, respectively, indicating that concentrations equal to or above these levels inhibit visible growth and kill these bacteria. (Graph 1,2)

Table 9.2 Observed growth rate and absorbance readings measured at 600nm in 1.5ml OPN treated and non-treated bacterial conditions to determine MIC as well as MBC

Sample Concentration (%)	S. mutans		S. salivarius	
	Growth rate	OD@600nm	Growth rate	OD@600nm
100	-	0.583	-	0.642
50	-	0.976	-	0.998
25	-	1.025	-	1.126
12.5	-	1.758	-	1.792
6.25	+	1.903	+	1.876
3.125	+++	1.954	+++	1.928
BC (Bacterial growth control)	+++	2.189	+++	1.976
PC (Positive control)	-	0	-	0
NC (Negative control)	-	0	-	0
MIC concentration (%)	12.5%		12.5%	
MBC concentration (%)	6.25%		6.25%	

Source: Author's compilation

3.2 3 ml OPN with Probiotic Paste

As shown in Table 9.3 shows that MIC and MBC values for S. mutans are 6.25% and 3.125%, respectively, indicating that concentrations equal to or above these levels inhibit visible growth and kill the bacteria. Whereas MIC and MBC values for S. salivarius are 12.5% and 6.25%, respectively. (Graph 1,2)

OPN is expressed in various human tissues and plays a role in mineralization, wound healing, and leukocyte recruitment (Sodek et al.,2006) (Mazzali et al.,2002). Schalfer S et al. found that OPN significantly reduced dental biofilm, likely due to its bactericidal effect or influence on biofilm formation mechanisms, enhancing mechanical cleaning efficacy (Schlafer et al.,2012). M. F. Kristensen et al. demonstrated OPN's potential to control biofilms by reducing bacterial adhesion without obstructing glycoconjugates on cell surfaces (Kristensen et al.,2017).

Probiotics are "live microorganisms that, when administered in adequate amounts, confer a health benefit on the host" (Bustamante et al.,2020). Probiotics counteract this through competitive exclusion, direct engagement, and immune modulation, preventing pathogen adherence and

Table 9.3 Observed growth rate and absorbance readings measured at 600nm in 3ml OPN treated and non-treated bacterial conditions to determine MIC as well as MBC

Sample Concentration (%)	S. mutans		S. salivarius	
	Growth rate	OD@600nm	Growth rate	OD@600nm
100	-	0.469	-	0.581
50	-	0.858	-	0.984
25	-	1.138	-	1.029
12.5	-	1.736	+	1.568
6.25	+	1.874	++	1.903
3.125	+++	2.087	+++	2.316
BC (Bacterial growth control)	+++	2.265	+++	2.483
PC (Positive control)	-	0	-	0
NC (Negative control)	-	0	-	0
MIC concentration (%)	6.25%		12.5%	
MBC concentration (%)	3.12%		6.25%	

Source: Author's compilation

Table 9.4 Depicts that MIC and MBC for both S. mutans and S. salivarius are 6.25% and 3.125%, respectively. These concentrations inhibit visible growth and kill the bacteria. (graph 1,2)

Sample Concentration (%)	S. mutans		S. salivarius	
	Growth rate	OD@600nm	Growth rate	OD@600nm
100	-	0.458	-	0.473
50	-	0.762	-	0.635
25	-	0.984	-	0.851
12.5	-	1.405	-	1.368
6.25	-	1.372	-	1.595
3.125	++	1.787	++	1.927
BC (Bacterial growth control)	+++	2.289	+++	2.435
PC (Positive control)	-	0	-	0
NC (Negative control)	-	0	-	0
MIC concentration (%)	6.25%		6.25%	
MBC concentration (%)	3.12%		3.12%	

Source: Author's compilation

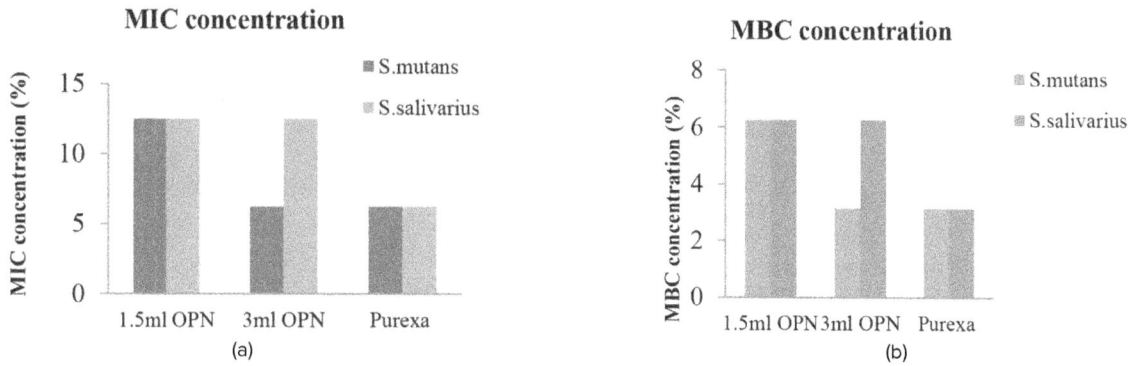

Fig. 9.1 (a) Overlaid bar graph represented the MIC concentration of 1.5ml OPN, 3ml OPN and Purexa molecules against the S. mutans and S. salivarius sps after the 24hours of treatment, (b) Overlaid bar graph represented the MBC concentration of 1.5ml OPN, 3ml OPN and Purexa molecules against the S. mutans and S. salivarius sps after the 24hours of treatment

Source: Author's compilation

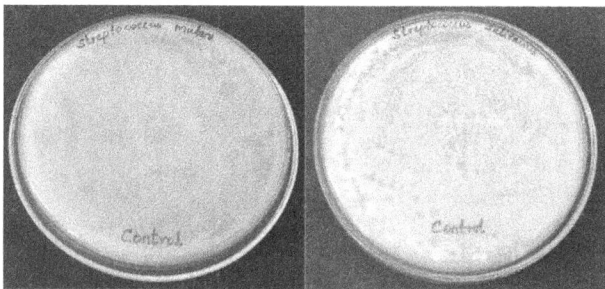

Fig. 9.2 Petri dishes represented the *S. mutans* & *S. salivarius* bacteria growth after the incubation period of 24hours in control groups

Source: Author's compilation

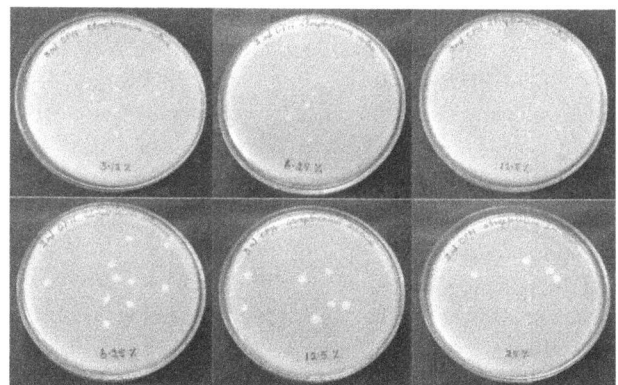

Fig. 9.3 Petri dishes represented the S. mutans and S. salivarius bacteria growth after the incubation period of 24hours in low one dose of MIC, MIC dose and higher dose of MIC of 1.5ml OPN compound after the incubation period of 24hours

Source: Author's compilation

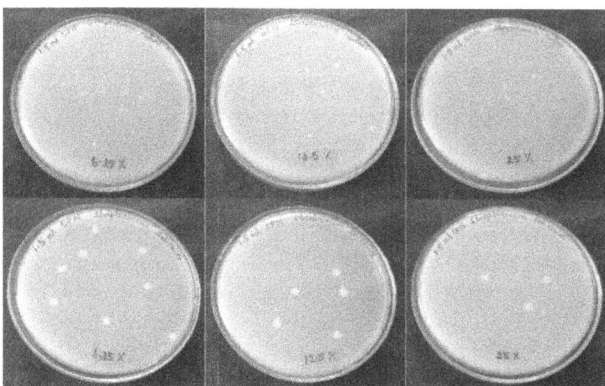

Fig. 9.4 Petri dishes represented the S. mutans and S. salivarius bacteria growth after the incubation period of 24hours in low one dose of MIC, MIC dose and higher dose of MIC of 3ml OPN compound after the incubation period of 24hours

Source: Author's compilation

enhancing host defence (Bustamante et al.,2020) (Reddy et al.,2011).

This study is the first in vitro investigation of OPN-probiotic synergy against oral microbes causing dental caries. We evaluated MIC & MBC of 1.5 ml OPN with probiotic, 3 ml OPN with probiotic, and pure probiotic (Purexa) against S. mutans and S. salivarius. MIC for S. mutans was highest for 1.5 ml OPN (12.5%) but lower for 3 ml OPN and Purexa (6.25%), indicating 3 ml OPN is more effective than 1.5 ml OPN and comparable to Purexa. For S. salivarius, MIC was 12.5% for both 1.5 ml and 3 ml OPN, while Purexa had a lower MIC (6.5%), showing greater efficacy.

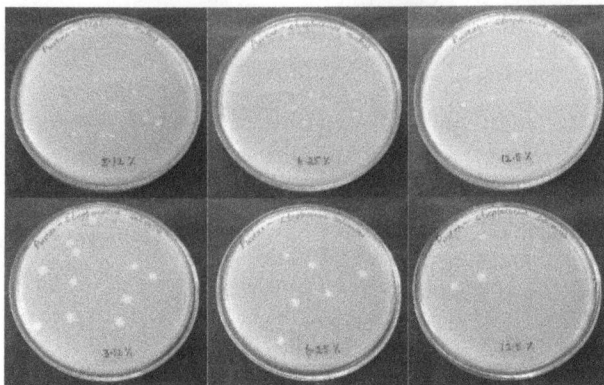

Fig. 9.5 Petri dishes Petri dishes represented the S. mutans and S. salivarius bacteria growth after the incubation period of 24hours in low one dose of MIC, MIC dose and higher dose of MIC of Purexa compound after the incubation period of 24hours

Source: Author's compilation

MBC for S. mutans was 6.25% for 1.5 ml OPN, 3.125% for 3 ml OPN, and 3.125% for Purexa, indicating 3 ml OPN matches Purexa in efficacy. For S. salivarius, MBC was 6.25% for both OPN concentrations but 3.125% for Purexa, making it more effective.

The test compound exhibited antibacterial activity comparable to Purexa, a commercial probiotic toothpaste. While Purexa showed better MIC and MBC, additional caries-protective factors like remineralization potential must be evaluated.

4. Conclusion

In conclusion, this study shows the potential of osteopontin incorporated probiotic paste as an antimicrobial agent against cariogenic bacteria. The dose-dependent antibacterial activity of OPN, along with its ability to enhance the effects of probiotics, suggests a promising role for OPN in dental caries management. The combination of OPN and probiotics is a novel and potentially effective approach to improving oral health and reducing the burden of dental caries.

Acknowledgments

The authors acknowledge M.S. Ramaiah University of Applied Sciences, Faculty of Pharmacy and Faculty of Dental Sciences.

References

1. Andrews, J.M., 2001. Determination of minimum inhibitory concentrations. *Journal of antimicrobial Chemotherapy*, 48(suppl_1), pp.5–16.

2. Bellahcène, A., Castronovo, V., Ogbureke, K.U., Fisher, L.W. and Fedarko, N.S., 2008. *Small integrin-binding ligand N-linked glycoproteins (SIBLINGs): multifunctional proteins in cancer*. Nature Reviews Cancer, 8(3), pp.212–226.

3. Bonaventura, A., Liberale, L., Vecchié, A., Casula, M., Carbone, F., Dallegri, F. and Montecucco, F., 2016. *Update on inflammatory biomarkers and treatments in ischemic stroke*. International journal of molecular sciences, 17(12), p.1967.

4. Bosshardt, D.D., Zalzal, S., Mckee, M.D. and Nanci, A., 1998. *Developmental appearance and distribution of bone sialoprotein and osteopontin in human and rat cementum*. The Anatomical Record: An Official Publication of the American Association of Anatomists, 250(1), pp.13–33.

5. Bustamante, M., Oomah, B.D., Mosi-Roa, Y., Rubilar, M., and Burgos-Díaz, C. (2020) *Probiotics as an Adjunct Therapy for the Treatment of Halitosis Dental Caries and Periodontitis*. Probiotics and Antimicrobial Proteins, 12, pp. 325–334.

6. Cagetti, M.G., Mastroberardino, S., Milia, E., Cocco, F., Lingström, P. and Campus, G., 2013. *The use of probiotic strains in caries prevention: a systematic review*. Nutrients, 5(7), pp.2530–2550.

7. Featherstone, J.D., 2000. *The science and practice of caries prevention*. The Journal of the American dental association, 131(7), pp.887–899.

8. Gajjeraman, S., Narayanan, K., Hao, J., Qin, C. and George, A., 2007. *Matrix macromolecules in hard tissues control the nucleation and hierarchical assembly of hydroxyapatite*. Journal of Biological Chemistry, 282(2), pp.1193–1204.

9. Gross, E.L., Beall, C.J., Kutsch, S.R., Firestone, N.D., Leys, E.J. and Griffen, A.L., 2012. *Beyond Streptococcus mutans: dental caries onset linked to multiple species by 16S rRNA community analysis*.

10. Ishizuka, H., Hamba, H., Nakamura, K., Miyayoshi, Y., Kumura, H. and Muramatsu, T., 2023. *Effects of bovine milk osteopontin on in vitro enamel remineralization as a topical application prior to immersion in remineralizing solutions with/without fluoride*. Dental Materials Journal, 42(1), pp.140–146.

11. Keller, M.K. and Twetman, S., 2012. *Acid production in dental plaque after exposure to probiotic bacteria*. BMC oral health, 12, pp.1–6.

12. Kristensen, M.F., Zeng, G., Neu, T.R., Meyer, R.L., Baelum, V., and Schlafer, S. (2017) *Osteopontin Adsorption to Gram-Positive Cells Reduces Adhesion Forces and Attachment to Surfaces Under Flow*. Journal of Oral Microbiology, 9, p. 1379826

13. Mazzali, M., Kipari, T., Ophascharoensuk, V., Wesson, J.A., Johnson, R., and Hughes, J. (2002) '*Osteopontin—a molecule for all seasons*', QJM: An International Journal of Medicine, 95, pp. 3–18

14. Mogana, R., Adhikari, A., Tzar, M.N., Ramliza, R. and Wiart, C.J.B.C.M., 2020. *Antibacterial activities of the extracts, fractions and isolated compounds from Canarium patentinervium Miq. against bacterial clinical isolates*. BMC complementary medicine and therapies, 20, pp.1–11.

15. Nurrohman, H., Carter, L., Barnes, N., Zehra, S., Singh, V., Tao, J., Marshall, S.J. and Marshall, G.W., 2022. *The Role of Process-Directing Agents on Enamel Lesion Remineralization: Fluoride Boosters.* Biomimetics, 7(2), p.54.

16. Ozturk, S. and Ercisli, S., 2006. *Chemical composition and in vitro antibacterial activity of Seseli libanotis.* World Journal of Microbiology and Biotechnology, 22, pp. 261–265.

17. Reddy, R.S., Swapna, L.A., Ramesh, T., Singh, T.R., Vijayalaxmi, N., and Lavanya, R. (2011) *Bacteria in Oral Health-Probiotics and Prebiotics: A Review.* International Journal of Biological Medical Research, 2, pp. 1226–1233.

18. Schlafer, S., Raarup, M.K., Wejse, P.L., Nyvad, B., Städler, B.M., Sutherland, D.S., Birkedal, H., and Meyer, R.L. (2012) '*Osteopontin Reduces Biofilm Formation in a Multi-Species Model of Dental Biofilm*', PLoS ONE, 7(e41534)

19. Sodek, J., Batista Da Silva, A.P. and Zohar, R., 2006. *Osteopontin and mucosal protection.* Journal of dental research, 85(5), pp.404–415.

20. Soderling, E., Isokangas, P., Pienihäkkinen, K. and Tenovuo, J., 2000. *Influence of maternal xylitol consumption on acquisition of mutans streptococci by infants.* Journal of Dental Research, 79(3), pp.882–887.

21. Twetman, S. and Keller, M.K., 2012. *Probiotics for caries prevention and control.* Advances in dental research, 24(2), pp.98–102.

22. Zaky, S.A., Al-Zoghbi, A.F., Taher, H.M., Abou-auf, E.A. and Mehanna, N.S. *Remineralization of Incipient Enamel Lesions with Probiotics Versus Casein Phosphopeptide Amorphous Calcium Phosphate with Fluoride for Treatment of Demineralization of Maxillary Bovine Anterior Teeth: An in Vitro Study*

Advances in Materials Science and Technology – Dr. Srikari Srinivasan et al. (eds)
© 2025 Taylor & Francis Group, London, ISBN 978-1-041-12342-2

10

In-Vitro Anti-Bacterial Effect of Lyophilized Umbilical Cord Blood Platelet Rich Plasma against *Streptococcus Mutans*

Ashmitha Kishan Shetty,
Serene Joy, Suraksha Shetty, Rutuja Biradar*

Department of Pediatric and Preventive Dentistry,
Faculty of Dental Sciences, M S Ramaiah University of Applied Sciences,
Bangalore, Karnataka, India

Anbu Jayaraman

Department of Pharmacology,
Faculty of Pharmacy, M S Ramaiah University of Applied Sciences,
Bangalore, Karnataka, India

Deveswaran Rajamanickam,
Sharon Caroline Furtado

Department of Pharmaceutics, Faculty of Pharmacy,
M S Ramaiah University of Applied Sciences,
Bangalore, Karnataka, India

ABSTRACT: Platelet lysate containing proteins and peptides show antimicrobial property and this can be affected by bacterial load and concentration platelet concentrate. Autogenous Platelet-Rich Plasma (PRP) containing microbicidal proteins was proved effective against nosocomial infections. The aim of study was to evaluate the antimicrobial effect of lyophilized Umbilical Cord Blood Platelet Rich Plasma (UCB-PRP) by Agar disc diffusion method against pathogenic bacteria *Streptococcus mutans* responsible for dental caries fresh full-term human Umbilical Cord Blood (UCB) was collected from 30 healthy mothers and was lyophilized via centrifugation. Agar disc diffusion method was used to evaluate the antibacterial activity. The results showed Zone of Inhibition (ZOI) of about 12±2mm for UCB-PRP (10µl) against *S mutans*. Lyophilized UCB-PRP suggested to have effective antimicrobial activity against *S mutans*

KEYWORDS: Antimicrobial activity, Platelet rich plasma, *Streptococcus mutans*, Umbilical cord blood

1. Introduction

Microorganisms in the oral cavity attaches to tooth surfaces and form biofilms. In 1924, J. Clarke isolated a bacterium from carious lesions, naming it *Streptococcus mutans,* primarily *S. mutans* inhabits the human oral cavity particularly within dental plaque, a multispecies biofilm on tooth surfaces (Assam et al.,2010). This bacterium synthesizes polysaccharides in dental plaque and is a leading cause of dental caries. *S. mutans*, known anaerobic bacterium produces lactic acid during metabolism and binds to tooth surfaces in the presence of sucrose by forming

---

*Corresponding author: rutujab1284@gmail.com

DOI: 10.1201/9781003664277-10

water-insoluble glucans which aids adhesion (Badade et al., 2016). Streptococcus mutans has been concerned like one of the main etiologies of dental caries. Managing deep carious lesions remains a challenge for clinicians to avoid pulp exposure, which may can compromise prognosis, conservative therapies such as application of calcium-based products which targets the dentine-pulp complex have been proposed (Tong et al., 2022). Over the last decade, there has been an increased interest in the use of biologics for regenerative medicine applications. Cord PRP is an excellent alternative to the autologous and allogenic preparations (Hashemi et al.,2017). UCB has great potential to be adopted for research and therapeutic purposes. Platelets naturally stimulate the secretion of growth factors that can initiate the physiological healing process in acute injuries (Samarkanova et al,2020).

Autogenous PRP is a concentrated form of autologous blood platelets containing 3 to 5 times the normal platelet count (Chai et al.,2019). Lyophilized PRP offers an appealing alternative to autologous PRP due to its availability and ready-to-use preparation (Clarke et al.,1924). Numerous studies have demonstrated the benefits of allogeneic PRP in various clinical settings. However, there is limited detailed assessment of its immunological characteristics for clinical use. Tissue regeneration that is mediated by macrophages exhibits both pro-inflammatory (M1) and anti-inflammatory (M2) phenotypes and M2 phenotype associates with tissue healing (Drago et., 2013). PRP preparation methods that can shift macrophages from the M1 to the M2 phenotype hold significant potential for enhancing tissue regeneration outcomes. Cord PRP derived from umbilical cord blood is gaining recognition as a superior alternative to adult blood PRPs due to its high levels of growth factors (Intravia et al., 2014). The present study aims to evaluate the in-vitro antimicrobial effect of lyophilized UCB-PRP against *Streptococcus mutans*.

2. Methodology

2.1 Preparation of Lyophilized UCB-PRP

The study was carried out in accordance with the recommendations for good clinical practice approved by the University Ethics Committee for Human trials of M S Ramaiah University of Applied Sciences.Cord blood was collected from 30 healthy mothers at 40 weeks of gestation who underwent Caesarean section delivery. Cord blood was promptly transferred into 5ml vacutainers containing EDTA. It underwent gentle centrifugation at 1400 rpm for 10 minutes, resulting in the formation of a buffy coat layer. This layer was then transferred into another vacutainer for a second centrifugation at 2400 rpm for 15 minutes to separate the PRP from Platelet-Poor Plasma. Subsequently,

all PRP samples were pooled together to obtain the final desired UCB-PRP which was then pre-frozen at -20°C and lyophilized using the freeze-drying method.

2.2 Antimicrobial Test by Disc Diffusion Method

The agar medium was prepared by mixing 4.5 ml of Mitis Salivarius Bacitracin (MSB) agar with 50 ml of distilled water. This mixture was then heated for sterilization and autoclaved at 120°C for 30 minutes at 15 lbs pressure. Once sterilized, the agar was transferred to a petri dish and allowed to solidify. Subsequently, the solidified agar plate was streaked with a KWIK STIK™ containing S. mutans ATCC strain bacteria. After streaking, the petri dish was inverted and placed in a hot air oven for incubation at 37°C for 24 hours. At the end of incubation, inhibition zones were examined around the disc and measured with transparent ruler in millimetres. The absence of ZOI was interpreted as the absence of activity. The activities are expressed as resistant, if the ZOI was less than 7 mm, intermediate (8-10 mm) and sensitive if more than 11 mm (Karan et al., 2023). Vancomycin with 10μg disc was used as a standard control. This study was done in triplicates.

3. Results and Discussion

The regenerative capabilities of platelet concentrates have been extensively studied over the past twenty years. However, there are only a few reports in the existing literature regarding their antimicrobial effects (Li et al, 2013).

Bacterial infections are among the most severe complications that hinder wound healing and tissue regeneration. Despite strict disinfection measures, bacteria can still infiltrate and colonize in the underlying tissues of the wound (Li et al.,1986).

The combination of proteolytic enzymes, toxin-laden bacterial exudates, and chronic inflammation can disrupt growth factors and metalloproteinases, impacting the cellular mechanisms essential for cell proliferation and wound healing (Loesche et al., 1986). This study has shown that lyophilized UCB-PRP was active against S. mutans.

Some studies compared the antibacterial effects of four different plasma and platelet preparations like PRP, activated PRP supernatant, solvent/detergent-treated platelet concentrate (S/D-PL), and virally inactivated platelet concentrate against various bacterial strains including S. aureus, E. coli, K. pneumoniae, E. faecalis, and P. aeruginosa (Mani et al.,2024). Over the past two decades, PRP has garnered attention due to its widespread off-label clinical use, driven by its hypothesized regenerative potential and antibacterial effects. However, the precise

Table 10.1 Representation of the results of the disc diffusion method used to assess the antibacterial activity of UCB-PRP against the bacterial strain *S. mutans*

Culture condition	Zone of Inhibition (mm)- *S. mutans*				
	Average	SD	SE	ZOI±SD	P value
Control	0	0	0	0	
Vancomycin-30µg	9.667	0.577	0.333	10±0.6	<0.000262*
UCB-PRP-10µl	1.967	0.058	0.033	2±0.06	

Source: Author's compilation

Standard Deviation (SD), Standard Error (SE), and the Zone of inhibition with Standard Deviation (ZOI±SD) (P<0.05) (*- Statistically significant)

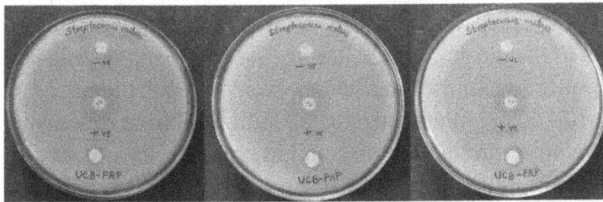

Fig. 10.1 Anti-bacterial activity of UCB-PRP (10µg/ml) against bacterial strain of *S.mutans* in comparison of positive control (Vacomycin-30µg) and negative control (distilled water)

Source: Author's compilation

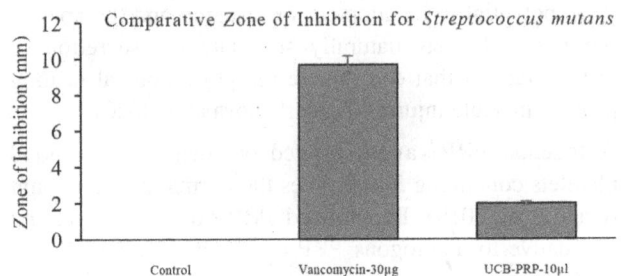

Fig. 10.2 Comparative zone of inhibition of *streptococcus mutans*

Source: Author's compilation

mechanism of dosage and efficacy of PRP against various bacterial strains remain unclear. The foundational research necessary to support specific treatment indications is still incomplete (Murphy et al., 2012). Autogenous PRP likewise inhibits the adherence and growth of periodontal pathogens like *Porphyromonas gingivalis* and *Aggregatibacter actinomycetemcomitans*, though PRP does not show bactericidal effect. Furthermore, Autogenous PRP also inhibits the growth of oral microorganisms such as *Enterococcus faecalis, Candida albicans, Streptococcus agalactiae, and Streptococcus oralis* (Napimoga et al., 2004).

The antibacterial activity of UCB-PRP at a concentration of 10µg/ml against *S. mutans*, measured by the diameter of the zone of inhibition after a 24-hour incubation period (see Table 10.1) (see Fig. 10.1). The negative control, distilled water showed no inhibition (0 mm). The positive control, Vancomycin at 30µg/ml exhibited a significant zone of inhibition with an average diameter of 10 mm demonstrating strong antibacterial activity (Pham et al.,2019) UCB-PRP displayed a moderate antibacterial effect with a minimal inhibition zone of 2 mm. This indicated UCB-PRP statistically significant antibacterial potential (P<0.00026) when compared to Vancomycin (see Fig. 10.2). The results suggest that UCB-PRP can inhibit the growth of *S. mutans*, supporting its potential use as an antibacterial effect.

PRP releases various antibacterial proteins such as connective tissue activating peptide 3, PF 4, RANTES, thymosin β-4, platelet basic protein, fibrinopeptide A and fibrinopeptide B. These proteins interact with bacterial cell membranes increasing permeability and disrupting protein synthesis (Sethi et al.,2021). Additionally, they may target intracellular proteins to affect DNA synthesis or inhibit enzyme activity. These molecules show stronger antimicrobial activity invitro against bacteria than fungi especially in acidic environments (Sheean et al., 2021). Autogenous PRP also inhibits the adherence and growth of periodontal pathogens like *Porphyromonas gingivalis* and *Aggregatibacter actinomycetemcomitans*. Besides, PRP inhibits the growth of oral microorganisms such as *Enterococcus faecalis, Candida albicans, Streptococcus agalactiae, and Streptococcus oralis* (Yeaman et al., 2010).

Platelet antimicrobial proteins, classified as classical chemokines possessed direct antimicrobial properties. These proteins work synergistically with conventional antibiotics and have a lower likelihood of promoting bacterial resistance. The in-vitro data regarding efficacy against specific bacteria were somewhat inconsistent (Zabidi et al., 2012). Most researchers concurred that platelet preparations exhibit varying degrees of activity against bacterial strains commonly found in wounds, including Methicillin-resistant *Staphylococcus aureus*,

Methicillin-Susceptible *Staphylococcus aureus, E. coli* (extended spectrum beta-lactamase), *K. pneumoniae, E. faecalis, P. aeruginosa, B. megaterium, P. mirabilis, E. cloacae, B. cereus, B. subtilis, S. epidermidis, and A. baumannii* (Zhang et al., 2019).

The present study was limited on a single bacterial strain, *S. mutans* and further studies are needed to explore its impact on other oral pathogens or mixed-species biofilms that are common in dental caries. The activation method of PRP which is critical for its antibacterial activity which needs further standardization to ensure consistency and reproducibility in clinical applications.

4. Conclusion

The study demonstrates that lyophilized UCB-PRP exhibits promising in-vitro antibacterial effects against *Streptococcus mutans*. UCB-PRP shows potential, further research is needed to compare its efficacy with adult peripheral PRP and other platelet preparations in clinical settings. Finally, long-term studies are required to evaluate the safety, optimal dosage, and clinical outcomes of using lyophilized PRP in dental and other medical applications. In scenarios where adult peripheral PRP is not available, UBC-PRP presents a viable alternative. Despite these promising findings, no study has yet explored the antibacterial efficacy of lyophilized allogenic PRP specifically against cariogenic bacteria of the oral cavity.

Acknowledgments

The authors acknowledge M.S. Ramaiah University of Applied Sciences, Faculty of Pharmacy and Faculty of Dental Sciences. The authors would like to thank Lakshmi Maternity, Bengaluru and Averin Biotech Laboratory for supporting the study.

References

1. Assam, J. P. A., Dzoyem, J. P., Pieme, C. A., and Penlap, V. B. (2010) 'In vitro antibacterial activity and acute toxicity studies of aqueous-methanol extract of *Sida rhombifolia* Linn. (Malvaceae)', *BMC Complementary and Alternative Medicine*, 10, pp. 1–7.

2. Badade, P. S., Mahale, S. A., Panjwani, A. A., Vaidya, P. D., and Warang, A. D. (2016) 'Antimicrobial effect of platelet-rich plasma and platelet-rich fibrin', *Indian Journal of Dental Research*, 27(3), pp. 300–304.

3. Chai, J., Jin, R., Yuan, G., Kanter, V., Miron, R. J., and Zhang, Y. (2019) 'Effect of liquid platelet-rich fibrin and platelet-rich plasma on the regenerative potential of dental pulp cells cultured under inflammatory conditions: A comparative analysis', *Journal of Endodontics*, 45(8), pp. 1000–1008.

4. Clarke, J. K. (1924) 'On the bacterial factor in the aetiology of dental caries', *British Journal of Experimental Pathology*, 5(3), pp. 141.

5. Drago, L., Bortolin, M., Vassena, C., Taschieri, S., and Del Fabbro, M. (2013) 'Antimicrobial activity of pure platelet-rich plasma against microorganisms isolated from oral cavity', *BMC Microbiology*, 13, pp. 1–5.

6. Hashemi, S. S., Mahmoodi, M., Rafati, A. R., Manafi, F., and Mehrabani, D. (2017) 'The role of human adult peripheral and umbilical cord blood platelet-rich plasma on proliferation and migration of human skin fibroblasts', *World Journal of Plastic Surgery*, 6(2), pp. 198.

7. Intravia, J., Allen, D. A., Durant, T. J., McCarthy, M. B., Russell, R., Beitzel, K., Cote, M. P., Dias, F., and Mazzocca, A. D. (2014) 'In vitro evaluation of the antibacterial effect of two preparations of platelet-rich plasma compared with cefazolin and whole blood', *Muscles, Ligaments and Tendons Journal*, 4(1), pp. 79.

8. Karan, C. L., Jeyaraman, M., Jeyaraman, N., Ramasubramanian, S., Khanna, M., and Yadav, S. (2023) 'Antimicrobial effects of platelet-rich plasma and platelet-rich fibrin: A scoping review', *Cureus*, 15(12).

9. Li, H., and Li, B. (2013) 'PRP as a new approach to prevent infection: Preparation and in vitro antimicrobial properties of PRP', *Journal of Visualized Experiments: JoVE*, 74, p. 50351.

10. Li, H., Hamza, T., Tidwell, J. E., Clovis, N., and Li, B. (2013) 'Unique antimicrobial effects of platelet-rich plasma and its efficacy as a prophylaxis to prevent implant-associated spinal infection', *Advanced Healthcare Materials*, 2(9), pp. 1277–1284.

11. Loesche, W. J. (1986) 'Role of *Streptococcus mutans* in human dental decay', *Microbiological Reviews*, 50(4), pp. 353–380.

12. Mani, R., Roopmani, P., Rajendran, J., Maharana, S., and Giri, J. (2024) 'Cord blood platelet-rich plasma (PRP) as a potential alternative to autologous PRP for allogenic preparation and regenerative applications', *International Journal of Biological Macromolecules*, 262, p. 129850.

13. Murphy, M. B., Blashki, D., Buchanan, R. M., Yazdi, I. K., Ferrari, M., Simmons, P. J., and Tasciotti, E. (2012) 'Adult and umbilical cord blood-derived platelet-rich plasma for mesenchymal stem cell proliferation, chemotaxis, and cryopreservation', *Biomaterials*, 33(21), pp. 5308–5316.

14. Napimoga, M. H., Kamiya, R. U., Rosa, R. T., Rosa, E. A., Höfling, J. F., de Oliveira Mattos-Graner, R., and Gonçalves, R. B. (2004) 'Genotypic diversity and virulence traits of *Streptococcus mutans* in caries-free and caries-active individuals', *Journal of Medical Microbiology*, 53(7), pp. 697–703.

15. Pham, T. A., Tran, T. T., and Luong, N. T. (2019) 'Antimicrobial effect of platelet-rich plasma against *Porphyromonas gingivalis*', *International Journal of Dentistry*, 2019, p. 7329103.

16. Samarkanova, D., Cox, S., Hernandez, D., Rodriguez, L., Casaroli-Marano, R. P., Madrigal, A., and Querol, S. (2020) 'Cord blood platelet-rich plasma derivatives for clinical

applications in non-transfusion medicine', *Frontiers in Immunology*, 11, p. 942.

17. Sethi, D., Martin, K. E., Shrotriya, S., and Brown, B. L. (2021) 'Systematic literature review evaluating evidence and mechanisms of action for platelet-rich plasma as an antibacterial agent', *Journal of Cardiothoracic Surgery*, 16(1), p. 277.

18. Sheean, A. J., Anz, A. W., and Bradley, J. P. (2021) 'Platelet-rich plasma: Fundamentals and clinical applications', *Arthroscopy*, 37(9), pp. 2732–2734.

19. Tong, H. J., Seremidi, K., Stratigaki, E., Kloukos, D., Duggal, M., and Gizani, S. (2022) 'Deep dentine caries management of immature permanent posterior teeth with vital pulp: A systematic review and meta-analysis', *Journal of Dentistry*, 124, p. 104214.

20. Yeaman, M. R. (2010) 'Platelets in defense against bacterial pathogens', *Cellular and Molecular Life Sciences*, 67, pp. 525–544.

21. Zabidi, M. A., Yusoff, N. M., and Kader, Z. S. (2012) 'Preliminary comparative analysis of antibacterial effects of activated and non-activated expired platelet concentrates by disc diffusion method', *Indian Journal of Pathology and Microbiology*, 55(1), pp. 47–51.

22. Zhang, W., Guo, Y., Kuss, M., Shi, W., Aldrich, A. L., Untrauer, J., Kielian, T., and Duan, B. (2019) 'Platelet-rich plasma for the treatment of tissue infection: Preparation and clinical evaluation', *Tissue Engineering Part B: Reviews*, 25(3), pp. 225–236.

Advances in Materials Science and Technology – Dr. Srikari Srinivasan et al. (eds)
© *2025 Taylor & Francis Group, London, ISBN 978-1-041-12342-2*

11

Modified Endodontic Bioceramic Sealer Comprising Gold Nanoparticles and PNIPAAm for Enhanced Antibacterial Ability against *Enterococcus Faecalis*

Sagarika Yadav*, Swaroop Hegde
Department of Conservative Dentistry and Endodontics,
Faculty of Dental Sciences, M S Ramaiah University of Applied Sciences,
Bangalore, Karnataka, India

Ashmitha Kishan Shetty
Department of Pediatric and Preventive Dentistry,
Faculty of Dental Sciences, M S Ramaiah University of Applied Sciences,
Bangalore, Karnataka, India

ABSTRACT: *Enterococcus faecalis* is one of the most frequently isolated species in case of secondary infection of endodontically treated teeth. The Bioceramic (Bio-C®) sealer is modified by addition of Gold Nanoparticles (AuNPs) hypothesizing that AuNPs can provide substantive antimicrobial activity. In the present in-vitro study Bio-C® sealer with addition of AuNPs and Poly (N-isopropylacrylamide) (PNIPAAm) was tried to increase the antibacterial property against pathogenic bacteria *E. faecalis*. The study groups were Bio-C®, AuNP, Bio- C + AuNP and Bio-C® + AuNP + PNIPAAm was tested against *E. faecalis* by agar disk diffusion method. The Zone of Inhibition (ZOI) demonstrated that AuNP+Bio-C®+PNIPAAm exhibited increased antibacterial effects against *E. faecalis*. It can be concluded that AuNPs bound to PNIPAAm when incorporated into bioceramic sealer increases the antibacterial ability.

KEYWORDS: *E. faecalis*, Bioceramic sealers, Gold nanoparticles, Thermoresponsive polymer

1. Introduction

Endodontic disease is a biofilm mediated infection (Rotstein and Ingle., 2019). Among the endodontic biofilm one such bacterial species is *E. faecalis* which is more prevalent in secondary infection in endodontically treated teeth. *E. faecalis* is a gram positive, anaerobic coccus that known to cause opportunistic infections. It forms resistant biofilms by adhering to the gutta percha and canal walls. It can also survive as a monoinfection of the root canal system. It shows resistance to intracanal medicaments such as calcium hydroxide. The high prevalence of *E. faecalis* is due the virulence factors such as esp, ace, asa, gelE, efaA and cylA. One of the relevant virulence factors is the ability to synthesize gelatinase which is necessary for the biofilm formation and bone resorption of endodontically treated teeth (Francisco et al., 2021).

Nanoparticles (NPs) such as silver, copper, titanium, gold, zinc oxide, magnesium oxide possess antibacterial property

*Corresponding author: sagarikay6@gmail.com

DOI: 10.1201/9781003664277-11

(Vimbela et al., 2017). Gold Nanoparticles (AuNPs) have unique properties such as adjustable size, shape, surface properties, optical properties, biocompatibility, low cytotoxicity, high stability making them useful in medical and dental field (Chen Su et al., 2020). Due to their versatile optical and photothermal properties causes disruption of bacterial cellular metabolism as a result of change in membrane potential further inhibiting ATPase activity. Alternative mechanism is that it inhibits binding of subunit of ribosome to tRNA which leads to collapse of cell metabolism (Vimbela et al., 2017). Antifungal action of AuNPs is by inhibition of ATPase action of candida species (Su et al., 2020).

Smart polymer is a type of material which undergoes a change in the physical property or chemical structure due to changes in the external environment reversibly or irreversibly(Mu and Ebara., 2020). The external stimuli for smart polymers include temperature, changes in pH, redox reaction, humidity, electric or magnetic field, light intensity (Huang et al., 2019). Poly (N-isopropylacrylamide) (PNIPAAm) is most widely used temperature-responsive polymer. Thermosensitive polymer has a Critical Solution Temperature (CST). This heat sensitive polymer undergoes a phase change in solution above or below a critical solution temperature (Mu and Ebara., 2020). PNIPAAm has various applications in cell culture, tissue engineering, enzymatic immobilization, drug delivery, wound dressing, biosensors, etc. (Yang et al., 2020).

Antibacterial property of bioceramic sealer is due to precipitation in-situ following setting of the material which causes bacterial sequestration. Nonetheless, in-vitro studies have found that antibacterial efficacy of the bioceramic sealers significantly decreases from third to seventh day, there is less antibacterial substantivity. Recent studies indicated that antibacterial activity of sealers is lost after setting (Simundić et al., 2019). This necessitates the use of materials with the inherent antibacterial property such as nanoparticles (Shrestha and Kishen, 2016). In endodontics when it comes to gold nanoparticles there are not enough studies. Thus, the present study aimed to evaluate the antibacterial property of bioceramic sealers by adding PNIPAAm - AuNP for enhanced antibacterial activity against *E. faecalis*. The null hypothesis was PNIPAAm - Au nanoparticles incorporated into bioceramic sealers will not result in enhanced antimicrobial activity against *E. faecalis*.

2. Methodology

The study design was approved by the University Ethics Committee for Human trials of M S Ramaiah University

of Applied Sciences, Bangalore. The preparation of AuNP (Aldrich 777137-15nm) + Bio-C® (Angelus, Brazil) 5 μL of AuNP was mixed with Bio-C® sealer. For preparation of PNIPAAm (Tokyo Chemical Industry, Chennai, India) + AuNP + Bio-C® about 5mg of PNIPAAm powder was weighed and mixed with 15μL of AuNP was mixed with Bio-C® sealer.E. faecalis was obtained from MTCC culture (Chandigarh, India) Microbial Type Culture Collection and Gene Bank 35550 strains. The bacterial strains were maintained on Nutrient Agar (NA).

2.1 Minimum Inhibitory Concentration (MIC) and Minimum Bactericidal Concentration (MBC) Determination

MIC was performed on the study groups determined by serial dilution, according to the Clinical and Laboratory Standards Institute (CLSI) guidelines. The serial dilutions from the stock solution were made ranging from 1mg/ mL to 0.003 mg/mL using HiMedium® in 96-well microplate. The bacterial suspension containing approximately 5×10^5 colony-forming units/mL was prepared from a 24 hours culture plate (Mueller–Hinton agar (MHA)). From this suspension, 100μl was inoculated into each well. A positive control (Norfloxacin 10 μg), negative control (only MHA) and a bacterial control (only *E. faecalis*) wells were also studied for the comparison. After incubation the micro titer plate was read at 600 nm using spectrophotometer. The wells were examined for visible growth (cloudy) and was recorded growth as (+) and no growth as (-). The concentration that inhibited bacterial growth completely (the first clear well) was taken as the MIC value. MBC was recorded as a lowest concentration kill 99.9% of the bacterial inoculum after 24 hours incubation at 37 °C for all samples. Ten microliters were taken from the well obtained from the MIC experiment and two wells above the MIC value well and spread on MHA plates. The number of colonies was counted after 18–24 hours of incubation at 37 °C. The concentration of sample that produces < 10 colonies was considered as MBC value. Each experiment was repeated at least three times.

2.2 Antimicrobial Test by Agar Disc Diffusion Method

The Mueller Hinton Agar Medium (HiMedium®) stock was autoclaved at 15 lbs pressure at 121°C for 15 min (pH 7.3) and poured on petri plates (25 ml/plate) after cooling and were swabbed with *E. faecalis* followed by 24 hours incubation. The samples were loaded in 6 mm diameter prepared wells with a concentration 100μl/ml were then placed in 4 wells and incubated at 37°C for 24 hours. The antibacterial activity was expressed as resistant, if the ZOI was less than 7 mm, intermediate (8-10 mm) and sensitive if more than 11 mm. Nearly 10µg of Norfloxacin disc was used as the standard control. The experiment was performed in triplicates.

3. Statistical Analysis

Statistical analysis was done with SPSS software (version 22). Data was analyzed statistically using ANOVA and Tukey's post hoc test. $P < 0.05$ was regarded as statistically significant.

4. Results and Discussion

In the present study, test groups were subjected to *E. faecalis* to determine MIC and MBC by microtitre broth dilution method. It was observed that no growth was seen of *E. faecalis* was seen at concentration a of 0.007mg/ml, 0.015mg/ml, 0.125mg/ml, 0.031mg/ml for Bio-C®, AuNP, AuNP + Bio-C®, PNIPAAm + AuNP + Bio-C® respectively (Fig. 11.1). The concentration of sample that produces < 10 colonies was considered as MBC value. MBC against *E. faecalis* was seen at a concentration of 0.003mg/ml, 0.007mg/ml, 0.062mg/ml and 0.015mg/ml Bio-C®, AuNP, AuNP + Bio-C®, PNIPAAm + AuNP + Bio-C® respectively (Fig. 11.2).

The results of the present study found that the AuNP, AuNP+Bio-C® and PNIPAAm + AuNP + Bio-C® may have effective antibacterial activity than the Bio-C® alone (Fig. 11.3). This may be because of the higher surface area to charge density of the AuNPs which allows them to have a greater degree of interaction to negatively charged

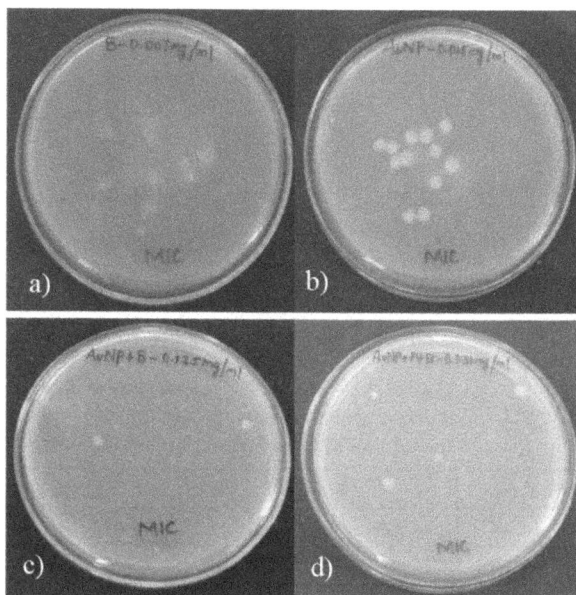

Fig. 11.1 The image (a), (b), (c) and (d) represented the *E. faecalis* bacteria growth with MIC for Bio-C® sealer (0.007mg/ml), AuNPs (0.015mg/ml), Bio- C sealer + AuNPs(0.125mg/ml), Bio-C® sealer + AuNPs + PNIPAAm (0.031mg/ml) concentrations after the incubation period of 24 hours.

bacterial cell surface. The ZOI of the groups Control, Norfloxacin-10µg, Bio-C®, AuNP, AuNP+Bio-C®, PNIPAAm + AuNP + Bio-C® was found to be statistically significant (P<0.001) (Table 11.2). The ZOI of the study groups was compared and was found that PNIPAAm + AuNP + Bio-C® group displayed statistically significant

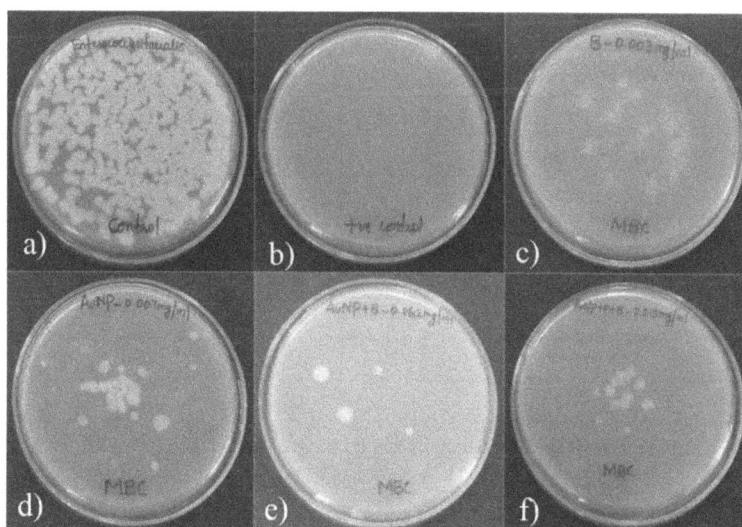

Fig. 11.2 The image (a), (b), (c), (d), (e)and (f) represented the MBC against *E. faecalis* growth at for Bio-C® (0.003mg/ml), AuNPs(0.007mg/ml), AuNPs + Bio-C® (0.062mg/ml), PNIPAAm + AuNPs + Bio-C® (0.015mg/ml) concentrations after the incubation period of 24 hours

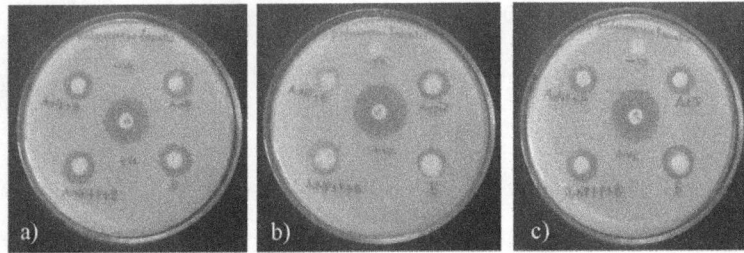

Fig. 11.3 Anti-bacterial activity showing ZOI against *E. faecalis* in comparison of positive control (Norfloxacin-10µg) and negative control (distilled water) and found that the AuNPs, AuNPs+Bio-C® and PNIPAAm + AuNPs + Bio-C® may have effective anti-bacterial activity than the Bio-C® with 1 dilution (1mg/ml) seen in (a), (b) and (c) image

Table 11.1 Observed growth rate and absorbance readings measured at 600nm in treated and untreated bacterial conditions to determine MIC for Bio-C®, AuNPs, AuNPs + Bio-C® and PNIPAAm + AuNPs + Bio-C®

(mg/ ml)	Bio-C®		AuNPs		AuNPs +Bio-C®		PNIPAAm + AuNPs + Bio-C®	
SC	GR	OD	GR	OD	GR	OD	GR	OD
1	-	0.001	-	0.001	-	0.001	-	0.001
0.5	-	0.001	-	0.001	-	0.001	-	0.001
0.25	-	0.001	-	0.001	-	0.001	-	0.001
0.125	-	0.001	-	0.001	-	0.029	-	0.001
0.062	-	0.001	-	0.001	-	0.037	-	0.001
0.031	-	0.001	-	0.001	-	0.112	-	0.206
0.015	-	0.001	+	0.348	+	0.221	+	0.348
0.007	+	0.390	+	0.570	+	0.277	+	0.530
0.003	+++	0.443	+++	0.703	+++	0.355	+++	0.605
BC	++++	0.948	++++	0.964	++++	0.997	++++	0.958
-ve C	-	0.001	-	0.001	-	0.001	-	0.001
+ve C	-	0	-	0	-	0	-	0

SC- Sample Concentration (mg/ ml), GR- Growth rate, OD- Optical Density and BC-Bacterial growth count

Table 11.2 Comparison of ZOI between Control, Norfloxacin-10µg, Bio-C®, AuNPs, AuNPs + Bio-C®, PNIPAAm + AuNPs + Bio-C® using One way ANOVA

Zone of inhibition (mm) -*E. faecalis*						
Culture condition	Average	SD	SE	ZOI±SD	F	P Value
Negative Control	0	0	0	0		
Norfloxacin-10µg	20.66667	1.527525	0.881917	21±1.53		
Bio-C®	9.666667	1.527525	0.881917	10±1.5	136.99	<0.001*
AuNPs	14	1	0.57735	14±1		
AuNPs+ Bio-C®	14.33333	0.57735	0.333333	14±0.57		
PNIPAAm + AuNPs + Bio-C®	18	1	0.57735	18±1		

Standard Deviation (SD), Standard Error (SE), and the Zone of Inhibition with Standard Deviation (ZOI±SD) (P < 0.001) (* - Statistically Significant)

higher ZOI when compared to AuNP and AuNP + Bio-C® (Table 11.3, Fig. 11.4).

AuNPs are effective against both Gram-negative and Gram-positive bacteria, including *E. coli, P. aeruginosa, S. typhi, Serratia species, K. pneumoniae, S. aureus, B. subtilis,* and *E. faecalis*, among others (Betancourt et al., 2020). In one study silver nanoparticles (AgNP) and AuNP were added into the root canal irrigants which resulted in increased

Table 11.3 Comparison of ZOI between control, Norfloxacin-10µg, Bio-C®, AuNPs, AuNPs + Bio-C®, PNIPAAm + AuNPs + Bio-C® using Tukey's post hoc analysis

Groups	Btw Groups	Mean Difference of ZOI	Std. Error	Sig.	95% Confidence Interval	
Control	Norfloxacin-10µg	-20.67	0.88	<.001*	-23.63	-17.70
	Bio-C®	-9.67	0.88	<.001*	-12.63	-6.70
	AuNPs	-14.00	0.88	<.001*	-16.96	-11.04
	AuNPs + Bio-C®	-14.33	0.88	<.001*	-17.30	-11.37
	PNIPAAm + AuNPs + Bio-C®	-18.00	0.88	<.001*	-20.96	-15.04
Norfloxacin-10µg	Bio-C®	11.00	0.88	<.001*	8.04	13.96
	AuNPs	6.67	0.88	<.001*	3.70	9.63
	AuNPs + Bio-C®	6.33	0.88	<.001*	3.37	9.30
	PNIPAAm + AuNPs + Bio-C®	2.67	0.88	0.087	-0.30	5.63
Bio-C®	AuNPs	-4.33	0.88	0.004*	-7.30	-1.37
	AuNPs + Bio-C®	-4.67	0.88	0.002*	-7.63	-1.70
	PNIPAAm + AuNPs + Bio-C®	-8.33	0.88	<.001*	-11.30	-5.37
AuNPs	AuNPs + Bio-C®	-0.33	0.88	0.999	-3.30	2.63
	PNIPAAm + AuNPs + Bio-C®	-4.00	0.88	0.007*	-6.96	-1.04
AuNPs + Bio-C®	PNIPAAm + AuNPs + Bio-C®	-3.67	0.88	0.013*	-6.63	-0.70

(P < 0.05) (* - Statistically Significant)

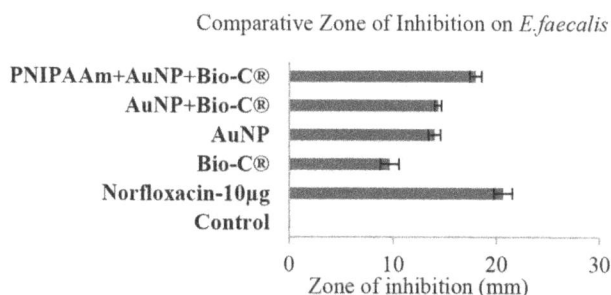

Fig. 11.4 Graph represents comparison of ZOI of control, Norfloxacin-10µg, Bio-C®, AuNPs, AuNPs + Bio-C®, PNIPAAm + AuNPs + Bio-C® against *E. faecalis*

optical properties of irrigants. Also, the NPs enhanced the ability of irrigation protocols to reduce residual bacteria in the root canal (Topala et al., 2020). Adding AgNP to TotalFill® BC sealer significantly enhanced its antimicrobial effectiveness against *E. faecalis*, but it adversely affects the adaptability of sealer to root dentin. Incorporating chitosan nanoparticles into TotalFill® BC sealer enhances its ability to adapt to root dentin, but it diminishes its antimicrobial effectiveness against *E. faecalis* (Magdy et al., 2022).

Incorporating AgNP at a volume concentration of 2.3% into Endosequence® BC sealer enhanced its antibacterial effectiveness, although the difference was not statistically

significant (El-Tayeb and Nabeel., 2023). Study reported that by adding Silica Doped Titanium Dioxide Nanoparticles (SiTiO2 NPs), to Bio-C® and MTA-Fillapex the bacterial count was significantly reduced compared to the unmodified sealer. MTA-Fillapex containing SiTiO2 nanoparticles exhibited greater bacterial viability when compared to Bio-C® containing SiTiO2 nanoparticles (Alekhya et al., 2023).

Incorporating a liposomal formulation of Chlorhexidine (CHX) loaded NPs into BioRoot™ RCS (BR) improved CHX release and its antimicrobial effectiveness. Adding 5% propolis nanoparticles to Ceraseal (Bioceramic sealer) and Adseal (Resin based sealer) enhanced their antibacterial properties against *E. faecalis* and improved their penetration into dentinal tubules. The bioceramic sealer exhibited greater antibacterial effects and better sealer penetration compared to the epoxy resin sealer (Raddi et al., 2024).

In this study AuNPs are combined with PNIPAAm as AuNPs have a tendency to aggregate to form larger particles. AuNPs need to be stable and fully dispersed in a solvent for their subsequent applications, thus polymers are commonly used as stabilizers (Tepale et al., 2019). Modifying polymers with AuNPs may offer numerous benefits, including improved sensitivity, specificity, speed, contrast, resolution, and penetration depth (Kumar and Mahajan., 2024). The antimicrobial effect against *E. faecalis* was more when polymer plus AuNP was incorporated into Bio-C® sealer in comparison to Bio-C® sealer alone in the present study.

5. Conclusion

It can be concluded that combination of PiPAAm + AuNP + Bio-C® showed effective antibacterial activity against *E. faecalis*. This modified Bio-C® sealer with AuNPs could be suggested as an endodontic sealer for secondary infections of endodontically treated teeth. In the future, further studies are required evaluate physiochemical properties of the Bio-C® sealer after adding AuNPs with regard to setting time, flow and pH of the sealer.

Acknowledgments

The authors would like to acknowledge M S Ramaiah University of Applied Sciences, Bangalore and Department of Conservative Dentistry and Endodontics, FDS, MSRUAS. The authors thank Averin Biotech Laboratory for their technical support.

References

1. Betancourt, J.A., Romero, C.C., Delgadillo, R.H., Villarreal, M.M., Rodriguez, L.E., Quintanilla, N.C., Kim, H. and Soto, J.M., 2020. Analysis of the antimicrobial and antibiotic activity of nanoparticles for endodontic use. Int. J. Appl. Dent. Sci, 6, pp.85–89.
2. El-Tayeb, M.N. and Nabeel, M., 2023. Antimicrobial efficacy of endosequence bioceramic sealer incorporated with silver and chitosan nanoparticles (an in-vitro study). Egyptian Dental Journal, 69(4), pp.3167–3177.
3. Francisco, P.A., Fagundes, P.I.D.G., Lemes-Junior, J.C., Lima, A.R., Passini, M.R.Z. and Gomes, B.P., 2021. Pathogenic potential of Enterococcus faecalis strains isolated from root canals after unsuccessful endodontic treatment. Clinical Oral Investigations, pp.1–9.
4. Huang, H.J., Tsai, Y.L., Lin, S.H. and Hsu, S.H., 2019. Smart polymers for cell therapy and precision medicine. Journal of Biomedical Science, 26, pp.1–11.
5. Kumar, P.P.P. and Mahajan, R., 2024. Gold Polymer Nanomaterials: A Promising Approach for Enhanced Biomolecular Imaging. Nanotheranostics, 8(1), p.64.
6. Mediboyina, A., Parvathaneni, K.P. and Raju, T.B.V.G., 2023. Comparative Evaluation of Antibacterial Efficacy of Two Bioceramic Root Canal Sealers Incorporated with Novel Silica Doped TiO₂ Nanoparticles: An In-vitro Study. Journal of Clinical and Diagnostic Research, 17(6), pp.ZC19-ZC24.
7. Mu, M. and Ebara, M., 2020. Smart polymers. In Polymer Science and Nanotechnology (pp. 257–279). Elsevier.
8. Raddi, S., El Karmy, B., Martinache, O., Richert, R., Colnot, C. and Grosgogeat, B., 2024. Development of Chlorhexidine-loaded Lipid Nanoparticles Incorporated in a Bioceramic Endodontic Sealer. Journal of Endodontics.
9. Rotstein, I. and Ingle, J.I. eds., 2019. Ingle's Endodontics. PMPH USA.
10. Shrestha, A. and Kishen, A., 2016. Antibacterial nanoparticles in endodontics: a review. Journal of Endodontics, 42(10), pp.1417–1426.
11. Šimundić Munitić, M., Poklepović Peričić, T., Utrobičić, A., Bago, I. and Puljak, L., 2019. Antimicrobial efficacy of commercially available endodontic bioceramic root canal sealers: A systematic review. PLoS One, 14(10), p.e0223575.
12. Su, C., Huang, K., Li, H.H., Lu, Y.G. and Zheng, D.L., 2020. Antibacterial properties of functionalized gold nanoparticles and their application in oral biology. Journal of Nanomaterials, 2020(1), p.5616379.
13. Tepale, N., Fernández-Escamilla, V.V., Carreon-Alvarez, C., González-Coronel, V.J., Luna-Flores, A., Carreon-Alvarez, A. and Aguilar, J., 2019. Nanoengineering of gold nanoparticles: Green synthesis, characterization, and applications. Crystals, 9(12), p.612.
14. Topala, F., Nica, L.M., Boariu, M., Negrutiu, M.L., Sinescu, C., Marinescu, A., Cirligeriu, L.E., Stratul, S.I., Rusu, D., Chincia, R. and Duma, V.F., 2021. En-face optical coherence tomography analysis of gold and silver nanoparticles in endodontic irrigating solutions: An in vitro study. Experimental and Therapeutic Medicine, 22(3), pp.1–6.
15. Vimbela, G.V., Ngo, S.M., Fraze, C., Yang, L. and Stout, D.A., 2017. Antibacterial properties and toxicity from metallic nanomaterials. International Journal of Nanomedicine, pp.3941–3965.
16. Yang, L., Fan, X., Zhang, J. and Ju, J., 2020. Preparation and characterization of thermoresponsive poly (N-isopropylacrylamide) for cell culture applications. Polymers, 12(2), p.389.
17. Yehia, N.M., Al-Ashry, S., Hashem, A. and Nabeel, M., 2022. Evaluation of Antimicrobial Efficacy and Adaptability to Root Canal Dentin of Bioceramic Sealer Containing Nanoparticles (In-vitro Study). Journal of Fundamental and Clinical Research, 2(1), pp.56–73.

Note: All the figures and tables in this chapter were made by the author.

Advances in Materials Science and Technology – Dr. Srikari Srinivasan et al. (eds)
© *2025 Taylor & Francis Group, London, ISBN 978-1-041-12342-2*

12

Bioactive Botanicals: Evaluating *Cynodon dactylon's* Antioxidant Role in Intracanal Treatment

Desai Simran Prakash*,
Shwetha G., Dhananjaya Gaviappa

Department of Pediatric and Preventive Dentistry,
Faculty of Dental Sciences, M S Ramaiah University of Applied Sciences,
Bangalore, Karnataka, India

Deveswaran Rajamanickam

Department of Pharmaceutics,
Faculty of Pharmacy, M S Ramaiah University of Applied Sciences,
Bangalore, Karnataka, India

ABSTRACT: *Cynodon dactylon* (CD) possesses anti-inflammatory, antiallergic, antibacterial, and antifungal properties and aids in wound healing. Its antioxidant effects in root canals may help reduce oxidative stress, promote tissue healing, and improve endodontic treatments The aim of the study was to determine the antioxidant activity of developed intracanal medicament containing *Cynodon dactylon* extract and to compare it with the standard quercetin group. The antioxidant potential of CD extract was assessed using the DPPH scavenging assay. A freshly prepared DPPH solution (4.3 mg in 3.3 ml of 80% methanol) was mixed with test compounds at various concentrations, vortexed, and incubated for 30 minutes. Quercetin served as the standard, and a control sample contained only DPPH. Absorbance at 517 nm was measured using a microplate reader. The results showed both Quercetin and CD Gel exhibited dose-dependent inhibition of DPPH RSA. The IC_{50} values were 56.86 µg/ml for Quercetin and 507.84 µg/ml for CD Gel, indicating lower potency for CD. To conclude the DPPH radical scavenging assay for *Cynodon dactylon* (CD) Gel suggest that it has a satisfactory dose-dependent scavenging effect.

KEYWORDS: Antioxidant, *Cynodon dactylon*, Quercetin

1. Introduction

Microorganisms are the crucial factors for any pulpal or periodontal disease to occur and eradication of these microorganisms fully during root canal treatment procedures plays a crucial role in endodontic therapy (Bouktaib et al,2002). The primary goal of intracanal medication is to eliminate post-instrumentation microorganisms, reduce periapical inflammation, neutralize the root canal, and prevent microleakage. It is a crucial part of endodontic therapy, especially in cases of apical periodontitis and pulpal necrosis. Common intracanal medications include calcium hydroxide, chlorhexidine, aldehydes, antibiotics, formocresol, cresol, and phenolic compounds.(Chong et

---

*Corresponding author: desaisimran47@gmail.com

DOI: 10.1201/9781003664277-12

al, 1992; Correia et al, 2022; Estrela et al ,2018).The rise in antibiotic resistance necessitates alternative therapies like intracanal medication (Furiga, A et al, 2009). Its clinical effectiveness depends on both physicochemical and biological properties, including toxicity (Hassan et al, 2006). Successful endodontic therapy requires complete elimination of intraradicular and extraradicular bacteria, influenced by factors like irrigation, access size, medication use, and treatment sessions. Intracanal medications are used between visits, especially for necrosis or periapical abscess, to enhance bacterial eradication (Karami et al, 2018; MacDonald-Wicks et al, 2006; Moon et al, 2009). Synthetic intracanal medicaments are common but have drawbacks like brittleness (calcium hydroxide) and staining (chlorhexidine). Gel-based forms offer better flow, sustained release, and targeting. Natural alternatives are preferred for biocompatibility, safety, and cost-effectiveness (Mozafari et al, 2018).

Cynodon dactylon (Doob/Bermuda grass) of the Poaceae family is globally distributed and has medicinal properties, including wound healing, antioxidant, diuretic, and antidiabetic effects (Murota et al, 2003; Nitu et al, 2021; Ordinola-Zapata et al, 2022). Its antibacterial properties stem from tannins, phenols, and quinines, while phenolic aldehydes act as bioactive, bacteriostatic agents. Its antioxidant potential may reduce oxidative stress, aid tissue healing, and enhance dental pulp treatment (Özden et al, 2017; Prada et al, 2019). This study aims to evaluate the DPPH free radical scavenging activity of an intracanal medicament containing *Cynodon dactylon* extract and compare it with quercetin.

2. Materials and Methods

The study was carried out in accordance with the recommendations for good clinical practice approved by the University Ethics Committee for Human trials of M S Ramaiah University of Applied Sciences

DPPH Assay: The DPPH (2, 2-diphenyl-1-picrylhydrazyl) assay is a simple and sensitive method widely used in natural product antioxidant studies. It measures the ability of compounds to act as radical scavengers, with the antioxidant effect correlating to the reduction of DPPH•. DPPH•, a stable nitrogen radical, absorbs UV light at 517 nm, appearing purple "Fig. 12.1". Upon accepting hydrogen from an antioxidant, its color shifts to yellow, indicating a reduction. This change is stoichiometric, allowing easy measurement of antioxidant activity by monitoring the decrease in absorbance at 517 nm (Shahriar et al, 2013; Shahi et al, 2019).

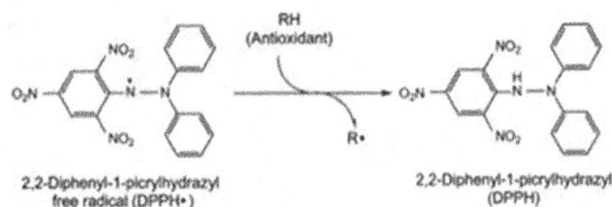

Fig. 12.1 DPPH• free radical conversion to DPPH by antioxidant compound

Source: Author's compilation

Materials Used: The materials and various equipment used in this study are, DPPH-2,2-Diphenyl-1-picrylhydrazyl (Cat No:D9132, Sigma), Quercetin as standard(Cat No: Q4951, Sigma), Adjustable multichannel pipettes and a pipettor (Benchtop, USA), Methanol (Cat No: 34860-1L-R, Sigma), 50 ml centrifuge tubes (# 546043 TARSON), 10ml Borosil Glass tubes (TARSON), 10 ml serological pipettes (TARSON), 10 to 1000ul tips (TARSON), Pipettes: 2-10µl, 10-100µl, and 100-1000µl and micro-plate reader (ELX-800, BIOTEK, USA).

3. Methodology

Freshly prepared DPPH solution (4.3mg in 3.3ml of 80% methanol) was taken in 1.5ml Eppendorf tubes and test compound with different concentrations of *Cynodon dactylon's* extract was added to all Eppendorf tubes in 1:1ratio to achieve the final volume of 300ul (150ul of Compound with desired concentration+150ul of DPPH). The tubes were vortexed and incubated for 30mins at room temperature in the absence of light by covering with aluminum foil. Each and every reaction mixtures were kept in separate tubes to keep them as a blank correction. Quercetin acid with different concentrations was used as standard and methanol as a blank. Control sample was prepared containing the same volume without any sample (DPPH alone). 100ul of reaction mixture was loaded to each well of 96 well plate and the absorbance was read on a microplate reader at 517nm. The %age radical scavenging activity of the given compound was calculated using the following formula:

$$\% \text{ DPPH Radical Scavenging Activity} = \frac{\text{Abs of Control} - \text{Abs of Sample}}{\text{Abs of Control}} * 100$$

Concentrations used in the Study: In this study, the given test compounds were evaluated to measure the DPPH Radical Scavenging activity. The used concentrations of the compounds for the study as follows

Table 12.1 Details of given test compound with different concentrations along with controls used for the study

Sl. No	Test Compounds	Concentration used
1	DPPH alone	4.3 mg in 3.3 ml of 80% methanol
2	Standard (Quercetin)	6 (31.25,62.5,125,250,500,1000µg/ml)
3	Blank	80% Methanol
4	CD Gel	6 (31.25,62.5,125,250,500,1000µg/ml)

Source: Author's compilation

4. Results

The observations in statistical data of DPPH RSA study by Spectrophotometer/ELISA reader suggesting us that given test compound viz Quercetin and *Cynodon dactylon* (CD) Gel

Table 12.2 Comparative % DPPH inhibition values of the quercetin on dose dependent fashion along with IC_{50} concentration

DPPH RSA activity-Quercetin	
Concentration (µg/ml)	% DPPH RSA inhibition
DPPH alone	0.00
Quercetin-31.25µg/ml	14.15
Quercetin-62.5µg/ml	45.70
Quercetin-125µg/ml	87.23
Quercetin-250µg/ml	90.33
Quercetin-500µg/ml	90.15
Quercetin-1000µg/ml	91.95
IC_{50} conc=56.86ug/ml	

Source: Author's compilation

showed DPPH RSA inhibition on a dose dependent manner with IC_{50} value of 56.86µg/ml and 507.84µg/ml respectively. Quercetin was used as a standard control for the study.

Table 12.3 Comparative % DPPH inhibition values of the CD Gel on dose dependent fashion along with IC_{50} concentration.

DPPH RSA activity-CD Gel	
Concentration (µg/ml)	% DPPH RSA inhibition
DPPH alone	0.00
CD Gel-31.25µg/ml	2.66
CD Gel-62.5µg/ml	4.29
CD Gel-125µg/ml	6.74
CD Gel-250µg/ml	19.89
CD Gel-500µg/ml	41.14
CD Gel-1000µg/ml	92.26
IC_{50} conc=507.84ug/ml	

Source: Author's compilation

Table 12.4 DPPH RSA activity of Quercetin with different concentrations in comparison to DPPH alone

Concentration (µg/ml)	% DPPH inhibition
DPPH alone	0.00
Quercetin-31.25µg/ml	14.15
Quercetin-62.5µg/ml	45.70
Quercetin-125µg/ml	87.23
Quercetin-250µg/ml	90.33
Quercetin-500µg/ml	90.15
Quercetin-1000µg/ml	91.95

Source: Author's compilation

Table 12.5 % DPPH inhibition of Quercetin with different concentrations in comparison to DPPH alone

Concentration Unit: µg/ml			30 mins					
Parameter	Blank	DPPH	Quercetin					
			31.25	62.5	125	250	500	1000
Abs Reading 1 @ 517nm	0.056	3.093	2.685	1.715	0.485	0.383	0.405	0.360
Abs Reading 2 @ 517nm	0.061	3.181	2.720	1.767	0.463	0.405	0.424	0.388
Abs Reading 3 @ 517nm	0.066	2.992	2.580	1.721	0.422	0.399	0.370	0.393
Mean abs	0.059	3.137	2.703	1.741	0.474	0.394	0.415	0.374
Reaction control		0.045	0.048	0.062	0.079	0.095	0.110	0.125
Corrected abs		3.092	2.655	1.679	0.395	0.299	0.305	0.249
Std deviation		0.062	0.025	0.037	0.016	0.016	0.013	0.020
Std error		0.044	0.018	0.026	0.011	0.011	0.009	0.014
% DPPH RSA Inhibition		0.000	14.149	45.699	87.225	90.330	90.152	91.947

Source: Author's compilation

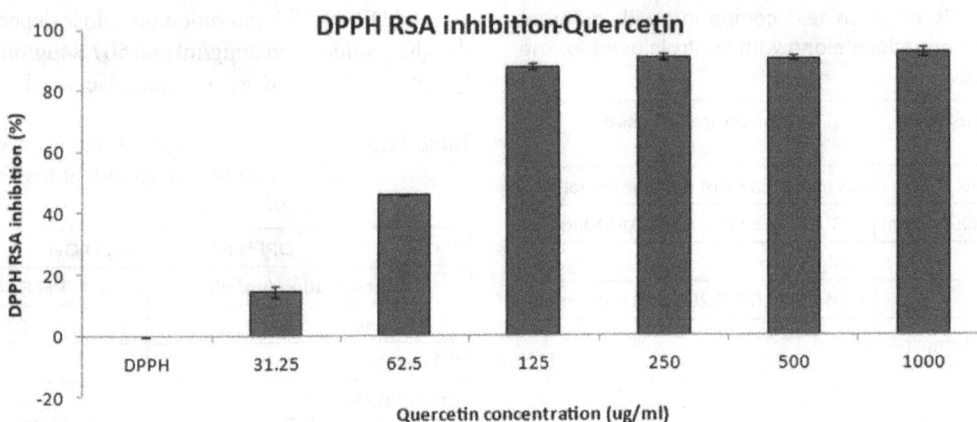

Fig. 12.2   Graph presentation: DPPH RSA activity of Quercetin with different concentrations in comparison to DPPH alone

Source: Author's compilation

Table 12.6   DPPH RSA activity of *Cynodon dactylon* (CD) Gel with different concentrations in comparison to DPPH alone

Concentration Unit: µg/ml		30mins						
Parameter	Blank	DPPH	Quercetin					
			31.25	62.5	125	250	500	1000
Abs Reading 1 @ 517nm	0.056	3.093	3.081	3.011	2.979	2.510	2.060	0.353
Abs Reading 2 @ 517nm	0.061	3.181	3.045	3.026	2.945	2.610	1.766	0.335
Abs Reading 3 @ 517nm	0.066	2.992	3.106	3.001	2.998	2.984	2.742	0.396
Mean abs	0.059	3.137	3.063	3.019	2.962	2.560	1.913	0.344
Reaction control		0.050	0.058	0.064	0.083	0.087	0.096	0.105
Corrected abs		3.087	3.005	2.955	2.879	2.473	1.817	0.239
Std deviation		0.062	0.025	0.011	0.024	0.071	0.208	0.013
Std error		0.044	0.018	0.007	0.017	0.050	0.147	0.009
% DPPH RSA Inhibition		0.000	2.656	4.292	6.738	19.890	41.140	92.258

Source: Author's compilation

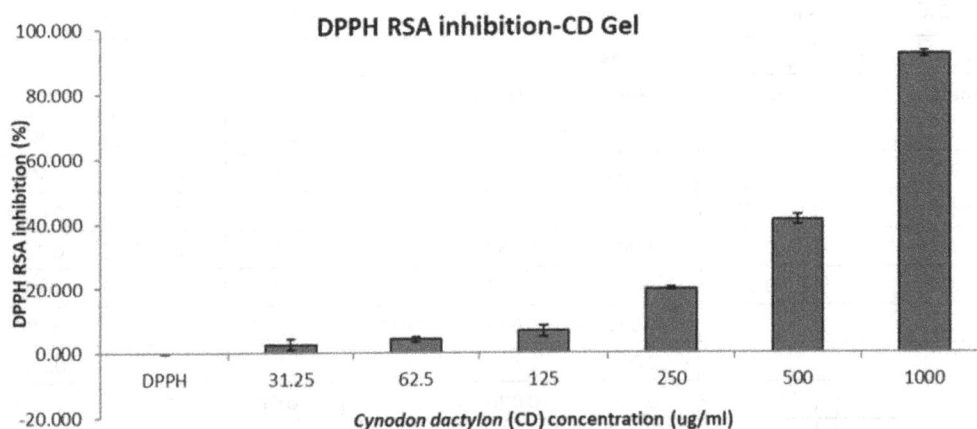

Fig. 12.3   Graph presentation: DPPH RSA activity of Quercetin with different concentrations in comparison to DPPH alone

Source: Author's compilation

Table 12.7 % DPPH inhibition activity of *Cynodon dactylon* (CD) Gel with different concentrations in comparison to DPPH alone

Concentration (µg/ml)	% DPPH inhibition
DPPH alone	0.00
Quercetin-31.25µg/ml	2.66
Quercetin-62.5µg/ml	4.29
Quercetin-125µg/ml	6.74
Quercetin-250µg/ml	19.89
Quercetin-500µg/ml	41.14
Quercetin-1000µg/ml	92.26

Source: Author's compilation

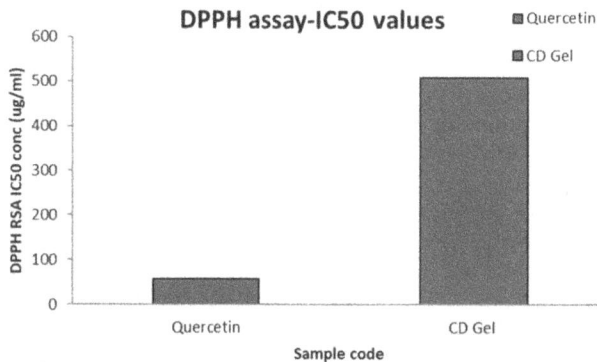

Fig. 12.4 DPPH RSA activity of Quercetin and *Cynodon dactylon* (CD) Gel with IC_{50} concentrations

Source: Author's compilation

This study focuses on the antioxidant properties of *Cynodon dactylon* (CD) extract when used as an intracanal medicament. The main aim of using intracanal medicament is to prevent secondary infection and to have a bactericidal action (Adikwu et al, 2022), so, the primary aim of this study was to compare the antioxidant activity of *Cynodon dactylon* (CD) extract with the standard antioxidant quercetin using the DPPH radical scavenging assay. Quercetin, a key flavonol, constitutes 60–75% of flavonoid intake and inhibits LDL oxidation, potentially preventing inflammation, atherosclerosis, and cancer. This study evaluated *Cynodon dactylon's* antioxidant activity using a DPPH assay, measuring absorbance at 517 nm with controls, including methanol (blank), DPPH alone (negative), and quercetin (positive) (Ordinola-Zapata et al, 2022; Biswas et al, 2017).

The results showed dose-dependent inhibition for both control and test compounds. While quercetin had stronger antioxidant potential (IC50: 56.86 µg/ml), *Cynodon dactylon* (CD) gel (IC50: 507.84 µg/ml) still exhibited satisfactory antioxidant activity. *Cynodon dactylon* (CD) extract shows potential as an intracanal medicament due to its antioxidant properties, which may help reduce oxidative stress and promote tissue healing in root canals. The results of the present study showed that *Cynodon dactylon* (CD) gel extract showed remarkable DPPH activity, though it was less compared to quercetin, suggesting that, *Cynodon dactylon* (CD) gel can be considered as an intracanal medicament because of its easy accessibility and availability.

4.1 Limitations

Cynodon dactylon (CD) extract's high IC_{50} value requires higher concentrations for comparable antioxidant effects. Future research should explore combining it with other antioxidants and conducting in vivo studies to validate its clinical efficacy.

5. Conclusion

The study confirms *Cynodon dactylon's* antioxidant potential in intracanal treatment. Though less potent than quercetin, it shows promise for reducing oxidative stress and aiding healing. Further research could enhance its clinical application.

Acknowledgment

The authors thank Dr Pushpalatha C, Head of Department of Pediatric and Preventive Dentistry for the support and motivation. The authors are thankful to Deans of Faculty of Dentistry and Faculty of Pharmacy for providing the necessary facilities.

References

1. Bouktaib, M., Atmani, A. and Rolando, C. (2002) 'Regio- and Stereoselective Synthesis of the Major Metabolite of Quercetin, Quercetin-3-O-β-D-Glucuronide', *Tetrahedron Letters*, 43(35), pp. 6263–6266.
2. Chong, B.S. and Ford, T.P. (1992) 'The Role of Intracanal Medication in Root Canal Treatment', *International Endodontic Journal*, 25(2), pp. 97–106.
3. Correia, B.L., Gomes, A.T., Noites, R., Ferreira, J.M. and Duarte, A.S. (2022) 'New and Efficient Bioactive Glass Compositions for Controlling Endodontic Pathogens', *Nanomaterials*, 12(9), p. 1577.
4. Estrela, C., Decurcio, D.D.A., Rossi-Fedele, G., Silva, J.A., Guedes, O.A. and Borges, Á.H. (2018) 'Root Perforations: A Review of Diagnosis, Prognosis and Materials', *Brazilian Oral Research*, 32, p. e73.
5. Furiga, A., Lonvaud-Funel, A. and Badet, C. (2009) 'In Vitro Study of Antioxidant Capacity and Antibacterial Activity on Oral Anaerobes of a Grape Seed Extract', *Food Chemistry*, 113(4), pp. 1037–1040.

6. Hassan, W., Noreen, H., Rehman, S., Gul, S., Kamal, M.A., Kamdem, J.P. and da Rocha, B.T. (2017) 'Oxidative Stress and Antioxidant Potential of One Hundred Medicinal Plants', *Current Topics in Medicinal Chemistry*, 17(12), pp. 1336–1370.

7. Karami, S., Rahimi, M. and Babaei, A. (2018) 'An Overview on the Antioxidant, Anti-Inflammatory, Antimicrobial and Anti-Cancer Activity of Grape Extract', *Biomed. Res. Clin. Pract.*, 3, pp. 1–4.

8. MacDonald-Wicks, L.K., Wood, L.G. and Garg, M.L. (2006) 'Methodology for the Determination of Biological Antioxidant Capacity in Vitro: A Review', *Journal of the Science of Food and Agriculture*, 86(13), pp. 2046–2056.

9. Moon, J.K. and Shibamoto, T. (2009) 'Antioxidant Assays for Plant and Food Components', *Journal of Agricultural and Food Chemistry*, 57(5), pp. 1655–1666.

10. Mozafari, A.A., Vafaee, Y. and Shahyad, M. (2018) 'Phytochemical Composition and In Vitro Antioxidant Potential of Cynodon Dactylon Leaf and Rhizome Extracts as Affected by Drying Methods and Temperatures', *Journal of Food Science and Technology*, 55, pp. 2220–2229.

11. Murota, K. and Terao, J. (2003) 'Antioxidative Flavonoid Quercetin: Implication of Its Intestinal Absorption and Metabolism', *Archives of Biochemistry and Biophysics*, 417(1), pp. 12–17.

12. Nitu, S.K., Tarique, H. and Islam, S.M.S. (2021) 'Leaf Epidermal Anatomy of Cynodon Dactylon (L.) Pers. in Relation to Ecotypic Adaptation', *Bangladesh Journal of Plant Taxonomy*, 28(1).

13. Ordinola-Zapata, R., Noblett, W.C., Perez-Ron, A., Ye, Z. and Vera, J. (2022) 'Present Status and Future Directions of Intracanal Medicaments', *International Endodontic Journal*, 55, pp. 613–636.

14. Özden, F.O., Sakallioğlu, E.E., Sakallioğlu, U., Ayas, B. and Erişgin, Z. (2017) *Journal of Applied Oral Science*, 25(2), pp. 121–129.

15. Prada, I., Micó-Muñoz, P., Giner-Lluesma, T., Micó-Martínez, P., Muwaquet-Rodríguez, S. and Albero-Monteagudo, A. (2019) 'Update of the Therapeutic Planning of Irrigation and Intracanal Medication in Root Canal Treatment: A Literature Review', *Journal of Clinical and Experimental Dentistry*, 11(2), pp. e185.

16. Pushpalatha, C., Stephen, A. and Deveswaran, R. (2020) 'Assessment of Antioxidant Activity of an Intracanal Medicament Containing Grape Seed Extract', *AIP Conference Proceedings*, 2274(1). AIP Publishing.

17. Shahriar, M., Hossain, I., Sharmin, F.A., Akhter, S., Haque, M.A. and Bhuiyan, M.A. (2013) 'In Vitro Antioxidant and Free Radical Scavenging Activity of Withania Somnifera Root', *IOSR Journal of Pharmacy*, 3, pp. 38–47.

18. Shahi, S., Özcan, M., Maleki Dizaj, S., Sharifi, S., Al-Haj Husain, N., Eftek-hari, A. and Ahmadian, E. (2019) 'A Review on Potential Toxicity of Dental Material and Screening Their Biocompatibility', *Toxicology Mechanisms and Methods*, 29(5), pp. 368–377.

19. Adikwu, P., Oyiwona, E.G., Johnson, A., Awua, Y., Hassan, A.O., Adenugba, T.O. and Ebiega, E.T. (2022) 'Antibacterial Activity of Psidium Guajava Leaf and Stem Bark Extracts on Selected Bacteria in Ugbokolo, Benue State, Nigeria', *Advances in Microbiology*, 12(10), pp. 569–578.

20. Biswas, T.K., Pandit, S., Chakrabarti, S., Banerjee, S., Poyra, N. and Seal, T. (2017) 'Evaluation of Cynodon Dactylon for Wound Healing Activity', *Journal of Ethnopharmacology*, 197, pp. 128–137.

Advances in Materials Science and Technology – Dr. Srikari Srinivasan et al. (eds)
© 2025 Taylor & Francis Group, London, ISBN 978-1-041-12342-2

13

Gas Chromatography Analysis of Bioactive Compounds from *Sargassum Wightii* Extract for Periodontal Therapy

Tanya Singh*,
Rohit Prasad, Akshatha Raj
Department of Periodontology,
Faculty of Dental Sciences, M S Ramaiah University of Applied Sciences,
Bangalore

Deveswaran Rajamanickam
Department of Pharmaceutics,
Faculty of Pharmacy, M S Ramaiah University of Applied Sciences,
Bangalore

Bhavya Shetty
Department of Periodontology,
Faculty of Dental Sciences, M S Ramaiah University of Applied Sciences,
Bangalore

ABSTRACT: Periodontitis is a multifactorial disease of the tooth's supporting tissues caused by specific microorganisms resulting in progressive destruction. Since plaque control by mechanical methods is not enough to keep the gingival health in many people, attention has been brought to many therapeutic agents adjunct to periodontal therapy. Therefore, assessing a biomaterial with natural archetype to support and facilitate periodontal therapy is of paramount importance. The proposed study aim was to assess the biological activity of *sargassum wightii* (brown seaweed) extract using gas chromatography analysis and mass spectroscopy. The extract was prepared using 30g of *Sargassum wightii* in 30% ethanol. The mass spectrometer was operated in the electron impact mode at 80 eV. The extract exhibited several beneficial properties such as dodecanoic acid also known as lauric acid which has bactericidal effect. Arachidonic acid acts as immunomodulatory in periodontitis. 3-acetylcoumarin coumarin derivatives possess antibacterial, antiinflammtory, antioxidant properties. It is evident from the present study that the ethanol extract of *Sargassum Wightii* could be utilized as a good source of agent antimicrobial, anti-inflammatory, anti-oxidant and anti-plaque as an adjunct to conventional periodontal therapy.

KEYWORDS: Gas chromatography analysis, Mass spectroscopy, Periodontitis, *Sargassum wightii* extract

1. Introduction

Historically, it has been found that seaweed has an incredibly ancient lineage, humans recognized the benefits of seaweed thousands of years ago, utilizing it for various purposes including food, medicine, and even as a fertilizer. This long history of use underscores its nutritional and therapeutic value, which is still being explored and utilized

*Corresponding author: tanyamasand24@gmail.com

DOI: 10.1201/9781003664277-13

today. Marine algae, commonly referred to as seaweed, encompass a diverse group of species classified into three main categories based on their pigmentation: Red Algae (Rhodophyta), Brown Algae (Phaeophyceae) and Green Algae (Chlorophyta). Sargassum wightii is a prominent species of brown seaweed found along the intertidal and subtidal zones of the Indian coast, particularly in the southern regions including the Gulf of Mannar, Palk Bay, and the coasts of Tamil Nadu and Andhra Pradesh. The anti-bacterial agents found in these algae include amino acids, terpenoids, phlorotannins, acrylic acid, phenolic compounds, steroids, halogenated ketones, alkanes, cyclic polysulphides and fatty acids. It is one of the marine jewels rich in sulphated polysaccharides that possess a wide range of biological properties including anti-oxidant, anti-inflammatory, anti- microbial and anti-carcinogenic properties. Recent studies have suggested that Fucoidans an extract from sargassum wightii has an anti- bacterial effect on periodontal pathogens including *Tannerella forsythia* and *Porphyromonas gingivalis.* Abundant flavonoids and terpenoids contributes to its greater anti-oxidant activity. Its great potency with cellular signal transduction interactions and ability to synthesize different pharmacophores with good nutritional values helps it to withstand against drug resistant bacteria Since, periodontal diseases are prevalent both in developed and developing countries and affect about 20-50% of global population which can result in tooth loss at an early stage. Microbial plaque is known to be the most important causative agent of periodontal disease. Since plaque control by mechanical methods is not enough to keep the gingival health in many people, attention has been brought to many therapeutic agents adjunct to supportive periodontal therapy. Crude fucoidan extract from *Sargassum polycystum* also has showed antibacterial activity. Brown seaweeds are rich in bioactive compounds such as polyphenols, carotenoids, polysaccharides, vitamins, minerals, and essential fatty acids. These components contribute to their high antioxidant activity, which can help neutralize free radicals and reduce

oxidative stress make them promising candidates for use in food, pharmaceuticals, and nutraceuticals, suggesting their potential role in preventing diseases associated with oxidative stress. In view of the above the present study is conducted to evaluate the potency of bioactive compounds extracted from sargassum wightii against periodontitis by the dint of gas chromatography and mass spectroscopy.

2. Methodology

Collection of *Sargassum Wightii* Extract: The samples of fresh brown marine algae, Sargassum wightii was collected from Mandapam, Ramanathapuram District, Tamil Nadu, Southeast coast of India. Immediately after collection, the seaweed samples were washed with marine water to remove surface salts and loosely attached particles. The seaweed was taxonomically identified and authenticated by Ramaiah Advance testing lab. The whole seaweed material was shade dried and was coarsely powdered.

Preparation of Extracts: 300 grams of shade dried coarse material was subsequently made in powder form. 15 grams of sample was measured and was diluted with 50 ml of ethanol. The prepared sample was then placed in microwave synthesiser at 50° Celsius for 15 mins. The solvent was filtered from seaweed debris through Whatman No.1 filter paper and the filtrates were concentrated in water bath for 48 hours.

2.1 Gas Chromatography and Mass Spectrometry Analysis

The Sargassum wightii extract was filtered on a Durapore-HV membrane filter disk with 2.5 cm diameter and 0.45 µm pore size by vacuum filtration. The sample was then freeze dried at at −80 °C The sample was diluted in 1 ml of methanol and vortexed for about 10 seconds. Later, the solution was centrifuged for 15000 rpm for 20 minutes at 4°C. The mass spectrometer was operated in the electron impact mode at 80 eV. The split ratio was set to 1:15, with

Fig. 13.1 15gram of sargassum wightii was measured through digital weighing scale. Two samples with same weight of 15gm sargassum wightii were diluted in 50ml of ethanol (99% v/v) and for further extraction was kept inside microwave synthesizer at 50°C for 15 minutes. Two samples with same weight of 15gm sargassum wightii were diluted in 50ml of ethanol (99% v/v) and for further extraction was kept inside microwave synthesizer at 50°C for 15 minutes. The prepared sample was filtered through Whatman no. 1 filter paper and was left undisturbed. 5 gram of crystalline form aqueous extract was obtained after 48hours of water bath for 2 hours

an injection volume of 1 μL. The injector temperature was maintained at 250°C. The oven temperature was maintained at 40°C for 10 minutes, then increased to 300°C over 20 minutes. Mass spectrometry data collection began at 2 minutes and ended at 30 minutes. For peak identification of the crude Sargassum wightii extract, retention times were compared with standards, and the mass spectra obtained were matched with the Mass Spectral Library using an acceptance criterion of a match factor above 75%.

3. Results

The compounds identified from S.wightii (brown algae) by interpreting the GCMS spectrum were 8,10-Dodecadien-1-ol acetate, Cumarin-3-carboxylic acid-7-methoxy, Z-(13,14-Epoxy)- tetradec-11-en-1-ol acetate, Hexadecenoic acid methyl ester, 9-Octadecenoic acid methyl ester, Nonadecanoic acid, 18-oxo-methyl ester, 13-Docosenoic acid methyl ester, 17-hydroxy-methyl ester with their retention time, molecular weight and molecular formula. Hexahydro farnesyl acetone has antibacterial, anti-nociceptive and anti-inflammation activities. Dodecanoic acid also known as lauric acid has bactericidal effect. Arachidonic acid, acts as immunomodulatory in periodontitis. The compound 9,12,15-Octadecatrienoic acid is known to possess several biological properties like analgesic, anaesthetic, anticonvulsant, anti-inflammatory, antioxidant, anti-pyretic, antibacterial, anticancer, antihistaminic, hepatoprotective, hypocholesterolemia, nematicide. Anthraquinones (9,10-dioxoanthracenes) constitute an important class of natural and synthetic compounds with a wide range of applications. Besides their utilization as colorants, anthraquinone derivatives have been used since centuries for medical applications, for example, as laxatives and antimicrobial and anti-inflammatory agents. Isoparvifuran is a benzofuran compound isolated from the heartwood of *Dalbergia odorifera*. Related research reported that isoparvifuran has antioxidant property. Studies indicate that the compound 6-hydroxy-4,4,7a-trimethyl-5,6,7,7a-tetrahydrobenzofuran-2(4H)-one (HTT) plays a promising role in anti-inflammatory processes. Specifically, HTT demonstrates effectiveness against inflammation induced by lipopolysaccharides (LPS) by reducing oxidative stress and modulating the Nuclear Factor-κappa beta and MAPK (Mitogen activated pathway) signaling pathways. This action helps suppress the production of pro-inflammatory cytokines and mitigates oxidative damage to cells. HTT, often derived from marine organisms like brown algae, exhibits bioactivity relevant to these cellular pathways, which can also contribute to reducing extracellular matrix degradation often associated with inflammatory responses. Hexahydro

farnesyl acetone (6,10,14-Trimethyl-2-pentadecanone), a sesquiterpene isolated from Impatiens parviflora, is the major constituents of the essential oil. Hexahydro farnesyl acetone has antibacterial, anti-nociceptive and anti-inflammation activities. 3-Acetylcoumarin coumarin derivatives also possess antibacterial, antiinflammatory, antioxidant properties.

4. Discussion

The marine brown algae Sargassum wightii has garnered attention in periodontal research due to its rich composition of bioactive compounds, including terpenoids, sterols, sulphated polysaccharides, polyphenols, carotenoids, vitamins, proteins, and essential minerals. Such compounds are widely recognized for their diverse biological activities, particularly those beneficial in managing inflammatory diseases like periodontitis which is characterized by chronic inflammation and progressive destruction of periodontium. The presence of polyphenols and carotenoids in Sargassum wightii offers significant antioxidant properties, which may help neutralize free radicals and reduce oxidative stress within periodontal tissues.This could be beneficial in mitigating the inflammatory response. Studies suggest that Sargassum wightii produces various antimicrobial compounds, including certain terpenoids and sulphated polysaccharides, that inhibit microbial growth. Given that periodontitis is primarily driven by a dysbiosis biofilm of pathogenic bacteria, these compounds may directly reduce or inhibit the growth of pathogenic microorganisms in the oral cavity. Sulphated polysaccharides, in particular, have demonstrated strong antimicrobial properties against bacterial biofilms, suggesting their potential as adjunctive agents for biofilm management in periodontal therapy. Some sulphated polysaccharides exhibit anti-quorum sensing properties, interfering with bacterial cell-to-cell communication. This disruption reduces biofilm formation and can inhibit the virulence of bacteria involved in periodontal diseases. The control of biofilm-forming bacteria is crucial for periodontal health, as plaque accumulation exacerbates inflammation and tissue degradation. The sulphated polysaccharides found in Sargassum wightii could play a role in inhibiting biofilm formation on tooth surfaces and reducing plaque build-up. By integrating extracts or derivatives of this algae into oral care products such as mouthwashes or gels its anti-inflammatory, antimicrobial, and antioxidant properties might enhance the outcomes of scaling, gingival curettage or other periodontal treatments. Further studies on its compatibility with existing chemical agents, such as chlorhexidine or fluoride, could pave the way for more comprehensive treatment strategies. Although the bioactive

Table 13.1 Compounds present in the sample with match factor

Component RT	Compound Name	CAS#	Formula	Component Area	Match Factor
3.1037	Succinic acid, hept-2-yl 2-fluoroethyl ester	1000390-88-0	C13H23FO4	166698.9	65.3
3.1272	Cyanogen chloride	506-77-4	CClN	796843.6	69.1
3.1446	Acetoin	513-86-0	C4H8O2	294624132.3	81.3
3.2230	Silanol, trimethyl-	1066-40-6	C3H10OSi	1422260164.0	61.3
3.2404	Propanoic acid, 2-hydroxy-, methyl ester, (.+/-.)-	2155-30-8	C4H8O3	334556783.2	60.6
3.2575	1,2,4,5-Tetroxane, 3,3,6,6-tetramethyl-	1073-91-2	C6H12O4	140935275.3	72.6
3.2904	Thioacetic acid	507-09-5	C2H4OS	8830507.6	61.5
3.3303	2-Propanol, 1-methoxy-	107-98-2	C4H10O2	829240905.1	63.6
3.3460	Methyl nitrate	598-58-3	CH3NO3	27332720.2	78.2
3.4049	2,3-Butanediol, dinitrate	6423-45-6	C4H8N2O6	35605297.5	63.7
3.5451	Acetoin	513-86-0	C4H8O2	230736619.6	83.6
3.6097	Dimethyl ether	115-10-6	C2H6O	259939894.4	92.6
3.6400	Ethane, fluoro-	353-36-6	C2H5F	391687182.5	66.1
3.7458	Nitric acid, 1-methylethyl ester	1712-64-7	C3H7NO3	119874459.9	78.8
3.7566	Silanol, trimethyl-	1066-40-6	C3H10OSi	44241220.3	62.1
3.7585	Acetic acid, dimethoxy-, methyl ester	89-91-8	C5H10O4	567166603.8	62.4
3.7596	Propanoic acid, 2-hydroxy-, ethyl ester	97-64-3	C5H10O3	1100668150.6	62.6
3.8590	Triethyl borate	150-46-9	C6H15BO3	10468470.1	62.2
3.9011	Ethoxycyclohexyldimethylsilane	1000375-96-9	C10H22OSi	3021018.4	66.3
3.9780	Acetic acid	64-19-7	C2H4O2	12414943.4	94.7
4.0987	Ethane, (chloromethoxy)-	3188-13-4	C3H7ClO	1289321.6	73.2
4.1840	Methane, diethoxy-	462-95-3	C5H12O2	1307554.1	72.9
4.2197	Acetic acid, methyl ester	79-20-9	C3H6O2	1095681.6	86.6
4.2210	Ethoxy(dimethyl)isopropylsilane	36850-66-5	C7H18OSi	658245.3	61.0
4.2329	Acetic acid, cesium salt	3396-11-0	C2H3CsO2	738976.0	75.3
4.3082	Methyl propyl ether	557-17-5	C4H10O	373825.7	63.5
4.4705	Silane, diethoxydimethyl-	78-62-6	C6H16O2Si	8607030.9	83.3
4.6339	Silane, diethoxydimethyl-	78-62-6	C6H16O2Si	8175606.9	94.3
4.9583	Hexanal	66-25-1	C6H12O	1567641.6	69.0
5.1125	Cyclotrisiloxane, hexamethyl-	541-05-9	C6H18O3Si3	4885456.6	93.6
5.3380	3-Furaldehyde	498-60-2	C5H4O2	3986344.1	88.8
5.5987	1,3-Difluoro-2-propanol	453-13-4	C3H6F2O	211133.4	63.0
5.5990	Cyanogen chloride	506-77-4	CClN	156197.3	61.3
5.6338	Dimethylsulfoxonium formylmethylide	31043-74-0	C4H8O2S	1977686.6	61.8
5.6464	4,6-Heptadiyn-3-one	29743-27-9	C7H6O	2974952.8	70.2
5.9236	4,6-Heptadiyn-3-one	29743-27-9	C7H6O	40219.6	63.5
6.3866	Ethanone, 1-(2-furanyl)-	1192-62-7	C6H6O2	1973663.6	77.4
6.5352	Dimethyl sulfone	67-71-0	C2H6O2S	148841.5	71.3
6.6363	Ethyl orthoformate	122-51-0	C7H16O3	3175143.7	69.1
7.6152	1H-Pyrazole, 5-methoxy-1,3-dimethyl-	53091-80-8	C6H10N2O	692491.2	79.0
7.7445	Ethyl orthoformate	122-51-0	C7H16O3	1729193.6	80.6
7.7585	Ethylphosphonic acid, dioctyl ester	6156-13-4	C18H39O3P	199646.5	60.1
7.8117	9,10-Anthracenedione, 1-phenyl-	1714-14-3	C20H12O2	10763479.6	60.1
7.8144	Cyclotetrasiloxane, octamethyl-	556-67-2	C8H24O4Si4	71634507.5	93.9
8.0572	1-Butanamine, N,N-dimethyl-	927-62-8	C6H15N	99607.2	61.6
8.1363	1-Butanone, 1-(2-furanyl)-	4208-57-5	C8H10O2	153739.4	65.6
8.2251	Acetic acid, diethoxy-, ethyl ester	6065-82-3	C8H16O4	299261.2	74.8
8.3423	Benzofuran	271-89-6	C8H6O	590192.7	64.8
8.3439	Indoline, 2-(hydroxydiphenylmethyl)-	1000164-32-0	C21H19NO	1732684.8	64.2
8.3893	Benzene, 1,2,4-trimethyl-	95-63-6	C9H12	760649.4	72.3
8.4259	Benzamidine	618-39-3	C7H8N2	135929.9	60.3
9.6586	Ethyl 3-thiopheneacetate	37784-63-7	C8H10O2S	535238.2	65.3
9.6761	2-Furaldehyde diethyl acetal	13529-27-6	C9H14O3	837775.6	70.6
9.7843	Silane, triethylmethoxy-	2117-34-2	C7H18OSi	38209.1	62.2
9.8545	4-Methylbenzoic acid, 2,5-dichlorophenyl ester	1000325-60-2	C14H10Cl2O2	141999.0	63.6
10.0326	Succinic acid, ethyl 2-norbornyl ester	1000330-19-4	C13H20O4	964500.2	60.3
10.0348	Hexane, 1,1-diethoxy-	3658-93-3	C10H22O2	3077498.6	89.0
10.5736	Mepivacaine	96-88-8	C15H22N2O	367805.1	62.3
11.6071	Cyclopentasiloxane, decamethyl-	541-02-6	C10H30O5Si5	16692836.7	98.1
11.7961	Hexane, 1,1-diethoxy-	3658-93-3	C10H22O2	1185879.4	80.5
11.9255	Ethyl orthoformate	122-51-0	C7H16O3	371283.0	82.5
12.0484	2-(2,2-Diethoxyethyl)furan	1000411-31-4	C10H16O3	896902.5	80.0
12.3867	cis-4-Hepten-1-al diethyl acetal	18492-65-4	C11H22O2	1591959.2	73.4

Component RT	Compound Name	CAS#	Formula	Component Area	Match Factor
12.6587	Heptane, 1,1-diethoxy-	688-82-4	C11H24O2	2484217.2	80.8
13.0886	Isoquinaldamide	1436-44-8	C10H8N2O	188569.8	67.0
13.4072	Isophthalaldehyde	626-19-7	C8H6O2	124467.9	72.1
13.9267	1H-Pyrrole-2,5-dione, 3-ethyl-4-methyl-	20189-42-8	C7H9NO2	1335600.0	71.5
14.0374	Diethylene glycol, bis(chlorodifluoroacetate)	1000375-79-1	C8H8Cl2F4O5	100490.2	62.6
14.3501	Silane, dimethyl(2-naphthoxy)tetradecyloxy-	1000347-22-0	C26H42O2Si	79913.5	61.9
14.4847	Diethylene glycol, bis(chlorodifluoroacetate)	1000375-79-1	C8H8Cl2F4O5	375074.8	60.4
14.6159	Benzeneacetic acid	103-82-2	C8H8O2	8395356.0	89.8
14.6169	1,3-Benzodioxol-2-one	2171-74-6	C7H4O3	4918483.3	64.2
14.8491	6-Bromo-2-hydroxyquinoline, tert-butyldimethylsilyl ether	1000463-48-8	C15H20BrNOSi	7644.7	61.5
15.5510	Phosphoric acid, diundecyl ethyl ester	1000308-89-0	C24H51O4P	146723.3	64.4
15.8499	2,2,2-Trifluoro-N-[2-(1-hydroxy-2,2,6,6-tetramethyl-piperidin-4-yl)-ethyl]-acetamide	1000278-26-3	C13H23F3N2O2	1013953.9	62.8
15.9275	5-[4-(Carboxyadamanthyl-3)-phenyl]-10,15,20-triphenyl-21H,23H-porphine	142230-18-0	C55H44N4O2	3769.0	72.1
16.2463	Cyclohexasiloxane, dodecamethyl-	540-97-6	C12H36O6Si6	5974005.5	72.0
16.4786	Pyrazine, 2-methoxy-3-(1-methylethyl)-	25773-40-4	C8H12N2O	731242.2	66.2
16.5298	Cyclohexasiloxane, dodecamethyl-	540-97-6	C12H36O6Si6	3319966.3	92.6
16.6883	2(3H)-Furanone, dihydro-4,4-dimethyl-5-(2-oxopropyl)-	89722-19-0	C9H14O3	215908.7	64.5
17.1393	Naphthalene, 1,2-dihydro-4,5,7-trimethyl-	53156-11-9	C13H16	289153.7	78.5
17.3090	1,3-Indandione, 2-acetyl-	1133-72-8	C11H8O3	594722.4	76.7
17.6357	Ethyl orthoformate	122-51-0	C7H16O3	3232195.4	72.6
17.7836	Ethyl orthoformate	122-51-0	C7H16O3	502866.9	74.7
17.9411	Sulfur hexafluoride	2551-62-4	F6S	6720825.3	66.6
18.1203	3,4,4,4-Tetrachloro-1-(2,4-dichloro-phenyl)-butan-1-one	301655-26-5	C10H6Cl6O	842586.7	71.5
18.2420	1-(3,6,6-Trimethyl-1,6,7,7a-tetrahydrocyclopenta[c]pyran-1-yl)ethanone	1000194-97-2	C13H18O2	1416640.7	70.6
18.2897	Alanylalanine, N,N'-dimethyl-N'-propargyloxycarbonyl-, ethyl ester	1000329-34-8	C14H22N2O5	327678.4	66.2
18.5598	2-Acetylcyclopentanone	1670-46-8	C7H10O2	2169998.5	72.9
18.7974	Ethanone, 1-(2,3-dihydro-1,1-dimethyl-1H-inden-4-yl)-	55591-10-1	C13H16O	4791346.5	82.6
19.1433	1,2-Bis(1,4,7-trioxa-10-azacyclododec-10-yl)-ethane	79645-07-1	C18H36N2O6	762046.4	62.4
19.1717	Naphthalene, 1,2,3,4-tetrahydro-1-methyl-8-(1-methylethyl)-	81603-43-2	C14H20	6735439.8	85.0
19.4344	Ethyltetramethylcyclopentadiene	57693-77-3	C11H18	2399003.2	72.9
19.5545	Formic acid, 2-bromomethyl-4,4-dimethyl-3-(3-oxobut-1-enyl)cyclohex-2-enyl ester	1000190-68-6	C14H19BrO3	5146095.1	63.8
19.8231	Benzene, 1,4-dimethyl-2,5-bis(1-methylethyl)-	10375-96-9	C14H22	1113422.5	65.0
20.0394	1-Acetyl-4,6,8-trimethylazulene	834-97-9	C15H16O	3170573.0	79.5
20.0934	Cycloheptasiloxane, tetradecamethyl-	107-50-6	C14H42O7Si7	1064185.6	89.1
20.1457	3-Phenylpropanoic acid, 2-(1-adamantyl)ethyl ester	1000282-93-7	C21H28O2	917730.3	65.6
20.2952	1,1,4,5,6-Pentamethyl-2,3-dihydro-1H-indene	16204-67-4	C14H20	10310168.6	82.5
20.3086	Pentanoic acid, 5-hydroxy-, 2,4-di-t-butylphenyl esters	166273-38-7	C19H30O3	1231647.8	76.3
20.5067	1,1'-Biphenyl, 4-(1-methylethyl)-	7116-95-2	C15H16	1186280.4	73.0
20.6171	2(4H)-Benzofuranone, 5,6,7,7a-tetrahydro-4,4,7a-trimethyl-, (R)-	17092-92-1	C11H16O2	2943190.2	85.2
20.7081	1,3-Benzodioxole-5-(4-keto-butyric acid)	41764-07-2	C11H10O5	871232.3	61.0
20.7380	Bis(heptamethylcyclotetrasiloxy)siloxane	17909-39-6	C14H42O9Si8	158045.3	61.6
20.7381	Bis(pentamethylcyclotrisiloxy)tetramethyldisiloxane	17909-18-1	C14H42O9Si8	239279.1	65.3
20.9560	Phosphorodichloridic acid, pentyl ester	1000309-09-3	C5H11Cl2O2P	205941.6	60.4
21.0158	4-Trifluoromethylbenzohydroxamic acid	40069-07-6	C8H6F3NO2	82914.1	61.7
21.0932	Dodecanoic acid	143-07-7	C12H24O2	4709531.6	90.1
21.1672	Sulfone, dichloromethyl m-(trifluoromethyl)benzyl	15894-29-8	C9H7Cl2F3O2S	1538488.0	68.4
21.2688	Fumaric acid, 2,4-dimethylpent-3-yl ethyl ester	1000348-54-1	C13H22O4	2317836.7	66.9
21.5512	Diethyl Phthalate	84-66-2	C12H14O4	1394425.2	80.1
21.6488	3-Methoxy-4-propoxybenzaldehyde	57695-98-4	C11H14O3	2636623.8	70.1
21.8240	1,1,4,5,6-Pentamethyl-2,3-dihydro-1H-indene	16204-67-4	C14H20	5515231.1	86.4
22.0047	Benzeneacetic acid, 3-methoxy-	1798-09-0	C9H10O3	1288692.8	65.8
22.0477	Benzophenone	119-61-9	C13H10O	600607.0	66.6
22.3387	1,1'-Biphenyl, 3,3',4,4'-tetramethyl-	4920-95-0	C16H18	1364439.3	76.2
22.3412	9,9-Dimethyl-9-silafluorene	13688-68-1	C14H14Si	589270.9	78.3
22.5748	4,4,5,8-Tetramethylchroman-2-ol	82391-05-7	C13H18O2	3785065.9	73.5
22.6041	Isophthalic acid, 2-formylphenyl pentyl ester	1000344-61-4	C20H20O5	397581.4	69.0
22.7516	2H-1-benzopyran-6-ol, 3,4-dihydro-2,2-dimethyl-7-(1,1,3,3-tetramethylbutyl)-	1000402-03-6	C19H30O2	244353.7	60.2
22.8784	5-Trimethylsilylpent-2-en-4-yne	18387-62-7	C8H14Si	1067941.8	77.2
23.1720	1H-Indene, 2,3-dihydro-1,1,3-trimethyl-3-phenyl-	3910-35-8	C18H20	297996.5	61.9
23.2056	6,6-Dimethyl-9-methylene-undec-3-ene-2,5,10-trione	1000193-23-0	C14H20O3	4802314.9	63.2

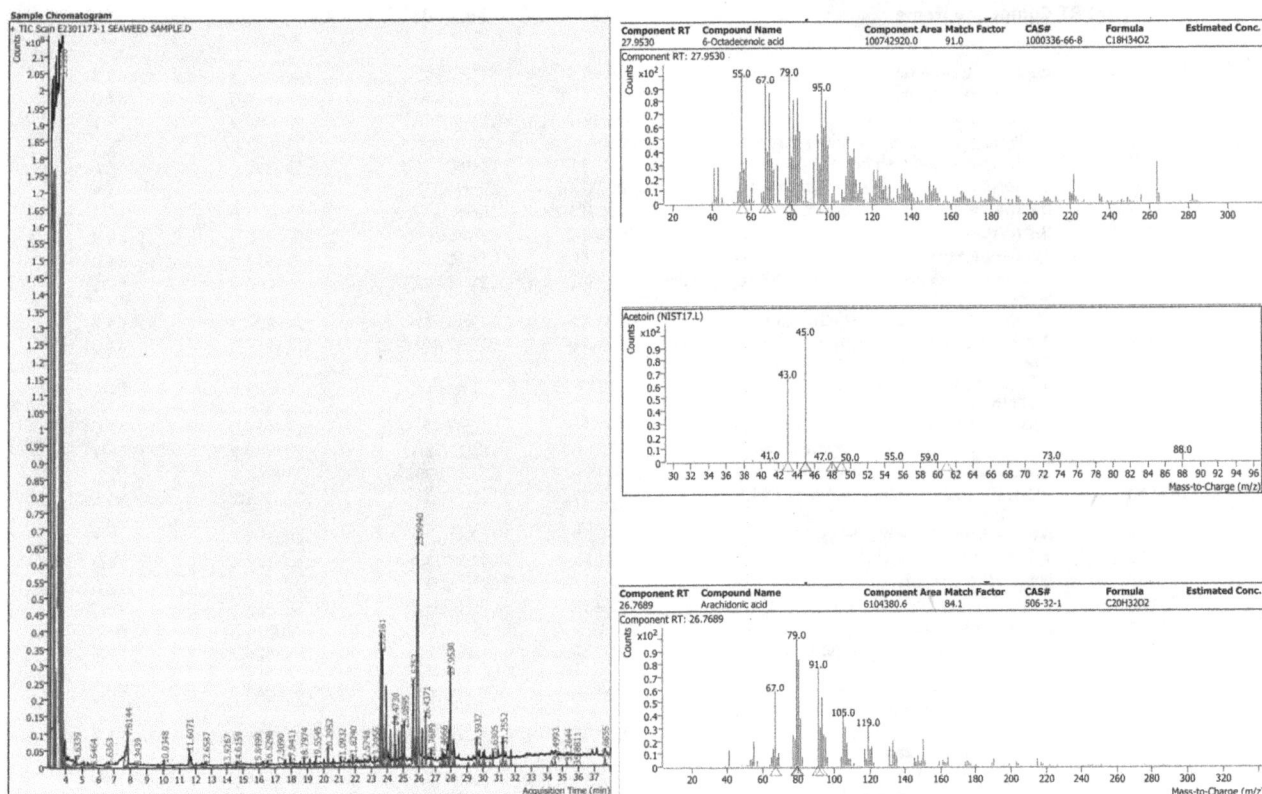

Fig. 13.2 Gas chromatography and mass spectroscopy spectrum of *sargassum wightii*

potential of Sargassum wightii is promising, it remains underexplored in clinical periodontal settings. Rigorous in vitro and in vivo studies are essential to validate its efficacy and safety in humans. Additionally, understanding the specific mechanisms through which Sargassum wightii compounds exert their therapeutic effects on periodontal tissues will provide insights into its potential role in clinical applications.

5. Conclusion

The Indian Ocean has abundant resources of brown seaweed Sargassum wightii, which have proven to have an innate effective defence system due to their adverse habitats. This study demonstrates the metabolite profiling and characterization of the bioactive compounds in Sargassum wightii which can be further isolated to be used in the production of pharmaceuticals and functional food supplements to treat several diseases such as periodontitis, hypertension, diabetes, and inflammatory disorders opening new frontiers in algal industry for this seaweed world-wide. In summary, Sargassum wightii presents a promising natural source of therapeutic agents for managing periodontal disease due to its anti-inflammatory, antioxidant, and antimicrobial properties. However,

while preliminary findings are encouraging, extensive clinical studies are needed to substantiate its application in periodontal therapy. Integrating Sargassum wightii into current periodontal treatment protocols could represent a significant advancement in both preventive and therapeutic oral health care, promoting a more holistic and natural approach to periodontal disease management.

Acknowledgment

The authors extend thankfulness to South India textile Research Association (SITRA) Lab, Coimbatore, Tamil Nadu, India for lab investigations and results. The authors have no funding or conflicts of interest to disclose for this study.

References

1. Cao, X., Cheng, X.W., Liu, Y.Y., Dai, H.W. and Gan, R.Y., 2024. Inhibition of pathogenic microbes in oral infectious diseases by natural products: Sources, mechanisms, and challenges. *Microbiological Research*, 279, p.127548.
2. Farvin, K.S. and Jacobsen, C., 2013. Phenolic compounds and antioxidant activities of selected species of seaweeds from Danish coast. *Food chemistry*, 138(2-3), pp. 1670–1681.

3. Hamrun, N., Oktawati, S., Haryo, H.M., Syafar, I.F. and Almaidah, A.N., 2020. Effectiveness of fucoidan extract from brown algae to inhibit bacteria causes of oral cavity damage. *Systematic Reviews in Pharmacy*, *11*(10), pp. 686–693.

4. Kim, T.H., Kim, S.C. and Jung, W.K., 2023. Therapeutic effect of marine bioactive substances against periodontitis based on in vitro, in vivo, and clinical studies. *Fisheries and Aquatic Sciences*, *26*(1), pp.1–23.

5. Magesh, K.T., Aravindhan, R., Kumar, M.S. and Sivachandran, A., 2020. Antibacterial Efficacy of the Extract of Sargassum Wightii Against Oral Pathogen– An In Vitro Study. *Journal of Orofacial Sciences*, *12*(2), pp.96–100.

6. Nagraj, B.K., Koregol, A.C., Puladas, H., Sulakod, K., Patil, K. and Gore, S., 2022. Antibacterial efficacies of brown sea weed Sargassum Wightii on Periodontal Pathogens: An In-Vitro Microbiological Analysis. *Journal of Ayurveda and Integrated Medical Sciences*, *7*(11), pp.52–58.

7. Nagai, T. and Yukimoto, T., 2003. Preparation and functional properties of beverages made from sea algae. *Food chemistry*, *81*(3), pp.327–332.

8. Nongpiur, C.G.L., Soh, C., Diengdoh, D.F., Verma, A.K., Gogoi, R., Banothu, V., Kaminsky, W. and Kollipara, M.R., 2023. 3-acetyl-coumarin-substituted thiosemicarbazones and their ruthenium, rhodium and iridium metal complexes: An investigation of the antibacterial, antioxidant and cytotoxicity activities. *Journal of Organometallic Chemistry*, *998*, p.122788.

9. Oka, S., Okabe, M., Tsubura, S., Mikami, M. and Imai, A., 2020. Properties of fucoidans beneficial to oral healthcare. *Odontology*, *108*, pp.34–42.

10. Pangal, A., Tambe, P. and Ahmed, K., 2023. Screening of 3-acetylcoumarin derivatives as multifunctional biological agents. *Current Chemistry Letters*, *12*(2), pp.343–352.

11. Panezai, J. and van Dyke, T., 2023. Polyunsaturated fatty acids and their immunomodulatory actions in periodontal disease. *Nutrients*, *15*(4), p.821.

12. Pramitha, V.S. And Sree Kumari, N., 2016. Anti-Inflammatory, Anti-Oxidant, Phytochemical And Gc-Ms Analysis of Marine Brown Macroalga, Sargassum Wighti. *International Journal of Pharmaceutical, Chemical & Biological Sciences*, *6*(1).

13. Ravi, C., Muthamil, R., & Karthiga, A. (2016). Biomedical Potential and Preliminary Phytochemistry of the Brown Seaweed Sargassum wightii Greville ex J. Agardh 1848. *Asian Fisheries Science*, *29*(1).

14. Syad, A.N., Shunmugiah, K.P. and Kasi, P.D., 2013. Antioxidant and anti-cholinesterase activity of Sargassum wightii. *Pharmaceutical biology*, *51*(11), pp.1401–1410.

15. Yuvaraj, N. and Arul, V., 2014. In vitro anti-tumor, anti-inflammatory, anti-oxidant, and antibacterial activities of marine brown alga Sargassum wightii collected from Gulf of Mannar. *Global Journal of Pharmacology*, *8*(4), pp. 566–577.

Note: All the figures and table in this chapter were made by the author.

Advances in Materials Science and Technology – Dr. Srikari Srinivasan et al. (eds)
© 2025 Taylor & Francis Group, London, ISBN 978-1-041-12342-2

14

Insilico Analysis of Compounds Isolated from Mulberry Plant Source for Anti Cariogenicity

Gaviappa Dhananjaya*
Department of Pediatric and Preventive Dentistry,
Faculty of Dental Sciences, M S Ramaiah University of Applied Sciences,
Bangalore

Sylvia Mathew
Department of Conservative Dentistry and Endodontics,
Faculty of Dental Sciences, M S Ramaiah University of Applied Sciences,
Bangalore

Deveswaran R.
Department of Pharmaceutics, Faculty of Pharmacy, MSRUAS,
Bangalore, Karnataka, India

ABSTRACT: Dental caries is a multifactorial disease impacting human health, primarily caused by S. mutans through biofilm formation. Compounds that inhibit S. mutans growth and disrupt biofilm can aid in treating or preventing caries. Molecular docking analysis evaluates the binding energies and positions of interacting molecules and ligands. Two compounds, Deoxyjiniromycin (DNJ) and Quercetin, isolated from V2 and S36 mulberry plant varieties, were evaluated for anti-cariogenic activity. Molecular Dynamic Simulation assessed the binding stability, conformation, and interaction modes between the bioactive compounds and receptors, using GROMACS software.Both isolated compounds showed strong inhibitory potential. Generally, in simulations, a higher SASA value of the protein indicates a relaxed structure and thus decreased stability. The highest SASA values were observed in the DNJ complex, followed by quercetin and ciprofloxacin.MDS revealed that DNJ had higher RMSD fluctuations compared to quercetin and ciprofloxacin. Overall, both quercetin and DNJ demonstrated strong potential as drug targets for treating dental caries.

KEYWORDS: Discontinuities, Joint properties, Plaxis, Sirovision, Slope stability

1. Introduction

Oral diseases, particularly dental caries, is a serious public health concern affecting humans (Gambhir et al.,2016). Dental caries is a biofilm-mediated, diet modulated, multifactorial, non-communicable, dynamic disease resulting in net mineral loss of dental hard tissues. Caries causing microorganism S. mutans (a major etiological agent) are mainly aciduric in nature (Machiulskiene, et al.,2020) (Valm et al.,2019). Regarding the virulence of S. mutans, it is the adhesive property that plays a significant role, and this adhesion can be mediated through sucrose-independent and sucrose dependent mechanisms. In a sucrose-dependent mechanism, the major mechanism involves the formation

*Corresponding author: dhanadarshil@gmail.com

DOI: 10.1201/9781003664277-14

of dextran (soluble) and glycans (insoluble) by GTFs (Miglani et al.,2020) (Islam et al.,2008). The activity and expression of S. mutans enzymes (lactate dehydrogenase) specific to glycolytic pathway have shown to be disrupted by plant crude extracts (Hirasawa et al.,2006) (Brighenti et al.,2008). The mulberry plant is a member of the genus and family, Morus and Moraceae, respectively. Alkaloid DNJ present in mulberry has the property of inhibiting elevated blood sugar levels (Sugiyama et al.,2013). Insilco analysis helps us to understand the interaction of the isolated compounds against the receptors, it strengthens the understanding of mechanism of action of isolated compounds. The Molecular Dynamics Simulation (MDS) is conducted to determine the dynamic behaviour of drug candidates. Mainly five different target proteins were considered for this study including 3AIE (Glucan sucrose), 3BBA (Cystein Protease), 6TZ6 (Calcineurin A), 6LOI (Undecarprenyl pyrophosphate synthase), 4TQX (Sortase A) because these proteins are potential targets for inhibitors to fight against infection causing bacteria.

2. Methodology

The study was carried out in accordance with the recommendations for good clinical practice approved by the University Ethics Committee for Human trials of M S Ramaiah University of Applied Sciences

The dynamic behaviour of test compounds was evaluated using molecular dynamics simulation (MDS). The MDS was evaluated to determine the binding stability, conformation and interaction modes between the selected bioactive compounds (ligands) and receptor. The test compounds used were DNJ and quercetin. The complex files were subjected to molecular dynamics studies using GROMACS software (Sugiyama et al.,20202). The selected ligands topology was downloaded from PRODRG server (Singh A et al.,20202). The PRODRG was used instead of PDF files as the former method can be used to study large complexes which are quite difficult when the latter method is used. Further, the PRODRG program is interfaced to WHAT IF (molecular modelling package) and the GROMOS (molecular dynamics package). This combination provides a quick path, where WHATIF was used to load a protein with a ligand, while the GROMOS was used to stimulate the protein ligand complex.

For MDS, first vacuum was minimized using the steepest descent algorithm for 5000 steps. The complex structure was solvated in a cubic periodic box of 0.5 nm with a simple point charge (SPC) water model. Trajectory analysis was performed by using GROMACS simulation package of Root Mean Square Deviation (RMSD), Root Mean Square Fluctuation (RMSF), Radius Of Gyration (RG)

and Solvent Accessible Surface Area (SASA) through the online server —WebGRO for Macromolecular Simulations (https://simlab.uams.edu/) (Singh A et al.,2014).

The MMPBSA (Molecular Mechanics Poisson–Boltzmann Surface Area) method was used to calculate the protein-ligand binding free energy of each complex. The free energy of binding was determined using the g_mmpbsa tool developed for GROMACS4. In the presence of explicit water, MDS creates a set of binding conformations in this case, while the MM-PBSA method evaluates the binding energy. The change in binding free energy was estimated using the Single 105 trajectory method (Valdés-Tresanco M.S et.,2021)

3. Results

3.1 RMSD

It was observed that the Ciprofloxacin complex revealed differences in the backbone RMSD till 40 ns in the range of 0.3 to 0.9 nm. In Fig. 14.1, DNJ showed considerable variations when compared to the control Ciprofloxacin throughout the simulation. The DNJ complex did not show any stability throughout the simulation period and the RMSD values ranged from ~0.15 nm to 3 nm. Quercetin complex showed similar variation to ciprofloxacin until 35ns but started at ~0.45ns. After 40ns, the structure showed higher RMSD values ranging from 0.9nm (at 40ns) to 5nm (at about 90ns). However, the fluctuations in quercetin were much smoother than DNJ. The RMSD values of DNJ lied between that of quercetin and ciprofloxacin, suggesting that the structural changes in DNJ were higher than ciprofloxacin but lower than quercetin. Thus, it can be inferred that the fluctuations clearly specified that DNJ underwent structural changes in the complex during molecular dynamics.

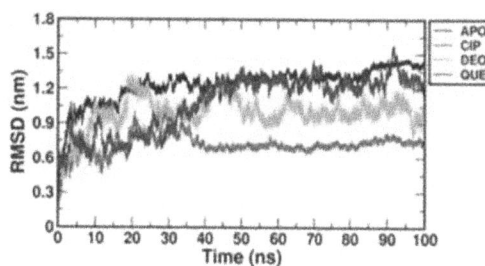

Fig. 14.1 RMSD activity

3.2 RMSF

In the present complex, minimal fluctuation of the amino acids was observed during the entire simulation process. The amino acid residues between 25 and 50 of quercetin interacted with Sortase A enzyme and showed higher fluctuation values of 6 nm (Fig. 14.2). However, ciprofloxacin and DNJ showed fluctuations at 2nm and 0.8nm, respectively at less than 50 residues. However, the fluctuations reduced among all three compounds between 50 and 100 amino acid residues and almost remained constant from 100 to 250 amino acids residues. Thus, results suggest that ligands interacted with minimal effects on the flexibility of the residues in the protein.

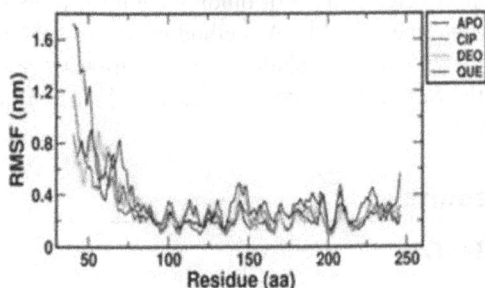

Fig. 14.2 RMSF activity

3.3 Radius of Gyrus

Rg is another important MDS parameter that illustrates the total fluctuations in protein structure complexes. The flexibility is determined by the Rg values. Proteins with higher Rg values are less compact and flexible, while proteins with lower Rg values are highly compact and rigid. These changes can be seen by plotting Rg values of protein backbone atoms against time. Findings also indicated that the trajectory showed peak values in the initial stage. However, in the later stages the Rg values for all the ligand remained constant from 20 to 100ns (Fig. 14.3). Thus, the data is suggestive of the stable nature of protein in the structural complex.

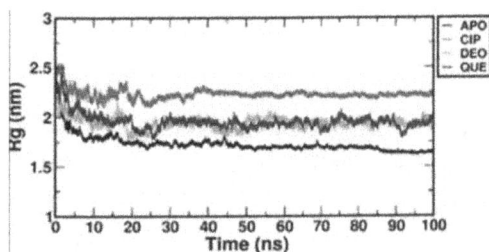

Fig. 14.3 Radius of gyrus values vs time

3.4 SASA

SASA examines the degree of receptors exposure to the solvent molecules presents in the surrounding during simulation. The changes in the surface area were estimated by plotting SASA values against time. Results indicated that until 20ns, the lower or equal SASA value was observed in DNJ when compared to ciprofloxacin and

quercetin (Fig. 14.4). Between 20ns to ~33ns, quercetin showed slightly higher SASA value than DNJ and ciprofloxacin. After 33ns until 100ns, it was found that DNJ had slightly higher SASA value, followed by quercetin and ciprofloxacin, respectively.

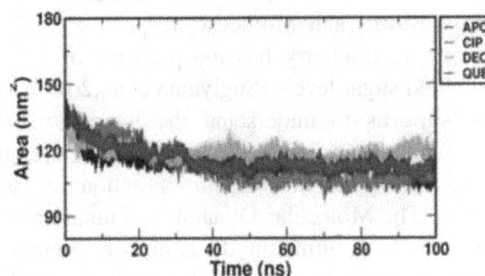

Fig. 14.4 SASA activity

3.5 H-BONDS

The stability of the protein structure can be determined by the hydrogen bonds count. Higher number of hydrogen bonds indicates higher stability. From the present study it can be observed that the count of hydrogen bonds for DNJ ranged from 2 to 5, while it was 1 to 4 for quercetin (Fig. 14.5). Further, the number of hydrogen bonds for ciprofloxacin also ranged from 1 to 4. Thus, suggesting that DNJ was found to be more stable than quercetin and ciprofloxacin as the number of hydrogen bonds formed by DNJ was higher compared to quercetin and ciprofloxacin.

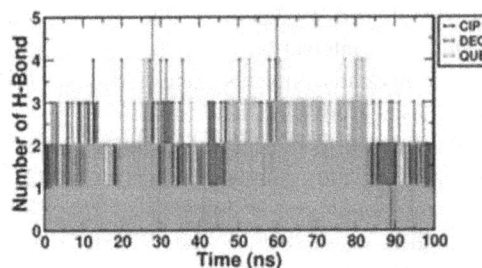

Fig. 14.5 H – Bonds

S. mutans plays a major role in the pathogenesis of caries and infection occur through various processes including adherence, biofilm formation, glucan synthesis, aciduracity, acidogenicity, hydrophobicity on the cell surface and through quorum sensing mechanism (Hasan S et al.,2014). For a phytochemical to be considered as a potential inhibitor, the quality and the stable nature of enzyme–ligand complex is extremely essential. To achieve this, MDS of docked complex was performed (Pahal V et al.,2016). In theory, higher values of RMSF measure the higher flexibility of the protein but lower protein stability (Pahal V et al.,2016). In this study, lower RMSF values were observed after 100 amino acid residues for all the three compounds, indicating higher stability. Further, the stability of the docked complex is directly linked to negative potential energy of this complex which is an indicative of the inhibitory potential of the phytochemical or test compound (Pahal V et al.,2016). Thus, the compounds were considered to have good inhibitory potential. In general, during simulation, higher SASA value of the protein indicates relaxed structure of the protein and hence decreased protein stability[16]. Higher SASA values were observed in DNJ complex, followed by quercetin and ciprofloxacin, respectively. Thus, ciprofloxacin-Sortase A complex was more stable than quercetin and DNJ complexes with Sortase A enzyme. Analysis of the number of hydrogen bonds showed that more positively charged residues were present (Pahal V et al.,2016). However, in terms of hydrogen bonds, DNJ was more stable than the other two compounds as it showed the highest number of hydrogen bonds.

4. Conclusion

DNJ was more stable than quercetin and ciprofloxacin as it showed higher number of hydrogen bonds which was evident in the docking analysis as well. Overall, both quercetin and DNJ showed great potential to be drug targets in treating dental caries.

Acknowledgments

This study was supported by the Department of Pharmaceutics, Faculty of Pharmacy, Ramaiah University of Applied Sciences, Bangalore -India.

References

1. Brighenti, F. L., Luppens, S. B., Delbem, A. C., Deng, D. M., Hoogenkamp, M. A., Gaetti-Jardim Jr, E., Dekker, H. L., Crielaard, W., & Ten Cate, J. M. (2008). Effect of Psidium cattleianum Leaf Extract on Streptococcus mutans Viability, Protein Expression and Acid Production. *Caries Research*, *42*(2), 148–154.
2. Chen, C., Mohamad Razali, U. H., Saikim, F. H., Mahyudin, A., & Mohd Noor, N. Q. (2021). Morus alba L. Plant: Bioactive Compounds and Potential as a Functional Food Ingredient. *Foods*, *10*(3), 689.
3. Gambhir, R. S., Kaur, A., Singh, A., Sandhu, A. R., & Dhaliwal, A. P. (2016). Dental Public Health in India: An Insight. *Journal of Family Medicine and Primary Care*, *5*(4), 747–751.
4. Hasan, S., Singh, K., Danisuddin, M., Verma, P. K., & Khan, A. U. (2014). Inhibition of Major Virulence Pathways of Streptococcus mutans by Quercitrin and Deoxynojirimycin:

5. A Synergistic Approach of Infection Control. *PLoS One*, *9*(3), e91736.
6. Hirasawa, M., Takada, K., & Otake, S. (2006). Inhibition of Acid Production in Dental Plaque Bacteria by Green Tea Catechins. *Caries Research*, *40*(3), 265–270.
7. Islam, B., Khan, S. N., Haque, I., Alam, M., Mushfiq, M., & Khan, A. U. (2008). Novel Anti-Adherence Activity of Mulberry Leaves: Inhibition of Streptococcus mutans Biofilm by 1-Deoxynojirimycin Isolated from Morus alba. *Journal of Antimicrobial Chemotherapy*, *62*(4), 751.
8. Kumari, R., Kumar, R., Open-Source Drug Discovery Consortium, & Lynn, A. (2014). g_mmpbsa--a GROMACS Tool for High-Throughput MM-PBSA Calculations. *Journal of Chemical Information and Modeling*, *54*(7), 1951–1962.
9. Machiulskiene, V., Campus, G., Carvalho, J. C., Dige, I., Ekstrand, K. R., Jablonski-Momeni, A., Maltz, M., Manton, D. J., Martignon, S., Martinez-Mier, E., & Pitts, N. B. (2020). Terminology of Dental Caries and Dental Caries Management: Consensus Report of a Workshop Organized by ORCA and Cariology Research Group of IADR. *Caries Research*, *54*(1), 7–14.
10. Miglani, S. (2020). Burden of Dental Caries in India: Current Scenario and Future Strategies. *International Journal of Clinical Pediatric Dentistry*, *13*(2), 155.
11. Pahal, V., Devi, U., & Dadhich, K. S. (2018). Quercetin, a Secondary Metabolite Present in Methanolic Extract of Calendula Officinalis, Is a Potent Inhibitor of Peptide Deformylase, Undecaprenyl Pyrophosphate Synthase, and DNA Primase Enzymes of Staphylococcus Aureus: An in Vitro and in Silico Result Analysis. *MOJ Drug Design Development & Therapy*, *2*, 216–225.
12. Páll, S., Zhmurov, A., Bauer, P., Abraham, M., Lundborg, M., Gray, A., Hess, B., & Lindahl, E. (2020). Heterogeneous Parallelization and Acceleration of Molecular Dynamics Simulations in GROMACS. *The Journal of Chemical Physics*, *153*(13).
13. Singh, A., & Mishra, A. (n.d.). Molecular Dynamics Free Energy Simulation Study to Investigate Binding Pattern of Isoliquiritigenin as PPARγ Agonist.
14. Sugiyama, M., Katsube, T., Koyama, A., & Itamura, H. (2013). Varietal Differences in the Flavonol Content of Mulberry (Morus spp.) Leaves and Genetic Analysis of Quercetin 3-(6-Malonylglucoside) for Component Breeding. *Journal of Agricultural and Food Chemistry*, *61*(38), 9140–9147.
15. Valdés-Tresanco, M. S., Valdés-Tresanco, M. E., Valiente, P. A., & Moreno, E. (2021). gmx_MMPBSA: A New Tool to Perform End-State Free Energy Calculations with GROMACS. *Journal of Chemical Theory and Computation*, *17*(10), 6281–6291.
16. Valm, A. M. (2019). The Structure of Dental Plaque Microbial Communities in the Transition from Health to Dental Caries and Periodontal Disease. *Journal of Molecular Biology*, *431*(16), 2957–2969.

Note: All the figures in this chapter were made by the author.

Advances in Materials Science and Technology – Dr. Srikari Srinivasan et al. (eds)
© 2025 Taylor & Francis Group, London, ISBN 978-1-041-12342-2

15

Comparative Evaluation of Anti-microbial Efficacy of *Ocimum Sanctum* (Tulsi) extract, *Ocimum Sanctum* with Silver Nanoparticles and 0.12% Chlorhexidine: An Invitro Study

Amulya Vishwanath[1]

Department of Periodontology and Implantology,
Bangalore Institute of Dental Sciences,
Bangalore

Umesh Yadalam[2]

Department of Periodontology and Implantology,
Sri Rajiv Gandhi College of Dental Sciences and Hospital,
Bangalore

ABSTRACT: Periodontitis is an infection of periodontal complex with severe forms of disease associated with specific bacteria colonizing the sub gingival area. Various microorganisms that have been known to be associated with periodontitis are P.gingivalis, P.intermedia, A.a comitants, Fusobacterium nucleatum, etc. Currently Chlorhexidine (CHX) is the most commonly used chemotherapeutic agent against periodontal disease-causing organisms. However, its widespread and prolonged use is limited by its side effects such as teeth staining, parotid gland swelling and taste disturbances. *Ocimum santum* or TULSI has been the pillar of Ayurvedic health system in India. It has been famously considered as "the queen of the herbs" for its medicinal and spiritual properties.In this in-vitro study the antimicrobial efficacy was evaluated against P.gingivalis and T.forsythia for 4%, 8%,and 12% of tusi extract, tulsiextract with silver nano particles, DMSO and 0.12% CHX. The mean zone of inhibition was calculated and external tooth staining was also calculated using spectrophotometry. The results of the present study demonstrated that tulsi extract and tulsi extract with silver-nano particles are effective against P.gingivalis, T.forsythia, hence can be used as an anti-microbial agent. 12% concentration of tulsi extract exhibited the least staining property.

KEYWORDS: Chlorhexidine, Mean zone of inhibition, P. gingivalis, Silver nano particles, Tulsi extract, T. forsythia

1. Introduction

Periodontal disease is a disease of complex etiology. Microorganisms play an important role in the etiology of periodontal disease (Mallikarjun et al., 2016). There is substantial evidence that suggests that bacteria like Porphyromonas gingivalis, Prevotella intermedia, Aggregatibacter actinomycetum comitants, Tannerella forsythia, Fusobacterium nucleatum are considered as periodontal pathogens that cause periodontal destruction (Jayanthi et al., 2018).

[1]amoolyavishwanath@gmail.com, [2]umeshyadalam@gmail.com

DOI: 10.1201/9781003664277-15

There are many treatment strategies which are robust, effective and feasible that are used to treat the periodontal disease. One such strategy would be to verify the enormous wealth of medicinal plants. Currently many pharmaceutical dispensed across the world are having plant origins and very few are intended for use as antimicrobials.

Tulsi scientifically known as *Ocimum Santum* is a time tested medicinal herb. It is a plant of Indian origin and is bestowed with enormous antimicrobial substances and is used to treat a variety of illnesses ranging from diabetes mellitus, arthritis ,bronchitis etc, recent studies have also demonstrated significant anticancer and anti gonnorial efficacy of *ocimum santum (*Ramurthy et al., 2019).

One of the recently introduced antimicrobial therapy is the use of metal nanoparticles and they are gaining the interest of researchers due to their distinctive potential which have applications in various field. Amongst many nano-particles, Silver nano-particles are efficient, non-specific antimicrobial agents against planktonic forms of a broad spectrum of bacterial, fungal and viral species. Their antimicrobial activities were attributed to the unique physiochemical characteristics, such as the high ratio of surface area to mass, high reactivity, and nanometer sizes (Charannya et al., 2018).

The impetus for the study is the non-availability of literature about antimicrobial activity of tulsi and metal nano-particles in combination with tulsi extract against periodontal pathogens mainly Tannerella forsythia and Porphyromonas gingivalis.

Chlorhexidine is used as a gold standard against which other antimicrobial agents are compared. Though, many studies have used 0.2% chlorhexidine, very scanty literature is available with use of 0.12% chlorhexidine. Consequently, chlorhexidine is also often used as positive control for assessing the antimicrobial potential of other agents hence an attempt is made to compare the antimicrobial activity of various concentrations of tulsi with 0.12% chlorhexidine aginst Tannerella forsytia and Porphyromonas gingivalis (Mallikarjun et al., 2016).

2. Methodology

Tulsi leaves were obtained from the local market in Bangalore city. Leaves were separated from the stem, washed and dried until they are adequately dry to be grounded. Dried leaves were powdered separately in an electric grinder until a homogenous powder is obtained. Ethanolic extract of the tulsi was prepared using cold extraction method.

Fig. 15.1 Tulsi leaves (ocimum sanctum)

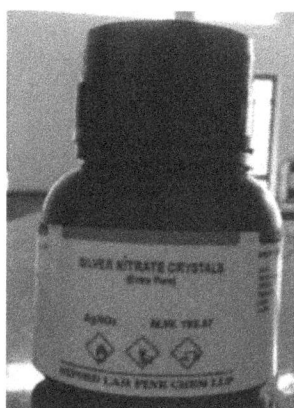

Fig. 15.2 Silver nano particles used for tulsi extract

2.1 Microbiological Assay

Agar well diffusion method was used to determine the antimicrobial activity of tulsi leaves extract in in-vitro method. The diameter of inhibition zone was measured to the nearest whole millimeter by using a vernier calliper. Microbiological procedure was repeated 3 times, and corresponding 3 values of zones of inhibition for each concentration of tulsi extract, tulsi extract with silver nano particles, chlorhexidine and dimethyl formamide was obtained for the bacteria. The values were compared within the group and with positive control (chlorhexidine) and negative control (dimethyl formamide).

3. Results and Discussion

3.1 Microbiological Assessment against the Individual Pathogen

Porphyromonas Gingivalis

On comparison, the mean zone of inhibition for 4%, 8%, 12% concentration of Tulsi extract against P. gingivalis, was statistically significant (p = 0.006). (Table 15.1). On comparison, the mean zone of inhibition for 4%, 8%, 12%

Table 15.1 Comparison of mean zone of inhibition (in mm) for P. Gingivalis b/w different concentrations of Tulsi extract using kruskal wallis test

Conc.	N	Mean	SD	Min	Max	p-value
12% Conc.	3	24.33	3.21	22	28	0.006*
8% Conc.	3	22.67	4.04	18	25	
4% Conc.	3	20.67	1.15	20	22	

concentration of Tulsi extract with silver nano particles was not statistically significant (p = 0.12). (Table 15.2). Comparison of mean zone of Inhibition In between different concentration of Tulsi extract, different concentration of tulsi extract with silver nano particles, Dimethyl formamide and chlorhexidine against P.gingivalis. (Table 15.3). At 12%, 8% and 4% concentrations and also for DMSO and CHX concentration, Tulsi Extract showed significantly higher mean Zone of Inhibition as compared to Tulsi extract with Silver Nano Particle group.

Table 15.2 Comparison of mean zone of inhibition (in mm) for P. Gingivalis b/w different concentrations of tulsi extract with silver nano Particle using kruskal wallis test

Conc.	N	Mean	SD	Min	Max	p-value
12% Conc.	3	11.00	5.20	8	17	0.12
8% Conc.	3	8.33	7.64	0	15	
4% Conc.	3	11.33	4.16	8	16	

Tannerella Forsythia

On comparison of mean zone of inhibition of concentration of 4%, 8%, 12% Tulsi extract against T.forsythia was statistically significant (p = 0.04). (Table 15.4). On comparison of mean zone of inhibition of 4%, 8%,

12% concentration of Tulsi extract with silver nano particles statistically significant (p < 0.001)(Table 15.5). Comparison of mean zone of Inhibition In between different concentration of Tulsi extract, different concentration of tulsi extract with silver nano particles, Dimethyl formamide and chlorhexidine against T.forsythia. (Table 15.6). At 12%, 8% and 4% concentrations and also for DMSO and CHX concentration, Tulsi Extract showed significantly higher mean Zone of Inhibition as compared to Tulsi extract with Silver Nano Particle group.

Table 15.4 Comparison of mean zone of inhibition (in mm) for Tannerella forsythia b/w different concentrations of tulsi extract using kruskal wallis test

Conc.	N	Mean	SD	Min	Max	p-value
12% Conc.	3	26.00	7.21	18	32	0.04*
8% Conc.	3	22.67	8.74	13	30	
4% Conc.	3	21.33	6.51	15	28	

Table 15.5 Comparison of mean zone of inhibition (in mm) for tannerella forsythia b/w different concentrations of tulsi extract with silver nano particle using kruskal wallis test

Conc.	N	Mean	SD	Min	Max	p-value
12% Conc.	3	13.00	2.65	11	16	<0.001*
8% Conc.	3	11.00	1.00	10	12	
4% Conc.	3	9.33	1.15	8	10	

Analysis of Extrinsic Stains of the Teeth using Spectophotometry

The evaluation of external teeth staining of different concentration of ocimum sanctum(tulsi) extract, Tulsi

Table 15.3 Comparison of mean zone of inhibition (in mm) for P. Gingivalis b/w tulsi extract with silver nano particle & tulsi extract at different conc. And DMSO and CHX using Mann whitney test

Conc.	Extract	N	Mean	SD	Mean Diff	p-value
12% Conc.	Tulsi extract with Silver Nano Particle	3	11.00	5.20	-13.33	0.02*
	Tulsi Extract	3	24.33	3.21		
8% Conc.	Tulsi extract with Silver Nano Particle	3	8.33	7.64	-14.33	0.04*
	Tulsi Extract	3	22.67	4.04		
4% Conc.	Tulsi extract with Silver Nano Particle	3	11.33	4.16	-9.33	0.02*
	Tulsi Extract	3	20.67	1.15		
DMSO	Tulsi extract with Silver Nano Particle	3	13.00	1.00	-1.00	0.48
	Tulsi Extract	3	14.00	2.00		
0.12% CHX	Tulsi extract with Silver Nano Particle	3	18.67	1.15	-0.67	0.52
	Tulsi Extract	3	19.33	1.15		

Table 15.6 Comparison of mean zone of inhibition (in mm) for tannerella forsythia b/w silver nano particle & tulasi extract at different conc. using mann whitney test

Conc.	Extract	N	Mean	SD	Mean Diff	p-value
12% Conc.	Tulsi extract with Silver Nano Particle	3	13.00	2.65	-13.00	0.04*
	Tulsi Extract	3	26.00	7.21		
8% Conc.	Tulsi extract with Silver Nano Particle	3	11.00	1.00	-11.67	0.08
	Tulsi Extract	3	22.67	8.74		
4% Conc.	Tulsi extract with Silver Nano Particle	3	9.33	1.15	-12.00	0.04*
	Tulsi Extract	3	21.33	6.51		
DMSO	Tulsi extract with Silver Nano Particle	3	8.67	1.15	-2.00	0.10
	Tulsi Extract	3	10.67	1.15		

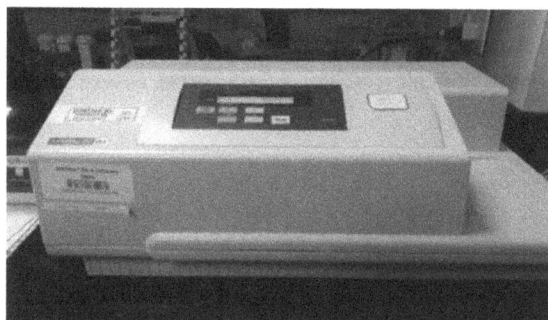

Fig. 15.3 Spectrophotometer

extract with silver metal nano-particles and 0.12% of Chlorhexidine using spectrophotometry showed the mean optical density (OD) for different concentration of tulsi extract with silver nano particles 4%, 8%, 12% is 0.666, 0.510, 0.300 respectively (graph 1). The mean OD values for different concentration of tulsi extract at 4%, 8%, 12% is 0.601, 0.499, 0.278 respectively (graph 2). And the mean OD for 0.12% CHX is 0.389(graph 3). Hence it reveals that, 4% concentration of tulsi extract with silver nano particles had higher staining property compared to other different concentration of tulsi extract and tulsi extract with silver nano particles and 0.12% Chlorhexidine.

In this study attempt was made to evaluate the antimicrobial efficacy of tulsi extract and Tulsi extract with silver nano particles against two periodontal pathogens, P. gingivalis and T. forsythia. Results in this invitro study showed that Tulsi extract at concentration of 4%, 8%, 12% can effectively inhibit the growth of P.gingivalis which is similar to the study done by Mallikarjun et al (Mallikarjun et al.,2016). 12% concentration had higher zone of inhibition in our study. In the study done by Ipsitha et al, 8% concentration of tulsi extract showed maximum zone of inhibition against P. gingivalis followed by the 4% concentration, which was similar to this study. The results were also similar to the studies done by Rathod et al, Shah et al (Shah et al., 2014), and Prasannabalaji et al (Jayanthi et al., 2018). This shows that higher the concentration of tulsi extract more the antimicrobial efficacy.

In present study, on comparison between the different concentration of tulsi extract and tulsi extract with silver nano particles, 12% of tulsi extract showed the maximum zone of inhibition against both P.gingivalis and T. forsythia. The study done by, Sirisha.P et al concluded that silver nanoparticles synthesized with Ocimum extract has superior sensitivity against P.gingivalis followed by ocimum sanctum extract (Sirisha et al., 2017).

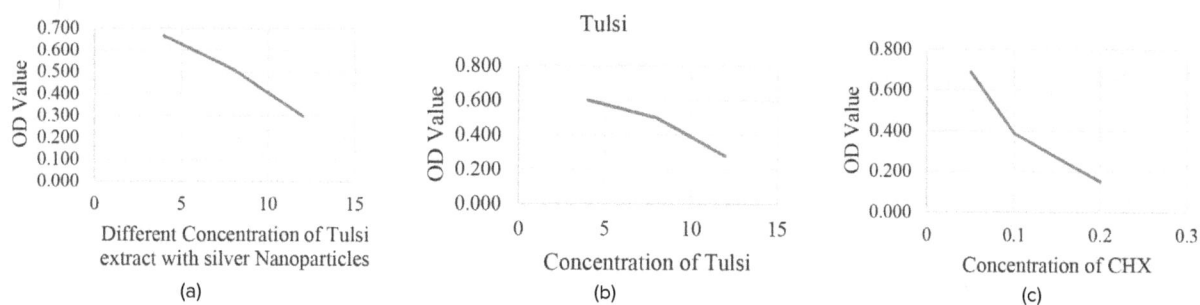

Fig. 15.4 Graph depicting (a) Different concentration of tulsi extract with silver nanoparticles. (b) Concentration of tulsi. (c) Concentration of CHX

This study showed that Tulsi extract and tulsi extract with silver nano particles at concentration of 4%, 8%, 12% can effectively inhibit the growth of T. forsythia. Comparision with previous studies are not justified here due to variation in the organisms tested against tulsi extract and tulsi extract with silver nano particles for its antimicrobial efficacy. Since there were scarce literature available that could depict the efficacy of Tulsi against periodontal microbes specifically, the present study encourages researchers to carry out further studies assessing toxicity, durability and other assessments followed by clinical trials to provide an insight into the activity of Tulsi against T. forsythia.

In the present study, DMSO depicted least zone of inhibition against both P.gingivalis and T. forsythia which was similar to the study done by Mallikarjun et al where DMSO showed least ZOI. On comparision of DMSO with different concentration of tulsi exract and tulsi extract with silver nano particles there was no statistical significance in the zone of inhibition.

0.12% CHX, also depicted least ZOI. In the study done by Eshwar .P et al (Eshwar et al., 2016), 0.2% chlorhexidine was found to be more effective against Actinobacillus actinomyctemcomitans when compared to Tulsi extract, though the microorganism tested in our study was different (Sirisha et al., 2017). In the study done by Ipsita Jayanti et al, 0.2% chlorhexidine was less effective for P. gingivalis, in comparison with the 8% tulsi extract, which was similar to the present study, where 0.12% CHX was less effective compared to the tulsi extract and tulsi extract with silver nano particles (Jayanti et al., 2018).

To evaluate the external staining of the teeth, Spectrophotometry was used for tulsi extract, tulsi extract with silver nano particles and 0.12% CHX. 12% tulsi extract had the least external staining of the teeth compared to other different concentration of tulsi and tulsi extract with silver nano particles and 0.12% CHX. The study done by, Mukhatar Ahmed Javali et al[11] concluded that mouthwashes containing TiO2 nanoparticles produced greater enamel discoloration compared to those of other antiseptic mouthwashes and dietary juices used in the study which is similar to the present study where, 4% concentration of tulsi extract with silver nano particles showed the highest staining (Eshwar et al., 2016).

4. Conclusion

It was concluded that 12% concentration of O. sanctum (tulsi) extract showed the maximal antimicrobial activity against T.forsythia and P.gingivalis. Both Tulsi extract and tulsi extract with silver nano particles at different concentrations showed antimicrobial activity against P. gingivalis and T.forsythia. The maximum Zone of inhibition was seen in tulsi extract compared to tulsi extract with silver nano particles, DMSO and CHX against both the pathogens. This study concluded that there was no significant change in the antimicrobial activity in using silver nano particles with the tulsi extract. The external staining of teeth was seen maximum at 4% tulsi extract with silver nano particles and the least was seen at 12% of tulsi extract. To conclude, Tulsi extract and tulsi extract with silver nano particles can to be used as an adjunct to mechanical therapy in the prevention and treatment of periodontal diseases at higher concentrations but further studies should be conducted for its clear implication.

Acknowledgments

The authors thanks to Bangalore Institute of Dental Sciences, Bangalore for supporting to necessary resources and analysis

References

1. Agarwal, P. and Nagesh, L., 2011. Comparative evaluation of efficacy of 0.2% Chlorhexidine, Listerine and Tulsi extract mouth rinses on salivary Streptococcus mutans count of high school children—RCT. Contemporary clinical trials, 32(6), pp.802–808.
2. Charannya, S., Duraivel, D., Padminee, K., Poorni, S., Nishanthine, C. and Srinivasan, M.R., 2018. Comparative Evaluation of Antimicrobial Efficacy of Silver Nanoparticles and 2% Chlorhexidine Gluconate When Used Alone and in Combination Assessed Using Agar Diffusion Method: An: In vitro: Study. Contemporary clinical dentistry, 9(Suppl 2), pp.S204–S209.
3. Eswar, P., Devaraj, C.G. and Agarwal, P., 2016. Antimicrobial activity of Tulsi {Ocimum sanctum (Linn.)} extract on a periodontal pathogen in human dental plaque: an invitro study. Journal of clinical and diagnostic research: JCDR, 10(3), p.ZC53.
4. Gupta, D., Bhaskar, D.J., Gupta, R.K., Karim, B., Jain, A., Singh, R. and Karim, W., 2014. A randomized controlled clinical trial of Ocimum sanctum and chlorhexidine mouthwash on dental plaque and gingival inflammation. Journal of Ayurveda and integrative medicine, 5(2), p.109.
5. Javali, M.A., Abdul Khader, M., Alqahtani, R.M., Almufarrij, M.J., Alqahtani, T.M. and Addas, M.K., 2020. Spectrophotometric analysis of dental enamel staining to antiseptic and dietary agents: In vitro study. International Journal of Dentistry, 2020(1), p.5429725.
6. Jayanti, I., Jalaluddin, M., Avijeeta, A., Ramanna, P.K., Rai, P.M. and Nair, R.A., 2018. In vitro Antimicrobial Activity of Ocimum sanctum (Tulsi) Extract on Aggregatibacter actinomycetemcomitans and Porphyromonas gingivalis. The journal of contemporary dental practice, 19(4), pp. 415–419.

7. Jayanti, I., Jalaluddin, M., Avijeeta, A., Ramanna, P.K., Rai, P.M. and Nair, R.A., 2018. In vitro Antimicrobial Activity of Ocimum sanctum (Tulsi) Extract on Aggregatibacter actinomycetemcomitans and Porphyromonas gingivalis. The journal of contemporary dental practice, 19(4), pp. 415–419.

8. Mallikarjun, S., Rao, A., Rajesh, G., Shenoy, R. and Pai, M., 2016. Antimicrobial efficacy of Tulsi leaf (Ocimum sanctum) extract on periodontal pathogens: An: in vitro: study. Journal of Indian Society of Periodontology, 20(2), pp.145–150.

9. Pai, K.R., Pallavi, L.K., Bhat, S.S. and Hegde, S.K., 2022. Evaluation of Antimicrobial Activity of Aqueous Extract of "Ocimum Sanctum-Queen of Herb" on Dental Caries Microorganisms: An In Vitro Study. International Journal of Clinical Pediatric Dentistry, 15(Suppl 2), p.S176.

10. Ramamurthy, J. and Jayakumar, N.D., 2019. Ocimum sanctum and its effect on oral health–A comprehensive review. Drug Invention Today, 11(4).

11. Shah, S., Trivedi, B., Patel, J., Dave, J.H., Sathvara, N. and Shah, V., 2014. Evaluation And Comparison Of Antimicrobial Activity Of Tulsi (Ocimum Sanctum) Neem (Azadirachta Indica) And Triphala Extract Against Streptococcus Mutans&Lactobacillus Acidophilus: An In Vitro Study. National Journal of Integrated Research in Medicine, 5(4).

12. Sirisha, P., Gayathri, G.V., Dhoom, S.M. and Amulya, K.S., 2017. Antimicrobial effect of silver nanoparticles synthesised with Ocimum sanctum leaf extract on periodontal pathogens. J Oral Health Dent Sci, 1, pp.1–7.

Note: All the figures and tables in this chapter were made by the author.

Advances in Materials Science and Technology – Dr. Srikari Srinivasan et al. (eds)
© 2025 Taylor & Francis Group, London, ISBN 978-1-041-12342-2

16

Demineralized Tooth as Bone Graft Material: A Systematic Review

Harshitha Gowda Bangalore Hanumanthe,
Pooja Gavanivari*, Ravishankar Krishna
Department of Prosthodontics and Crown & Bridge,
Faculty of Dental Sciences, M S Ramaiah University of Applied Sciences,
Bangalore, Karnataka, India

Srikari Srinivasan
Department of Automotive and Aerospace,
Faculty of Engineering, M S Ramaiah University of Applied Sciences,
Bangalore, Karnataka, India

ABSTRACT: Demineralized tooth has been proposed as an alternative to bone graft materials due to its biocompatibility and similarity in composition to alveolar bone. It contains bone morphogenic proteins (BMPs) and growth factors (GFs) that assist in new bone formation. This systematic review evaluates its effectiveness as a bone graft material for socket preservation and sinus lift procedures. A comprehensive search of PubMed and Google Scholar (January 2017–September 2023) identified 195 records, of which 19 met the inclusion criteria (RCTs, meta-analyses, systematic reviews). Case reports, retrospective studies, and non-peer-reviewed articles were excluded. The selected studies demonstrated that demineralized tooth materials have comparable osteogenic potential similar to other autografts and xenografts. However, there were significant heterogeneity in study designs and outcome measures. To conclude, demineralized tooth in different forms can be used as an alternative graft material in dentistry because of its ability for new bone formation.

KEYWORDS: Demineralized dentin matrix, Demineralized tooth, Guided bone regeneration

1. Introduction

Bone graft materials play a crucial role in reconstructive surgery and dentistry due to their reliability in repairing bony defects, promoting bone formation, and ensuring osseointegration. Among various options, demineralized tooth materials have gained attention over the past two decades for their biocompatibility and osteogenic properties [1,2]. Demineralized teeth serve as a valuable autogenous bone graft source, containing BMPs, GFs, calcium, and phosphorus, all aiding bone regeneration [3,4]. However, comprehensive analysis of its clinical application remains limited. Previous reviews have focused on isolated properties or lacked systematic evaluation of clinical outcomes. Variability in study designs, sample sizes, and methodologies has further led to inconsistencies in assessing its effectiveness. This review aims to bridge these gaps by providing an updated and thorough evaluation.

The need for biocompatible and cost-effective bone grafts underscores this review's significance. Conventional autografts pose risks like donor site morbidity and disease transmission, allografts lack osteogenic cells, synthetic grafts are costly and lack osteoinductivity, and xenografts

*Corresponding author: gavanivari.pooja@gmail.com

DOI: 10.1201/9781003664277-16

raise ethical concerns [5]. Demineralized tooth material offers a promising alternative. By systematically analyzing existing evidence, this review seeks to enhance clinical guidelines and improve patient outcomes.

Most systematic reviews focus on traditional graft materials, often overlooking demineralized tooth material [4]. Updating the knowledge base and assessing its unique properties is essential. This review provides a comprehensive understanding of its applications, efficacy, and influencing factors such as patient conditions, surgical techniques, and post-operative care in bone grafting procedures.

2. Methodology

2.1 Objectives

To evaluate the benefits of using demineralized tooth material as a bone graft substitute and assess its effectiveness in promoting bone regeneration and integration compared to traditional bone graft materials.

2.2 Research Questions

The following focus questions were developed according to the problem, intervention, comparison, and outcome (PICOS) design.

1. What are the clinical outcomes of using demineralized tooth as a bone graft material in terms of bone regeneration and integration?

2. How does the demineralized tooth material behave compare to other commonly used bone graft materials in terms of osteoinductive and osteoconductive properties?

3. What are the potential risks or complications associated with the use of demineralized tooth material as a bone graft?

4. What are the effects of incorporating Platelet-Rich Fibrin (PRF) and Injectable Platelet-Rich Fibrin (I-PRF) with demineralized tooth material on bone healing and regeneration outcomes?

2.3 Literature Search Strategy

Following PRISMA guidelines electronic databases; PubMed and Google scholar were searched in order to locate articles concerning the use of demineralized tooth for socket preservation. The Keywords used for the search included: ("demineralized tooth") AND (("bone regeneration") OR ("demineralized dentine matrix") OR ("demineralized tooth matrix") OR ("autologous tooth")) AND ("platelet rich fibrin") OR ("injectable platelet rich fibrin") OR ("PRF") OR ("i-PRF"). The search was

restricted to English language only. Articles published from January 2017 to September 2023 were searched.

2.4 Selection of Studies

Only one researcher independently searched and screened each record's title and abstract to assess eligibility based on the inclusion criteria. Full-text articles were then retrieved and assessed for inclusion. Data extraction was done manually using a standardized data extraction. Any discrepancies or uncertainties during the data extraction process were resolved through discussion with a second reviewer. In cases where specific data were not reported, it was assumed that the study did not measure or report those variables.

2.5 Inclusion Criteria

Randomized controlled trials, systematic reviews and articles published from January 2017 to September 2023.

2.6 Exclusion Criteria

Case reports, case series, studies with no biomaterials used for socket preservation at-least in one group, studies with no statistical data.

2.7 Information Sources

PubMed: June 2017 to September 2023 and Google Scholar: last searched on September 2023

2.8 Sequential Search Strategy

During initial literature search, all articles were screened and excluded based on titles and abstracts. The following stage of screening involved reading full text articles to evaluate and confirm study's eligibility based on selected inclusion and exclusion criteria.

2.9 Data Extraction

The data were independently extracted from studies in the form as variables, according to the aims and themes of present review as follows; Study design (RCT or systematic review), population characteristics (age, sex, health status), intervention details (type of bone graft material, use of PRF or i-PRF), outcomes measured (bone regeneration, graft integration), follow-up duration, results for primary and secondary outcomes, study quality indicators (risk of bias, funding sources).

2.10 Data Items

Data were collected from selected articles and arranged in the following fields: "Year" - describes the date of publication, "Type of study" - indicates the type of study, "Socket preservation technique" - described what biomaterial was used for socket preservation and "Results" – described the results of the study done.

3. Results and Discussion

3.1 Study Selection

Article review and data extraction were performed according to the PRISMA flow diagram as shown in Fig. 16.1. An electronic literature search was performed on the PubMed and google scholar databases. A total of 195 search results were filtered. After inclusion and exclusion criteria were applied, a number of 19 articles were selected. Finally, 19 full text articles were included in this study.

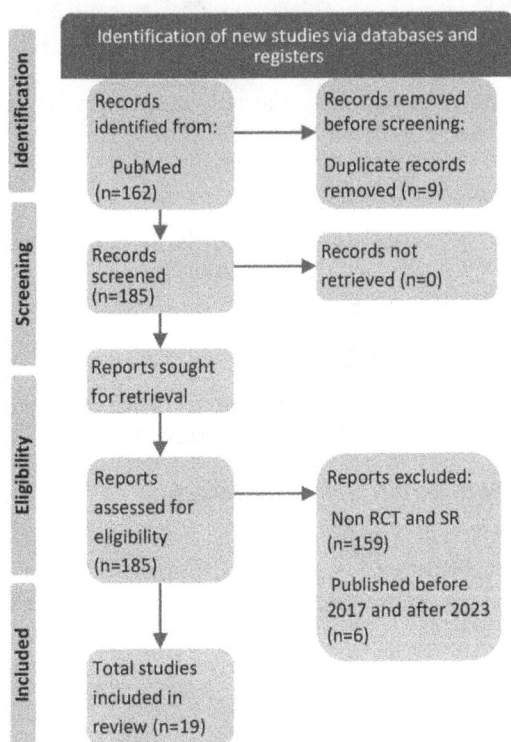

Fig. 16.1 PRISMA flow diagram

Source: Author

3.2 Study Characteristics

19 articles were included in systemic literature review out of which 9 were randomized controlled trials and 10 were systematic reviews. The summarized included studies characteristics are presented in Table 16.1 and Table 16.2.

Demineralized tooth material (DTM) has shown significant promise in clinical applications for bone regeneration and repair, particularly in the context of alveolar ridge preservation (ARP) and augmentation following tooth extraction. In a study done to check the rate of bone loss, the use of DTM demonstrated considerable efficacy in reducing alveolar bone loss compared to untreated

Table 16.1 List of selected systematic reviews and their summary

Year	Research question
2018 [2]	In patients exhibiting alveolar ridge deficiencies and being in need of an implant retained restoration, what is the efficacy of reconstructive procedures employing autogenous teeth (AT) on changes in ridge dimensions compared with control measures?
2019 [13]	What biomaterials are used for socket preservation after the tooth extraction and which of those show the best results regarding alveolar dimensional changes and quality of newly formed bone?
2021 [14]	Not mentioned
2022 [4]	Not mentioned
2022 [15]	Are there any differences in implant treatment outcome following lateral alveolar ridge augmentation with an autogenous demineralized tooth block graft compared with autogenous bone block graft?
2022 [16]	In patients scheduled for restoration with dental implants after dental extraction(s), is the use of APD appropriate for ARP, in comparison with spontaneous healing or the use of other bone substitutes, in order to maintain bone volume and so avoid structural and compositional changes in the overlying soft tissue?
2022 [17]	Not mentioned
2023 [18]	Not mentioned
2023 [19]	Based on histology and cone-beam computed tomography (CBCT) scan results, what are the effects of different graft materials used in socket preservation in medically fit individuals with an indication for teeth extraction?
2023 [20]	Not mentioned

extraction sites. The structured scaffold provided by DTM facilitated the formation of vascularization leading to effective mineralization and bone remodeling [1]. This process is essential for maintaining the structural integrity of the alveolar ridge, which is critical for subsequent dental implant placement.

Similarly, another study, highlighted the biological benefits of DTM in ARP. The inherent growth factors and collagen present in DTM promote osteogenesis and healing. The study found that by the incorporation of DTM in ARP, there was significant reduction in horizontal and vertical bone loss post-extraction, thereby, supporting its role as an effective biomaterial for maintaining alveolar ridge dimensions [1]. Furthermore, it was noted that DTM's ability to eliminate risks associated with cross-infection and immunogenicity makes it a safer alternative to allografts and xenografts, which often carry these risks [2].

Table 16.2 List of randomized controlled trials and their summary

Year	Socket preservation technique	Results
2016 [6]	*Group 1:* Autogenous tooth graft (ATG) (n=5) *Group 2:* Beta-tricalcium phosphate (β-TCP) alloplast (n=5) *Group 3:* Ungrafted socket (n=5)	ATG has shown more promising results as compared to β-TCP in achieving minimum volumetric alveolar bone loss when it is grafted immediately in post-extraction socket.
2017 [3]	*Group 1:* Autogenous tooth graft material (AutoBT) (n=21) *Group 2:* Anorganic bovine bone (Bio-Oss) (n=12)	AutoBT grafted to extraction sockets for the vertical augmentation of alveolar defect exhibited minimal tissue response and bone regenerative potential similar to Bio-Oss.
2018 [7]	*Group 1:* Immediate implantation + GBR with autogenous DDM granules from the extracted tooth (DDM group) (n=23) *Group 2:* Immediate implantation + GBR with Bio-Oss granules (Bio group) (n=22)	Similar clinical and radiographic performance of DDM and traditional osseous powder in immediate placement of implants in periodontal post extraction sites
2018 [8]	*Group 1:* Demineralised dentine matrix (DDM graft) (n = 10) *Group 2:* DDM graft combined with recombinant human bone morphogenetic protein-2 (rhBMP-2) (rhBMP-2/DDM) (n = 10) *Group 3:* deproteinized Bovine bone with collagen (DBBC - Bio-Oss)(n = 10)	The application of rhBMP-2/DDM into an extraction socket results in higher new bone formation compared with the use of DDM alone or DBBC at 4 months after ridge
2018 [9]	*Group 1:* Autogenous demineralised tooth roots (TR) (n=15) *Group 2:* Cortical autogenous bone blocks harvested from the retromolar area (AB) (n=15)	TR grafts were associated with significantly higher crestal ridge width values than AB grafts
2019 [10]	*Group 1:* Autologous demineralized tooth matrix (aDTM) in combination with platelet-rich fibrin (PRF) membrane (aDTM/PRF) (n=20) *Group 2:* PRF membrane alone (n=20)	Application of aDTM with PRF membrane is useful for alveolar ridge preservation by reducing horizontal alveolar ridge collapse and promoting bone healing
2021 [1]	*Group 1:* Autogenous whole tooth (AWTG) (n=10) *Group 2:* Autogenous demineralized dentin graft (ADDG) (n=10)	AWTG or ADDG employed in ARP is equally effective at reducing dimensional losses after 6 months, with ADDG seeming to exert higher osteoinductive properties.
2022 [11]	*Group 1:* Autogenous demineralised tooth roots (TRS) (n = 14) *Group 2: Autogenous bone blocks (ABS) (n = 14)I*	TR grafts were associated with a significantly higher ridge width gain

DTM's osteoinductive and osteoconductive properties distinguish it from other commonly used bone graft materials. The growth factors embedded in DTM, such as TGF-β and BMP-2, play a crucial role in promoting mesenchymal cell differentiation and new bone matrix formation [8]. This intrinsic property enhances the osteogenic potential of DTM, making it a superior choice of material compared to synthetic grafts that lack bioactivity [3].

Additionally, a study done in 2011, discussed the osteoconductive nature of DTM, which provides an ideal matrix for new bone growth, unlike synthetic grafts, which may not integrate as effectively with natural bone. DTM's organic composition facilitates better cellular adhesion and proliferation, leading to improved osseointegration and biomechanical stability, which is vital for the long-term success of dental implants [7].

Despite its advantages, the use of DTM is not without potential risks. It includes risks related to variable graft resorption rates and the possibility of inflammatory responses. These issues can affect the long-term stability of the graft and patient outcomes [10]. It is emphasized,

the importance of patient specific considerations and meticulous surgical techniques to minimize these risks. They recommended thorough monitoring and follow-up to ensure successful bone healing and graft integration. Moreover, the chairside preparation of DTM, involved cleaning, grinding, and disinfection, which requires significant time and effort; and only non-carious tooth can be used as an autograft. These limitations underscore the need for alternative sources or synthetic analogs that can mimic the properties of DTM while addressing these practical challenges [21].

The incorporation of Platelet-Rich Fibrin (PRF) with DTM has been shown to enhance bone healing and regeneration outcomes significantly. PRF provides a rich source of growth factors, which when used in conjunction with DTM, provides a favorable microenvironment for bone formation and graft integration [10,13]. The combination of PRF with DTM in ARP procedures resulted in superior bone regeneration outcomes compared to DTM alone. It was found that the synergistic effects of PRF's growth factors and DTM's osteoinductive properties led to enhanced new bone formation and healing [13]. Similarly, it was reported that i-PRF, when combined with other autografts and allografts, improved the quality and quantity of new bone formed, highlighting the potential of this combination to optimize clinical results in bone grafting procedures [10].

Overall, the evidences suggests that DTM alone or when used with platelet concentrates, offers significant benefits in bone regeneration. Its osteoinductive and osteoconductive properties, combined with the biological enhancement, makes it a highly effective material for dental applications. Future research should focus on standardizing protocols and elucidating the optimal clinical applications of DTM in diverse forms of bone grafting scenarios.

4. Conclusion

The demineralized tooth material represents a valuable innovation as a bone graft substitute, offering significant advantages in terms of biological compatibility, osteoinductive properties, and clinical efficacy. The evidence synthesized from the reviewed articles supports DTM's role in promoting bone regeneration and integrating seamlessly with host tissues. DTM's unique biological composition, enriched with growth factors and collagen, enhances its osteogenic potential and distinguishes it from conventional graft materials. Furthermore, the adjunctive use of platelet concentrates with DTM demonstrates synergistic benefits in enhancing bone healing outcomes.

Acknowledgments

The authors acknowledge to M S Ramaiah University of Applied Sciences, Bangalore, Karnataka, India for providing resources and testing facilities.

References

1. Elfana, Ahmed, Samar El-Kholy, Heba Ahmed Saleh, and Karim Fawzy El-Sayed. "Alveolar ridge preservation using autogenous whole-tooth versus demineralized dentin grafts: A randomized controlled clinical trial." *Clinical Oral Implants Research* 32, no. 5 (2021): 539–548.
2. Ramanauskaite, Ausra, D. Sahin, R. Sader, J. Becker, and F. Schwarz. "Efficacy of autogenous teeth for the reconstruction of alveolar ridge deficiencies: A systematic review." *Clinical Oral Investigations* 23 (2019): 4263–4287.
3. Pang, Kang-Mi, In-Woong Um, Young-Kyun Kim, Jae-Man Woo, Soung-Min Kim, and Jong-Ho Lee. "Autogenous demineralized dentin matrix from extracted tooth for the augmentation of alveolar bone defect: a prospective randomized clinical trial in comparison with anorganic bovine bone." *Clinical oral implants research* 28, no. 7 (2017): 809–815.
4. Li, Yanfei, Wanhang Zhou, Peiyi Li, Qipei Luo, Anqi Li, and Xinchun Zhang. "Comparison of the osteogenic effectiveness of an autogenous demineralised dentin matrix and Bio-Oss® in bone augmentation: a systematic review and meta-analysis." *British Journal of Oral and Maxillofacial Surgery* 60, no. 7 (2022): 868–876.
5. Zhao, Rusin, Ruijia Yang, Paul R. Cooper, Zohaib Khurshid, Amin Shavandi, and Jithendra Ratnayake. "Bone grafts and substitutes in dentistry: A review of current trends and developments." *Molecules* 26, no. 10 (2021): 3007.
6. Joshi, Chaitanya Pradeep, Nitin Hemchandra Dani, and Smita Uday Khedkar. "Alveolar ridge preservation using autogenous tooth graft versus beta-tricalcium phosphate alloplast: A randomized, controlled, prospective, clinical pilot study." *Journal of Indian Society of Periodontology* 20, no. 4 (2016): 429–434.
7. Li, Peng, HuiCong Zhu, and DaHong Huang. "Autogenous DDM versus Bio-Oss granules in GBR for immediate implantation in periodontal postextraction sites: A prospective clinical study." *Clinical implant dentistry and related research* 20, no. 6 (2018): 923–928.
8. Jung, Gyu-Un, Tae-Hyun Jeon, Mong-Hun Kang, In-Woong Um, In-Seok Song, Jae-Jun Ryu, and Sang-Ho Jun. "Volumetric, radiographic, and histologic analyses of demineralized dentin matrix combined with recombinant human bone morphogenetic protein-2 for ridge preservation: a prospective randomized controlled trial in comparison with xenograft." *Applied Sciences* 8, no. 8 (2018): 1288.
9. Schwarz, Frank, Didem Hazar, Kathrin Becker, Robert Sader, and Jürgen Becker. "Efficacy of autogenous tooth roots for lateral alveolar ridge augmentation and staged implant placement. A prospective controlled clinical study." *Journal of clinical periodontology* 45, no. 8 (2018): 996–1004.
10. Ouyyamwongs, Warisara, Narit Leepong, and Srisurang Suttapreyasri. "Alveolar ridge preservation using autologous demineralized tooth matrix and platelet-rich fibrin versus platelet-rich fibrin alone: A split-mouth randomized controlled clinical trial." *Implant Dentistry* 28, no. 5 (2019): 455–462.
11. Schwarz, Frank, Karina Obreja, Stephanie Mayer, Ausra Ramanauskaite, Robert Sader, and Puria Parvini. "Efficacy of autogenous tooth roots for a combined vertical and horizontal alveolar ridge augmentation and staged implant placement. A prospective controlled clinical study." *Journal of clinical periodontology* 49, no. 5 (2022): 496–505.
12. Oguić, Matko, Marija Čandrlić, Matej Tomas, Bruno Vidaković, Marko Blašković, Ana Terezija Jerbić Radetić, Sanja Zoričić Cvek, Davor Kuiš, and Olga Cvijanović Peloza. "Osteogenic potential of autologous dentin graft compared with bovine xenograft mixed with autologous bone in the esthetic zone: radiographic, histologic and immunohistochemical evaluation." *International journal of molecular sciences* 24, no. 7 (2023): 6440.
13. Stumbras, Arturas, Povilas Kuliesius, Gintaras Januzis, and Gintaras Juodzbalys. "Alveolar ridge preservation after tooth extraction using different bone graft materials and autologous platelet concentrates: a systematic review." *Journal of oral & maxillofacial research* 10, no. 1 (2019).

14. Hazballa, D., A. D. Inchingolo, A. M. Inchingolo, G. Malcangi, L. Santacroce, E. Minetti, D. Di Venere et al. "The effectiveness of autologous demineralized tooth graft for the bone ridge preservation: A systematic review of the literature." *J. Biol. Regul. Homeost. Agents* 35, no. Suppl. S1) (2021): 283–294.

15. Starch-Jensen, Thomas, Julie Vitenson, Daniel Deluiz, Kimie Bols Østergaard, and Eduardo Muniz Barretto Tinoco. "Lateral alveolar ridge augmentation with autogenous tooth block graft compared with autogenous bone block graft: a systematic review." *Journal of Oral & Maxillofacial Research* 13, no. 1 (2022).

16. Sánchez-Labrador, Luis, Santiago Bazal-Bonelli, Fabian Pérez-González, Luis Miguel Sáez-Alcaide, Jorge Cortés-Bretón Brinkmann, and José María Martínez-González. "Autogenous particulated dentin for alveolar ridge preservation. A systematic review." *Annals of Anatomy-Anatomischer Anzeiger* 246 (2023): 152024.

17. Noronha Oliveira, Miguel, Hugo A. Varela, João Caramês, Filipe Silva, Bruno Henriques, Wim Teughels, Marc Quirynen, and Júlio CM Souza. "Synergistic benefits on combining injectable platelet-rich fibrin and bone graft porous particulate materials." *Biomedical Materials & Devices* 1, no. 1 (2023): 426–442.

18. Solyom, Eleonora, Eszter Szalai, Márk László Czumbel, Bence Szabo, Szilárd Váncsa, Krisztina Mikulas, Zsombor Radoczy-Drajko et al. "The use of autogenous tooth bone graft is an efficient method of alveolar ridge preservation–meta-analysis and systematic review." *BMC Oral Health* 23, no. 1 (2023): 226.

19. Madi, Marwa, Ibrahim Almindil, Maria Alrassasi, Doha Alramadan, Osama Zakaria, and Adel S. Alagl. "Cone-beam computed tomography and histological findings for socket preservation techniques using different grafting materials: a systematic review." *Journal of Functional Biomaterials* 14, no. 5 (2023): 282.

20. Inchingolo, Angelo Michele, Assunta Patano, Chiara Di Pede, Alessio Danilo Inchingolo, Giulia Palmieri, Elisabetta de Ruvo, Merigrazia Campanelli et al. "Autologous Tooth Graft: Innovative Biomaterial for Bone Regeneration. Tooth Transformer® and the Role of Microbiota in Regenerative Dentistry. A Systematic Review." *Journal of Functional Biomaterials* 14, no. 3 (2023): 132.

21. Nelson, Aaron C., and Brian L. Mealey. "A randomized controlled trial on the impact of healing time on wound healing following ridge preservation using a 70%/30% combination of mineralized and demineralized freeze-dried bone allograft." *Journal of periodontology* 91, no. 10 (2020): 1256–1263.

Advances in Materials Science and Technology – Dr. Srikari Srinivasan et al. (eds)
© 2025 Taylor & Francis Group, London, ISBN 978-1-041-12342-2

17

Assessment of Proliferative Effect of Human Adult Peripheral and Umbilical Cord Blood Platelet-Rich Fibrin Enhanced with Gold Nanoparticles on Human Dental Pulp Stem Cells: An In-Vitro Study

Sophia Saud[1], Indiresha Narayana
Department of Conservative Dentistry and Endodontics,
Faculty of Dental Sciences, M S Ramaiah University of Applied Sciences,
Bangalore, Karnataka, India

Ashmitha Kishan Shetty[2]
Department of Pediatric and Preventive Dentistry,
Faculty of Dental Sciences, M S Ramaiah University of Applied Sciences,
Bangalore, Karnataka, India

Sowmya S. V., Dominic Augustine
Department of Oral & Maxillofacial Pathology and Oral Microbiology,
Faculty of Dental Sciences, M S Ramaiah University of Applied Sciences,
Bangalore, Karnataka, India

ABSTRACT: Platelet Rich Factor (PRF) scaffolds containing growing factors can help stimulate tissue development and revascularize non-vital young permanent teeth suggesting when PRF extraction requirements are met, the texture of the PRF membrane appears to be more robust than the collagen membrane. Gold nanoparticles (AuNPs) and other biomaterials are forming a new therapeutic paradigm, offering great promise for applications in regenerative medicine and bone tissue engineering. In the present study adult peripheral PRF (pPRF) and cord blood PRF (cbPRF) was prepared via centrifugation at 1500 rpm for 14 minutes and doped with 3 different concentrations (3,6 and 9μl) of AuNPs by centrifuged at 1000 rpm for 1 minute followed by MTT assay to evaluate the proliferative effect on human Dental Pulp Stem Cells (hDPSCs) with 5 different concentrations (6.25,12.5,25,50 and 100 μl) of the prepared combination. Results showed the addition of AuNPs to pPRF and cbPRF offers proliferative effect on human dental pulp stem cells (hDPSc). The proliferative effect was maximum and comparable with pPRF and cbPRF when 3 μl of AuNPs added. The present study signifies AuNPs could be useful in delivering potential proteins of PRF suggesting nano bioconjugation and could be used for future regenerative endodontic therapies.

KEYWORD: Umbilical cord blood, Platelet-rich fibrin, Gold nanoparticles, Human dental pulp stem

Corresponding author: [1]sophiasaud5@gmail.com, [2]ashmitha.pe.ds@msruas.ac.in

DOI: 10.1201/9781003664277-17

1. Introduction

Pulp revascularization leverages the remaining pulp tissue, stem cells at the tooth apex, and stem cells in the periodontal area to differentiate and form living tissue rich in blood vessels and connective elements (Jung et al.,2019; Amrollahi et al.,2016; de Souza et al.,2017). A recent trend in regenerative endodontics involves using autologous platelet concentrates as scaffolds, showing promising clinical and radiographic results. Platelet-Rich Plasma (PRP) and Platelet-Rich Fibrin (PRF) are two concentrated platelet sources currently in use, containing a dense suspension of growth factors that promote wound healing when applied locally (Arshad et al.,2021; Paul et al.,2020).

PRF, a second-generation platelet concentrate offers significant advantages over PRP in regenerative endodontics6. PRF is produced by the homogenization of plasma, initiated by the formation of a fibrin clot during the centrifugation of human-derived blood. This clot contains abundant cytokines, growth factors, and platelets from a polymerization reaction, without requiring foreign enzymes or anticoagulants (Khiste and Naik, 2013). Umbilical cord blood, derived from the umbilical cord and placenta vessels, contains various stem cell populations with distinct characteristics. Initially used primarily for hematopoietic stem cell transplantation in haematology, its application was later focused on blood-related disorders (Khademhosseini et al.,2020; Devi et al.,2023).

Gold nanoparticles (AuNPs) are a promising tool in nanomedicine, widely used as therapeutic agents, drug delivery systems, and for photothermal therapy, diagnosis, and imaging. Their nanoscale size aligns with biological compounds, and they offer easy preparation, high surface area, and (Vial et al.,2017). AuNPs promote osteogenic differentiation and bone mineralization, significantly improving osteoblastic differentiation and bone formation around dental implants. The size and shape of AuNPs are crucial for differentiating Human Periodontal Ligament Stem Cells (hPDLSCs) into osteoblasts10. Adding nanoparticles (NPs) to PRF formulations enhances regenerative potential (Khorshidi et al., 2018). In the present study we have hypothesized that the addition of AuNPs to PRF obtained from Human Adult Peripheral and Umbilical Cord Blood will enhance the proliferative properties of Human dental pulp stem cells.

2. Methodology

2.1 Preparation of pPRF

The ethical clearance was obtained from the institutional committee to obtain sample. 5-6 ml of intravenous blood was drawn from the cubital fossa of the healthy adult volunteers and collected in a 10-ml sterile glass tube without anticoagulant and immediately centrifuged at 3000 rpm for 10 minutes. Blood centrifugation results in separation of blood into a structured fibrin clot in the middle of the tube, just between the red corpuscles at the bottom and acellular plasma or Platelet-Poor Plasma (PPP) at the top. After removal of PPP, PRF can be easily separated from red corpuscles base, preserving a small red blood cell (RBC) layer using sterile tweezers and scissors (Fig. 17.1).

2.2 Preparation of cbPRF

Six young and healthy pregnant women scheduled for Caesarean sections, were chosen for this procedure. After childbirth, the umbilical cord was secured using two clamps. Using a sterile syringe, 5-6 ml of cord blood was drawn from umbilical cord. This blood was then transferred into separate vacutainers that did not contain any anticoagulants. The absence of anticoagulants is crucial for the subsequent formation of a fibrin matrix. The collected blood samples were centrifuged at a speed of 3000 rpm for 10 minutes. This process allows the blood components to separate based on their densities. The supernatant (plasma) is carefully removed and discarded. The middle segment, which is rich in platelets and leukocytes and forms a fibrin matrix, is extracted. placed into a sterile tube (Fig. 17.1).

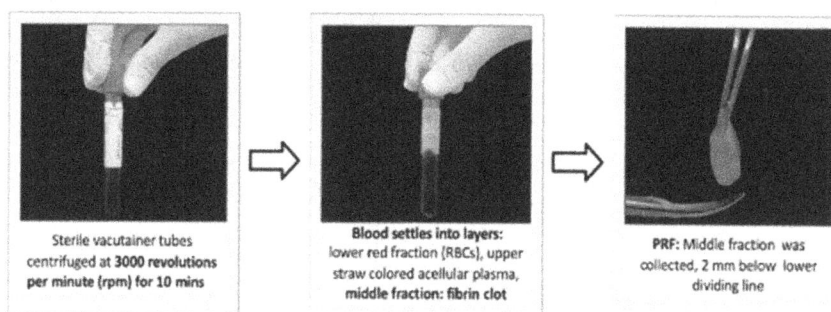

Sterile vacutainer tubes centrifuged at **3000 revolutions per minute (rpm)** for 10 mins

Blood settles into layers: lower red fraction (RBCs), upper straw colored acellular plasma, middle fraction: fibrin clot

PRF: Middle fraction was collected, 2 mm below lower dividing line

Fig. 17.1 Preparation protocol of pPRF/cbPRF

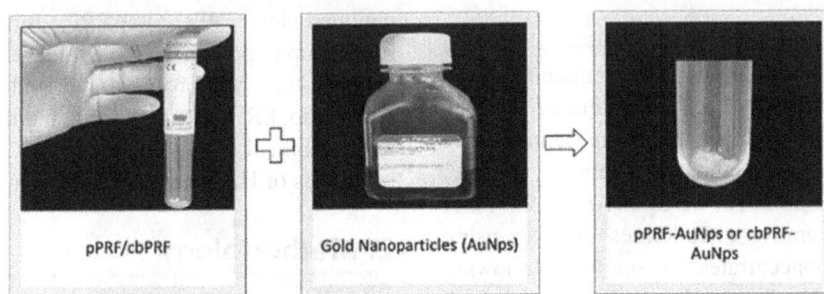

Fig. 17.2 Preparation protocol of pPRF-AuNps/cbPRF-AuNps

2.3 Preparation of pPRF-AuNPs and cbPRF-AuNPs

The preparation of pPRF-AuNPs/ cbPRF-AuNPs involved dividing the sample into three groups, with each group receiving a gradual addition of 15 nm AuNps to the obtained sample of pPRF. Using a micropipette, 3 μl, 6 μl, and 9 μl of the liquid AuNps were added to the respective groups. This was followed by centrifugation (Gel Eltek Tc 650 D –Multispin Centrifuge, Elektrocraft India Pvt.LTd.) for 1 minute at 1000 rpm (Fig. 17.2).

2.4 Histological Examination of pPRF-AuNPs and cbPRF-AuNPs

Histological examination of pPRF-AuNPs and cbPRF-AuNPs involved fixing samples in 10% formalin for 24 hours, dehydrating through graded alcohol solutions, and embedding in paraffin. Thin sections (~5mm) were cut, placed on glass slides, and stained with Haematoxylin and Eosin. Sections were deparaffinized, stained for nuclei and cytoplasm, and examined under a bright-field microscope (CKX415F, Olympus, Japan) at 10x, 40x, and 100x magnifications. The analysis focused on tissue structure, fibrin network, and the presence of AuNPs within the fibrin strands.

2.5 Characterize pPRF-AuNPs and cbPRF-AuNPs under Polarised Light Microscope

The darkfield illumination mode was activated on the Polarisation Microscope BA310POL trinocular, the prepared slides of pPRF-AuNPs 6 μl and cbPRF-AuNPs 6 μl were placed on the microscope stage to observe the overall distribution of AuNps over the pPRF and cbPRF matrix.

2.6 MTT Assay

HiFi hDPSCs (HiFi Human Dental Pulp Stem cells) were maintained in mesenchymal stem cell expansion media supplemented with 10% FBS and 1% antibiotic-antimycotic solution in a CO^2 incubator at 37°C with 5% CO2. Cells were sub-cultured every 2 days. A suspension with 20,000

cells per well was seeded in 200 μl per well in a 96-well plate. After 24 hours, culture medium was replaced with fresh medium containing test agents or control medium. Following another 24-hour incubation, MTT reagent was added at 0.5 mg/mL for 3 hours, allowing viable cells to form purple formazan crystals. Solubilization was done with Dimethylsulfoxide, and absorbance of the solubilized formazan was measured at 570 nm using an ELISA reader to assess cytotoxicity effects (Table 17.1).

Table 17.1 Details of drug treatment to respective cell line used for the study

Sl. No	Culture condition	Cell lines	Concentration treated to cells
1	Untreated	hDPSCs	No treatment
2	Blank	-	Only Media without cells
3	Doxorubicin	hDPSCs	1 μM/ml
4	Test compounds	hDPSCs	(6.25,12.5, 25, 50, 100 μg/ml)

3. Statistical Analysis

Statistical Package for Social Sciences (SPSS) for Windows Version 22.0 Released 2013. Armonk, NY: IBM Corp., will be used to perform statistical analyses. Descriptive analysis includes expression of Absorbance values in terms of Mean & SD for each group. Independent Student t Test was used to compare the mean Absorbance values between pPRF& cbPRF groups with different levels of AuNPs. One-way ANOVA Test followed by Tukey's post hoc Test was used to compare the mean Absorbance values between different Au NPs levels in each concentration of pPRF & cbPRF group. The level of significance was set at $P < 0.05$.

4. Results and Discussion

4.1 Histological Examination of AuNPs Trapped by Fibrin Strands

Histological examination revealed AuNPs trapped within fibrin strands, confirming the presence of AuNPs in fibrin clots derived from pPRF-AuNPs 6 µl and cbPRF-AuNPs 6 µl. This demonstrates the successful identification of AuNPs within the fibrin clots confirms their effective incorporation and entrapment (Fig. 17.3 and 17.4).

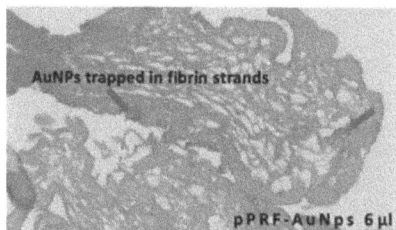

Fig. 17.3   Histological examination of pPRF-AuNps 6µl

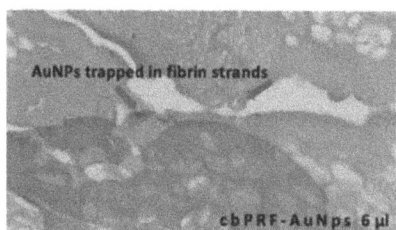

Fig. 17.4   Histological examination of cbPRF-AuNps 6 µl

4.2 Characterize pPRF-AuNPs and cbPRF-AuNPs under Polarised Light Microscope

Characterization of pPRF-AuNPs 6 µl and cbPRF-AuNPs 6 µl was performed using a polarised light microscope. Dark field polarization, revealed small-sized, luminescent dense particles interspersed within the fibrin strands. This detailed imaging provided clear evidence of the AuNPs being effectively captured within the fibrin matrix. The luminescent nature of the particles, enhanced by the dark field polarization technique, allowed for precise visualization and confirmation of their presence. This further supports the incorporating AuNPs into the fibrin clot, highlighting the successful integration and distribution of the nanoparticles within the clot structure (Fig. 17.5 and 17.6).

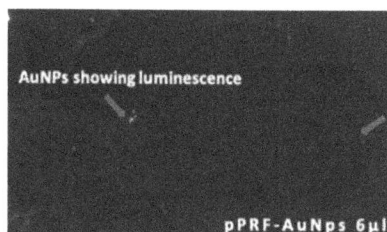

Fig. 17.5   Dark field polarization- pPRF-AuNps 6 µl

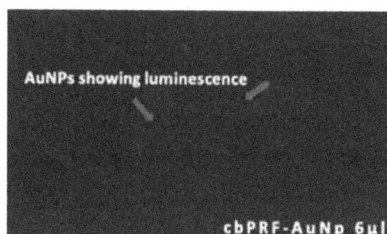

Fig. 17.6   Dark field polarization- cbPRF-AuNps 6 µl

4.3 MTT Assay

Microphotographic Analysis

Lower and moderate concentrations (6.25, 12.5, and 25 µg/mL) maintained cells with normal spindle-shaped morphology akin to untreated hDPCs, suggesting minimal impact on cellular health. Both pPRF-AuNps and cbPRF-AuNps demonstrated similar results across these concentrations. At higher concentrations (50 and 100 µg/mL), cells displayed morphological changes such as shrinkage, rounding and detachment, indicative of stress and potential damage. This pattern was consistently observed in both pPRF-AuNps and cbPRF-AuNps groups highlighting a more significant impact on cell viability and health at elevated concentrations (Fig. 17.7 and 17.8).

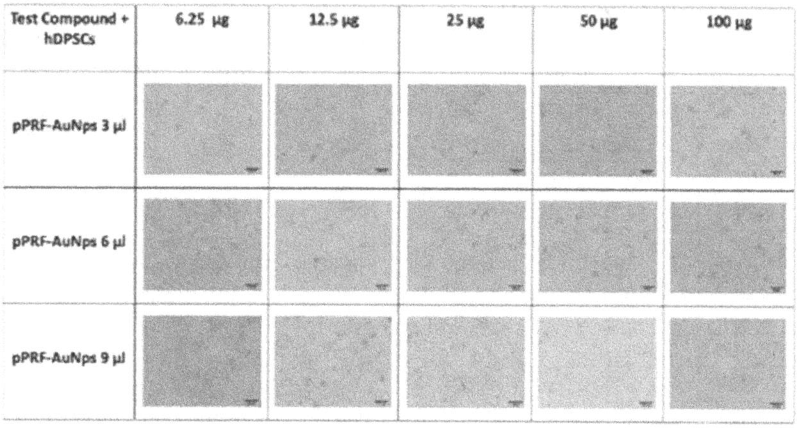

Test Compound + hDPSCs	6.25 µg	12.5 µg	25 µg	50 µg	100 µg
pPRF-AuNps 3 µl					
pPRF-AuNps 6 µl					
pPRF-AuNps 9 µl					

Fig. 17.7   Characteristics of hDPSCs at 24hr exposed to pPRF- AuNps 3/6/9 µl at 5 concentrations (6.25,12.5, 25, 50, 100µg/ml)

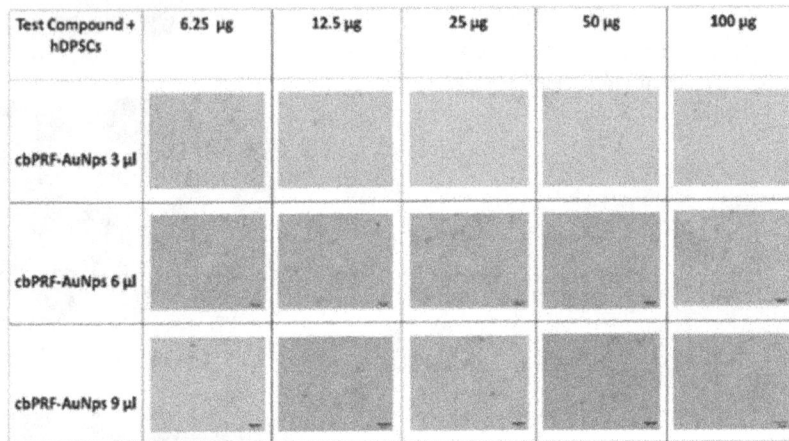

Fig. 17.8 Characteristics of hDPSCs at 24hr exposed to cbPRF- AuNps 3/6/9 µl at 5 concentrations (6.25,12.5, 25, 50, 100µg/ml)

Mean Absorbance Value and Percentage of Cell Viability

Table 17.2 Comparison of mean absorbance values between pPRF and cbPRF groups with 3 µl AuNPs using independent student t test

Conc.	Groups	N	Mean	SD	Mean Diff	p-value
6.25 µg	pPRF	3	1.015	0.005	0.199	<0.001*
	cbPRF	3	0.817	0.005		
12.5 µg	pPRF	3	1.005	0.006	0.201	<0.001*
	cbPRF	3	0.804	0.002		
25 µg	pPRF	3	0.999	0.000	0.207	<0.001*
	cbPRF	3	0.792	0.002		
50 µg	pPRF	3	0.996	0.001	0.239	<0.001*
	cbPRF	3	0.758	0.006		
100 µg	pPRF	3	0.993	0.002	0.274	<0.001*
	cbPRF	3	0.719	0.009		

Table 17.3 Comparison of mean absorbance values between pPRF and cbPRF groups with 6 µl AuNPs using independent student t test

Conc.	Groups	N	Mean	SD	Mean Diff	p-value
6.25 µg	pPRF	3	1.005	0.006	0.007	0.10
	cbPRF	3	0.998	0.002		
12.5 µg	pPRF	3	1.004	0.007	0.015	0.03*
	cbPRF	3	0.989	0.004		
25 µg	pPRF	3	0.994	0.001	0.020	0.007*
	cbPRF	3	0.975	0.007		
50 µg	pPRF	3	0.989	0.002	0.009	0.03*
	cbPRF	3	0.980	0.004		
100 µg	pPRF	3	0.984	0.002	0.094	<0.001*
	cbPRF	3	0.890	0.003		

Table 17.4 Comparison of mean absorbance values between pPRF and cbPRF groups with 9 µl AuNPs using independent student t test

Conc.	Groups	N	Mean	SD	Mean Diff	p-value
6.25 µg	pPRF	3	0.995	0.002	0.019	0.02*
	cbPRF	3	0.976	0.009		
12.5 µg	pPRF	3	0.992	0.001	0.038	<0.001*
	cbPRF	3	0.955	0.004		
25 µg	pPRF	3	0.980	0.005	0.062	<0.001*
	cbPRF	3	0.918	0.008		
50 µg	pPRF	3	0.976	0.008	0.096	<0.001*
	cbPRF	3	0.880	0.005		
100 µg	pPRF	3	0.952	0.002	0.145	<0.001*
	cbPRF	3	0.807	0.004		

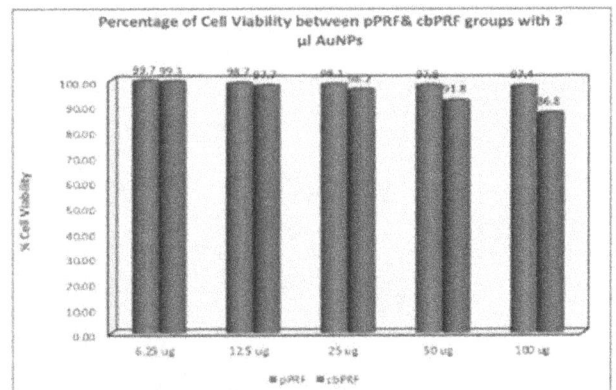

Fig. 17.9 Percentage of cell viability between pPRF & cbPRF group with 3 µl AuNPs

The MTT assay assessed the impact of varying concentrations of AuNPs combined with cbPRF and pPRF on hDPSC morphology. At lower and moderate

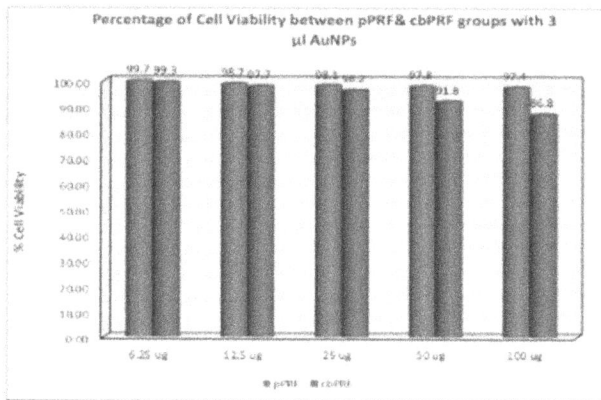

Fig. 17.10 Percentage of cell viability between pPRF & cbPRF group with 6 μl AuNPs

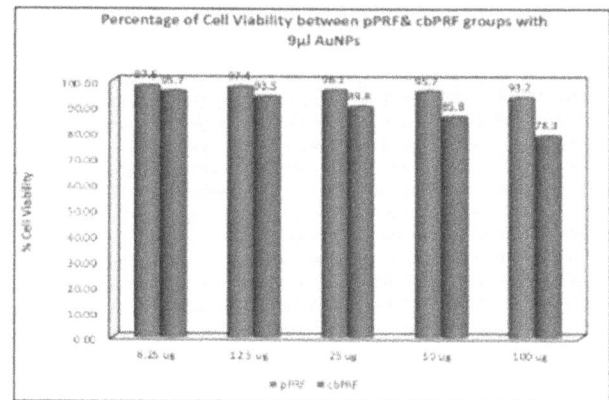

Fig. 17.11 Percentage of cell viability between pPRF & cbPRF group with 9 μl AuNPs

Table 17.5 Comparison of mean absorbance values between different AuNPs levels in each concentration of pPRF group using one-way ANOVA test & tukey's post hoc test

Conc.	AuNPs	N	Mean	SD	p-value[a]	Sig. Diff	p-value[b]
6.25 μg	3 μl	3	1.015	0.005		3 vs 6	0.04*
	6 μl	3	1.005	0.006	0.004*	3 vs 9	0.003*
	9 μl	3	0.995	0.002		6 vs 9	0.07
12.5 μg	3 μl	3	1.005	0.006		3 vs 6	0.97
	6 μl	3	1.004	0.007	0.04*	3 vs 9	0.04*
	9 μl	3	0.992	0.001		6 vs 9	0.07
25 μg	3 μl	3	0.999	0.000		3 vs 6	0.13
	6 μl	3	0.994	0.001	<0.001*	3 vs 9	<0.001*
	9 μl	3	0.980	0.005		6 vs 9	0.001*
50 μg	3 μl	3	0.996	0.001		3 vs 6	0.18
	6 μl	3	0.989	0.002	0.004*	3 vs 9	0.003*
	9 μl	3	0.976	0.008		6 vs 9	0.03*
100 μg	3 μl	3	0.993	0.002		3 vs 6	0.002*
	6 μl	3	0.984	0.002	<0.001*	3 vs 9	<0.001*
	9 μl	3	0.952	0.002		6 vs 9	<0.001*

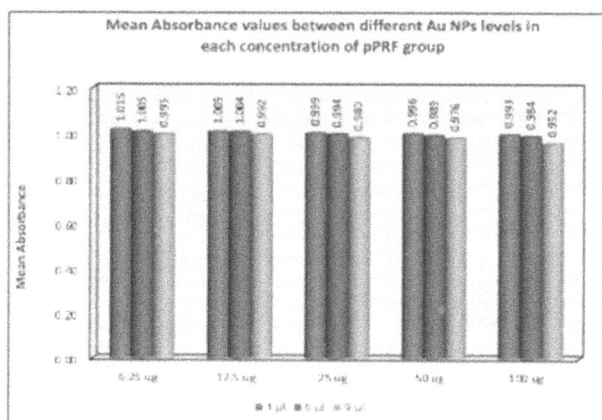

Fig. 17.12 Mean absorbance values: AuNPs levels in each concentration of pPRF

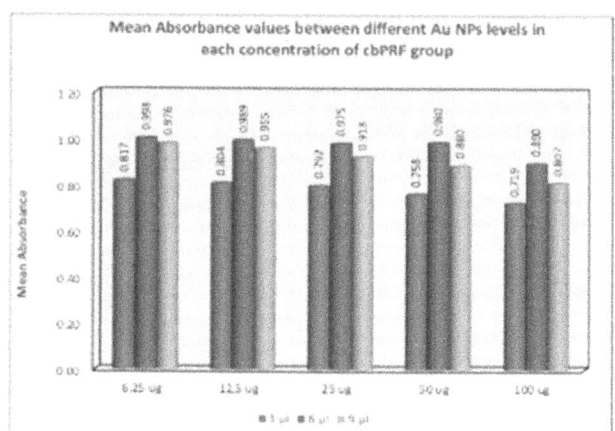

Fig. 17.13 Mean absorbance values: AuNPs levels in each concentration of cbPRF

Table 17.6   Comparison of mean absorbance values between different AuNPs levels in each concentration of cbPRF group using one-way ANOVA test & tukey's post hoc test

Conc.	AuNPs	N	Mean	SD	p-value[a]	Sig. Diff	p-value[b]
6.25 µg	3 µl	3	0.817	0.005		3 vs 6	<0.001*
	6 µl	3	0.998	0.002	<0.001*	3 vs 9	<0.001*
	9 µl	3	0.976	0.009		6 vs 9	0.007*
12.5 µg	3 µl	3	0.804	0.002		3 vs 6	<0.001*
	6 µl	3	0.989	0.004	<0.001*	3 vs 9	<0.001*
	9 µl	3	0.955	0.004		6 vs 9	<0.001*
25 µg	3 µl	3	0.792	0.002		3 vs 6	<0.001*
	6 µl	3	0.975	0.007	<0.001*	3 vs 9	<0.001*
	9 µl	3	0.918	0.008		6 vs 9	<0.001*
50 µg	3 µl	3	0.758	0.006		3 vs 6	<0.001*
	6 µl	3	0.980	0.004	<0.001*	3 vs 9	<0.001*
	9 µl	3	0.880	0.005		6 vs 9	<0.001*
100 µg	3 µl	3	0.719	0.009		3 vs 6	<0.001*
	6 µl	3	0.890	0.003	<0.001*	3 vs 9	<0.001*
	9 µl	3	0.807	0.004		6 vs 9	<0.001*

concentrations (6.25, 12.5, and 25 µg/mL), cells maintained a healthy spindle-shaped appearance, suggesting minimal cytotoxic effects. However, at higher concentrations (50 and 100 µg/mL), cells showed morphological changes like shrinkage, rounding, and detachment from the culture surface, indicating potential cell stress and damage. Both pPRF-AuNPs and cbPRF-AuNPs groups exhibited these changes consistently at higher concentrations, underscoring the need for cautious dosage in cell applications. Higher nanoparticle concentrations likely induce cellular stress and damage, resulting in cell shrinkage, rounding, and detachment (Cameron et al., 2022; Tabari et al.,2017).

The pPRF consistently showed higher cell viability than cbPRF across all AuNPs concentrations (3 µl, 6 µl, 9 µl). For instance, at 3 µl AuNPs, pPRF maintained viability from 99.7% to 97.4%, while cbPRF decreased from 99.3% to 86.8% ($p<0.001$). This trend was similarly observed at higher AuNPs levels (6 µl and 9 µl), reaffirming superior effectiveness of pPRF in maintaining hDPSCs viability compared to cbPRF.Among the variations tested within the pPRF group, 3 µl AuNPs consistently demonstrated the highest mean absorbance values, indicating superior cell viability across all concentrations (6.25 µg, 12.5 µg, 25 µg, 50 µg, and 100 µg). Significant differences were observed, with 3 µl AuNPs consistently outperforming 6 µl and 9 µl AuNPs at various concentrations ($p=0.004$ to $p<0.001$). This highlights 3 µl AuNPs as the most effective concentration within the pPRF group for enhancing cell viability.

In the cbPRF group, significant variations in mean absorbance were observed across all tested concentrations (6.25 µg, 12.5 µg, 25 µg, 50 µg, and 100 µg) among different AuNP levels ($p<0.001$). Consistently, the 6 µl AuNP group showed the highest mean absorbance at each concentration, significantly surpassing both the 3 µl and 9 µl AuNP groups ($p<0.001$ and $p=0.007$ at 6.25 µg; $p<0.001$ for all other concentrations). Additionally, the 9 µl AuNP group exhibited significantly higher mean absorbance than the 3 µl AuNP group across all concentrations ($p<0.001$). Therefore, within the cbPRF group, 6 µl AuNPs proved most effective, followed by 3 µl AuNPs, with 9 µl AuNPs demonstrating the least effectiveness. The comparison between pPRF and cbPRF consistently showed higher cell viability percentages with pPRF at all tested concentrations. This finding is significant, especially considering the assumption that cord blood, known for its abundance of stem cells and growth factors, might demonstrate equal or superior viability.

The percentage of cell viability decreased with increasing concentrations of AuNPs in both pPRF and cbPRF groups. However, pPRF consistently exhibited higher cell viability percentages compared to cbPRF across most concentrations, indicating stronger interaction or compatibility with AuNPs. This suggests that pPRF may offer enhanced biocompatibility when combined with AuNPs. The decrease in cell viability, as measured by MTT assay, could imply that cells underwent apoptosis (Tabari et al.,2017). AuNPs can be internalized into cells, potentially causing ultrastructural changes. Molecules with positively charged surfaces typically have higher uptake ratios but lower intracellular stability compared to neutral or negatively charged molecules (Landgraf et al.,2015).

The study found that cell viability decreased as the concentration of AuNPs increased from 3 µl to 9 µl in both pPRF and cbPRF groups, consistent with previous research indicating higher AuNP concentrations correlate with increased cytotoxicity (Chueh at el.,2015; Skalska and Struzynska, 2015). Lower concentrations (e.g., 3 µg/mL) may be tolerated or beneficial to cells, while higher concentrations (e.g., 6 or 9 µg/mL) can induce toxicity. Higher concentrations are prone to forming aggregates, which reduces their surface area-to-volume ratio and alters their interaction with cells, potentially increasing toxicity (Nasrullah et al.,2023). The optimal performance of 3 µl AuNPs in both groups at all tested concentrations suggests this volume achieves a balance of nanoparticle density, surface area-to-volume ratio, and cellular uptake efficiency (Khlebtsov and Dykman, 2011). This study marks the first exploration into generating platelet-rich fibrin (PRF) from cord blood, revealing potential differences in fibrin architecture compared to PRF derived from peripheral blood. Peripheral blood PRF may offer a more supportive matrix for sustained growth factor release and cell interactions whereas fibrin structure of cbPRF influenced by platelet activation and clotting factors may affect cell viability outcomes. Limitations of current research include significant variability in cord blood quality due to differences in collection, storage and handling procedures, necessitating standardization for consistent outcomes.

5. Conclusion

The integration of AuNPs into PRF was confirmed through histological and stereomicroscopic examinations, showing successful incorporation and distribution of nanoparticles within the fibrin matrix. At lower concentrations, both pPRF-AuNPs and cbPRF-AuNPs maintained typical stem cell morphology. The findings reveal that pPRF exhibits higher cell viability compared to cbPRF when combined with AuNPs, indicating superior biocompatibility and lower cytotoxicity of pPRF. However, the effect of AuNPs on cell viability is dose-dependent, with higher concentrations leading to decreased cell viability. This suggests the need for careful optimization of AuNP concentrations to balance therapeutic benefits.

Acknowledgments

We would also like to thank the Faculty of Dental Sciences, Ramaiah University of Applied Sciences for providing the necessary resources and facilities and Averin Pvt lab for the support.

Funding

This study was funded by Office of Research and Innovation, M S Ramaiah University of Applied Sciences (Project Reference No.: ORI-SG/003/2023).

References

1. Amrollahi, P., Shah, B., Seifi, A. and Tayebi, L., (2016). Recent advancements in regenerative dentistry: A review. *Materials Science and Engineering: C*, 69, pp.1383–1390.
2. Arshad, S., Tehreem, F., Ahmed, F., Marya, A. & Karobari, M.I., (2021). Platelet-rich fibrin used in regenerative endodontics and dentistry: current uses, limitations, and future recommendations for application. *International Journal of Dentistry*, 1, p.4514598.
3. Cameron, S.J., Sheng, J., Hosseinian, F. & Willmore, W.G., (2022). Nanoparticle effects on stress response pathways and nanoparticle-protein interactions. *International Journal of Molecular Sciences*, 23, p.7962.
4. Chueh, P.J., Liang, R.Y., Lee, Y.H., Zeng, Z.M. & Chuang, S.M., (2015). Differential cytotoxic effects of gold nanoparticles in different mammalian cell lines. *Journal of Hazardous Materials*, 6, pp.303–312.
5. de Souza Araujo, P.R., Silva, L.B., dos Santos Neto, A.P., de Arruda, J.A., Alvares, P.R., Sobral, A.P. & Sampaio, G.C., (2017). Pulp revascularization: a literature review. *The Open Dentistry Journal*, 10, p.48.
6. Devi, S., Bongale, A.M., Tefera, M.A., Dixit, P. & Bhanap, P., (2023). Fresh umbilical cord blood—A source of multipotent stem cells, collection, banking, cryopreservation, and ethical concerns. *Life*, 13, p.1794.
7. Ghaznavi, D., Babaloo, A., Shirmohammadi, A., Zamani, A.R., Azizi, M., Rahbarghazi, R. & Ghaznavi, A., (2019). Advanced platelet-rich fibrin plus gold nanoparticles enhanced the osteogenic capacity of human mesenchymal stem cells. *BMC Research Notes*, 12, p.721.
8. Jung, C., Kim, S., Sun, T., Cho, Y.B. and Song, M., (2019). Pulp-dentin regeneration: current approaches and challenges. *Journal of tissue engineering*, 10, p.2041731418819263.
9. Khademhosseini, A., Ashammakhi, N., Karp, J.M., Gerecht, S., Ferreira, L., Annabi, N., Darabi, M.A., Sirabella, D., Vunjak-Novakovic, G. and Langer, R., (2020). Embryonic stem cells as a cell source for tissue engineering. In *Principles of tissue engineering*. Pp.467–490. Academic Press.
10. Khiste, S.V. & Naik Tari, R., (2013). Platelet-rich fibrin as a biofuel for tissue regeneration. *International Scholarly Research Notices*, 1, p.627367.
11. Khlebtsov, N. & Dykman, L., (2011). Biodistribution and toxicity of engineered gold nanoparticles: a review of in

vitro and in vivo studies. *Chemical Society Reviews*, 40, pp.1647–1671.

12. Khorshidi, H., Haddadi, P., Raoofi, S., Badiee, P. & Dehghani Nazhvani, A., (2018). Does adding silver nanoparticles to leukocyte-and platelet-rich fibrin improve its properties? *Biomedical Research International*, 1, p.8515829.

13. Landgraf, L., Muller, I., Ernst, P., Schafer, M., Rosman, C., Schick, I., Kohler, O., Oehring, H., Breus, V.V., Basche, T. & Sonnichsen, C., (2015). Comparative evaluation of the impact on endothelial cells induced by different nanoparticle structures and functionalization. *Beilstein Journal of Nanotechnology*, 6, pp.300–312.

14. Nasrullah, M., Meenakshi Sundaram, D.N. & Uludag, H., (2023). Nanoparticles and cytokine response. *Frontiers in Bioengineering and Biotechnology*, 11, p.1243651.

15. Paul, M.P., Swathi Amin, D.A.M. & Naik, R., (2020). Platelet rich fibrin in regenerative endodontics. *International Journal of Applied Dental Sciences*, 6, pp.25–29.

16. Skalska, J. & Struzynska, L.,(2015). Toxic effects of silver nanoparticles in mammals—does a risk of neurotoxicity exist? *Folia Neuropathologica*, 53, pp.281–300.

17. Tabari, K., Hosseinpour, S., Parashos, P., Khozestani, P.K. & Rahimi, H.M., (2017). Cytotoxicity of selected nanoparticles on human dental pulp stem cells. *Iranian Endodontic Journal*, 12, p.137.

18. Vial, S., Reis, R.L. & Oliveira, J.M., (2017). Recent advances using gold nanoparticles as a promising multimodal tool for tissue engineering and regenerative medicine. *Current Opinion in Solid State and Materials Science*, 21, pp. 92–112.

Note: All the figures and tables in this chapter were made by the author.

Advances in Materials Science and Technology – Dr. Srikari Srinivasan et al. (eds)
© 2025 Taylor & Francis Group, London, ISBN 978-1-041-12342-2

18

"Knocking on Nature's Door"— An in Vitro Validation of the Anti-Inflammatory, Anti-Microbial Efficacy of Convolvulus Pluricaulis

Soumya N Sajjan[1],
Greeshma Chandrashekar[2],
Ashwini Shivananje Gowda
Department of Periodontology,
Faculty of Dental Sciences, Ramaiah University of Applied Science,
Bangalore, Karnataka, India

Rajamanickam Deveswaran
Department of Pharmaceutics,
Faculty of Pharmacy, Ramaiah University of Applied Science,
Bangalore, Karnataka, India

M. K. Akhila
Department of Periodontology,
Faculty of Dental Sciences, Ramaiah University of Applied Science,
Bangalore, Karnataka, India

ABSTRACT: Inflammation is the body's response to injury or infection, which can be acute or chronic. Chronic inflammation, such as periodontitis, is due to plaque and bacteria, leading to tissue destruction. Non-steroidal anti-inflammatory drugs (NSAIDs) are commonly used for treatment but have limitations, including side effects, cost, and rebound effects. Phytopharmaceuticals, derived from medicinal herbs, offer a promising alternative. *Convolvulus pluricaulis* (Shankhpushpi), a plant native to India, contains bioactive compounds with various therapeutic properties. These include neuroprotective, antioxidant, analgesic, antidiabetic, and antimicrobial effects. Scientific validation of these herbs is crucial for their safe and effective use in treating inflammatory diseases. To evaluate the cytotoxicity, anti-bacterial and anti-inflammatory effect of *Convolvulus pluricaulis* (CP).25g of Shankhpushpi powder was mixed with 100ml of 70% ethanol and subjected to microwave extraction at 50°C for 60sec at 55W. The mixture was filtered, and the filtrate dried in a water bath. The dried extract was stored in a sterile container. The invitro, cytotoxicity was evaluated using the Methyltiazolyldiphenyl-tetrazolium bromide (MTT) assay with different extract concentrations, antimicrobial test by disc diffusion method and anti-inflammatory effect was assessed using the Raw 264.7 cells. The extract of Shankhpushpi was found to possesses, effective anti-inflammatory and antimicrobial effect with least cytotoxic effect. Shankhpushpi can be considered as an alternative to conventional anti-inflammatory agents in chronic periodontitis.

KEYWORDS: Anti-inflammatory, Chronic periodontitis, *Convolvulus pluricaulis*, Cytotoxicity and Microwave extraction

Corresponding author: [1]soumyasajjan28@gmail.com, [2]greeshma.c17@gmail.com

DOI: 10.1201/9781003664277-18

1. Introduction

Inflammation is the physiological response to a variety of injuries or insults, including heat, chemical agents or bacterial infection. The acute phase of inflammation, is of short duration. Resolving and elimination of the noxious stimulus is necessary. If stimulus is not resolved, inflammation remains unresolved and enters the chronic phase, which can be considered as non-physiologic or pathologic. Periodontitis is a chronic oral disease common in adults primarily, due to plaque accumulation and pathogenic-bacterial interactions with host immune response. Systemic factors and environmental factors play a role in the degradation and periodontal tissue destruction (Preshaw et al., 2012). The pathogenesis of periodontal diseases is mediated by the inflammatory response to bacteria in the dental biofilm. Inflammation serves as a double-edged sword in periodontal disease, producing both protective and detrimental effects. It is key to defending against infections and in tissue repair, simultaneously leading to tissue damage, pain and discomfort. Inflammation becomes non selective of its targets, thus affecting pathogenic and host cells.

Anti-inflammatory therapy is therefore one of the treatment modalities routinely employed in the treatment for chronic disease. Non-steroidal anti-inflammatory drugs (NSAIDs) are for treating and relieving the inflammatory effect. But the effect of NSAIDS is transient, the condition ricocheting to pretreatment status upon discontinuation. Prolonged use of these drugs causes serious health hazards and their expense must also be considered. 'Rebound effect, noted upon withdrawal of the anti-inflammatory drugs, is defined as "the increased production of negative symptoms when the effect of a drug has passed or the patient no longer responds to the drug. If a drug produces a rebound effect, the condition it was used to treat may return even more strongly when the drug is discontinued or loses effectiveness" (Teixeira ., 2013).

Therefore, addressing only the inflammation of periodontitis, without removal of the bacterial etiology is redundant. It is thus necessary to look for medication that possess both anti-inflammatory and antibiotic effect. More recently, foray has been made into phytopharmaceuticals which involves research and development of medicinal herbs. These may be used as alternatives or adjunct to conventional allopathic medication. This requires a scientific approach in validating these herbs to assertion their usefulness in the safe treatment of inflammatory diseases. The use of phytopharmaceuticals is more desirable due to them being compatible, cheaper, and less harmful effects to the system. Natural anti-inflammatory, antibiotic compounds have been found in plant parts such as flowers, stems, and leaves. One such plant, known to possess these properties is *Convolvulus pluricaulis*, which is commonly known as Shankhpushpi, an indigenous plant (Bhowmik et al., 2012). This medicinal herb has been reported to contain many bioactive phytoconstituents, such as, alkaloid (convolamine), flavonoid (kaempferol) and phenolics (scopoletin, b-sitosterol and ceryl alcohol), that have been ascribed to the observed medicinal properties. According to the ancient literature, this herb has been attributed with several therapeutic properties, such as anti-inflammatory, anxiolytic, neuroprotective, antioxidant, analgesic, immunomodulatory, antimicrobial, antidiabetic and cardioprotective activities (Malik et al., 2021, Balkrishna A et al., 2020). Given its multifaceted chemistry it might prove to be the drug to address immuno-infective condition such as periodontitis. But the efficacy and safety of the Shankhpushpi leaf and stem, is yet to be ascertained. The antibacterial property of this herb has not yet been tested against the periodontitis causing bacteria's such as *Porphyromonas gingivalis, Tannerella forsythia & Treponema denticola*. And also, its use in the management of gingival and periodontal conditions has as yet not been attempted. Therefore, the aim of the study was to evaluate the in vitro ability of *Convolvulus pluricaulis* (Shankhpushpi) extract for its cytotoxicity, anti-bacterial, and anti-inflammatory.

2. Methodology

2.1 Plant Materials

Fresh Shankhpushpi plants were collected from Hunsur taluk, Mysore, Karnataka, India. The leaves and stems were separated from the roots and dried in shade for 1 week. Shade dried leaves and stems were ground into a uniform powder using a mechanical blender and stored in sterile container at room temperature.

2.2 Preparation of Plant Extracts

25 grams of Shankhpushpi powder was weighed using a digital balance and transferred to a beaker, with 100 ml 70% ethanol and the sample was subjected to microwave (Nu Wave Pro) (Fig. 18.1. Microwave (Nu Wave Pro)) extraction. The beaker was placed in microwave and heated at 50° C for 60 seconds at 55 watts, to obtain the ethanolic extract of Shankhapushpi (Alara et al., 2020).

The sample was collected and filtered. The filtrate was transferred to a porcelain dish and subject to evaporation of the solvent with a boiling water bath. The concentrated extract was scraped with the stainless-steel spatula, the final weight of the concentrated extract, 2g, was collected and stored in a closed container.

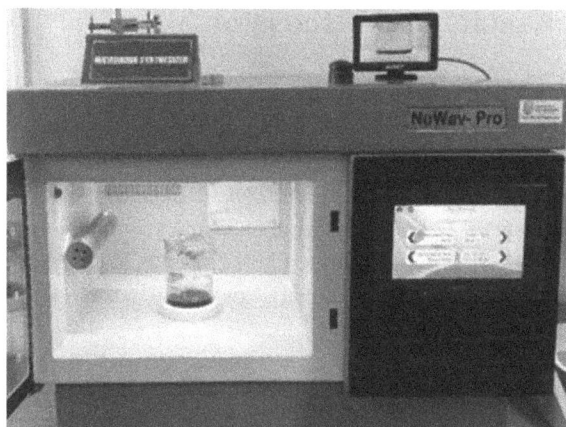

Fig. 18.1 Microwave (Nu Wave Pro)

2.3 Cytotoxicity Study

Cell Line

The Human dermal fibroblast cell line (HDF) was purchased from Sigma Aldrich, USA. The HDF cells were maintained in Fibroblast growth medium supplemented with 10% Fetal bovine serum (FBS) along with the 1% antibiotic-antimycotic solution in the atmosphere of 5% CO_2, 18-20% O_2 at 37^0C temperature in the CO_2 incubator and sub-cultured for every 2-3days.

Procedure

Inflammation was induced by stimulating the cells with 1µg/ml of Lipopolysaccharide (LPS) for 2 hours. Different concentrations of test agent Shankhpushpi extract (SPE) (62.5, 125, 250, 500,1000ug/ml) were added to wells in the culture, leaving one well untreated as a control (Table 18.1).

Table 18.1 Drug treatment with different concentrations to the Raw 264.7 cells used for the study

Sl. No	Culture condition	Cell lines	Concentration treated to cells
1	Untreated	Raw 264.7	No treatment
2	Blank	-	Only media without cells
3	LPS alone	Raw 264.7	1ug/ml
4	Shankhpushpi extract	LPS induced Raw 264.7	LPS+5 (62.5, 125, 250, 500,1000ug/ml)
5	LPS+Std		LPS+1mM/ml

The plate was incubated for 24 hours at 37°C in a 5% CO2 atmosphere. After incubation, the spent media was discarded and MTT reagent was added at a concentration of 0.5 mg/ml. The plate was protected from light with aluminum foil and incubated for an additional 3 hours. The MTT reagent was removed and 100µl of Dimethyl sulfoxide (DMSO) was added to each well, with gentle stirring to dissolve the formazan crystals. The absorbance was measured at 570 nm using a spectrophotometer or Enzyme linked immunosorbent assay (ELISA) reader.

Antimicrobial Analysis

The microorganisms used for antimicrobial analysis were purchased from American Type Culture Collection, USA. The bacterial strains were maintained on Nutrient Agar (NA). Pure cultures from the plate were inoculated into Nutrient Agar plate and sub cultured at 37°C for 24hours. Inoculum was prepared by aseptically adding the fresh culture into 2 ml of sterile 0.145 mol/L saline tube and the cell density was adjusted to 0.5 McFarland turbidity standard to yield a bacterial suspension of 1.5×10^8cfu/ml. Standardized inoculum was used for antimicrobial test.

Antimicrobial Test by Disc Diffusion Method

The medium was prepared by dissolving 38 g of Muller Hinton Agar Medium (Hi Media) in 1000 ml of distilled water. The dissolved medium was autoclaved medium was cooled, mixed well and poured into Petriplates (25ml/plate). The plates were swabbed with desired bacterial cultures. viz., *Porphyromonas gingivalis, Tannerella forsythia & Treponema denticola*. Three wells of 6 mm were bored in the inoculated media with the help of sterile cork-borer (6 mm). The sample loaded discs with 100ug/ml were then placed on the surface of Muller-Hinton medium. The standard drug, Ciprofloxacin with 3ug/ml concentration disc was used as a positive control and empty sterile disc loaded with double distilled water was used as a negative control. The plates were kept for incubation at 37°C for 24 hours. At the end of incubation (24hrs), inhibition zones were examined around the disc and measured with transparent ruler in millimeters. The size of the zone of inhibition (ZOI) was measured in millimeters. The absence of zone inhibition was interpreted as the absence of activity [Kohner. P.C et al., 1994, Mathabe, M.C et al., 2006]. The activities are expressed as resistant, if the zone of inhibition (ZOI) was less than 7 mm, intermediate (8-10 mm) and sensitive if more than 11 mm (Assam et al, 2010).

2.4 Anti-inflammatory

Cell Lines

The Raw 264.7 (Mouse macrophages) is purchased from NCCS, Pune, India. The cells were maintained in Dulbecco's Modified Eagle Medium (DMEM) with high glucose media supplemented with 10 % Fetal Bovine Serum (FBS) along with the 1% antibiotic-antimycotic solution in the atmosphere of 5% CO_2, 18-20% O_2 at 37^0C temperature in the CO_2 incubator and sub-cultured for every 2days.

Treatment of Shankhpushpi to LPS induced Raw 264.7 cells

0.5 x 106 cells/ml of Raw 264.7 were cultured in a 6 well plate and incubated for 48 hours for cell attachment and to reach required cell density. Inflammation was induced in the cells by stimulating with 1ug/ml of LPS for 2hours followed by treating the cells with required concentrations of test molecule (Shankhpushpi extract) and incubation for 24hours. Cells treated with LPS alone served as a Positive or disease control and cells without any treatment were considered as Control. LPS followed by Diclofenac was considered as Std control and medium/assay buffer alone was considered as a Blank control for all the ELISA studies (Table 18.2).

Table 18.2 Drug treatment to respective cell lines used for the study

Sl. No	Culture condition	Cell lines	Concentration treated to cells
1	Untreated	Raw 264.7	No treatment
2	Blank	-	Only Media without cells
3	LPS	Raw 264.7	1ug/ml
4	LPS+SPE	Raw 264.7	LPS-1ug+62.5, 125, 250, 500, 1000ug/ml
5	Diclofenac	Raw 264.7	LPS-1ug+1mM/ml

IL-1 Beta and TNF-alpha Estimation by ELISA

Proinflammatory cytokines play an important role in the mediation of inflammation. In the present study, we evaluated the expression of Pro-inflammatory cytokines. Interleukin-1 beta (IL1 beta) and Tumor necrosis factor alpha (TNF-alpha) were measured by quantitative ELISA analysis. IL-1 beta and TNF-alpha ELISA Kits were procured from RayBiotech Labs, Peachtree corners, GA and used for the estimation in cell culture supernatants. The assay employs an antibody specific for Mouse IL-1 beta or Mouse TNF-alpha coated on a 96-well plate. Standards and samples were pipetted into the wells and IL-1 beta/TNF-alpha present in a sample was bounded to the wells by the immobilized antibody. The wells were washed and biotinylated anti-Mouse IL-1 beta/TNF-alpha antibody was added. After washing away unbound biotinylated antibody, Horseradish peroxidase (HRP) - conjugated streptavidin was pipetted to the wells. The wells were again washed, a TMB (3,3',5,5'-Tetramethylbenzidine) substrate solution was added to the wells and color develops in proportion to the amount of IL-1 beta or TNF-alpha bound. The Stop Solution changes the color from blue to yellow, and the intensity of the color was measured at 450 nm.

3. Results and Discussion

3.1 Cytotoxicity

Given test compound, Shankhpushpi extract shows the moderate cytotoxicity against Raw 264.7 cells with 85.38% cell viability, at the concentration of 500ug/ml after the treatment of 24 h of incubation at 37°C temperature (Table 18.3: % cell viability values of given sample, Shankhpushpi extract against LPS induced Raw 264.7 cells after the treatment period of 24hrs).

Table 18.3 % cell viability values of given sample, shankhpushpi extract against LPS induced raw 264.7 cells after the treatment period of 24hrs

Culture condition	% cell viability	Max protective dose
Untreated	100	
LPS-1ug	98.50	
LPS+Diclofenac-1mM	90.22	
LPS+Shankhpushpi-62.5ug/ml	97.88	
LPS+Shankhpushpi-125ug/ml	93.51	500ug/ml
LPS+Shankhpushpi-250ug/ml	93.46	
LPS+Shankhpushpi-500ug/ml	85.38	
LPS+Shankhpushpi-1000ug/ml	44.19	

3.2 Anti-Bacterial Study

Anti-bacterial activity of Shankhpushpi with 1 dilution (100ug/ml) against 3 bacterial strains i.e., *Porphyromonas gingivalis* (Fig. 18.2. *Porphyromonas gingivalis*), *Tannerella forsythia* (Fig. 18.3. *Tannerella forsythia*) and *Treponema denticola* (Fig. 18.4. *Treponema denticola*) was done in comparison of Ciprofloxacin with 3ug (Std control) and Negative control (Phosphate Buffered Saline - PBS). It was found that the Shankhpushpi extract shows effective anti-bacterial activity on all species, based on the zone of inhibition measured (Table 18.4: Diameter of

Table 18.4 Diameter of zones of inhibition (mm) of given extract, Shankhpushpi extract against microorganisms at 100ug/ml concentration after the incubation period of 24hrs

Zone of inhibition (ZOI) in mm			
Sample code	Porphyromonas gingivalis	Tannerella forsythia	Treponema denticola
Control	0	0	0
Ciprofloxacin-5ug	24±1.73	28±2.64	28±2
Shankapushpi-100u	17±1	21±2.64	21±2

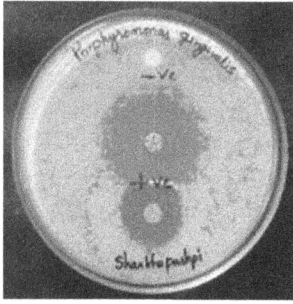

Fig. 18.2 Porphyromonas gingivalis

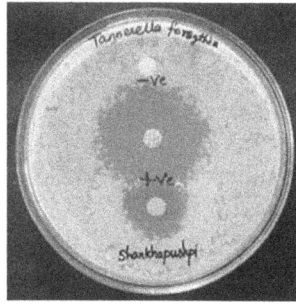

Fig. 18.3 Tannerella forsythia

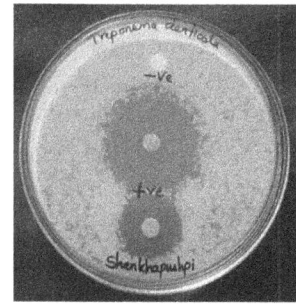

Fig. 18.4 Treponema denticola

zones of inhibition (mm) of given extract, Shankhpushpi extract against microorganisms at 100ug/ml concentration after the incubation period of 24hrs).

3.3 Anti-Inflammatory Study

In LPS alone group, IL-1 beta and TNF-alpha cytokines were effectively expressed and gradually inhibited in SPE on dose dependent fashion till the maximum dose of 500ug respectively. 1000ug dose caused toxicity and increased levels of both cytokines. In control group, low concentration of Cytokine expression was observed. Diclofenac with 1ug was used as a standard (std) control for the study (Fig. 18.5 and Fig. 18.6)

Periodontitis is a common inflammatory disease of infectious origins that often evolves into a chronic condition (Thomson et al., 1994), which affects the supporting structures of teeth and characterized by the destruction of the gingiva, periodontal ligament, and alveolar bone. Untreated periodontal disease can lead to tooth loss. It is highly prevalent (severe periodontitis affects 10–15% of adults) and negatively impacts the quality of the life of the affected individuals (Preshaw et al., 2012).

When the host's defense mechanisms are compromised even intraoral commensal bacteria become pathogenic and initiate the process of periodontal infection and inflammation. Therefore, elimination of the pathogenic load has been traditionally ben prioritized in the treatment of periodontitis. But the identification of the inflammatory component role in localized tissue destruction has birthed concept of host modulation to be applied to treat these

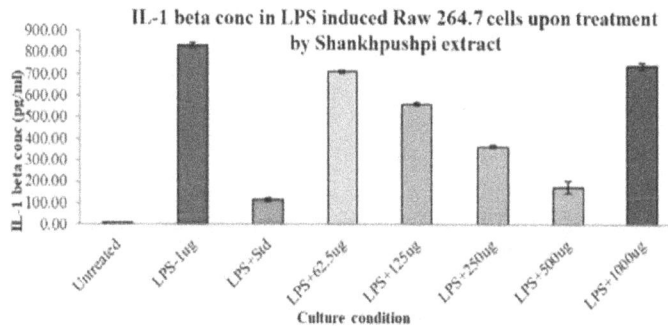

Fig. 18.5 IL-1 Overlay bar graph

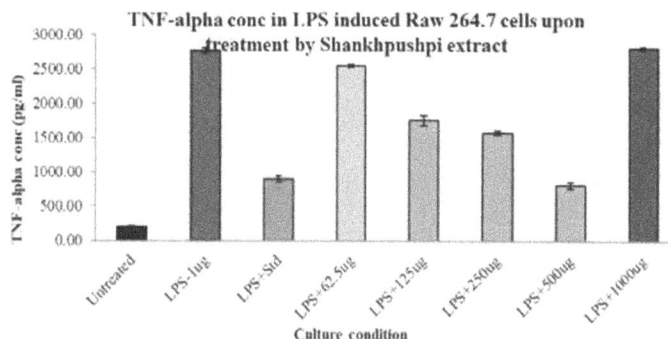

Fig. 18.6 TNF-alpha overlay bar graph

conditions. Iatrogenic or unintentional tissue damage results in inflammation, which affects the tissue enzyme, leukocyte, coagulation, and microcirculation systems. Both healing and host protection are the goals of these procedures. On the other hand, severe damage and infection may cause organisms or their toxins to accumulate to dangerous levels, which may result in systemic sepsis. The inflammatory response may become detrimental in some circumstances (Rogers., 2008). Host modulatory therapy is directed at pro-inflammatory mediators i.e. interleukin-1alpha, IL-6, IL-8 are activated. Others like tumor necrosis factor-alpha, reactive oxygen species, nitric oxide and prostaglandins. Over production in long time cause degenerative diseases i.e. asthma, cancer, arthritis, atherosclerosis and chronic periodontitis (Rogers., 2008). This is accomplished by other adjunctive use of anti-collagenase, proresolving molecules, OMEGA 3 fatty acids etc.

But in the field of phytopharmaceuticals it has been noted that antimicrobial, anti-inflammatory properties may be present in the same source. Generally, plants are the factory of metabolites in which phenolic components play vital role. Whereas, phenolic components have multiple activities in which anti-inflammatory activity is very important. They inhibit either the production or the action of pro-inflammatory mediator (Garima et al 2020).

Chakravarthi KK et al. studied and gave suggestions that the memory enhancing effects of Glyceriza glabra may be due to its antioxidant and anti-inflammatory activities. Thus, it can be used in the management of inflammatory condition like chronic periodontitis (Chakravarthi et al., 2013). In the study of Laxmidutta Shukla (2018), the ethanolic extract of the Shankhpushpi was found to have prominent anti-bacterial activity against bacteria such as Pseudomonas aeruginosa, Bacillus subtilis, and Escherichia coli. In the study conducted by Lai Teng Ling et al., Assessed for the antioxidant capacity and cytotoxicity of the thirteen Malaysian Plants and concluded that the extracts were found not cytotoxic at concentrations as high as 100µg/mL and also the ethanolic extracts were free radical scavengers (Ling et al., 2010).U. Santo Grace et al. studied and concluded that ethanolic extract of ginger is best effective against *S. aureus* when compared with *E. faecalis* when diluted up to 15 µl (Santo et al., 2017). In the present study, ethanolic extraction was performed of a variety of Shankhpushpi plant. In vitro assessment of the extraction was performed to determine cytotoxicity by using HDF, because fibroblasts are the primary cells to come in contact during the treatment. Maximum protective dose is 500µg/mL at which 85.38% of cell viability was observed. Anti-bacterial assessment was done by disc diffusion method against commonest periodontal pathogens such as *Porphyromonas gingivalis, Tannerella forsythia and* *Treponema denticola* at 500µg/mL and achieved effective anti-bacterial activity and anti-inflammatory efficacy was assessed by Raw 264.7 (Mouse macrophages), showed significant Anti-inflammatory activity by inhibiting the pro-inflammatory cytokines expression viz IL-1 beta and TNF-alpha. But the limitation of our study is that the test component that is Shankhpushpi extract is not compared with the standard chlorhexidine.

4. Conclusion

In conclusion, results indicates that the ethanolic extract of Shankhpushpi stem and leaves (test compound) demonstrates no cytotoxicity at doses below 500ug/ml, with greater than the 80% cell viability value after the 24hours of incubation. The Shankhpushpi extract was notes to possess effective anti-microbial activity at 100ug/ml and anti-inflammatory efficacy at 500ug/ml. As Shankhpushpi has been shown in this study to possess all three of the effects required to combat an infectious, inflammatory condition such as periodontitis, it can be considered an alternative to conventional pharmacotherapy in chronic periodontitis. Therefore, it can be concluded that 500ug/ml of Shankhpushpi is the optimal concentrate for future formulations aimed at the treatment of periodontitis, in further in vivo studies.

Acknowledgments

The authors thank Ms. Arpita (Post Graduate, Department of Pharmaceutics, Faculty of Pharmacy, Ramaiah University of Applied Sciences) for her valuable inputs.

References

1. Alara, O.R., Abdurahman, N.H. and Olalere, O.A., 2020. Ethanolic extraction of flavonoids, phenolics and antioxidants from Vernonia amygdalina leaf using two-level factorial design. *Journal of King Saud University-Science*, 32(1), pp.7–16.

2. Assam JP, A., Dzoyem, J.P., Pieme, C.A. and Penlap, V.B., 2010. In vitro antibacterial activity and acute toxicity studies of aqueous-methanol extract of Sida rhombifolia Linn.(Malvaceae). *BMC complementary and alternative medicine*, 10, pp.1–7.

3. Balkrishna, A., Thakur, P. and Varshney, A., 2020. Phytochemical profile, pharmacological attributes and medicinal properties of convolvulus prostratus–A cognitive enhancer herb for the management of neurodegenerative etiologies. *Frontiers in pharmacology*, 11, p.171.

4. Bhowmik, D., Kumar, K.S., Paswan, S., Srivatava, S. and Dutta, A., 2012. Traditional Indian herbs Convolvulus pluricaulis and its medicinal importance. *Journal of Pharmacognosy and Phytochemistry*, 1(1), pp.44–51.

5. Chakravarthi, K.K. and Avadhani, R., 2013. Beneficial effect of aqueous root extract of Glycyrrhiza glabra on learning and memory using different behavioral models: An experimental study. *Journal of natural science, biology, and medicine, 4*(2), p.420.

6. Kohner, P.C., Rosenblatt, J.E. and Cockerill 3rd, F.R., 1994. Comparison of agar dilution, broth dilution, and disk diffusion testing of ampicillin against Haemophilus species by using in-house and commercially prepared media. *Journal of clinical microbiology, 32*(6), pp.1594–1596.

7. Ling, L.T., Radhakrishnan, A.K., Subramaniam, T., Cheng, H.M. and Palanisamy, U.D., 2010. Assessment of antioxidant capacity and cytotoxicity of selected Malaysian plants. *Molecules, 15*(4), pp.2139–2151.

8. Malik, J. and Choudhary, S., 2021. Indian Traditional Herbs and Alzheimer's Disease: Integrating Ethnobotany and Phytotherapy. *Evidence Based Validation of Traditional Medicines: A comprehensive Approach*, pp.1129–1151.

9. Mathabe, M.C., Nikolova, R.V., Lall, N. and Nyazema, N.Z., 2006. Antibacterial activities of medicinal plants used for the treatment of diarrhoea in Limpopo Province, South Africa. *Journal of ethnopharmacology, 105*(1-2), pp. 286–293.

10. Preshaw, P.M., Alba, A.L., Herrera, D., Jepsen, S., Konstantinidis, A., Makrilakis, K. and Taylor, R., 2012. Periodontitis and diabetes: a two-way relationship. *Diabetologia, 55*, pp.21–31.

11. Rogers, J., 2008. The inflammatory response in Alzheimer's disease. *Journal of periodontology, 79*, pp.1535–1543.

12. Santo Grace, U. and Sankari, M., 2017. Antimicrobial activity of ethanolic extract of Zingiber Officinale-an in vitro Study. *Journal of Pharmaceutical Sciences and Research, 9*(9), p.1417.

13. Teixeira, M.Z., 2013. Rebound effect of modern drugs: serious adverse event unknown by health professionals. *Revista da Associação Médica Brasileira, 59*, pp.629–638.

14. Thomson, P. D., & Smith Jr, D. J., 1994. What is infection? *The American journal of surgery, 167*(1), S7-S1

Note: All the figures and tables in this chapter were made by the author.

Advances in Materials Science and Technology – Dr. Srikari Srinivasan et al. (eds)
© 2025 Taylor & Francis Group, London, ISBN 978-1-041-12342-2

19

Formulation and Assessment of Cytotoxicity of Mg-Cu Alloy in Combination with Xenogenic Bone Graft—An In-Vitro Study

Sanjana Hemashree, Lavanya Ramamurthy, Greeshma Chandrashekar*
Department of Periodontology,
Faculty of Dental Sciences, Ramaiah University of Applied Science,
Bangalore, Karnataka, India

Niranjana Prabhu
Department of Chemistry,
Faculty of Mathematical and Physical Sciences,
MS Ramaiah University of Applied Sciences,
Bengaluru, Karnataka, India

Premkumar
Department of Periodontology,
Faculty of Dental Sciences, Ramaiah University of Applied Science,
Bangalore, Karnataka, India

Department of Physics,
Faculty of Mathematical and Physical Sciences,
MS Ramaiah University of Applied Sciences,
Bengaluru, Karnataka, India

Akshay Arjun
Department of Physics,
Faculty of Mathematical and Physical Sciences,
MS Ramaiah University of Applied Sciences,
Bengaluru, Karnataka, India

ABSTRACT: Bone grafts are widely used in the management of periodontal defects and are always accompanied by a protocol of systemic antimicrobial therapy to prevent post operative infection. At present, there are no bone graft materials that possess inherent antibacterial property. Magnesium copper alloy (Mg-Cu) is known to possess antimicrobial properties and has been used in orthopedic implants with significant success. Recently, use of Mg-Cu as a standalone grafting material has received tremendous attention, as it has demonstrated mechanical properties similar to bone[1]. At present, in the field of dentistry, there have been no studies regarding the use of Mg-Cu alloy in conjunction with bone grafts to manage periodontal osseous defects. Prior to its incorporation as an antimicrobial with bone grafts, its chemical stability and cytotoxicity needs to be ascertained. The aim of this study was to formulate of Mg-Cu alloy of varying proportions of magnesium and copper, combine it with the xenogenic bone graft and assess their compatibility and potential cytotoxicity. Four different proportions of magnesium and copper were combined using the Solution combustion method to formulate

---

*Corresponding author: greeshma.c17@gmail.com

DOI: 10.1201/9781003664277-19

Mg-Cu alloy. The stability of the chemical structure following its combination with the graft was assessed using FTIR analysis. MTT assay was performed to evaluate the cytotoxicity levels of these alloys upon combination with the xenogenic bone graft. The FTIR analysis of the alloy with the xenogenic bone graft showed minimal variations in peaks implying that the functional groups remained intact. The MTT assay results show that that Mg-Cu alloy combined with the xenogenic bone graft showed minimal cytotoxicity against human bone cells. Mg-Cu alloys offer a promising vista in the surgical management of periodontal defects, combining the beneficial properties of both metals to enhance healing and support bone regeneration. However, further vitro studies and clinical trials are required to fully establish their efficacy and safety in periodontal therapy.

KEYWORDS: Mg-Cu, Bone graft, Xenograft, Cytotoxicity. Alloy, Magnesium-copper alloy, Xenogenic bone graft, In vitro study, Biocompatibility, Bone regeneration, Cell viability, Bone grafting

1. Introduction

Periodontal disease is one of the most common diseases in humans (Pihlstrom et al.,2005). Periodontitis is a multifactorial, chronic inflammatory disease affecting the oral cavity characterized by progressive destruction of the tooth supporting apparatus, mainly the alveolar bone. Management of periodontal disease is of utmost importance to restore the form, function and aesthetics of the tissues. Both non-surgical and surgical methods are used to manage the above condition.

Regeneration of lost tissues is one of the primary goals of periodontal regenerative therapy. Bone grafts are the most commonly used regenerative material in Periodontology. Autogenous bone graft, which is the gold standard, requires a second surgical procedure at the donor site. The limited volume of the harvested bone graft and unpredictable resorption rate among patients are its main disadvantages. Alternatively, allografts and xenografts may be used, but maybe associated with disease transmission and bear risks of infection[3]. In order to overcome the potential infection of the graft, antimicrobial therapy is prescribed in all recipients. Despite these measures, the use of antimicrobials following bone grafting cannot be foregone as grafts themselves have no antibacterial properties. Recently, addition of biodegradable magnesium copper (Mg-Cu) alloys in bone grafts have received tremendous attention, since they have demonstrated similar mechanical properties to bone and antimicrobial activity (Hong et al.,2015). Degradable magnesium and copper-based metals have been widely investigated both *in vitro* and *in vivo* due to their unique degradation property in physiologic environment. Besides having similar mechanical properties as human bones, magnesium and copper alloys have positive effects on the growth of bone tissues. Hence in this study, we will be formulating a novel Mg-Cu alloy combined with xenogenic bone graft.

2. Methodology

In this study, we aim to explore the properties of different combinations Mg-Cu alloy combined with the xenogenic bone graft to evaluate its effectiveness in managing periodontal defects. The objectives of the study were to formulate Mg-Cu alloy and combine it with the xenogenic bone graft and evaluate its cytotoxicity using MTT assay. Materials used: Magnesium nitrate hexahydrate Mg $(NO_3)_2$ (98%/), Cupric nitrate trihydrate Cu Mg $(NO_3)_2$ (95%), procured from Merck India Pvt Limited, Osseograft, Oxalyl dehydrate (ODH) and Distilled water. Various combinations of Mg-Cu alloy concentrations were formulated to identify appropriate concentration of Mg-Cu alloy (Table 19.1) and assessed through FTRI analysis to check for its compatibility with the xenogenic bone graft. Cytotoxicity of the formulated Mg Cu alloy combined with the xenogenic bone graft was assessed using MTT Assay (Table 19.3).

Table 19.1 Concentrations of Mg and Cu were formulated in gm

Composition	Mg(NO3)$_2$	Cu(No3)$_2$	ODH
MgO	10.2564	0.0000	4.7238
Mg$_{0.9}$Cu$_{0.1}$O	9.2308	0.9664	4.7238
Mg$_{0.8}$Cu$_{0.2}$O	8.2051	1.9328	4.7238
Mg$_{0.7}$Cu$_{0.3}$O	7.1795	2.8992	4.7238
Mg$_{0.6}$Cu$_{0.4}$O	6.1538	3.8656	4.7238

2.1 Formulation of MgCu alloy using Solution Combustion Method

Mg(NO3)$_2$ and Cu(NO3)$_2$ were measured as mentioned in Table 19.1 and the components were combined in a petri dish. ODH which serves as a fuel, helps in the combustion reaction of Mg Cu, along with 40 mL of distilled water, was added to the contents in the petri dish. The resulting blend was stirred using a magnetic stirrer for around 10-15 minutes until it achieved a uniform consistency. Following this, the mixture was transferred to a crucible and placed in a furnace preheated to 400°C for approximately 10 minutes to facilitate the combustion reaction. After cooling, the resultant mixture was in powder form. The obtained mixture in powder form was allowed to cool for it to come down to room temperature following which the contents were transferred to another crucible and kept in the furnace for 1 day for the process of calcination to be completed. After calcination, the powder was transferred into vials for subsequent combination with the xenogenic bone graft for further testing.

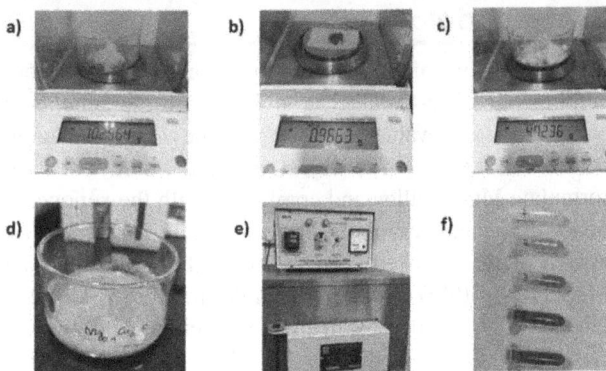

Fig. 19.1 Formulation of Mg-Cu alloy. a) Mg measured in gm b) Cu measured in gm c) ODH added as flux. d) MgCu mixture after addition of distilled water. e) Mg Cu mixture kept in furnace for solution combustion. f) Final MgCu alloy powder after calcination

2.2 Fourier Transform Infrared Spectroscopy (FTIR)

FTIR analysis is a characterization study done to identify organic or inorganic materials that could be a source of product contamination or cause of malfunction. It works by measuring the absorbance of Infrared Radiation by a sample. The resulting sample can then be used to identify the functional groups present in the compound.

Machine used: Bruker Alpha II-P FTIR Spectrometer

Technical specifications: Platinum diamond ATR Sampling Module

Wavelength range 350-8000cm^{-1}

Spectral resolution: 1cm^{-1}

Mode of operation: Transmittance mode

2.3 Cytotoxicity Study

MTT assay is a colorimetric assay used for the determination of cell proliferation and cytotoxicity, based on reduction of the yellow-coloured water-soluble tetrazolium dye MTT to formazan crystals. Mitochondrial lactate dehydrogenase produced by live cells reduces MTT to insoluble formazan crystals, which upon dissolution into an appropriate solvent exhibits purple colour, the intensity of which is proportional to the number of viable cells and can be measured spectrophotometrically at 570nm (Alley MC et al.,1986)

2.4 Cell Line

The MG-63 (Human bone osteosarcoma cell line) was purchased from NCCS, Pune, India. The cells were maintained in DMEM with high glucose media supplemented with 10 % FBS along with the 1% antibiotic-antimycotic solution in the atmosphere of 5% CO_2, 18-20% O_2 at 37^{0}C temperature in the Co2 incubator and sub-cultured for every 2days. Passage number of MG-63 cells was 37 was used for the current study

2.5 Procedure

Cytotoxicity in this study was assessed using MTT Assay. MTT assay is a colorimetric assay used for the determination of cell proliferation and cytotoxicity, based on reduction of the yellow-coloured water-soluble tetrazolium dye MTT to formazan crystals. Mitochondrial lactate dehydrogenase produced by live cells reduces MTT to insoluble formazan crystals, which upon dissolution into an appropriate solvent exhibits purple colour, the intensity of which is proporti47onal to the number of viable cells and can be measured spectrophotometrically at 570nm.

In this study, given test compounds were evaluated to measure the cytotoxicity study against MG-63 cells. The used concentrations of the compound used to treat the cells are as follows:

Table 19.2 Details of drug treatment to respective cell lines used for the study

Sl. No	Test	Cell line	Concentration treated to cells
1	Untreated	MG-63	No treatment
2	Std Control	MG-63	20uM/ml
3	Blank	-	Only Media without cells
4	Test compounds	MG-63	500µg/ml

3. Results and Discussion

3.1 Fourier Transform Infrared Spectroscopy (FTIR)

FTIR analysis was done for the formulated material (Mg Cu alloy), bone graft and the combination of the Mg Cu

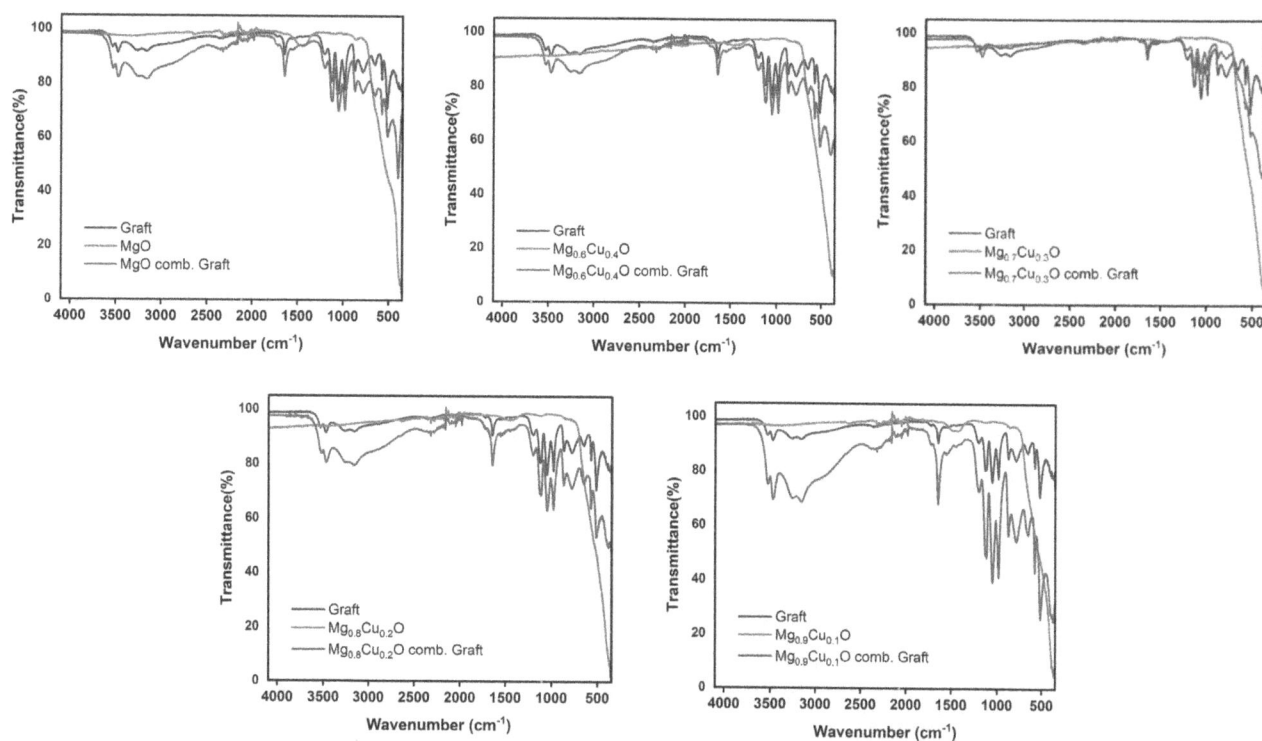

Fig. 19.2   FTIR analysis

alloy with the xenogenic bone graft. The bands at 1500-2000 cm^{-1} correspond to the amide group (CH_3). The collagen properties of bone are seen in the bands from 1548-1634 cm^{-1} (amide group). The bands at 1747-1773 cm^{-1} correspond to the C=O group (carbonyl) group. The closest peaks were observed with respect to these combinations: $Mg_{0.6}Cu_{0.4}O$ (2nd image), MgO (1st image) and $Mg_{0.7}Cu_{0.3}O$ (3rd image). The flowing combinations of Mg Cu alloy ($Mg_{0.6}Cu_{0.4}O$, MgO and $Mg_{0.7}Cu_{0.3}O$) combined with the xenogenic bone graft do not show much variation in peaks implying that there is minimal variation in the functional groups when the combination is used. Even after the combination of Mg Cu alloy with the xenogenic bone graft, the FTIR analysis showed that the functional groups have remained unaltered concluding that Mg Cu alloy is compatible with the xenogenic bone graft and its combination can be further tested in vitro and in vivo to be used in treatment periodontal defects.

3.2 Cytotoxicity

The MTT assay results suggested us that the given test compounds were minimally toxic against Human bone (MG-63) cells after the 24hours of incubation.

This study was designed to explore the properties Mg-Cu combined with xenogenic bone graft to treat periodontal defects where the Mg-Cu alloy was formulated using the

Table 19.3   Table shows the % cell viability values of MG-63 cells treated with the given test compounds with 500ug/ml against the human bone cells (MG-63) after the treatment period of 24hrs

MTT ASSAY-Summary	
Drug conc (µg/ml)	% cell viability
Untreated	100
MG-63 (Control)	100
PN-37 (Toxic Control)	72.584
(1) MgO+ Bone graft	89.449
(2) $Mg_{0.9}Cu_{0.1}O$+ Bone graft	86.006
(3) $Mg_{0.8}Cu_{0.2}O$+ Bone graft	91.353
(4) $Mg_{0.7}Cu_{0.2}O$+ Bone graft	88.402
(5) $Mg_{0.6}Cu_{0.3}O$+ Bone graft	84.045

solution combustion method. The formulated material was the combined with the xenogenic bone graft and subjected to characterization studies where FTRI analysis and cytotoxicity tests were done.

The treatment of periodontitis with bone grafting demonstrates significant benefits in promoting periodontal regeneration and improving clinical outcomes. Bone grafts enhance the healing process, support the regeneration of lost periodontal tissues and help restore bone architecture.

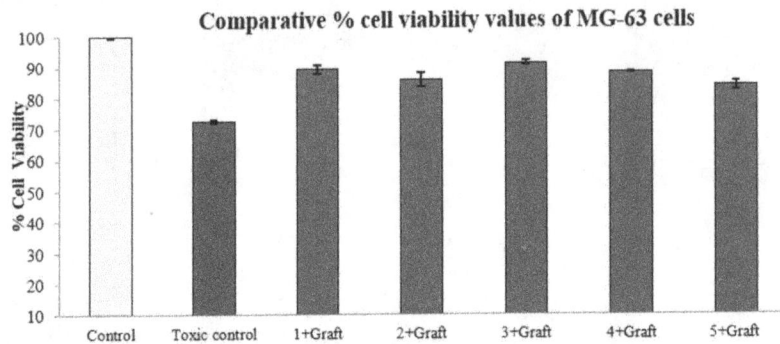

Fig. 19.3 Bar graph depicting the % cell viability values of MG-63 cells treated with various compounds with 500ug/ml after the incubation period of 24hours

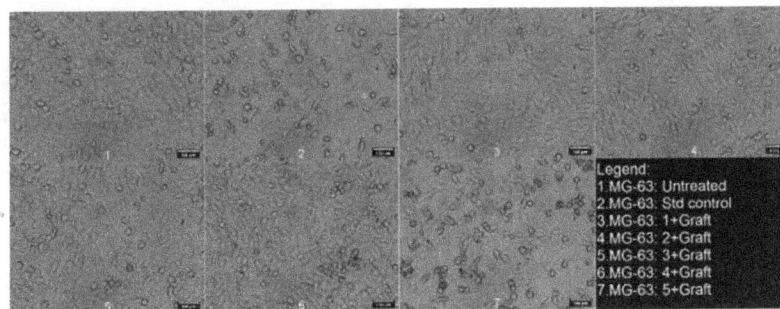

Fig. 19.4 Overlaid montage photo representing the morphology of MG-63 cells treated with various compounds with 500ug/ml along with controls after the incubation period of 24hours

When combined with other periodontal therapies, they can effectively reduce pocket depths, improve attachment levels and ultimately contribute to the preservation of tooth structures. Long-term studies indicate that these interventions can lead to sustained improvements in periodontal health, making bone grafting a valuable option in the management of periodontitis. However, individualized treatment plans and ongoing maintenance are essential for achieving optimal results. Magnesium alloys are gaining attention as materials for orthopedic implants due to their distinctive properties. These include, their favorable mechanical properties that match those of bone, and their ability to be customized for specific needs. Additionally, magnesium alloys can help reduce inflammation and their corrosion rates are being actively managed to ensure optimal performance. Copper (Cu) is widely utilized in medical and biomedical fields primarily for its strong antibacterial properties. Its applications span various areas including orthopedic implants, where it contributes to durability and strength; wound healing, where it promotes tissue repair; dental applications, where it enhances the properties of dental materials; and bioelectronic devices, due to its excellent electrical conductivity. However, the use of copper must be carefully controlled to address potential toxicity and ensure biocompatibility.

Bone grafts are frequently used to address periodontal defects and systemic antibiotics are prescribed adjunctively to prevent infection. Currently, no bone graft materials have inbuilt antibacterial properties. To address this, Mg Cu has been combined with xenogenic bone grafts, endowing them with antimicrobial properties and potentially eliminating the need for systemic antibiotics. The antibacterial properties of Mg Cu were investigated in previous studies (Li et al.,2014). Previous research has demonstrated that the combination of Mg and Cu provides sustained antibacterial effects against Staphylococcus aureus (Liu et al.,2016). Previous research has demonstrated the vital role of magnesium in regulating microvascular functions (Bernardini et al.,2004). Extracellular magnesium ions are known to act as receptor-mediated chemo attractants for endothelial cells. A deficiency in magnesium may impede endothelial cell migration and proliferation, likely by disrupting certain signal transduction pathways activated by angiogenic factors (Wolf et al.,2009). Previous studies have demonstrated that the bioactive magnesium and copper ions released from Mg-Cu alloys promote bone vasculature formation and enhance the angiogenesis process (Sato et al.,1993). Magnesium-based metals are recognized for stimulating bone cell growth and accelerating bone tissue healing. Additionally, magnesium is a crucial element for human metabolism. When used in vivo, the degradation of magnesium is typically non-toxic and is excreted through urine (Saris et al.,2000). Mg-Cu alloys have been used as a regenerative medicine as grafting materials and have shown promising results invitro. Further studies validating their efficacy and safety in the field of periodontal therapy are required.

4. Conclusion

Mg-Cu alloys offer a promising advancement in the treatment of periodontal defects, combining the beneficial properties of both metals to enhance healing and support bone regeneration. The combination of Mg-Cu alloy with the xenogenic bone graft paves way for further research in the management of periodontal defects. However, further vitro studies and clinical trials assessing antibacterial properties against dental plaque organisms, characterization studies on the formulated Mg-Cu alloy and Mg-Cu alloy combined with the xenogenic bone graft are required fully establish their efficacy and safety in periodontal therapy.

Acknowledgments

Sincere gratitude to all the authors of this study for their invaluable contributions and continuous support throughout the research process.

References

1. Alley, M.C., Scudiero, D.A., Monks, A., Hursey, M.L., Czerwinski, M., Fine, D.L., et al., (1986). Cancer cell line screening panel: A new in vitro testing panel for anticancer drug discovery. Cancer Research, 46(11), pp.4487–4496.
2. Bernardini, D., Nasulewicz, A., Mazur, A. and Maier, J.A., (2005). Magnesium and microvascular endothelial cells: A role in inflammation and angiogenesis. *Frontiers in Bioscience*, 10, pp. 1177–1182.
3. Chai H et al., (2012). In vitro and in vivo evaluations on osteogenesis and biodegradability of a beta-tricalcium phosphate coated magnesium alloy. *Journal of Biomedical Materials Research Part A*, 100, pp. 293–304.
4. Chou D.T., Hong, D., Saha, P., (2013). In vitro and in vivo corrosion, cytocompatibility and mechanical properties of biodegradable Mg-Y-Ca-Zr alloys as implant materials. *Acta Biomaterialia*, 9(10), pp. 8518–8533.
5. Figueiredo M.M., et al., (2012). Characterization of bone and bone-based graft materials using FTIR spectroscopy. In: *Infrared Spectroscopy – Life and Biomedical Sciences*. pp. 315–338.
6. Hong D., Saha, P., Chou, D.T., (2013). In vitro degradation and cytotoxicity response of Mg-4% Zn-0.5% Zr (ZK40) alloy as a potential biodegradable material. *Acta Biomaterials*, 9(10), pp. 8534–8547.
7. Li R.W., et al., (2014). The influence of biodegradable magnesium alloys on the osteogenic differentiation of human mesenchymal stem cells. *Journal of Biomedical Materials Research Part A*, 102, pp. 4346–4357.
8. Liu C., Fu, X., Pan, H., (2016). Biodegradable Mg-Cu alloys with enhanced osteogenesis, angiogenesis, and long-lasting antibacterial effects. *Scientific Reports*, 6.
9. Pihlstrom B.L., Michalowicz, B.S. and Johnson, NW, (2005). Periodontal diseases. *The Lancet*, 366(9499), pp. 1809–1820.
10. Qin H., Zhao, Y., An, Z., (2015). Enhanced antibacterial properties, biocompatibility, and corrosion resistance of degradable Mg-Nd-Zn-Zr alloy. *Biomaterials*, 53, pp. 211–220.
11. Saris N.E.L., Mervaala, E., Karppanen, H., Khawaja, J.A. and Lewenstam, A., (2000). Magnesium: An update on physiological, clinical and analytical aspects. *Clinical Chemistry and Laboratory Medicine*, 294, pp. 1–26.
12. Sato T.N., Qin, Y., Kozak, C.A. and Audus, K.L., (1993). Tie-1 and Tie-2 define another class of putative receptor tyrosine kinase genes expressed in early embryonic vascular system. *Proceedings of the National Academy of Sciences of the United States of America*, 90, pp. 9355–9358.
13. Wolf, F., et al., (2007). Magnesium and neoplasia: From carcinogenesis to tumor growth and progression or treatment. *Archives of Biochemistry and Biophysics*, 458, pp. 24–32.
14. Wolf, F.I., Cittadini, A.R. and Maier, J.A., (2009). Magnesium and tumors: Ally or foe? *Cancer Treatment Reviews*, 35, pp. 378–382.
15. Zhao, X., Wan, P., Wang, H., (2020). An antibacterial strategy of Mg-Cu bone grafting in infection-mediated periodontics. *Biomed Research International*, pp. 1–9.

Note: All the figures and tables in this chapter were made by the author.

Advances in Materials Science and Technology – Dr. Srikari Srinivasan et al. (eds)
© 2025 Taylor & Francis Group, London, ISBN 978-1-041-12342-2

20

A Novel Formulation of Collagen and Albumin-Based Nanoparticles: Optimization of the Methodology and in Vitro Validation

Greeshma Chandrashekar*

Department of Periodontology,
Faculty of Dental Sciences, Ramaiah University of Applied Science,
Bangalore, Karnataka, India

Deveswaran Rajamanickam,
Jahanavi S.

Department of Pharmaceutics,
Faculty of Pharmacy, Ramaiah University of Applied Science,
Bangalore, Karnataka, India

Kranti Konuganti,
Ashwini Shivananjegowda

Department of Periodontology,
Faculty of Dental Sciences, Ramaiah University of Applied Science,
Bangalore, Karnataka, India

ABSTRACT: Targeted drug delivery systems are currently a widely researched vista across multiple specialties. The advantages to this modality, as opposed to conventional oral administration, are use of lower drug concentrations and fewer systemic adverse effects, among others. Several vehicles have been formulated using synthetic or natural polymers. Collagen and albumin are natural proteins in the human body which can potentially be used to carry drugs for targeted delivery, as they are biologically inert and ubiquitous. Bovine variants of these proteins are close in structure to that of humans. Therefore, the aim of this study was to formulate and characterize a novel drug delivery vehicle by combining the two proteins. Fourier Transformation Infrared (FTIR) spectroscopy of bovine serum albumin (BSA) and bovine collagen (BC) were done to assess their compatibility. The nanoparticle formulation was done using two methods; BSA and BC were ultrasonicated in the first method and combined using desolvation in the second method. Fourier transform infrared spectroscopy (FTIR), Scanning Electron Microscopy (SEM) and X Ray Diffraction (XRD) were performed on both nanoparticles prepared and the results of these tests were compared. The functional groups of BC and BSA were noted to be present in the nanoparticles, upon performing FTIR. The SEM images showed nanoparticles to have formed via both methods of synthesis, but with marked difference in the morphology. XRD showed the same peaks to be present in the nanoparticles as the pure polymers. The two proteins were found to be compatible with each other for formulation of nanoparticles, implying that there were no alterations in the structural chemistry of BC and BSA upon combining. The BC and BSA nanoparticles synthesised by ultrasonication and desolvation retain the physical and structural chemistry of the pure polymers, and can thus be use to carry several drugs, owing to their docking sites.

KEYWORDS: Bovine serum albumin, Bovine collagen, Nanoparticles, Targeted drug delivery vehicles

*Corresponding author: greeshma.c17@gmail.com

DOI: 10.1201/9781003664277-20

1. Introduction

Nanotechnology has permeated into and had significant impact in the field of health sciences. It has found application in diagnosis as well as treatment of several conditions. But nowhere has its use been more pronounced than in drug delivery, where it boasts several advantages over conventional modalities (Mitchell et al., 2021). Chief among them is targeted delivery of the drug, with lower drug concentration required for treatment, thus reducing adverse effects. A wide variety of drugs are currently delivered using nanotechnology, mainly in the treatment of cancer and for gastrointestinal disorders (Liu.,2023).

There are certain requirements a drug delivery vehicle must possess to be clinically effective. It must be biocompatible, entrap the drug, concentrate at the site of action, and release the drug over a prolonged time span. Over the years synthetic bioresorbable polymers have been successfully used to formulate nanoparticles. They have been proven quite effective in drug delivery, but have certain drawbacks. Primarily, the byproducts released following the systemic disintegration of these materials may have a detrimental effect on the tissues. Therefore, the use of biopolymers for the formulation of nanoparticles for drug delivery is on the rise. Biopolymers are derived from living organisms and represent possible materials for the replacement of synthetic plastics. These materials are biodegradable and have a bearing on developing environmental sustainability through green synthesis. Biopolymers have a structural backbone of carbon, oxygen, and nitrogen atoms, making these the metabolic byproducts. Chitosan, gelatin, agar, and starch are examples of biopolymers that have been used as vehicles for the delivery of active molecules (Syed et al., 2023). Proteins such as albumin, zein etc. have emerged as drug delivery vehicles more recently. They demonstrate certain unique functions and properties, and allow for surface modification of particles, and ease of particle size control (Hong et al., 2020).

Collagen is one of the most ubiquitous proteins found in the human body. It has a triple helical structure, assembled into fibers, which accounts for its high tensile strength. It is biodegradable, biocompatible, biomimetic, demonstrates weak antigenicity, and possesses remarkable safety profile. Its limitations of rapid enzymatic degradation, weak mechanical strength and low thermal stability, can be overcome by crosslinking, grafting polymerization, blending and covalent conjugation, with other polymers. As a result, collagen can be formulated into discrete structures such as shields, films, sponges, hydrogels, microspheres, sheets, coatings, liposomes, disks, nanofibers, tablets, pellets and nanoparticles (Arun et al.,2021).

Albumins are a family of globular proteins, the most common of which are serum albumins. Its position as an exogenous or endogenous carrier protein can be extrapolated in the treatment of various diseases. It can be used to encapsulate several drugs due to the presence of Sudlow's sites on its structure, which are binding sites for various ligands. Bovine Serum Albumin (BSA) is widely accepted and applied in research because of its low cost, and easy availability and purification (Hornok et al., 2021). Although both proteins have been independently used in the formulation of drug delivery vehicles, they have hereto not been combined to form a nanoparticle. Therefore, there is no literature to validate their compatibility, method of combining or characterization of the formulation. This is the first study to address this research gap. Thus, the aim of this study is to assess the compatibility of bovine collagen (BC)and bovine serum albumin (BSA) in the formulation of a nanoparticle, to combine them using two different methods and the physical characterization of the novel formulations.

2. Methodology

Lyophilized bovine collagen type I (CF090) and bovine serum albumin (MB083) were purchased from HiMedia Laboratories Pvt Ltd, Bangalore, Karnataka. The bovine collagen (BC) was in liquid state and the bovine serum albumin (BSA) was in the form of flakes. Dichloromethane (028096) was procured from Central Drug House (P) Ltd. Ethanol solution (48075) was purchased from Sigma-Aldrich.

2.1 Fourier Transform Infrared Spectroscopy (FTIR) (Pre Formulation)

As this is the first instance where BC and BSA are being combined to formulate a nanoparticle, assessment of their compatibility was a prerogative (Derkach et al., 2020). This was done by Fourier transform infrared spectroscopy (FTIR), performed individually on BC (Fig. 20.1a) and BSA (Fig. 20.1b), using a model BRUKER Alpha II FTIR spectrometer. Following this, spectroscopy was also performed on a simple 1:1 mixture of BC and BSA, prior to any processing. The IR images thus obtained were evaluated and the peaks were compared. It was observed that there was no drastic alteration in the functional groups between the images, indicating the compatibility of BC and BSA (Fig. 20.1c).

2.2 Nanoparticle Formulation

Upon ascertainment of their compatibility, two different methods were selected to combine BC and BSA, to formulate nanoparticles.

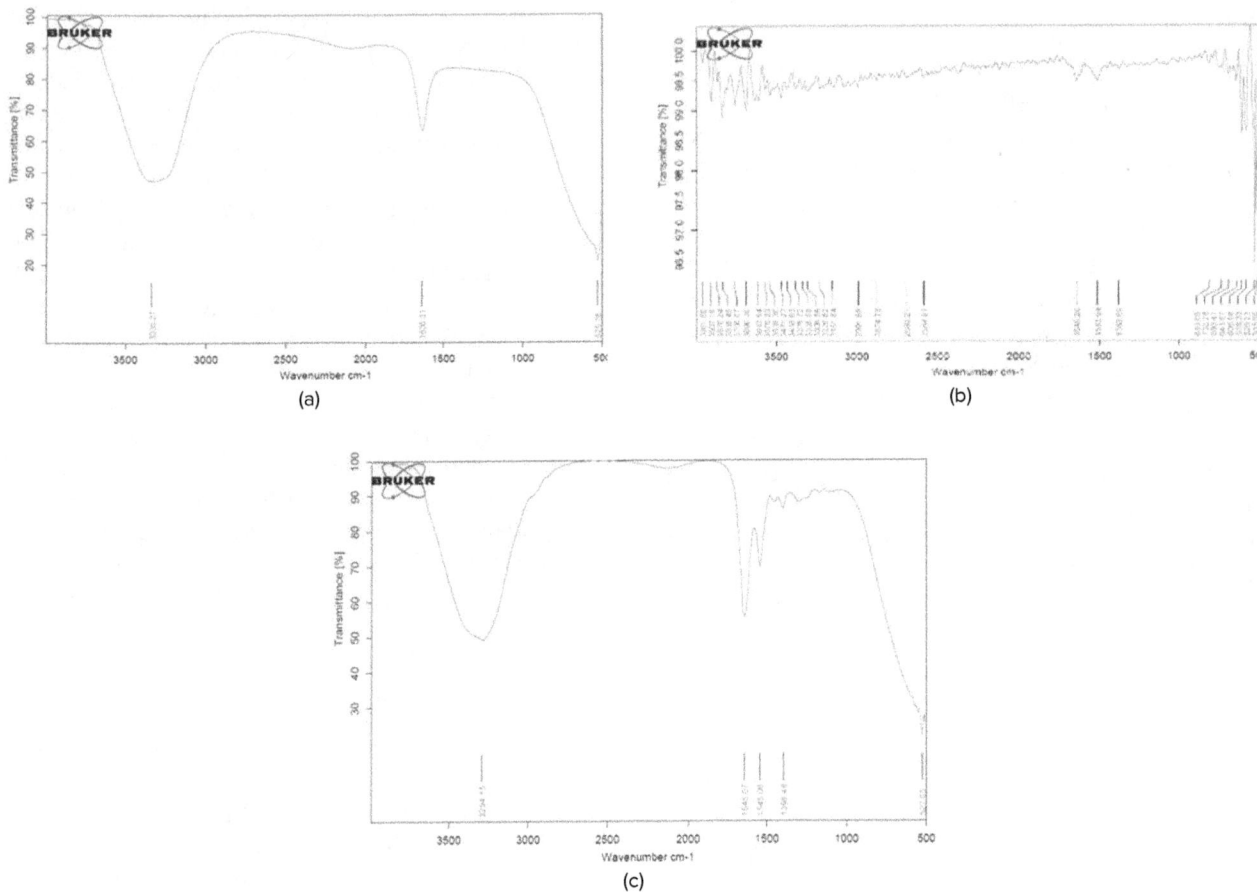

Fig. 20.1 Pre formulation FTIR images. (a) bovine collagen. (b) bovine serum albumin. (c) BC and BSA mixture

Method 1: Ultrasonication (Figure 20.2)

Equal proportions of BC and BSA were used in this method. Using a digital balance 1mg of BSA was weighed. 10 ml of 90 % ethanol was measured and the pre-weighed BSA was slowly added to the same with continuous stirring. 1 ml of BC was slowly added to 10 ml of 90% ethanol with continuous stirring. The two solutions were combined and subject to probe sonication for 1 hour (Ahmad et al., 2020). The resultant mixture was cloudy, with sediments. The mixture was filtered, the residue obtained was dried and further crushed using a mortar and pestle to form a powder. An organic phase of the powder was prepared by dissolving it in 10 ml of dichloromethane. This was subject to lyophilisation. (Fig. 10.4)

Method 2: Desolvation (Figure 20.3)

1ml of BC was dissolved in 9 ml distilled water with continuous stirring. 1mg BSA flakes was weighed with a digital balance and slowly dissolved in 10ml of distilled water under continuous stirring. Both solutions were subject to magnetic stirring for 30 minutes. The two solutions were mixed slowly under constant stirring, creating a uniform

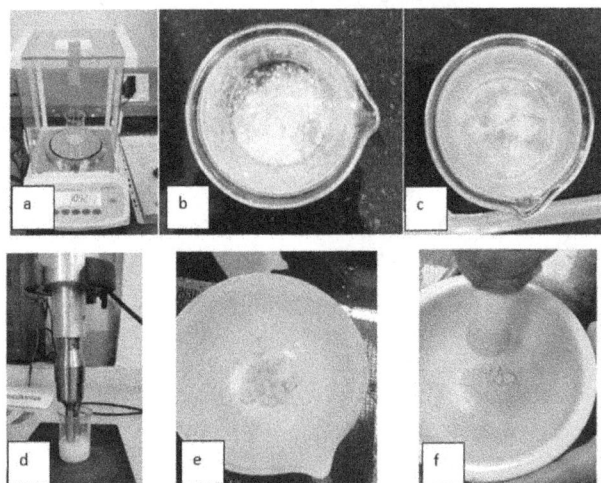

Fig. 20.2 Ultrasonication (a) weighing of pure polymers, (b) Ethanolic mixture of BSA, (c) Ethanolic mixture of BC, (d) Ultrasonication of b and c, (e) Residue of BSA+BC, (f) Trituration

solution, which was then lyophilized (Zhong et al., 2023) (Fig. 20.4).

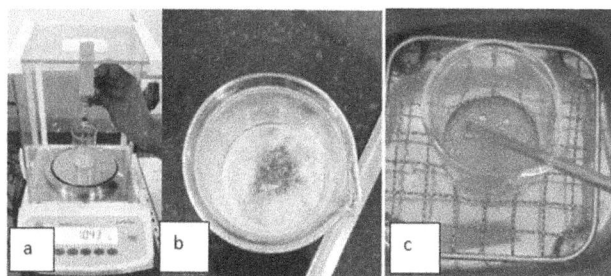

Fig. 20.3 Desolvation (a) Weighing of pure polymers, (b) Aqueous solution of polymers, (c) Magnetic stirring

Fig. 20.4 Lyophilisation

2.3 Characterization Tests

Attenuated Total Reflection-Fourier Transform Infrared Spectrometry (ATR-FTIR)

Attenuated total reflection-Fourier transform infrared spectrometry (ATR-FTIR) Infrared spectra of all the samples were recorded in Bruker ATR alpha kept at an ambient temperature of $25.0 \pm 0.5°$ C. The analytical procedure was simple and did not need any special sample preparation. Few mg of sample was placed on the Zinc solenoid crystal plate; Anvil was rotate to fix the sample and the spectra were recorded by scanning the samples in region of 4000-400 cm -1 to determine various functional groups.As mentioned preciously, FTIR was performed pre formulation, on BC and BSA, on the mixture of BC and BSA, as well as on the nanoparticles formulated with the two (Fig. 20.5b, 20.5c).

Scanning Electron Microscopy Imaging

The lyophilised powders were imaged by a scanning electron microscope (SEM) run at an accelerating voltage of 10kV using Hitachi SU 3500. Few micrograms of the

powder were fixed on to stub by a double-sided sticky carbon tape and kept inside the SEM chamber. Analysis was performed at different magnification such as 60X, 200X, 500X. 1.10X and 2.50X respectively to obtain better clarity on the particle morphology/ topology (Fig. 20.6a, 20.6b).

X-Ray Diffraction

A focused X-Ray beam was shot at the sample at a specific angle of incidence, which deflects depending on the crystal structure (inter-atomic distances) of the sample. The locations (angles) and intensities of the diffracted X-Rays were measured (Fig. 20.7a, 20.7b). Identification of phases was achieved by comparison of the acquired data to that in reference databases.

3. Results and Discussion

3.1 FTIR

The FTIR spectroscopic images of collagen showed its main functional groups. The observed spectral peaks were

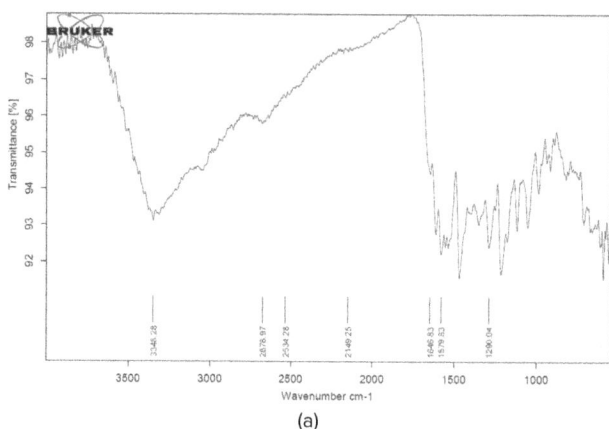

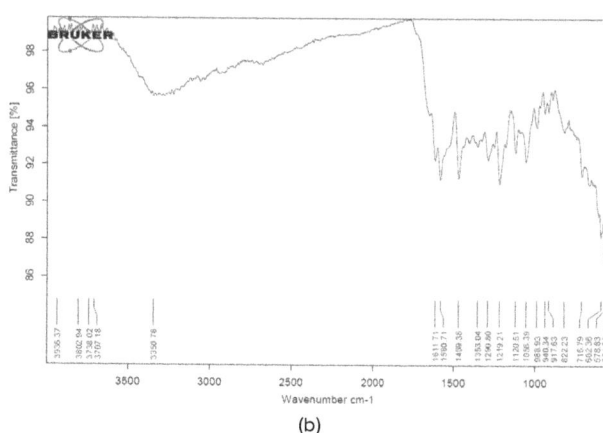

Fig. 20.5 (a) FTIR of BC+BSA nanoparticles formulated by ultrasonication, (b) FTIR of BC+BSA nanoparticles formulated by desolvation

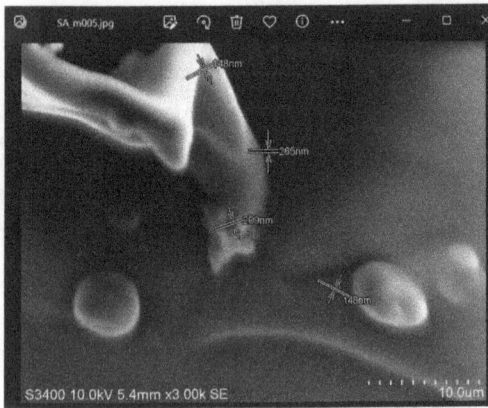 

(a)   (b)

Fig. 20.6 (a) SEM of ultrasonicated nanoparticles, (b) SEM of nanoparticles formulated through desolvation

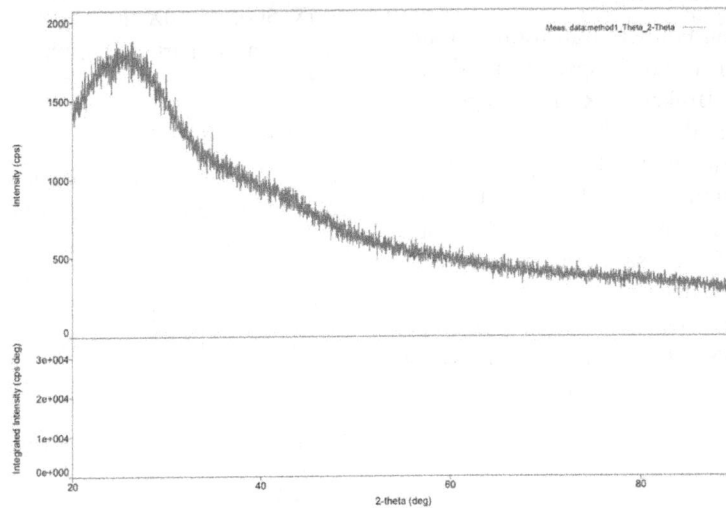

(a)

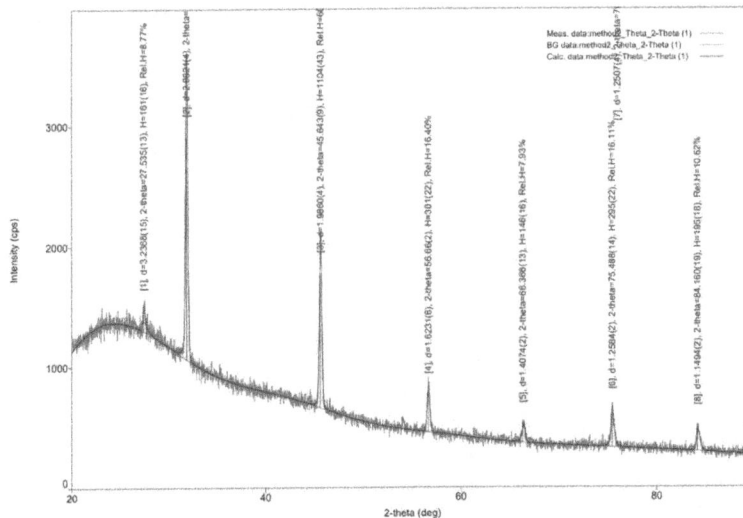

(b)

Fig. 20.7 (a) XRD of nanoparticles formulated with ultrasonication, (b) XRD of nanoparticles formulated with desolvation

amide I (1680–1620 cm¡1), amide II (1580–1480 cm⁻¹), and amide III (1300 1200 cm⁻¹). The strong hydroxyl band (3200–3600 cm⁻¹) was found to be overlapped with the amide A band (3360–3320 cm⁻¹)(Fig. 20.1a). These results match the FTIR of collagen in other studies as well (Stani, C et al., 2020). The FTIR of BSA showed bands which corresponded to amide A (3280 cm⁻¹), amide B (2970 cm⁻¹), amide I (1643 cm⁻¹), amide II (1515 cm⁻¹), CH2 bending groups (1392 cm⁻¹) and amide III (1260 cm⁻¹)(Fig. 20.1b) (Solanki, R et al., 2021).Combination : The peaks obtained in the pure polymers and in the combination were found to be similar at 3294.15 cm⁻¹, 1645.07 cm⁻¹, 1545.06 cm⁻¹ and 1398.48 cm⁻¹(Fig. 20.1c). Method 1 : The peaks formed in the nanoparticles formulated via ultrasonication matched the peaks noted in the pure polymers (Fig. 20.5a).

Method 2 : The peaks corresponding to those of the pure polymers were less pronounced in the formulation (Fig. 20.5b) (Ju, H et al., 2020)

3.2 SEM

Scanning electron microscopy was performed to observe the morphology and particle size of the materials formulated using both methodologies. Nanoparticles were observed to have formed through both methods.

Method 1: SEM of material showed a relatively uniform and amorphous structure. Nanoparticles of the size 148-265 nm were noted. Spherical particles were found to be distributed throughout, with even porosity (Fig. 20.6a).

Method 2: SEM showed a granular appearance, with well-formed particles. The size of these particles varied from 314-397 nm. Porosities were evenly distributed through material (Fig. 20.6b)

3.3 XRD

X-Ray Diffraction analysis was performed to assess the crystallographic structure of the nanoparticles formulated and to compare it to that of the pure BC and BSA. Following the formulation, XRD was performed on the nanoparticles. (Solanki et al., 2021; Florkiewicz et al.,2020; Maxwell et al., 2021) [11, 13, 14]

The XRD characteristics of the nanoparticles synthesised using method 1 (ultrasonication) showed a relatively smooth curve (Fig. 20.7a). Sharp peaks were evident in the image of nanoparticles synthesised using method 2 (desolvation)(Fig. 20.7b). The use of biopolymers in the formulation of nanoparticles has gathered momentum over the past few years. The materials developed have shown remarkable versatility of application, biocompatibility, and methodological feasibility. They have been used chiefly as scaffolds in tissue engineering and as drug delivery vehicles. Their primary drawback of rapid enzymatic degradation

can be overcome by cross-linkage with other polymers (Syed et al., 2023). The objective of the present study was to evaluate the compatibility of bovine collage and bovine serum albumin in the synthesis of a novel nanoparticle, to optimize the methodology for the formulation and characterize the nanoparticles thus formulated.There are several known methods to formulate nanoparticles such as ultrasonication, block co-polymerization, desolvation, electrospinning etc (Verma et al., 2020). As albumin and collagen have thus far not been combined to formulate a nanoparticle, two different methods were attempted to develop the same.

The first method (Method 1) used to synthesize the nanoparticle was ultrasonication. Ultrasonication is a tried and tested method for nanoparticle formulation. It is simple and reliably produces particles of a desired size (Souza, H. K et al., 2013) Ultrasonics refers to application of 20 kHz to 1 GHz frequency. The process works on the principle of cavitation, wherein high frequency soundwaves produce microbubbles in the solution subject to ultrasonic frequency. Upon the collapse of these microbubbles, the mechanical force generated breaks the covalent bonds in polymeric materials due to intense shear stress. In the present study, the ethanolic solutions of BC and BSA were mixed. The resultant mixture displayed phase separation by forming clumps immediately after being combined. This mixture was ultrasonicated for one hour, but the clumps did not dissipate upon probe sonication. To obtain a powder, the mixture was filtered, and the filtrate triturated. The triturate did not dissolve completely in the organic solvent. The fibrous content was evident and the mixture itself remained cloudy. In a previous study (Portier et al., 2017), gelatin and collagen were combined, where similar phase separation was noted. It was hypothesised that the polydispersity differences between the polymers, alteration in the configurational entropy or the tertiary protein structures may explain the phenomenon.

The second method (Method 2) employed desolvation to synthesize nanoparticles. It is a technically simple and economical process, requiring no specialized equipment (Zhong et al., 2023)]. It is therefore a more feasible, method than most. The solvent that is used may vary based on the polymers being used for the nanoparticle formulation. In this case, both BC and BSA were completely soluble in water, thus making distilled water the solvent of choice. This is the most common process used for the formulation of BSA nanoparticles (Tanjung et al., 2024). In the present study, the aqueous solutions of BC and BSA, upon being mixed with each other, were completely miscible. The solution was clear, with no deposits or phase separation noted. This indicates that the choice of methodology, and the thermodynamic variations have an impact on the phase separation. Akbarzadeh et al in 2014 discussed the impact of processing on phase separation, which may also be a factor in the present study.

Prior to combining the two proteins, FTIR Spectroscopy was performed to assess their potential compatibility (Segall et al., 2019, Stani et al., 2020). FTIR spectroscopy is not only useful to study the behavior a material but also as a compatibility screening tool, since the vibrational changes detected by this method serve as evidence for potential intermolecular interaction. The images obtained of BC and BSA matched those noted in previous studies, demonstrating the predominant functional groups of each. The analysis was repeated after combining the proteins, which closely resembled that of BC and BSA. Physical mixture of BC and BSA was, therefore noted to be compatible with each as the peaks of each were intact in the infrared images. This may be interpreted as the retention of most functional groups in BC and BSA. Following lyophilization, FTIR spectrograph of the nanoparticles demonstrated peaks which coincided with the pure polymers. The nanoparticles obtained by method 2 demonstrated less pronounced peaks than the one with ultrasonication, but were evidenced at the same loci. This indicated that the functional groups of BC and BSA remained undisturbed post formulation of nanoparticles. It was possible to interpret that minimal interaction occurred between the two proteins.

The SEM confirmed the formation of nanoparticles via both methods. The lyophilised powder of method 1 upon SEM analysis appeared more amorphous whereas that of method 2 demonstrated a more crystalline morphology. Well defined particles were noted to have formed with the desolvation method as compared to the ultrasonication. The results of this study match that of other formulations of collagen nanoparticles and albumin nanoparticles (Novaes et al., 2020). Both nanoparticles also showed evenly distributed pores.

XRD analysis, by way of the study of the crystal structure, is commonly used to identify the crystalline phases present in a material and thereby reveal chemical composition information. The XRD analysis of BC and BSA depict 2 theta value of approximately 22.5 degrees. Both graphs show broad peaks denoting the amorphous nature of the proteins (Solanki et al.,2021; Florkiewicz et al., 2021; Maxwell et al., 2006). The XRD images of the formulations show greater crystallinity in the nanoparticles obtained using desolvation as compared to ultrasonication. This corresponds to the SEM images of the two nanoparticles, which showed a more granular appearance in the latter. The peak matches that of the BC and BSA, implying their compatibility. It can be interpreted that following

combining the two polymers, there are no integral changes in the crystalline properties of BC and BSA

Based on the comparison of in vitro test results, it can be concluded that despite being similar in several characteristics, the nanoparticles obtained via different methods vary in certain aspects. In the present study, desolvation was a a more favourable method for combining BC and BSA to obtain crystalline particles and ultrasonication for an amorphous output.

Therefore we can validate the combination of Bovine collagen and Bovine serum albumin to formulate nanoparticles. These may be utilised in the targeted delivery of many a drug, at lower doses, with fewer systemic adverse effects. The limitation of this study is that drug entrapment efficacy of the nanoparticles has not been assessed. Other methods of crosslinking the two proteins may be evaluated as well.

4. Conclusion

The application of biopolymers in the field of nanotechnology has increased exponentially with new vistas being explored every day. Given their high safety profile, they have been used for many purposes, targeted drug delivery being one. Collagen, in its native and cross-linked forms, has been used as a scaffold in tissue engineering, due to its fibrous nature, excellent tensile strength and biocompatibility. Albumin has seen application in the form of a nanoparticle, usually conjugated with another polymer. But these two proteins have not been combined thus far, for the formulation of nanoparticles. The present study is the first where a nanoparticle was synthesized using bovine serum collagen and bovine serum albumin. Bovine collagen and bovine serum albumin were found to be compatible with each other for the synthesis of a nanoparticle, as confirmed by FTIR analysis. Two different methods of formulation, ultrasonication and desolvation, were used and the efficacy of the processes were compared. It was noted that nanoparticles were formed in both, but their microscopic structures were different. Ultrasonication yielded a more amorphous material which was evidenced by both SEM and XRD, whereas desolvation produced a granular material with well-formed particles. But in both methods of preparation the functional groups were intact.The results of the study showed that ultrasonication resulted in a more uniform, amorphous material, whereas desolvation yielded a granular material. Based on the requirement either method of formulation may be followed to obtain nanoparticles. Further investigations in terms of drug loading efficiency and biologic properties of the nanoparticles need to be evaluated.

Acknowledgments

We acknowledge the support of Faculty of Pharmacy in the formulation of the nanoparticles, and the financial assistance by RUAS in the form of Seed Grant Funding.

Funding

This study is funded by the Seed Grant received from RUAS. (Grant Number ORI/SG/FDS/007/2023)

References

1. Ahmad, M., Gani, A., Hassan, I., Huang, Q. and Shabbir, H., 2020. Production and characterization of starch nanoparticles by mild alkali hydrolysis and ultra-sonication process. *Scientific Reports*, *10*(1), p.3533.
2. Arun, A., Malrautu, P., Laha, A., Luo, H. and Ramakrishna, S., 2021. Collagen nanoparticles in drug delivery systems and tissue engineering. *Applied Sciences*, *11*(23), p.11369.
3. Derkach, S.R., Voron'ko, N.G., Sokolan, N.I., Kolotova, D.S. and Kuchina, Y.A., 2020. Interactions between gelatin and sodium alginate: UV and FTIR studies. *Journal of Dispersion Science and Technology*.
4. Florkiewicz, W., Słota, D., Placek, A., Pluta, K., Tyliszczak, B., Douglas, T.E. and Sobczak-Kupiec, A., 2021. Synthesis and characterization of polymer-based coatings modified with bioactive ceramic and bovine serum albumin. *Journal of functional biomaterials*, *12*(2), p.21
5. Hong, S., Choi, D.W., Kim, H.N., Park, C.G., Lee, W. and Park, H.H., 2020. Protein-based nanoparticles as drug delivery systems. *Pharmaceutics*, *12*(7), p.604.
6. Hornok, V., 2021. Serum albumin nanoparticles: problems and prospects. *Polymers*, *13*(21), p.3759.
7. Ju, H., Liu, X., Zhang, G., Liu, D. and Yang, Y., 2020. Comparison of the structural characteristics of native collagen fibrils derived from bovine tendons using two different methods: modified acid-solubilized and pepsin-aided extraction. *Materials*, *13*(2), p.358
8. Liu, R., Luo, C., Pang, Z., Zhang, J., Ruan, S., Wu, M., Wang, L., Sun, T., Li, N., Han, L. and Shi, J., 2023. Advances of nanoparticles as drug delivery systems for disease diagnosis and treatment. Chinese chemical letters, 34(2), p.107518..
9. Mitchell, M.J., Billingsley, M.M., Haley, R.M., Wechsler, M.E., Peppas, N.A. and Langer, R., 2021. Engineering precision nanoparticles for drug delivery. Nature reviews drug discovery, 20(2), pp.101–124.
10. Maxwell, C.A., Wess, T.J. and Kennedy, C.J., 2006. X-ray diffraction study into the effects of liming on the structure of collagen. *Biomacromolecules*, *7*(8), pp.2321–2326.
11. Novaes, J., Silva Filho, E.A.D., Bernardo, P.M.F. and Yapuchura, E.R., 2020. Preparation and characterization of Chitosan/Collagen blends containing silver nanoparticles. *Polímeros*, *30*(2), p.e2020015
12. Portier, F., Teulon, C., Nowacka-Perrin, A., Guenneau, F., Schanne-Klein, M.C. and Mosser, G., 2017. Stabilization of collagen fibrils by gelatin addition: a study of collagen/gelatin dense phases. *Langmuir*, *33*(45), pp.12916–12925.
13. Segall, A.I., 2019. Preformulation: The use of FTIR in compatibility studies..
14. Solanki, R., Patel, K. and Patel, S., 2021. Bovine serum albumin nanoparticles for the efficient delivery of berberine: Preparation, characterization and in vitro biological studies. *Colloids and Surfaces A: Physicochemical and Engineering Aspects*, *608*, p.125501..
15. Souza, H.K., Campiña, J.M., Sousa, A.M., Silva, F. and Gonçalves, M.P., 2013. Ultrasound-assisted preparation of size-controlled chitosan nanoparticles: Characterization and fabrication of transparent biofilms. *Food Hydrocolloids*, *31*(2), pp.227–236..
16. Stani, C., Vaccari, L., Mitri, E. and Birarda, G., 2020. FTIR investigation of the secondary structure of type I collagen: New insight into the amide III band. *Spectrochimica Acta Part A: Molecular and Biomolecular Spectroscopy*, *229*, p.118006.
17. Syed, M.H., Zahari, M.A.K.M., Khan, M.M.R., Beg, M.D.H. and Abdullah, N., 2023. An overview on recent biomedical applications of biopolymers: Their role in drug delivery systems and comparison of major systems. Journal of drug delivery science and technology, 80, p.104121.
18. Tanjung, Y.P., Dewi, M.K., Gatera, V.A., Barliana, M.I., Joni, I.M. and Chaerunisaa, A.Y., 2024. Factors affecting the synthesis of bovine serum albumin nanoparticles using the desolvation method. *Nanotechnology, Science and Applications*, pp.21–40.
19. Verma, M.L., Dhanya, B.S., Rani, V., Thakur, M., Jeslin, J. and Kushwaha, R., 2020. Carbohydrate and protein based biopolymeric nanoparticles: current status and biotechnological applications. *International journal of biological macromolecules*, *154*, pp.390–412..
20. Zhong, W., Li, J., Wang, C. and Zhang, T., 2023. Formation, stability and in vitro digestion of curcumin loaded whey protein/hyaluronic acid nanoparticles: Ethanol desolvation vs. pH-shifting method. *Food Chemistry*, *414*, p.135684.

Note: All the figures in this chapter were made by the author.

Advances in Materials Science and Technology – Dr. Srikari Srinivasan et al. (eds)
© 2025 Taylor & Francis Group, London, ISBN 978-1-041-12342-2

21

Bridging the Gap Between Phytopharmaceuticals and Periodontology: Coleus Amboinicus as a Potential Topical Treatment Modality

Melethu Krishnakumari Akhila[1],
Ashwini Shivananje Gowda[2], Greeshma Chandrashekar

Department of Periodontology,
Faculty of Dental Sciences, Ramaiah University of Applied Science,
Bangalore, Karnataka, India

Rajamanickam Deveswaran

Department of Pharmaceutics,
Faculty of Pharmacy, Ramaiah University of Applied Science,
Bangalore, Karnataka, India

Soumya Nagappa Sajan

Department of Periodontology,
Faculty of Dental Sciences, Ramaiah University of Applied Science,
Bangalore, Karnataka, India

ABSTRACT: Medicaments derived from natural sources have been used to treat a multitude of ailments. The rich herbal diversity of India has a significant influence on health care practices, which continues till date. Systematic research methods have paved the way and integrated these into modern dental health care, inciting the present study. *Coleus ambonicus* is a succulent folkloric medicinal herb that has proven anti-microbial, anti-diabetic and neuro pharmacological properties. To test the antimicrobial activity of ethanolic extract of *Coleus ambonicus* on subgingival plaque microbiota by estimating the reduction in colony forming units 100uL of ethanolic extract of *Coleus amboinicus* at the concentration of 400ug was tested against subgingival microbiota isolated from subgingival plaque samples by assessing colony forming units by streak plate method at 4-hour intervals for a total duration of 12 hours. This study exemplified the potential antimicrobial activity of Coleus ambonicus on subgingival microbiota as the total bacterial colony forming units reduced from 87 colonies to 2 colonies in the presence of the ethanolic extract of *Coleus Amboinicus*. Measurable reduction in colony forming units of subgingival microbiota depicted the significant antibacterial activity of the crude plant extract.

KEYWORDS: Coleus amboinicus, Microwave extraction, Chronic periodontitis, Phytopharmaceuticals

1. Introduction

Coleus Amboinicus (CA) is a large succulent herb found in the Indian subcontinent used as an anti-diarrhea and antifungal agent, and possess antidiabetic and neuropharmacological properties. The essential oil of Coleus Aromaticus/Amboinicus, is rich in carvacrol, thymol, eugenol, chavicol, ethyl salicylate (Dutta et

Corresponding author: [1]akhilamk15@gmail.com, [2]ashwiniperio1973@gmail.com

DOI: 10.1201/9781003664277-21

al.,1959).The herb has carvacrol and thymol as the major components responsible for the flavor (Rathod et al., 2023); while chlorogenic acid and rosmarinic acid are the phenolic components (Wadikar et al.,2016).The presence of alkaloids, carbohydrates, glycosides, proteins amino acids, terpenoids, quinine, tannins and flavonoids were found in CA leaves (Arumugam et al.,2016). They further found out the ethanolic leaf extract of CA showed more antibacterial activity against all the bacterial strains than the aqueous leaf extract of CA (Patel et al., 2010. Antibacterial activity of CA was analyzed against six bacterial strains and the results proved that the use of the leaves of CA to cure several illnesses, especially those caused by microbes, is valid (Ramalakshmi et al., 2014). The antibacterial property of this herb, has not been tested against subgingival microflora as a whole to the best of our knowledge, thus inciting the need for the present study.

2. Methodology

2.1 Preparation of the Plant Extract

The fresh leaves of CA were harvested (500g) from Bengaluru urban district of South Karnataka, India. The leaves were shredded, dried and powdered with a mechanical grinder to a coarse consistency and was then passed through a sieve numbered 44. The dry weight of the leaves was 20 grams, which was processed further to obtain the extract.

2.2 Microwave Assisted Extraction of Crude Plant Extract

200 ml of 99.9% pure ethanol was used as the solvent to extract the active components of CA. The coarse powder was submerged in ethanol contained in a glass beaker.

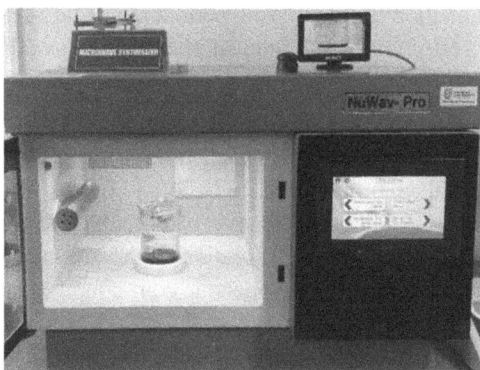

Fig. 21.1 Microwave synthesizer (Nu Wave Pro)

CA ethanolic extract was obtained by microwave synthesis (Nu Wave Pro) at 50°C, 55W for 1 minute. The solvent was evaporated using a boiling water bath, and the resulting 2-gram extract was transferred to a sterile container for further analysis.

2.3 MTT Assay

MTT assay is a colorimetric assay used for the determination of cell proliferation and cytotoxicity [Mosmann et al., 1983]. The cytotoxicity of the CA plant extract was tested against human fibroblast cell line (Cat No:106-05A, Sigma). Assay controls used were medium control (medium without cells), Negative control (medium with cells but without the experimental drug/compound), Positive control (medium with cells and 1% of Dimethyl sulfoxide-DMSO) [Alley et al.,1988].

Table 21.1 Test compounds evaluated to measure the cytotoxicity study against the human dermal fibroblast cell line against different concentrations

Sl no	Test	Cell line	Concentration treated to cells
1	Untreated cell line	HDF	No treatment
2	Toxic control	HDF	1% DMSO
3	Blank (No cells)	-	Only media without cells
4	Test compound	HDF	300,350,400,450,500 ug/ml

2.4 Microbiological Analysis

Five patients with chronic periodontitis (pocket probing depth >4mm) who reported to the Department of Periodontology, Faculty of Dental Sciences, Ramaiah university of Applied Sciences were included in the study. After supragingival plaque and calculus removal, subgingival plaque was collected from the deepest pockets using a Gracey curette, avoiding bleeding. The plaque thus collected was transferred into a sterile container with 10 mL of sterile saline solution. The semi quantitative method was used to assess the effect of CA extract on the colony forming units (CFUs) on sheep blood agar medium. The crude plant extract was diluted to 400 ug/ml in dimethyl sulfoxide (2% DMSO). The sub-gingival plaque sample(test) suspension was diluted up to 100uL with tryptic soy broth media which acts as a growth media for the bacteria. The growth control was made with 100ug each of 2% DMSO and tryptic soy broth and 10ul of bacteria. The samples were then placed in a shaker incubator at a speed of 150 RPM at 37°C for a period of one hour. The subgingival plaque sample was inoculated onto 5% Sheep blood agar plates with streak plate method at 1hr,4hr and 8hr intervals. The plates were then incubated at 37°C in bacteriological incubator for a period of 24 hours. The antimicrobial activity was determined by counting the colonies formed in the growth control in comparison to the test specimen.

3. Results and Discussion

3.1 Rapid Colorimetric Assay

The observations of cytotoxicity study by MTT assay suggests that against HDF (Human Dermal Fibroblast) cells, given test compound CA showed the moderate toxicity with IC50 concentration at 429.89ug/ml after the incubation period of 24hrs.

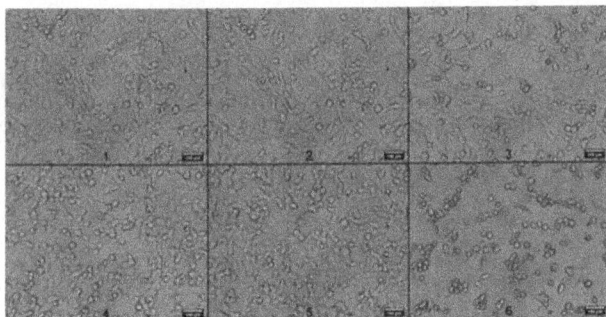

Fig. 21.2 Microscopic images of human dermal fibroblast cells treated with vary-ing concentrations of CA extract (1-untreated, 2-300ug, 3-350ug, 4-400ug, 5-450ug, 6-500ug) after 24 hours. The maximum non-toxic dose was 300ug/ml (with cell viability greater than 80%), while cell viability at 400ug/ml was 64.87%.

3.2 Antimicrobial Analysis

The Ethanolic extract of CA showed significant reduction in the colony forming units from 87 colonies to 2 colonies at 8 hours proving its antimicrobial efficacy against subgingival microflora.

India is recognized as one of the top 12 megadiversity's in the world with over 8,000 species [Chandra et al., 2016]. The use of plant extracts and phytochemicals with proven antimicrobial properties can be of great significance in the treatment of various infections (Alviano et al., 2009). The vital biologics in the medical sector fall into three main categories: Plant derived active substances, small bio molecules, phytopharmaceutical drugs, which are a newly developed third category. The ethnopharmacological method is the foundation for the development of phytopharmaceutical medications. This resulted in the inception of a novel idea known as polypharmacology, which claims that natural substances from various sources can interact with several targets within human physiology (Nasim et al., 2022). The minimum bactericidal concentration (MBC) of 80–150 μL/L and the minimum

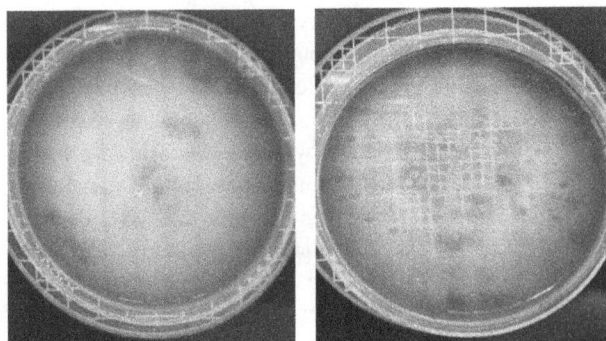

Test at 8 hours Growth control at 8 hours

Fig. 21.3 Colony forming units (CFU) at 8 hours in test and growth control on blood agar medium. The colony count in all the growth controls was >105. The colony forming unit (CFU) count in the test specimen was 87 at 1 hour, 5 at 4 hours and at 2 colonies at 8 hours

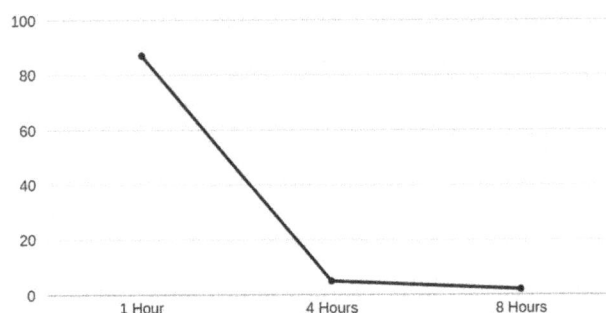

Fig. 21.4 Screening of antimicrobial activity of CA leaf extract against subgingival microflora by evaluation of colony forming units against incubation time.

inhibitory concentration (MIC) of 40–80 μL/L were found in the in vitro antibacterial studies (Koba et al., 2011). The typical method for identifying periodontopathic bacteria is by the collection of subgingival plaque samples (Haririan et al., 2014).

Hence the aim of this study was to determine the antimicrobial property of the plant extract of CA against subgingival microbiota as a whole. High antioxidant capacity was a defining feature of all ethanolic extracts. The 40% and 60% ethanol solutions were adequate for the efficient extraction of polyphenolic com-pounds generally, but the 80% ethanol solution was thought to be the optimal solvent for the extraction of flavonoids (Warycha et al., 2022). The Observations of cytotoxicity, in this study by MTT assay (Riss et al.,2004) suggests that against HDF cells, given test compound CA, showed the moderate toxicity with IC50 concentration at 429.89ug/ml after an incubation period of 24hrs.

The ethanolic crude extract of CA leaf showed the highest antimicrobial activity against Salmonella typhi 11 mm and Staphylococcus aureus 8 mm zone of inhibitions at 1000 µg/ml concentration (Ramalakshmi et al., 2014). There was significant reduction in the colony forming units from 87 colonies at to 2 colonies showcasing the effectiveness of the crude plant extract against subgingival microbiota.

4. Conclusion

CA is used for a variety of ailments and is extensively researched for its antibacterial and antioxidant potential. There are limited human studies to test the efficacy of its ethanolic extract on oral subgingival microbiota which warranted the current study. The present study unveils the remarkable antimicrobial action of the crude plant extract of CA against subgingival plaque microbiota. The exemplary antimicrobial action warrants further investigation to study the impact of the extract against oral microbiota, its use in various interventional or novel targeted treatment options and to determine the long-term efficacy with well-structured clinical trials.

Acknowledgments

Sincere gratitude to Dr. Tushar Shaw (Assistant Professor and Head of Department. Faculty of Life and Allied Health Sciences, Ramaiah University of Applied Sciences) for his valuable inputs.

References

1. Alley, M.C., Scudiero, D.A., Monks, A., Hursey, M.L., Czerwinski, M.J., Fine, D.L., Abbott, B.J., Mayo, J.G., Shoemaker, R.H. and Boyd, M.R., (1988). Feasibility of drug screening with panels of human tumor cell lines using a microculture tetrazolium assay. *Cancer research*, 48(3), pp.589–601.
2. Alviano, D.S. and Alviano, C.S., (2009). Plant extracts: search for new alternatives to treat microbial diseases. Current pharmaceutical biotechnology, 10(1), pp.106–121.
3. Arumugam, G., Swamy, M.K. and Sinniah, U.R., (2016). Plectranthus amboinicus (Lour.) Spreng: botanical, phytochemical, pharmacological and nutritional significance. *Molecules*, 21(4), p.369.
4. Chandra, L.D., 2016. Bio-diversity and conservation of medicinal and aromatic plants. *Adv Plants Agric Res*, 5(4), p.00186.
5. Dutta, S., (1959). Essential oil of Coleus aromaticus of Indian origin. *Indian Oil Soap J*, 25(1), pp.120–123.
6. Haririan, H., Andrukhov, O., Bertl, K., Lettner, S., Kierstein, S., Moritz, A. and Rausch-Fan, X., (2014). Microbial analysis of subgingival plaque samples compared to that of whole saliva in patients with periodontitis. *Journal of periodontology*, 85(6), pp.819–828.
7. Koba, K., Nénonéné, A.Y., Sanda, K., Garde, D., Millet, J., Chaumont, J.P. and Raynaud, C., (2011). Antibacterial activities of Coleus aromaticus Benth (Lamiaceae) essential oil against oral pathogens. *Journal of Essential Oil Research*, 23(1), pp.13–17.
8. Kulbat-Warycha, K., Oracz, J. and Żyżelewicz, D., (2022). Bioactive properties of extracts from Plectranthus barbatus (Coleus forskohlii) roots received using various extraction methods. *Molecules*, 27(24), p.8986.
9. Mosmann, T., (1983). Rapid colorimetric assay for cellular growth and survival: application to proliferation and cytotoxicity assays. *Journal of immunological methods*, 65(1-2), pp.55–63.
10. Nasim, N., Sandeep, I.S. and Mohanty, S., (2022). Plant-derived natural products for drug discovery: current approaches and prospects. *The Nucleus*, 65(3), pp.399–411.
11. Patel, R.D., Mahobia, N.K., Singh, M.P., Singh, A., Sheikh, N.W., Alam, G. and Singh, S.K., (2010). Antioxidant potential of leaves of Plectranthus amboinicus (Lour) Spreng. *Der Pharmacia Lettre*, 2(4), pp.240–245.
12. Ramalakshmi, P., Subramanian, N. and Saravanan, R., (2014). Antimicrobial activity of coleus amboinicus on six bacterial strains. *International Journal of Current Research*, 6(11), pp.9909–9914.
13. Rathod, N.B., Elabed, N., Punia, S., Ozogul, F., Kim, S.K. and Rocha, J.M., 2023. Recent developments in polyphenol applications on human health: A review with current knowledge. *Plants*, 12(6), p.1217.
14. Riss, T.L., Moravec, R.A., Niles, A.L., Duellman, S., Benink, H.A., Worzella, T.J. and Minor, L., (2016). Cell viability assays. *Assay guidance manual [Internet]*.
15. Wadikar, D.D. and Patki, P.E., (2016). Coleus aromaticus: a therapeutic herb with multiple potentials. *Journal of food science and technology*, 53, pp.2895–2901.

Note: All the figures and table in this chapter were made by the author.

Advances in Materials Science and Technology – Dr. Srikari Srinivasan et al. (eds)
© 2025 Taylor & Francis Group, London, ISBN 978-1-041-12342-2

22

Lemongrass oil Infused Dental Wafers: A Natural Shield against Oral Pathogens

Dhanya Shri Mahendran*,
Lalitha Shanka Jairam
Department of Pediatric and Preventive Dentistry,
Faculty of Dental Sciences, Ramaiah University of Applied Sciences,
Bangalore, Karnataka, India

Deveswaran Rajamanickam
Department of Pharmaceutics,
Faculty of Pharmacy, Ramaiah University of Applied Sciences,
Bangalore, Karnataka, India

ABSTRACT: This study aimed to formulate, characterize, and evaluate the antibacterial activity of a dental wafer containing lemongrass oil. The wafer was prepared using Hydroxypropyl Methylcellulose (HPMC), croscarmellose sodium, aspartame, and lemongrass oil, followed by freeze-drying. It was assessed for physical properties, disintegration time, and antibacterial efficacy against S. mutans using the disc diffusion method. Cytotoxicity was analyzed via MTT assay, and MIC/MBC values were determined by broth dilution. Results indicated that the lemongrass wafer exhibited significant antibacterial activity comparable to Ciprofloxacin, suggesting its potential as an adjunctive treatment for dental caries.

KEYWORDS: Dental caries, Drug delivery, Essential oil, Lemongrass oil, Natural agents

1. Introduction

Dental caries is a bacterial disease caused by the demineralization of enamel and dentin due to acidic by-products from bacterial metabolism (Toda et al.,20008). The condition remains highly prevalent, affecting 36.7% of adolescents and up to 83% of young children in developing nations. Globally, dental caries is the most common oral health issue, impacting billions. While chemical agents like triclosan, fluoridation, and chlorhexidine have been used for prevention, concerns over their side effects have led to interest in natural alternatives such as lemongrass oil (Pandey et al.,2021).

Lemongrass oil contains bioactive compounds, including vitamins, phenols, and flavonoids, which exhibit antioxidant, anti-inflammatory, and antimicrobial properties. It has shown effectiveness against various bacteria and fungi, making it a potential substitute for synthetic drugs (Meenapriya et al., 2017). Though widely used in medicine for treating infections and inflammation, its application in dentistry is limited. Existing drug delivery systems like mouthwash and capsules have drawbacks, such as irritation and ingestion risks. This study aims to formulate a dental wafer containing lemongrass oil and evaluate its cytotoxicity and antibacterial activity.

*Corresponding author: dhanyashri1996@gmail.com, dhanyashri1996@gmail.com

DOI: 10.1201/9781003664277-22

2. Methodology

2.1 Formulation of the Wafer Containing Lemongrass Oil

The reagents used for the preparation of the wafer were Hydroxypropyl Methylcellulose (HPMC), Croscarmellose sodium, Aspartame and Lemongrass oil (Table 22.1) was procured from Vasa Scientific Company, Bangalore, Karnataka. The wafer was formulated by adding 1g of Hydroxypropyl Methylcellulose (HPMC) in 25ml of ethanol in a beaker and it was stirred completely to obtain a clear solution. The obtained solution was transferred to beaker A. In beaker B, 2 ml of ethanol was taken, to which 50mg of croscarmellose sodium and 25mg of aspartame was added and then 0.5 ml of lemongrass oil was incorporated and then the mixture was stirred thoroughly. After which, the solution in beaker B was transferred to beaker A and was mixed thoroughly. The mixture was then transferred to petri plate for casting. The formulation was subjected to a freeze-phase in a freeze-dryer at-58°C for 2 hours. Then the obtained wafer was stored in an air tight container until further analysis (Fig. 22.1).

Table 22.1   Ingredients for the preparation the wafer

S. No	Ingredients	Quantity
1.	Hydroxypropyl Methylcellulose (HPMC)	1 g
2.	Ethanol	25 ml
3.	Croscarmellose sodium	25g
4.	Aspartame	25mg
5.	Lemongrass oil	0.5ml

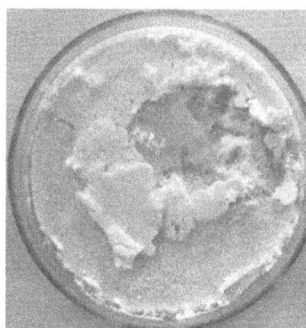

Fig. 22.1   Developed wafer

The developed wafer was characterized for its thickness, folding endurance, disintegration time and surface analysis were done.

Thickness

The thickness of the film was measured by micrometer screw gauge at different strategic locations (at least 5 locations). This was essential to determine uniformity in the thickness of the film as this is directly related to the accuracy of dose in the film (Baljeet et al., 2010).

Folding Endurance

It was determined by repeated folding of the wafer at the same place till it broke (Monton et al., 2024).

Disintegration Time (DT)

A wafer tablet was placed on a glass Petri dish containing 2 mL of water, gently agitating it every 5 s. The DT was recorded when the complete disintegration was observed

Structural Analysis

It was determined by Fourier Transformed Infrared (FTIR) Spectroscopy.

Antibacterial Activity of the Developed Wafer

The antibacterial activity was determined using the disc diffusion method and the minimum inhibitory concentration (MIC) determination assays, following the guidelines of the National Committee for Clinical Laboratory Standards (NCCLS) and the Clinical and Laboratory Standards Institute (CLSI).

2.2 Disc Diffusion Method

Microorganisms and Culture Preparation

The microorganisms were obtained from the Microbial Type Culture Collection and Gene Bank (MTCC), Chandigarh. Pure bacterial cultures were inoculated into Nutrient Agar plates and incubated for 24 hours at 37°C. Subsequently, fresh cultures were aseptically transferred to 2 ml of sterile 0.145 mol/L saline, and cell density was standardized to a 0.5 McFarland turbidity standard, yielding a bacterial suspension of approximately 1.5×10^{-8} colony-forming units per milliliter (CFU/ml). A standardized inoculum was utilized for antimicrobial testing.

Medium Preparation and Inoculation

Muller Hinton Agar Medium (Hi Media) was prepared by dissolving in 1000 mL of distilled water, followed by autoclaving at 121°C and 15 psi for 15 minutes to achieve sterility (pH 7.3). After autoclaving, the medium was allowed to cool, mixed thoroughly, and poured into petri plates (25 ml/plate). Each plate was inoculated with the standardized pathogenic bacterial culture.

Antimicrobial Testing

A concentration of 0.5 mg/ml lemongrass oil was incorporated into wafers. Using a sterile cork borer, three wells of 6 mm diameter were created in the inoculated Muller-Hinton medium. A sample loaded disc containing 1 mg/ml of the test substance was placed onto the surface of the medium. The plates were incubated for 24 hours at 37°C. Ciprofloxacin was used as a positive control.

Measurement of Antibacterial Activity

Post-incubation, the diameter of the zone of inhibition (ZOI) around each well was measured to assess antibacterial activity. The absence of a ZOI indicated no antibacterial activity. The antibacterial efficacy was classified as follows: resistant (ZOI < 7 mm), intermediate (ZOI 8-10 mm), and sensitive (ZOI > 11 mm).

Minimum Inhibitory Concentration (MIC)

The MIC was determined using the microtiter broth dilution method, following Andrews et al.,2001 and CLSI guidelines. Serial dilutions of the stock solution were prepared in sterile Nutrient broth, ranging from 4 mg/ml to 0.1 mg/ml, in test tubes. A bacterial suspension containing approximately 5×10^{-5} CFU/ml was prepared from a 24-hour culture. Using a standard wire loop (Merck), 10 µl of this culture was inoculated into test tubes containing 1 ml of the sample concentrations in Nutrient broth. Controls included positive (Ciprofloxacin at 5 µg/ml), negative (Nutrient broth only), and bacterial (microbe only) controls. The microtiter plate was incubated at 37°C for 24 hours. Post-incubation, absorbance at 600 nm was measured using a spectrophotometer. Tubes were examined for visible growth (cloudiness), recording growth as (+) and no growth as (-). The lowest concentration that completely inhibited bacterial growth (first clear tube) was recorded as the MIC value.

Minimum Bactericidal Concentration (MBC)

The Minimum Bactericidal Concentration (MBC) was determined using the microtiter broth dilution method as described by Ozturk[12]. The MBC was defined as the lowest concentration that achieved a 99.9% reduction in the bacterial inoculum after 24 hours of incubation at 37°C. Ten microliters were sampled from the wells corresponding to the MIC value and the next higher concentration and were plated on Mueller-Hinton agar (MHA). Colony counts were performed after 18–24 hours of incubation at 37°C. The MBC was identified as the concentration yielding fewer than 10 colonies. Each test was performed in triplicates.

Cytotoxicity Activity

The HOrF (Human oral fibroblasts) was purchased from ScienCell Research labs, CA, USA. The HOrF cells were maintained in DMEM with high glucose media supplemented with 10% FBS along with the 1% antibiotic-antimycotic solution in the atmosphere of 5% CO_2, 18-20% O_2 at 37°C temperature in the CO_2 incubator and sub-cultured for every 2-3days.

MTT [3-(4,5-dimethylthiazol-2-yl)-2,5-diphenyl-2,4-tetrazolium bromide] Assay

The cytotoxicity was measured using MTT assay developed by Mosmann[13] with slight modifications. After being harvested from culture flasks, the cells were seeded at 1.5×10^4 cells in each well of 96-well plate containing 100 µL of fresh growth medium per well and cells were permitted to adhere for 24 hours. The cells were treated with the lemongrass oil which were serially diluted with growth medium to obtain various concentration (12.5, 25, 50, 100, 200µg/ml). Then, 100 µl of each concentration was added to each well. After 48 hours of treatment the medium was aspirated and the cells were washed once with sterile phosphate buffered saline (PBS). To each well 5 mg/mL of MTT in PBS was added at 10% v/v and the plate was incubated at 37°C in 5% CO_2 for 3 hours. The medium was discarded and 200 µL of dimethyl sulfoxide (DMSO) was added to each well to dissolve the dark blue crystals of formazan salt. After incubation at 37°C for 10 minutes, the absorbance was measured at a primary wavelength of 570 nm and a reference wavelength of 650 nm, using a Multiskan Ascent micro plate reader. Each plate contained the lemongrass oil, a negative control (0.1% DMSO) and a blank. All tests and analyses were run in triplicate. Cell viability was calculated as a percentage using the formula:

Percentage cell viability = [Mean absorbance of treated cells / Mean absorbance of untreated cells] x 100

The IC_{50} value, which represents the concentration of the test agent that reduces cell viability by 50%, was determined using the linear regression equation

$$Y = Mx + C,$$

where Y equals 50 and M and C are derived from the viability graph.

3. Statistical Analysis

One-way ANOVA Test followed by Tukey's post hoc Test was used to compare Zone of inhibition between different groups. The level of significance was set at $P < 0.05$.

4. Results and Discussion

4.1 Characterization of the Developed Wafer
Thickness of the Wafer

The wafer used in the oral cavity had a thickness of approximately 100 micrometers (µm), which is comparable to the thickness of other wafers. This minimal thickness ensured that the wafer quickly dissolved upon contact with saliva, making it ideal for efficient drug delivery.

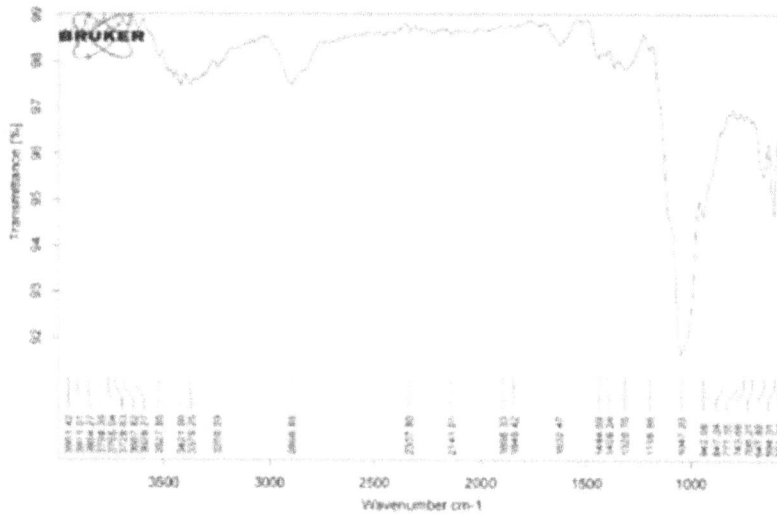

Fig. 22.2   FTIR analysis of the wafer

Folding Endurance

The folding endurance of the oral wafer was evaluated by repeatedly folding the wafer at the same location until it broke. The wafer demonstrated excellent folding endurance, withstanding over 30 folds before breaking. This high folding endurance indicated that the wafer is highly flexible and durable, making it suitable for manipulation and handling in the oral cavity without risk of breakage.

Disintegration Time

The wafer demonstrated a disintegration time of 2 minutes and 30 seconds, which ensured rapid dissolution and onset of action.

Structural Analysis

The FTIR spectrum of the wafer containing lemongrass oil shows characteristic peaks that align well with the known components of lemongrass oil, including citral, geraniol, and other terpenes. Key functional groups identified include O-H, C-H, C=O, and C=C, along with C-O stretching vibrations. These peaks confirm the presence of alcohols, aldehydes, and hydrocarbons typically found in lemongrass oil (Fig. 22.2).

4.2 Antibacterial Property

The results showed potential antibacterial potency of lemongrass (LG) molecule against the bacterial strain tested, *S.mutans* similar to the standard control used for the study. Ciprofloxacin was used as a standard control for the study (Fig. 22.3).

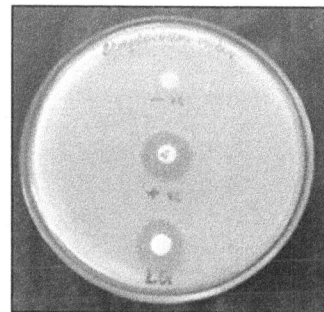

Fig. 22.3   Antibacterial activity test by disc diffusion method

The zone of inhibition was observed with the positive control (+ ve) and the lemongrass oil sample, indicating that these samples exhibit antimicrobial activity against *S.mutans* (Table 22.2).

Table 22.2   Diameter of zone of inhibition (Mm) of lemongrass oil against the microorganisms at 1mg/ml concentration after the incubation period of 24 hours

Culture condition	Zone of inhibition (mm)						
	Exp-1	Exp-2	Exp-3	Average	SD	SE	ZOI±SD
Control	0	0	0	0.00	0	0	0
Ciprofloxacin-5µg	17	19	21	19.00	2	1.154701	19 ± 2
Lemongrass oil	15	18	19	17.33	2.081666	1.20185	17.33 ± 2.08

MIC & MBC

The MBC was complementary to the MIC; whereas the MIC test demonstrated the lowest level of antimicrobial agent that greatly inhibited the growth, the MBC demonstrated the lowest level of antimicrobial agent resulting in microbial death [Table 22.3]. In the present study MIC and MBC values of lemongrass oil was 2mg/ml and 4mg/ml respectively which confirmed the antibacterial activity (Fig. 22.4).

Table 22.3 Growth pattern of *S.mutans* was observed after 24 hours of treatment with LG with desired concentrations

Sample Concentration (mg/ml)	Growth rate	OD@600nm
4	-	0.736
2	-	1.025
1	+	1.379
0.5	++	1.538
0.25	++	1.737
0.1	+++	1.905
Negative control	+++	2.025
NC (Blank)	-	0
Positive control	-	0

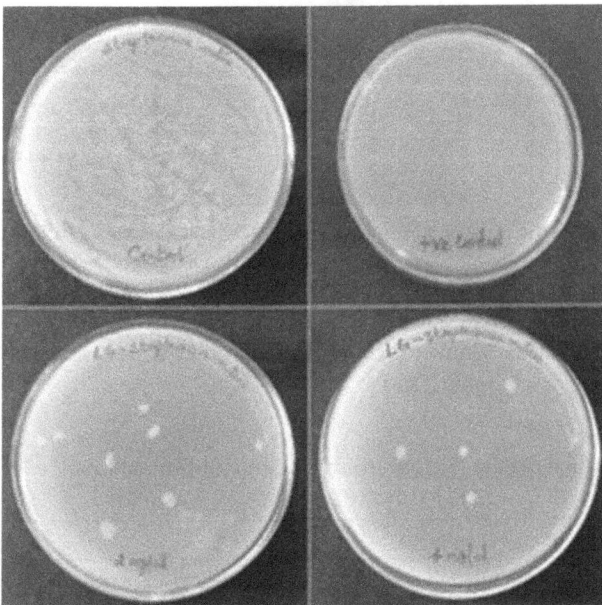

Fig. 22.4 Petri-dishes represented the *S.mutans* bacteria growth after the incubation period of 24 hours in control

MTT Assay

The MTT assay results suggested that the given test compound, lemongrass oil was non-toxic in nature with % cell viability values greater than 90% at the highest concentration of 200ug/ml after the incubation period of 24 hours (Fig. 22.5). So, lemongrass oil molecule proved to be safer against the Human oral fibroblasts till the 24 hours of incubation. Camptothecin (CPT) with 20uM was used as a toxic control for the study. As the molecule was safe to use on Non-malignant Human oral cells it's suggested to use the molecule for therapeutic purpose without any adverse or side effects on normal human health (Table 22.4).

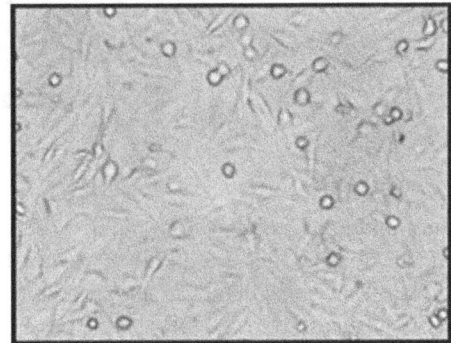

Fig. 22.5 Microscopic view of cultured cells post MTT assay treated with lemongrass

Table 22.4 Percentage of cell viability values of lemongrass treated on HOrF cells after the treatment period of 24 hours

Condition	Percentage cell viability
Untreated	100.00
CPT-20uM	46.16
LG-12.5ug	99.29
LG-25ug	98.76
LG-50ug	96.96
LG-100ug	92.74
LG-200ug	90.48

Ciprofloxacin has the highest mean ZOI at 19.40 mm, indicating strong antibacterial activity against *S. mutans* LG also demonstrates significant antibacterial activity with a mean ZOI of 17.60 mm [Fig. 22.6].

Dental caries is a dynamic, non-communicable disease influenced by biofilm and diet, leading to mineral loss in dental tissues. With untreated decay being highly prevalent in low- and middle-income countries, modern dentistry emphasizes preventive measures over traditional treatments

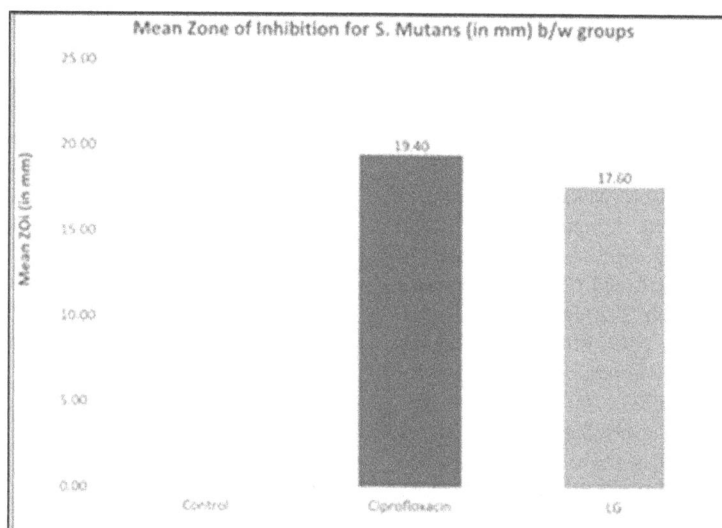

Fig. 22.6 Mean zone of inhibition for *S. mutans* (in mm) between groups

(Machiulskiene et al., 2020). Lemongrass oil (LGO), rich in citral, has demonstrated strong antibacterial activity against *S. mutans*, a key bacterium in caries formation. Our study confirmed that LGO-infused dental wafers effectively reduced *S. mutans* growth, aligning with previous research on its biofilm-disrupting potential and highlighting its promise as a natural alternative for caries prevention (da Silva et al., 2021). Numerous studies have documented the antimicrobial properties of LGO, supporting its potential in dental care (Palambo et al., 2011). Our MIC and MBC tests validated its efficacy at concentrations safe for oral use, reinforcing its suitability for dental applications. However, the study was limited to testing against a single microorganism. Further research and clinical trials are needed to explore its broader therapeutic benefits, optimize formulation, and determine ideal dosages for clinical application.

5. Conclusion

The dental wafer containing lemongrass oil demonstrated significant inhibitory effects on bacterial growth, indicating its potential efficacy as an adjunctive treatment for dental caries. The observed antibacterial activity suggests that lemongrass oil could help in reducing the bacterial load in the oral cavity, thereby aiding in the prevention and management of dental caries. This could be particularly beneficial in enhancing the effectiveness of standard dental treatments and providing an additional layer of protection against cariogenic bacteria.

Acknowledgments

The authors are thankful to MS Ramaiah University of Applied Sciences- Faculty of Dental Sciences and Faculty of Pharmacy for providing continual supervision for the work. The authors are also thankful to Averin Biotech Laboratories for helping to carry out the antibacterial testing.

Funding

This study was funded by Indian Society of Pedodontics and Preventive Dentistry, under "ISPPD Student Research Assistance Grant 2022".

References

1. Ambade SV, Deshpande NM. Antimicrobial and antibiofilm activity of essential oil of Cymbopogon citratus against oral microflora associated with dental plaque. European Journal of Medicinal Plants. 2019 Sep 2;28(4):1–1.
2. Andrews JM. Determination of minimum inhibitory concentrations. Journal of antimicrobial Chemotherapy. 2001 Jul 1;48(suppl_1):5–16.
3. Baljeet SY, Ritika BY, Roshan LY. Studies on functional properties and incorporation of buckwheat flour for biscuit making. International Food Research Journal. 2010 Nov 1;17(4).
4. Clementino MA, Gomes MC, Pinto-Sarmento TC, Martins CC, Granville-Garcia AF, Paiva SM. Perceived impact of dental pain on the quality of life of preschool children and their families. PloS one. 2015 Jun 19;10(6):e0130602.
5. da Silva Martins W, de Araújo JS, Feitosa BF, Oliveira JR, Kotzebue LR, da Silva Agostini DL, de Oliveira DL, Mazzetto SE, Cavalcanti MT, da Silva AL. Lemongrass (Cymbopogon citratus DC. Stapf) essential oil microparticles: Development, characterization, and antioxidant potential. Food Chemistry. 2021 Sep 1;355:129644.

6. Kumar R, Chopra S, Choudhary AK, Mani I, Yadav S, Barua S. Cleaner production of essential oils from Indian basil, lemongrass and coriander leaves using ultrasonic and ohmic heating pre-treatment systems. Scientific Reports. 2023 Mar 17;13(1):4434.

7. Machiulskiene V, Campus G, Carvalho JC, Dige I, Ekstrand KR, Jablonski-Momeni A, Maltz M, Manton DJ, Martignon S, Martinez-Mier E, Pitts NB. Terminology of dental caries and dental caries management: consensus report of a workshop organized by ORCA and Cariology Research Group of IADR. Caries research. 2020 Jan 29;54(1):7–14.

8. Meenapriya, M. and Priya, J., 2017. Effect of lemongrass oil on rheumatoid arthritis. *Journal of Pharmaceutical Sciences and Research*, 9(2), p.237.

9. Monton C, Kulvanich P, Chankana N, Suksaeree J, Songsak T. Fabrication of Orally Fast-Disintegrating Wafer Tablets Containing Cannabis Extract Using Freeze-Drying Method. Medical Cannabis and Cannabinoids. 2024 Jan;7(1):51.

10. Palombo EA. Traditional medicinal plant extracts and natural products with activity against oral bacteria: potential application in the prevention and treatment of oral diseases. Evidence-based Complementary and Alternative Medicine. 2011;2011(1):680354.

11. Pandey P, Nandkeoliar T, Tikku AP, Singh D, Singh MK. Prevalence of dental caries in the Indian population: A systematic review and meta-analysis. Journal of International Society of Preventive and Community Dentistry. 2021 May 1;11(3):256–65.

12. Toda S, Featherstone JD. Effects of fluoride dentifrices on enamel lesion formation. Journal of dental research. 2008 Mar;87(3):224–7.

Note: All the figures and tables in this chapter were made by the author.

Advances in Materials Science and Technology – Dr. Srikari Srinivasan et al. (eds)
© 2025 Taylor & Francis Group, London, ISBN 978-1-041-12342-2

23

Comparative 3D Finite Element Analysis of Stress Distribution and Displacement Patterns in Craniofacial Structures Using Hyrax and Memorax Maxillary Expansion Appliances

Selvi Vijayakumar*,
Prashantha Govinakovi Shivamurthy, Silju Mathew
Department of Orthodontics and Dentofacial Orthopaedics,
Faculty of Dental Sciences, M S Ramaiah University of Applied Sciences,
Bangalore, Karnataka, India

ABSTRACT: Evaluating the interplay of biology with the mechanics applied during Rapid Maxillary Expansion (RME), could shed light on the way that the teeth and bone respond to the applied force and this can be studied with time-dependent Finite Element Method (FEM) models that take those effects into consideration1. Studies on FEM show that stress distribution values with RME were found to be the highest with the Hyrax expansion screw 2. The magnitude of stress distribution patterns and displacements on dentofacial structures between RME and Memory Palatal Split Screw (Memorax) using FEM has not been addressed so far. In this present study the stress distribution pattern and the displacement of skeletal and dentofa-cial structures using Memorax and Hyrax appliance via the 3D FEM was evaluated. The 3D FEM was based on the computed to-mography data of a 16-year-old female patient with a constricted maxilla. The Hyrax model included 4,32,280 tetrahedral elements with 77,276 nodes. The Memorax model included 4,39,616 tetra-hedral elements with 79282 nodes. Activated the screw two quar-ter-turns a day ($0.2 \times 2 = 0.4$ mm) and von Mises stress distribu-tions and displacements were evaluated. A wedge-shaped opening was observed with both the appliances. Greater lateral displace-ment was observed in dentoalveolar region (Hyrax= 0.717mm, Memorax = 0.706mm) and the periodontal ligament of anterior teeth (Hyrax = 0.414mm, Memorax = 0.403) and craniofacial su-tures with Hyrax. All the anatomical structures showed slightly greater anterior displacement with Hyrax appliances when com-pared to Memorax and in the vertical plane, greater inferior dis-placement was recorded with the Hyrax (Hyrax= - 0.221mm, Memorax = -0.214). While both appliances exhibit nearly equiva-lent lateral, anteroposterior, and superoinferior displacement val-ues with minimal variance, the Hyrax appliance demonstrates greater displacement and desirable maxillary expansion.

KEYWORDS: Finite element method, Rapid maxillary expansion, Stresses, Displacement

1. Introduction

The history of Rapid Maxillary Expansion (RME) traces back to Angell's initial description in 1860, with a resurgence by Haas nearly a century later, marking significant advancements in orthodontics. Traditionally, RME involves jack screw appliances that exert substantial forces, as detailed by Isaacson and Ingram (Isaacson & Ingram,1964). These forces, escalating up to 22 pounds (10 kg) over the treatment period, are known to compress

*Corresponding author: selvijohn80@gmail.com

DOI: 10.1201/9781003664277-23

the periodontal ligament, facilitating orthopedic midpalatal suture opening and orthodontic movement of the posterior teeth (Oliveira et al.,2021; Haas, 1961; de Silva et al.,1991).

In orthodontics, Rapid Maxillary Expansion (RME) seeks to achieve minimal dental and maximal skeletal effects (Haas,1961), but the high initial forces often lead to discomfort and adverse reactions. Researchers like Isaacson and Ingram (Isaacson & Ingram, 1964) have suggested that reducing these forces could stabilize expansion earlier, potentially shortening treatment duration. To address these issues, recent studies have explored alternatives such as light, continuous forces to minimize trauma and enhance patient comfort.

Innovations in appliance design, such as the W appliance for midpalatal sutural opening and nickel-titanium (Ni-Ti) expanders activated by oral temperature, illustrate ongoing efforts to refine RME techniques. Additionally, magnetic sys-tems for maxillary expansion demonstrate varying force applications, reflecting a trend towards more effective and patient-friendly procedures. A recent study (Wichelhaus et al., 2004) introduced the "Memory Palatal Split Screw" by Forestadent, which was evaluated using an Instron universal testing machine. This screw showed consistent forces of 12-14 N (1,224-1,428 g) during activations, significantly lower than traditional screws. This aligns with Isaacson and Ingram's advocacy for con-stant, low-load deflection rates, aiming for rapid, physiological expansion forces with improved patient tolerance.

In 2012 (Halicioğlu et al.,2012), an initial study aimed at evaluating the dentoskeletal impacts of the memory screw over a 6-month follow-up period with 5 cases. Their findings suggested that the newly developed memory expansion screw offers benefits characteristic of both rapid and slow expansion procedures. It effectively widens the midpalatal suture and expands the maxilla using comparatively lighter forces within a shorter timeframe. Moreover, the resulting increases in maxillary apical base and intermolar width remained stable even after the retention period. Further research (Halicioğlu et al., 2010) indicated that the Memory screw resulted in significant increases in interpremolar and intermolar distances and reduced nasal airway resistance. A comparative study (Halicioğlu and Yavuz ,2014), with Hyrax expander and memorax using CBCT, plaster models, and cephalograms at pretreatment, postexpansion, and after retention, found that the Memory screw might offer advantages such as shorter expansion periods, additional expansion during reten-tion, and lower forces compared to the conventional Hyrax screw. However, more research is needed to fully evaluate the comparative effects of these screws on dentofacial structures.

In orthodontics, the Finite Element Method (FEM) is essential for analyzing stress distribution in biological systems like the periodontal ligament and alveolar bone. FEM allows for the accurate prediction of biomechanical responses through virtual models, enabling researchers to simulate various clinical conditions without exposing patients to risk, thereby ensuring ethical experimentation. This method's capacity to model complex structures with varying material properties provides a versatile platform for analyzing intricate force systems in orthodontic treatments (Jafari et al., 2003).

As recent research indicates that while both Memorax and Hyrax appliances induce dentoalveolar tipping in anchored teeth, Memorax generally promotes bodily movement expansion with less tipping, particularly in the premolars. In contrast, the Hyrax appliance tends to cause greater tipping in anchor premolars and molars. The newly developed Memory screw, known for its low and constant force application, might accelerate the maxillary expansion phase to approximately one week. These findings underscore the need to further investigate stress distribution patterns in sutures and dentofacial structures during the active expansion phase.

This study aims to compare the magnitude and distribution of stress and displacement changes during maxillary expansion using Memorax and Hyrax appliances in a 16-year-old female patient, employing FEM. Currently, no research has directly compared the stress distribution patterns and magnitudes, nor the displacement magnitudes, between these two appliances using FEM. The findings from this research are expected to offer important biomechanical insights for enhancing orthodontic treatment protocols.

2. Methodology

2.1 Construction of an FE Model

The 3D FEM was based on the computed tomography data of a 16-year-old female patient with a constricted maxilla. Spiral C.T Scan Machine - An X-force/SH spiral C.T scan machine was used for taking the C>T scan images of the cranium. Using the data extracted from the archival dental volumetric tomography scan images of a 16-year-old female patient with maxillary transverse deficiency, 3D finite element models were generated with institutional review board approval. The parents/legal guardian of the patient previously signed an informed consent form stating that his archival data could be used for scientific purposes (Table 23.1). The Hyrax model included 4,32,280 tetrahedral elements with 77,276 nodes. The MEMORAX model included 4,39,616 tetrahedral elements with 79282 nodes. The teeth, cortical and cancellous bones, sutures,

Table 23.1 Number of nodes and elements of the hyrax and memorax

Material		HYRAX		MEM-ORAX	
	No of Nodes	No. of Elements	No. of Nodes	No. of Elements
Cortical bone	19817	38293	19785	38229
Cancellous bone	47138	190211	47123	190191
Teeth	38374	174843	38374	174843
Sutures	1856	2840	1856	2840
Appliance	3278	6800	4547	12533
PDL	10753	20957	10753	20957

periodontal ligament, and stainless steel were considered to be homogenous and isotropic. The material properties (Young's modulus and Poisson's ratio) of the tooth, cortical bone, and PDL (periodontal ligament) were entered in the pre-processing stage (Table 23.2). The assembled finite element model of the skull, tooth, and PDL was then imported into ANSYS software for analysis.

Table 23.2 Young's modulus and poisson's ratio of various materials used in this study

	Young's modulus N/mm^2	Poisson's Ratio
Periodontal Ligament	50	0.49
Cancellous bone	7.9×10^3	0.30
Cortical bone	1.37×10^4	0.30
Suture	7	0.40
Tooth	2.07×10^4	0.30
Stainless Steel	2.1×10^5	0.3
Nickel Titanium	44×10^3	0.33

ANSYS software was used to solve the mathematical equation and to calculate the stress and displacement pattern of the skull. Post processing was the last stage of the FEM in which contour plots of the displacement and stresses was obtained from the results of the analysis performed. The von Mises stress distributions after 0.2 mm of expansion and displacement patterns after 5 mm of expansion were evaluated.

The Study will be Carried Out using a Three-Dimensional Finite Element Analysis using Following Resources

1. Workstation computer with following configuration 11th Gen Intel core i5 with 2.42 GHz 8 GB of RAM 4GB Graphics card 500GB hard Disc 17" Monitor. 2. Spiral C.T Scan Machine: An X-force/SH spiral C.T scan machine was used for taking the C>T scan images of the cranium. 3. Software's used was Mimics 8.11, Materialize 's Interactive Medical Image Control Sys-tem (MIMICS) is a medical modeling software used for the visualization and segmentation of CT/MRI images. Hyper mesh 2019.0 was the software used for converting geometric model into finite element model Altair HyperWorks is the Engineering Framework for Product Design for maximizing product performance, automating design processes, and improving profitability within an open and flexible environment. ANSYS 2017.2: Analysis System Software. This software is used for carrying out Finite element Analysis of structures and Fluids for various applications like Automotive, Civil, Manufacturing, Aerospace and Bio-medical fields, etc

Procedure

The CT scan data is read into MIMICS software and processed further to ex-tract only the region of interest for the study. Region of interest like maxilla, upper teeth, PDl, are extracted. The extracted DICOM data is then exported to Hypermesh software.Geometric model is created in Hypermesh software, also the physical models of the appliances are converted to geometric models by reverse engineering method (measuring the dimensions of the physical parts and creating the CAD models). Finite element model is created and physical parts are assembled on maxilla and material properties are assigned for each part like PDL, Teeth, Bones, sutures, appliances, etc. Loads and boundary conditions are applied to the model as per appliance and then the model is exported to ANSYS software. Run the model using ANSYS software and extract the results like movement of teeth, stress and strains, etc. Post processing the results and documentation is the last step

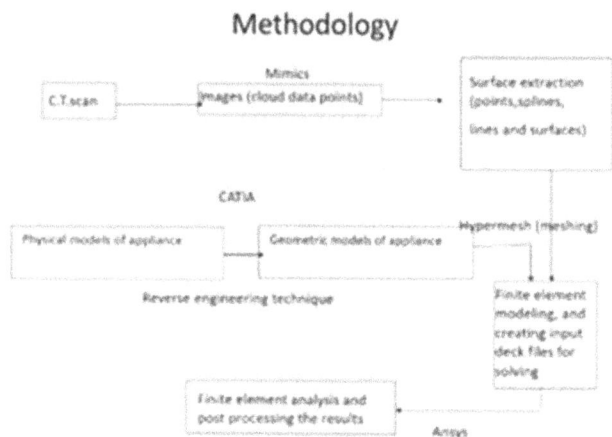

Fig. 23.1 The digital imaging and communications in medicine images on mimics software

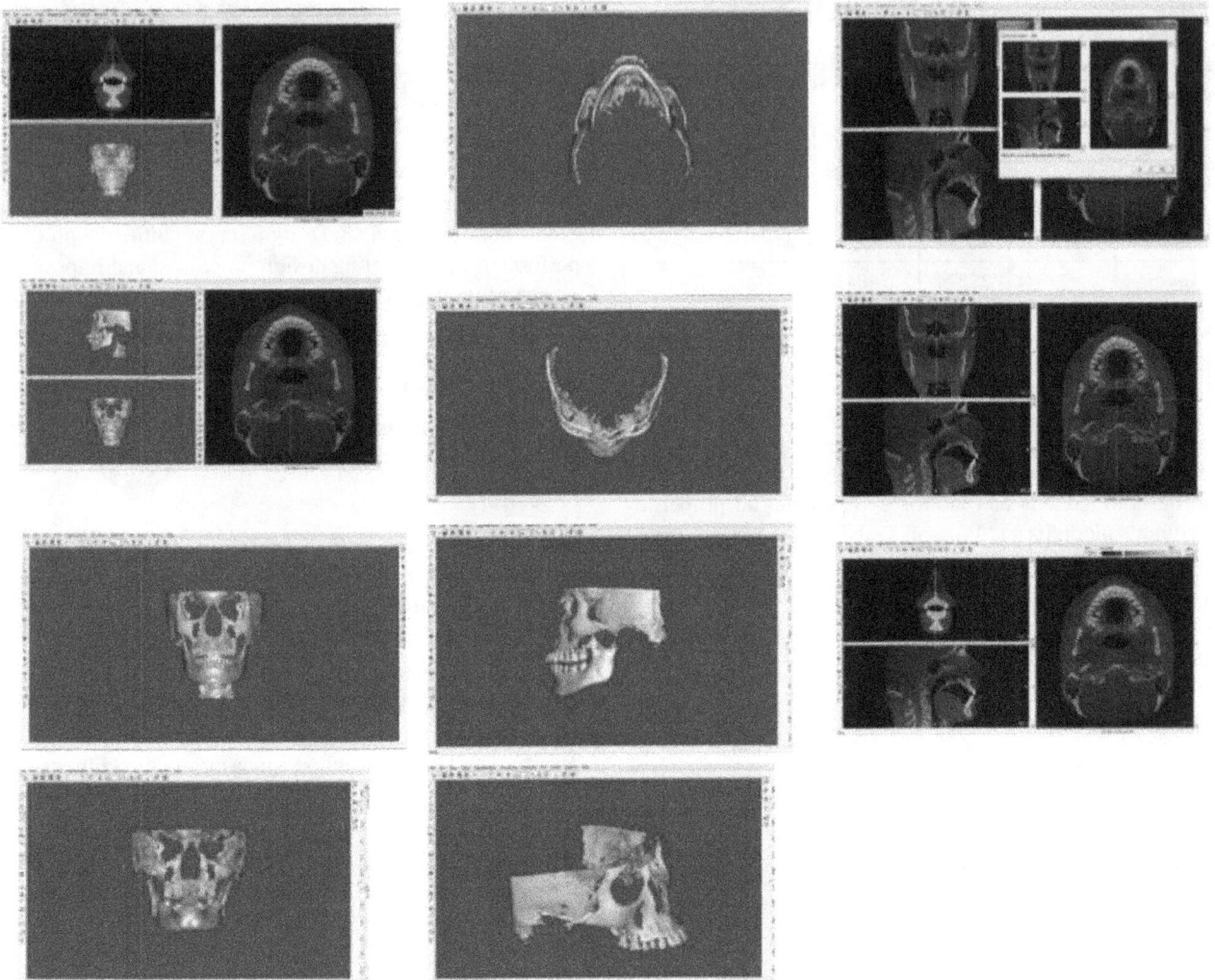

Fig. 23.2 CT scan data of 16y female

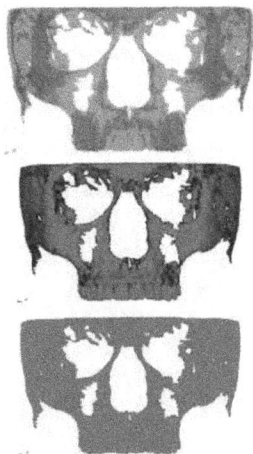

Fig. 23.3 Rapid form images showing surface creation for maxilla

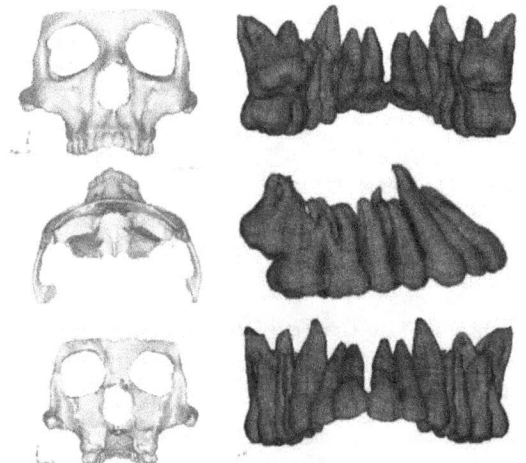

Fig. 23.4 Geometric model image of maxilla and upper teeth

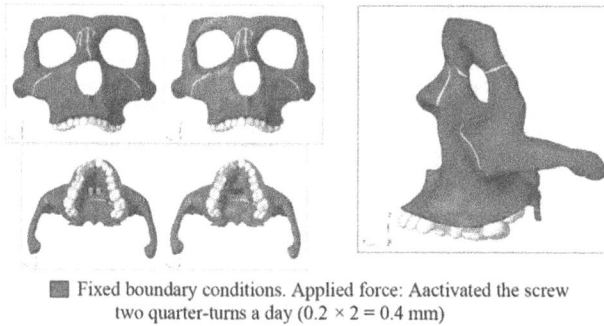

Fixed boundary conditions. Applied force: Aactivated the screw two quarter-turns a day (0.2 × 2 = 0.4 mm)

Fig. 23.5 Boundary conditions

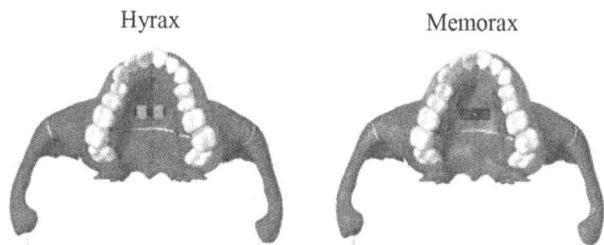

Hyrax   Memorax

Fig. 23.6 Designs of the RME types

Boundary Conditions and Reference Point (Fig. 23.5)

The boundary limitation was as follows: the Zygomatic arch and the forehead as a fixed point, the upward and downward, forward and backward, and right and left displacement was constrained; the shape and load was made symmetric around the X – Y axis (vertical and central section). The 3D co-ordinates were X- transverse direction plane; Y- antero-posterior di-rection plane; and Z- vertical direction. The midpoints of the buccal and lingual alveolar ridge of each tooth were used as reference points to evaluate alveolar bone displacement. Positive values indicate forward, outward, and upward dis-placements on the X, Y, and Z planes, respectively. The models were sectioned at canines and first molar by YZ plane and reference points were placed to assess teeth displacement.

Design of the Appliance

Hyrax expander and "Memory Palatal Split Screw" described by Wichelhaus et al.12 was used in the present study. Maxillary first premolars and first molars were banded and four-armed expansion screws were soldered to the bands in both groups. It incorporates superelastic Ni-Ti open-coil springs in the screw bed, which reduce excessive expansion forces. Special precaution was taken to position the screw body parallel to the occlusal plane and as close as possible to the palatal mucosa. Anterior arms of the screw were soldered to the first premolar bands, and the posterior arms were soldered to the first molar bands. In addition, a piece of stainless-steel wire (diameter, 1 mm) was soldered between the first premolar and first molar bands.

3. Results

The 3D coordinates were recorded for various craniofacial structures before and after the screw activation in all 3 dimensions (X-axis positive value: lateral movement, X-axis- negative value: medial movement, Y-axis-positive value: anterior movement, Y-axis-negative value: posterior movement, Z-axis -positive value: superior movement, Z-axis -negative value: inferior movement). Positive changes indicated lateral, anterior and superior displacements.

3.1 The Von Mises Stress Distribution (see Table 23.3)

In this study, the screw was activated two quarter-turns a day (0.2 × 2 = 0.4 mm). The stress distribution with the initial 0.2 mm of expansion is presented in a color band, with different colors representing various stress levels, where red indicates areas with the highest stress and blue indicates the lowest stress. In the Fig. 23.7 & 23.8, areas of elevated stress were observed around the teeth and periodontal ligament when using the Hyrax appliance (72.36 MPa and 1.89 MPa, respectively), compared to the Memorax. The highest stress in the cortical bone of the maxilla was found in the anterior cervical regions, with the Hyrax device registering 23.35 MPa and the Memorax device showing 22.65 MPa. At the mid-palatal suture, the Hyrax appliance exhibited a maximum stress of 2.82 MPa, while the Memorax showed 1.698 MPa. Similarly, higher stress fields were noted with the Hyrax appliance around various sutures such as zygomaticomaxillary, frontomaxillary, nasomaxillary, zygomaticotemporal, and internasal (0.094 MPa, 0.091 MPa, 0.087 MPa, 0.081 MPa, and 0.033 MPa, respectively) compared to the Memorax (0.086 MPa, 0.078 MPa, 0.075 MPa, 0.071 MPa, and 0.07 MPa). Consequently, increased stress levels were observed in the maxilla, mid-palatal sutures, and periodontal ligament of all teeth up to the first molar. In both the Hyrax and Memorax appliances, maximum stress was concentrated in the anterior part of the midpalatal suture at the position of the incisive papilla, emphasizing the critical role of this location in stress distribution. So, overall maximum stress was concentrated on the location of incisive papilla below the junction of two central incisors.

The von Mises distribution results reveal that the HYRAX appliance exerts higher stress compared to MEMORAX for most teeth. Specifically, the stress values for the HYRAX are 72.36 for the central incisor (vs. 70.502

HYRAX   MEMORAX   HYRAX   MEMORAX

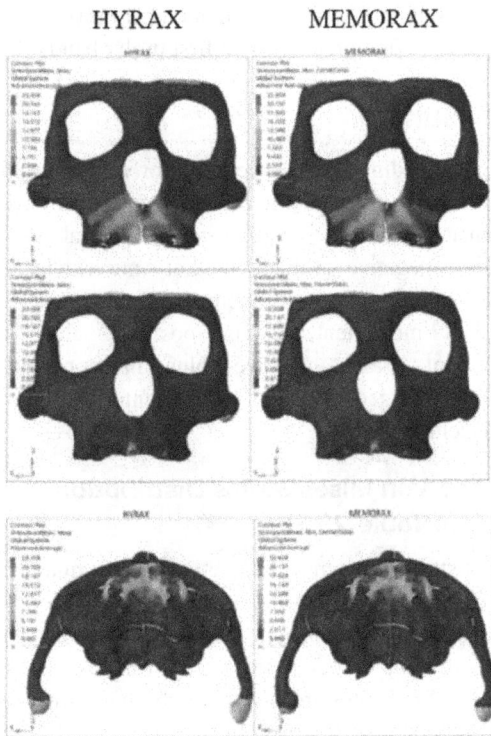

Fig. 23.7 Stress distribution in cortical and cancellous bone (Mpa) - frontal & occlusal view

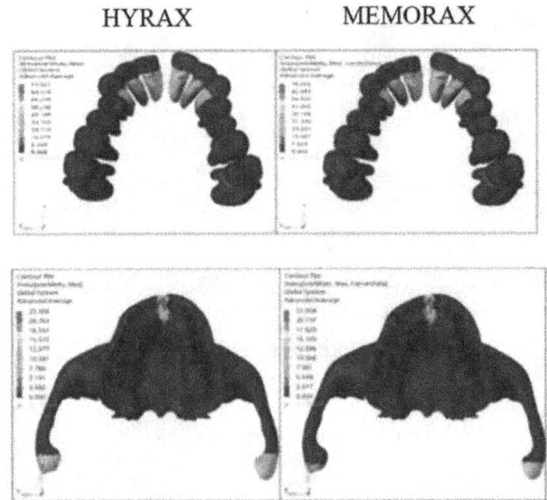

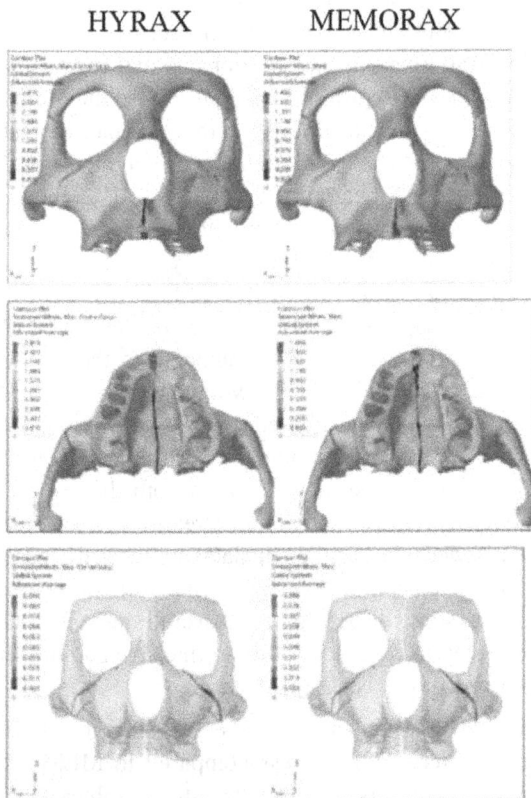

HYRAX   MEMORAX

Fig. 23.8 Stress distribution in various craniofacial sutures

Table 23.3 A comparison on stress distribution between hyrax and memorax

Site	HYRAX Von-Mises Stress (Mpa)	MEMORAX Von-Mises Stress (Mpa)
Teeth	72.36	70.5
PDL	1.89	1.77
Mid palatal Suture	2.82	1.69
Nasomaxillary Suture	0.087	0.075
Zygomatico maxillary Suture	0.094	0.086
Frontomaxillary suture	0.091	0.078
Internasal Suture	0.033	0.027
Pterygomaxillary Suture	0.118	0.113
Zygomatico Temporal suture	0.081	0.071
Maxilla	23.36	22.65

for MEMORAX), 71.55 for the canine (vs. 66.53), and 66.51 for the first premolar (vs. 60.08), suggesting a more aggressive expansion effect. In contrast, the stress values for the second premolar are nearly identical (10.62 for HYRAX vs. 10.28 for MEMORAX), and the first molar and second molar show only slight differences (4.91 vs. 4.79 and 18.71 vs. 17.63, respectively) (Table 23.4).

Table 23.4 Comparison of stress distribution in individual maxillary teeth between Hyrax and Memorax * VS = Von Mises stress (Mpa)

Teeth	Memorax			
	X	Y	Z	VS
Central Incisor	0.321	0.195	-0.203	70.502
Lateral Incisor	0.321	0.186	-0.128	32.21
Canine	0.349	0.186	0.184	66.53
First Premolar	0.171	0.169	-0.044	60.08

Teeth	HYRAX	MEMORAX
	VS	VS
Central Incisor	72.36	70.502
Lateral Incisor	34.85	32.21
Canine	71.55	66.53
First Premolar	66.51	60.08
Second Premolar	10.62	10.28
First Molar	4.91	4.79
Second Molar	18.71	17.63

3.2 Displacement Pattern between Hyrax and Memorax

Table 23.5 Displacement along each axis at the skeletal points (mm)

Site	HYRAX			MEMORAX		
	X	Y	Z	X	Y	Z
Teeth	0.717	0.195	-0.24	0.706	0.198	-0.235
PDL	0.414	0.151	-0.223	0.403	0.157	-0.219
Mid palatal suture	0.322	0.171	-0.216	0.325	0.162	-0.221
Nasomaxillarysuture	0.069	0.025	-0.013	0.065	0.018	-0.017
Zygomaticomaxillary suture	0.164	0.037	-0.036	0.139	-0.026	-0.035
Frontomaxillarysuture	0.012	0.019	-0.011	0.01	0.014	-0.009
Internasal suture	0.009	0.032	0.014	0.007	0.024	0.009
Pterygomaxillary Suture	0.137	0.155	-0.187	0.106	0.132	-0.155
ZygomaticoTemporal Suture	0.098	0.028	-0.017	0.088	0.028	-0.014
Maxilla	0.414	0.162	-0.221	0.403	0.171	-0.214

Table 23.6 Displacement along each axis at the maxillary teeth (mm)

Dentoalveolar and Dental Structures	HYRAX			MEMORAX		
	X	Y	Z	X	Y	Z
Apical region of incisor	-0.041	0.072	-0.198	-0.039	0.063	-0.203
Apical region of canine	-0.051	0.025	-0.07	-0.053	0.031	-0.06
Apical region of premolar	0.029	0.052	-0.063	0.022	0.051	-0.049
Apical region of molars	0.011	0.04	-0.096	0.002	0.036	-0.076
Incisal edge of incisor	0.326	0.198	0.023	0.321	0.195	0.021
Palatine cusp tip of Canine	0.352	0.195	0.181	0.349	0.186	0.184
Palatine cusp tip of molar	0.151	0.162	-0.057	0.125	0.148	-0.044

3.3 Displacement Patterns along the X-axis

Displacement of maxilla in X-axis-positive value: lateral movement, X-axis-negative value: medial movement as seen in Fig. 23.9 and in Table 23.5 & 23.6

Skeletal Finding

The study illustrates the X-axis displacement of the maxilla using the Hyrax and Memorax appliances. When viewed from the front, both appliances caused noticeable

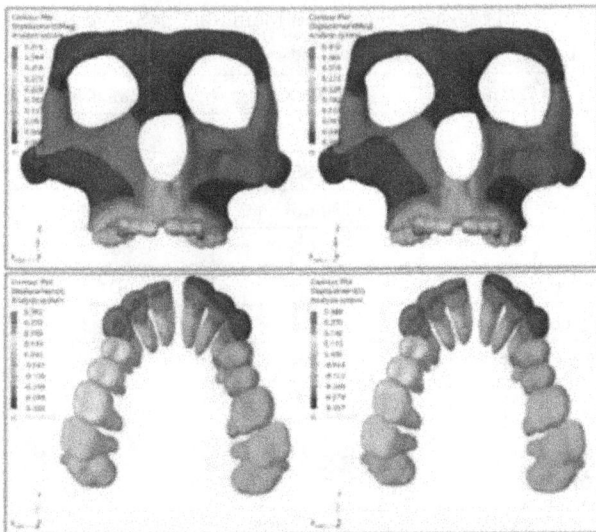

Fig. 23.9 Displacement contours of craniofacial structures and teeth in X- direction

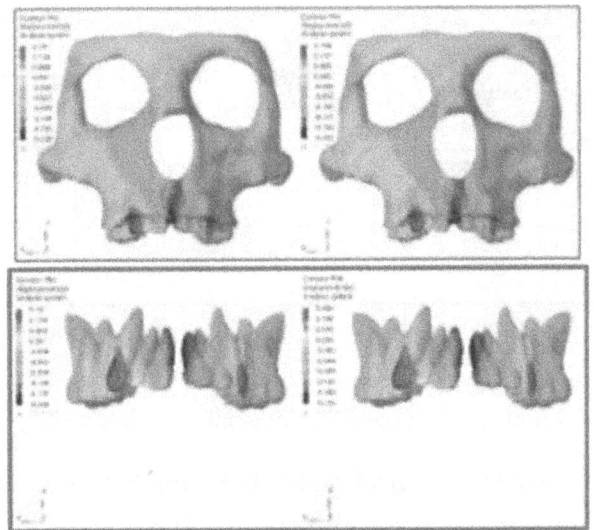

Fig. 23.11 Displacement contours of craniofacial structures and teeth in Z- direction

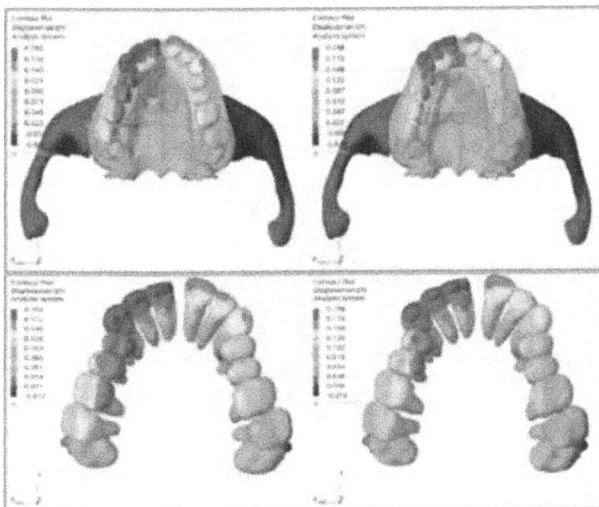

Fig. 23.10 Displacement contours of craniofacial structures and teeth in Y- direction

pyramidal displacement of the maxilla, resulting in a wedge-shaped opening. The midpalatal suture showed greater lateral displace-ment with the Hyrax appliance compared to Memorax. Additionally, the zygo-maticomaxillary, pterygomaxillary, and zygomaticotemporal sutures exhibited more lateral displacement with the Hyrax appliance. Conversely, minimal lateral displacement was observed in the nasomaxillary, internasal, and frontomaxillary sutures with both appliances, showing negligible differences between them. The apex of the pyramid faced the nasal bone, while the base was positioned on the oral side. Both appliances achieved a total expansion. From an occlusal view, both halves of the maxillary dentoalveolar complex and the basal maxilla dis-played increased anterior separation, alongside wider lateral walls of the nasal cavity. Minimal lateral movement was observed in the upper posterior part of the nasal cavity, while the width of the nasal cavity at the floor of the nose notably expanded.

Dental Finding

Hyrax appliance resulted in significantly greater lateral dis-placement in the dentoalveolar region and the periodontal ligament of anterior teeth compared to the Memorax appliance. The maximum lateral displacement at the incisal edge and apical region of the upper central incisor was slightly higher with the Hyrax appliance than with the Memorax appliance. Anchor teeth (pre-molars and molars) at the apical region also showed greater lateral displacement with the Hyrax appliance compared to Memorax. Additionally, the palatine cusp tips of the anchor teeth exhibited slightly more lateral displacement with the Hyrax appliance than with the Memorax appliance.

3.4 Displacement Pattern in the Y-axis

Displacement of the maxilla anteriorly (labially) in the y axis was indicated as a positive sign and posteriorly was indi-cated as a negative sign as shown in Fig. 23.10 and in Table 23.5 & 23.6.

Skeletal Finding

Both appliances resulted in anterior movement of the maxilla, with the MEMORAX appliance showing the greatest displacement compared to for the Hyrax appliance. Across all sutures, there was slightly greater anterior displacement observed with the Hyrax appliance compared to MEMORAX. Specifically, the zygomaticomaxillary suture exhibited slight posterior displacement with MEMORAX (-0.026 mm), whereas it showed anterior displacement with Hyrax (0.037 mm)."

Dental Findings

When comparing the displacement of the dentoalveolar com-plex between the Hyrax and Memorax appliances, the Hyrax appliance shows slightly greater anterior displacement. Both appliances primarily affect the central incisors, lateral incisors, and canines, with diminishing effects as you move towards the posterior teeth. Specifically, the Hyrax appliance causes forward displacement of the incisal edge and apical region of incisors, which is slightly higher compared to the Memorax appliance. Anchor teeth, such as premolars and molars in the apical region, also exhibit slightly greater displacement with the Hyrax appliance. Additionally, the palatine cusp tip of molars shows slightly more anterior displacement with the Hyrax appliance compared to Memorax.

3.5 Displacement Pattern in the Z-axis

Displacement of the maxilla in the caudal (occlusal) direction was indicated as negative and in the cephalic direction was indicated as positive in Fig. 23.11 and in Table 23.5 & 23.26.

Skeletal Finding

Both the Hyrax and Memorax appliances caused inferior movement of the maxilla. The Hyrax appliance resulted in slightly greater overall inferior displacement of the maxilla compared to Memorax. Specifically, the midpalatal suture showed more inferior displacement with Memorax than with Hyrax. The pterygomaxillary and zygomaticomaxillary sutures exhibited slightly greater inferior displacement with the Hyrax appliance. Conversely, the zygo-maticotemporal and internasal sutures showed slightly more inferior displace-ment with Hyrax compared to Memorax. The nasomaxillary suture demonstrated slightly more inferior displacement with Memorax compared to Hyrax. Overall, superoinferior displacement values were similar between both appliances with minimal differences.

Dental Finding

The incisal edge and apical region of incisors in Memorax (-0.021mm & -0.203mm) shows slightly more inferior displacement in the apical region than Hyrax (-0.023mm & -0.198mm). The anchor teeth such as premolars and molars at the apical region shows greater occlusal displacement (-0.063mm and -0.096mm respectively) with Hyrax as compared to Memorax (-0.049mm and -0.076mm

respectively). Palatine cusp tip of molars in Hyrax exhibits -0.057mm which is greater inferior displacement than Memorax with -0.044mm.

4. Conclusion

FEM is a valid method for comparing the effects of maxillary expansion appliances on the craniofacial complex. FEM modelling software now allows for the incorporation of sutures, for a more accurate and convincing model. Hyrax achieves slightly increased widening of the midpalatal suture with higher magnitude of stress and thus increasing maxillary transverse dimensions as compared to Memorax. The Memorax appliance tends to result in lower levels of stress concentration and displacement compared to the Hyrax appliance. This suggests that the Memorax may offer advantages in terms of minimizing adverse effects on tooth alignment and stability during treatment. Hyrax shows more buccal tipping and extrusion of anchored premolar and molar teeth when compared to Memorax, which is an adverse effect of maxillary expansion. Clinicians should consider the specific treatment goals, patient characteristics, and potential adverse effects such as buccal tipping and tooth extrusion when selecting between these appliances. Accurately transferring material properties and applying loads precisely can be challenging. The study concentrated on initial displacements and stress distributions, which may not fully address dynamic clinical scenarios. The effects of retention periods were not assessed, representing a constraint of the study.

Acknowledgments

The authors would like to thank Faculty of Dental Sciences, MSRUAS for approving the study/ giving an opportunity to conduct the study. We would also thank Mr. Rupesh kumar, FEM engineer for assisting us with the FEM study.

References

1. de Silva, F. O. G., Boas, C. V., & Capelozza, L. F. (1991). Rapid maxillary expansion in the primary and mixed dentitions: A cephalometric evaluation. *American Journal of Orthodontics and Dentofacial Orthopedics, 100*(2), 171–179.

2. Haas, A. J. (1961). Rapid expansion of the maxillary dental arch and nasal cavity by opening the midpalatal suture. *Angle Orthodontist, 31*, 73–90.

3. Halicioğlu, K., & Yavuz, İ. (2014). Comparison of the effects of rapid maxillary expansion caused by treatment with either a memory screw or a Hyrax screw on the dentofacial structures—transversal effects. *European Journal of Orthodontics, 36*(2), 140–149.

4. Halicioğlu, K., Kiki, A., & Yavuz, İ. (2012). Maxillary expansion with the memory screw: A preliminary investigation. *Korean Journal of Orthodontics, 42*(2), 73–79.

5. Halicioğlu, K., Kiliç, N., Yavuz, İ., & Aktan, B. (2010). Effects of rapid maxillary expansion with a memory palatal split screw on the morphology of the maxillary dental arch and nasal airway resistance. *European Journal of Orthodontics, 32*(6), 716–720.

6. Isaacson, R. J., & Ingram, A. H. (1964). Forces produced by rapid maxillary expansion: II. Forces present during treatment. *Angle Orthodontist, 34*(4), 261–270.

7. Jafari, A., Shetty, K. S., & Kumar, M. (2003). Study of stress distribution and displacement of various craniofacial structures following application of transverse orthopedic forces—a three-dimensional FEM study. *Angle Orthodontist, 73*(1), 12–20.

8. Oliveira, P. L. E., Soares, K. E. M., Andrade, R. M. D., Oliveira, G. C. D., Pithon, M. M., Araújo, M. T. D. S., & Sant'Anna, E. F. (2021). Stress and displacement of mini-implants and appliance in Mini-implant Assisted Rapid Palatal Expansion: Analysis by finite element method. *Dental Press Journal of Orthodontics, 26.*

9. Wichelhaus, A., Geserick, M., & Ball, J. (2004). A new nickel titanium rapid maxillary expansion screw. *Journal of Clinical Orthodontics, 38*(12), 677–682.

Note: All the figures and tables in this chapter were made by the author.

Advances in Materials Science and Technology – Dr. Srikari Srinivasan et al. (eds)
© 2025 Taylor & Francis Group, London, ISBN 978-1-041-12342-2

24

SPT based Liquefaction Prediction of Soils Employing AI Techniques

S. Sharika[1]

Research Scholar, Department of CE,
M.S. Ramaiah University of Applied Sciences,
Bangalore, Karnataka, India

P. Bora[2]

Research Scholar, Department of CSE,
M.S. Ramaiah University of Applied Sciences,
Bangalore, Karnataka, India

S. D. Anitha Kumari[3]

Professor, Department of CE,
M.S. Ramaiah University of Applied Sciences,
Bangalore, Karnataka, India

S. Chatterjee[4]

Associate Professor, Department of CSE,
M.S. Ramaiah University of Applied Sciences,
Bangalore, Karnataka, India

ABSTRACT: Soil, a material with its intricate and varied nature, requires advanced methods like AI for effective behavior analysis. AI has proven its efficacy in diverse challenges, can effectively able to solve soil complexities also. This paper presents the application of AI for soil characterization in the context of well-known phenomena, 'Liquefaction. Liquefaction refers to the transition of a soil mass from its solid to liquid state under dynamic loading conditions. Researchers have identified this as the prime factor for ascertaining the structural stability and having a direct link with the major earthquake damages. To identify the potential of the same, an extensive dataset of Standard Penetration tests (SPT) with 46 attributes from 202 different seismic prone sites are given in the AI classifier for the validation of results. The study includes data analysis and visualizations showing interdependencies among crucial parameters and predictions using tree based and neural network classifications. The evaluations of the models are done using various classification criteria like accuracy, precision, recall, F1 score and AUC where it is observed that tree-based classification performed better than neural models with an accuracy 0.88. In future, AI methods can be implemented for various other tests to predict the characteristics of soil as a material. Additionally, if image dataset for soil is captured and added with this classifier, this study can be extended for automated analysis and characterization of soil at various sites.

KEYWORDS: AI, Liquefaction, ML, SPT, Soil

[1]sharika2091@gmail.com, [2]borahparinita123@gmail.com, [3]anithakumari.ce.et@msruas.ac.in, [4]subarna.cs.et@msruas.ac.in

DOI: 10.1201/9781003664277-24

1. Introduction

Soil, being a highly heterogeneous material, exhibits complex and inconsistent behavior. complexities in soil not only affects the building foundation, but also creates catastrophic failures in the entire structure. The varied composition and structure of soils contribute to the uneven response of soil layers under dynamic loading like earthquakes, and this research addresses "liquefaction," a secondary effect of earthquakes. Liquefaction refers to the phenomena where solid soil mass, transitions to a liquid state under sudden and transient loading conditions, particularly during dynamic events such as earthquakes. This event is critical in saturated undrained soils, as the rapid loading causes excess pore water pressure to build up in the soil voids. This pressure buildup reduces the soil's effective stress, loses its strength and stiffness, and ultimately changing its state from solid to liquid. Identifying the possibility of liquefaction is crucial in seismically active regions, leading to the development of various techniques for assessment. Traditional methods like Standard penetration tests (SPT) (Goren et al., 2017; Patel and Kumar, 2017; Tokimatsu and Yoshimi, 1981), Cone penetration tests (CPT) (Robertson and Campanella, 1985; Zhang et al., 2002) and Shear wave velocity tests(Andrus et al., 2004; Baziar and Alibolandi, 2024), analytical techniques including soil modelling (Andrianopoulos et al., 2010; Boukpeti et al., 2002; Dafalias and Taiebat, 2016)and Numerical methods using Finite element softwares(Kanth and Maheshwari, 2021; Kumari and Sawant, 2021; Yang and Kavazanjian, 2021) are generally adopted to evaluate the potential of liquefaction. However, traditional methods often depend on empirical correlations and site-specific tests, which can introduce uncertainties due to soil heterogeneity, potentially affecting the accuracy of these assessments. Thus, researches are extended to the domain of Artificial intelligence (AI) which is emerged as a promising frontier in geotechnical engineering (Gao, 2024; Ghani et al., 2022; Jas and Dodagoudar, 2023; Kumar et al., 2024; Samui and Sitharam, 2011). By leveraging machine learning algorithms, AI facilitates the analysis and prediction of soil liquefaction potential more efficiently than traditional methods. AI models can integrate diverse data sources including dynamic soil properties, seismic characteristics, and liquefaction events to generate predictive models. These models capture complex nonlinear relationships and offer enhanced accuracy in identifying liquefaction-prone areas. AI techniques enable rapid evaluation and mapping of liquefaction risks, supporting proactive mitigation strategies and improving overall resilience in geotechnical engineering practices. The present study aims to identify the probability of liquefaction from an SPT dataset with the help of Machine learning algorithms.

2. Materials and Methods

2.1 Collection of Soil Data

The current research used SPT (Standard penetration test) data, as SPT stands out as a reliable field method specifically noted for its effectiveness in assessing liquefaction potential (Goren et al., 2017; Patel and Kumar, 2017). It is a standard insitu soil testing method where the number of blows required to penetrate a sampler for a depth of 300mm is recorded as the SPT N-value. This value reflects the soil resistance to penetration and provides valuable information about its density and consistency. It is instrumental in liquefaction analysis, particularly in assessing the susceptibility of soils to liquefy during seismic events. By measuring the SPT N-values at various depths within a site, engineers can identify layers of loose, saturated soils that are prone to liquefaction under cyclic loading. Lower N-values indicate softer, more liquefiable soils, while higher values suggest denser, more stable conditions. This data helps in characterizing the soil profile and determining the potential for liquefaction. In this study, data of 202 sites from USA, Japan, Argentina, China, and Philippines for SPT is collected (Cetin et al., 2018). The distribution of liquefiable and non-liquefiable soils in the collected dataset is shown in Fig. 24.1.

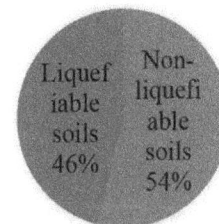

Fig. 24.1 Data distribution in current dataset

2.2 AI and Machine learning on SPT data

Artificial intelligence (AI) is about getting a machine to do the things in an intelligent way. A human intelligence knows the surrounding environment, learn various factors and make some decision to take next action involuntarily. Machines are programmed to achieve certain such goals using various technologies and mathematical algorithms. Technology like computer vision, natural language processing, neural network, machine learning is adopted to make decision makings. Machine learning (ML) is the set of mathematical algorithms that learns from data and can find inferences from data. It considers the specific characteristic of data items (exemplars), and predicts the probability of the closest category (label) which it might belongs to. In data, when the category or label is known, then it is a supervised learning. Otherwise, it is an

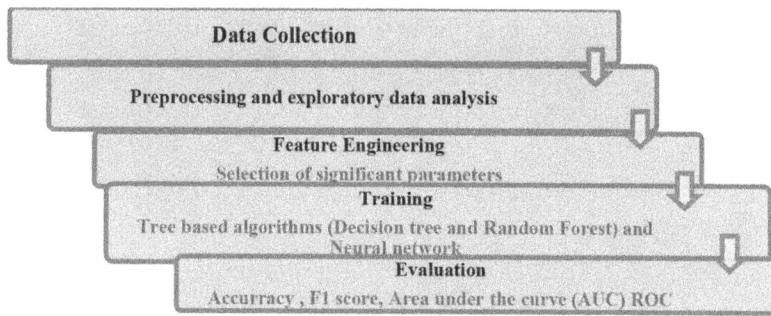

Fig. 24.2 Machine learning phases

unsupervised classification where the learning algorithm must uncover different categories or labels existing in the data. For the current study, supervised classification is done as the dataset has a target variable "Liquefies "which predicts whether the sample in a particular site undergoes liquefaction. The original contributor (Cetin et al., 2018) has done an excel based analysis for the dataset using maximum likelihood estimate (MLE). However, MLE has disadvantages like sensitivity to outliers, susceptibility to overfitting and computationally intensive numerical iterations. Considering adaptability and flexibility to handle complex data, other ML methods are preferred by researchers. From literatures, it is found that tree based and neural network ML methods are widely adapted for soil classification (Ali et al., 2012; Chala and Ray, 2023; Hateffard et al., 2019; Pham et al., 2021). Analyzing the efficiency, these techniques are applied in the current dataset.

2.3 Machine Learning Phases

Machine learning method in general involves two phases: (I) training and (II) testing. In the training phase, ML methods is applied on training data so that a learned model (relating input to known label) is produced. For the current study, 80% of data is given in the first phase to train the model and the remaining 20% is given in the testing phase to predict for the unknown input data. The flowchart shown in Fig. 24.2 **Error! Reference source not found.** exhibits the phases of AI model where the training phase involves data collection, pre-processing and training of the model using proposed algorithms and the testing phase involves applying the trained model on unknown samples from same distribution.

Data Pre-Processing and Feature Selection

In machine learning, the weightage of attributes in the dataset determines how effectively the trained model perform in classification task. For the current dataset, an exploratory data analysis was conducted to determine the importance of various features in predicting the probability

of liquefaction. The original SPT dataset has 46 attributes. For the computation purposes, null values of the dataset are replaced with the number zero. The data is then subjected to analysis to determine the most significant ones which influences the soil behaviour. To do this, the correlation between each pair of attributes is calculated. A correlation namely 'Pearson correlation coefficient' which is the linear relationship between two random variables is a good measure. This works on interval scale with resultant values in the interval between -1 to +1. The values when plotted as a digital image, the brightness of the pixels make it easily interpretable.

Figure 24.3 is a sample correlation matrix where the parameters 'sv' (total stress) and 'sv_dash' (effective stress) is having very high correlation. Thus, the total stress factor is omitted and the effective stress parameter is taken for the model simulations. Similarly, to obtain the relation between remaining attributes, Pearson correlation is adopted. In the correlation matrix shown in Fig. 24.4 every element in a square, located at a row and a column intersection shows the correlation coefficient between the two attributes one positioned as row and the other as column. When this value is near to 1, they are very highly correlated. This indicates

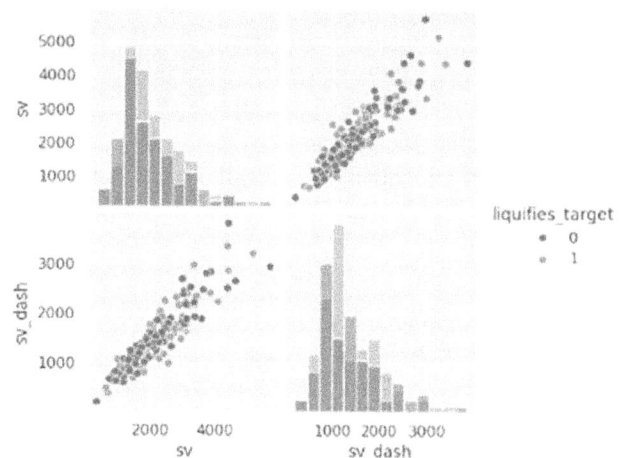

Fig. 24.3 Sample correlation matrix

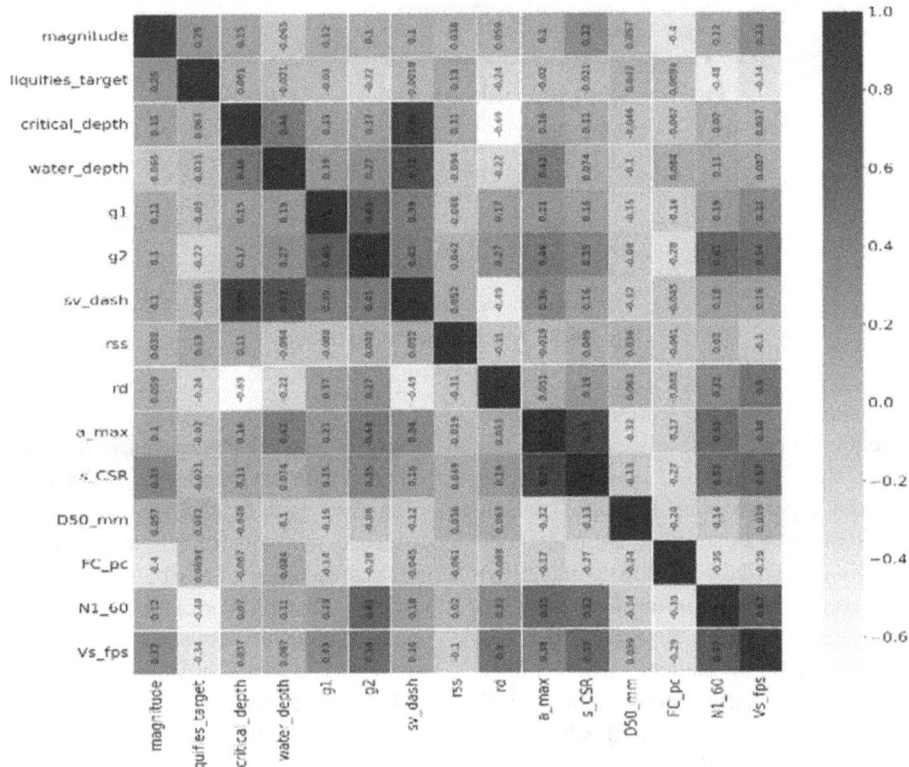

Fig. 24.4   Correlation matrix for significant parameters

that the features contribute similar information while training. To remove the redundancy and multicollinearity, if two attributes are highly correlated, one of them can be ignored. Here, the elimination criteria are considered for the attribute pairs with correlation value greater than 0.9. Hence 14 most significant attributes are considered for training whose description is given below:

The proposed AI models are trained using 14 parameters from the SPT data which highly influences liquefaction and has 1 parameter i.e., target variable, indicating whether a site liquefies or not. The description of the 14 parameters (unit weight of soil above and below water table is clubbed together) is as follows:

a. Magnitude: Earthquake magnitudes play a crucial role in triggering liquefaction. For the current study, magnitudes varying from 5.8 to 7 is taken for the analysis.

b. Critical depth: Critical depth is the soil depth where seismic shear stress triggers liquefaction and rapid pore pressure buildup.

c. Location of ground water table: The location of ground water table plays a crucial role in the prediction of pore pressure generation and there by the intensity of liquefaction.

d. Unit weight of soil (Y): Unit weight above (Y1) and below (Y2) water table are taken for the model simulations. As saturated samples are more prone to liquefaction, effect of these two factors is of prime importance.

e. Effective stress (sv_dash) (σ'): Effective stress is the difference between total stress and porewater pressure; liquefaction occurs if it reaches zero.

f. Correlation coefficient (ρσσ'): The correlation coefficient (ρσσ') quantifies the relationship between total and effective stress, aiding in liquefaction prediction. A strong correlation helps assess soil behavior, improving risk analysis and seismic safety in construction.

g. Cyclic shear stress reduction factor (rd): The depth-dependent reduction in shear stress during an earthquake is accounted for by (rd), crucial for seismic soil analysis. It reflects the impact of seismic waves on soil behavior.

h. Peak ground acceleration (PGA): PGA is the maximum ground acceleration experienced during an earthquake, often expressed as a fraction of gravity (g). Higher value of PGA results in greater cyclic shear stresses which increases the likely hood of liquefaction.

i. Cyclic stress ratio (CSR): It is the ratio of cyclic shear stress from an earthquake to the initial effective vertical stress. A higher cyclic stress ratio increases pore pressure, reducing soil strength and causing liquefaction.

j. Grain diameter D50 (mm): The median grain size D50, is key in liquefaction risk assessment. Smaller D50 educes permeability, increasing pore pressure buildup, while larger D50 improves drainage and reduces susceptibility.

k. Percentage of Fines (% FC): Fines content significantly impacts liquefaction probability. Higher fines reduce permeability, slowing pore pressure dissipation and increasing risk. Soils with more fines have higher void ratios and lower density, reducing stability. A fines content below 12% enhances liquefaction resistance.

l. Number of SPT blow counts (N1_60): SPT blow counts represents soil density, with higher values signifying denser, stronger soils less prone to liquefaction. Lower N1_60 suggest looser, more compressible soils with higher void ratios, increasing liquefaction risk. SPT values also aid in liquefaction potential correlations.

m. Shear wave velocity (Vs_fps): Shear wave velocity reflects soil stiffness and resistance to shear deformation. Higher values indicate denser soils less prone to liquefaction by resisting shear stress and limiting pore pressure buildup.

2.4 Selection of ML Algorithms

For the current study, Neural network and Tree based algorithms (Decision tree and Random Forest) are applied in training. The target attribute to be predicted is set as "liquefies" which has two values 'yes' and 'no'. In the preprocessing phase, the 'yes' and 'no' have been updated as 1 and 0. A brief description of algorithms used in current research is given below:

Neural Network

The concept of neural networks is inspired by the functioning of the human brain in decision-making. In artificial neural networks (ANNs), a threshold-based processing is mathematically represented using activation functions. Nodes in ANNs are interconnected with edges that have weights and the output is calculated based on the inputs and these weights. Learning in the network involves adjusting the weights to produce the most accurate outputs for given inputs. Training occurs over several iterations, each consisting of forward and backward propagation. Initially, weights are random; during the forward pass,

outputs are observed, and if incorrect, errors are calculated and used to adjust weights in the backward pass from the output layer back to the input layer. The layers between the input and output layers are known as hidden layers. Figure 24.5(a) represents a basic neuron/perceptron and Fig. 24.5(b) shows the multilayer neural network/perceptron.

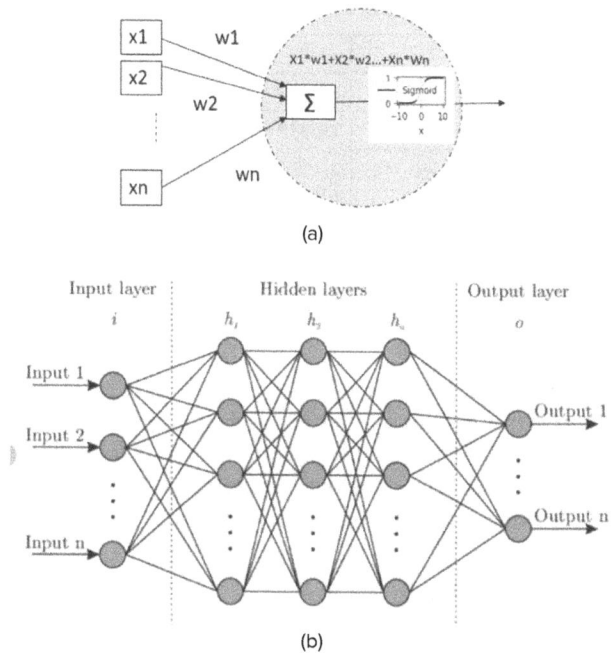

(a)

(b)

Fig. 24.5 (a) Sample neuron, (b) Multilayer neural network

Tree Based Methods - Decision Tree and Random Forest

A decision tree is a structure with nodes and leaf nodes used for classification, where each internal node represents an input feature, and leaf nodes provide classification outcomes. The algorithm has two phases: construction, where data is split based on criteria like Gini impurity, and prediction, where data is passed through the tree to a leaf. In this study, CART is used with Gini impurity to split the data. Random Forest builds multiple decision trees to improve predictive accuracy and reduce overfitting by averaging their predictions. It also handles large datasets, provides feature importance, and is resistant to outliers.

3. Results and Discussion

The evaluation of the models is done using the performance parameters accuracy, F1 score, Confusion matrix and Area under the curve (AUC) ROC (Receiver Operating Characteristics)(Sokolova et al., n.d.). Table shows the Metrics and the formula used for evaluation of the trained models where: TP stands for Total number of predicted samples as True positive; TN stands for Total number of predicted samples as True negative; FN stands for Total number of predicted samples as False negative; FP stands for Total number of predicted samples as False Positive and AUC stands for Area under the curve Receiver Operating characteristic (ROC) (Ling and Huang, n.d.)

Table 24.1 Evaluation metric

Metric	Formula
Precision	TP/ (TP + FP)
Recall or true positive rate	TP/ (TP + FN)
F measure or f1 score	2 * (Precision * Recall)/(Precision + Recall)
Accuracy	(TP + TN)/ (TP + TN + FP + FN)
AUC	Area under the curve TPR vs FPR

3.1 Decision Tree

Figure 24.6(a) shows the confusion matrix, and Fig. 24.6(b) displays the ROC curve. Out of 42 test samples, 4 are false negatives (liquefied but classified as non-liquefied), 3 are false positives (non-liquefied but predicted as liquefied), 16 are true negatives, and 19 are true positives. The ROC

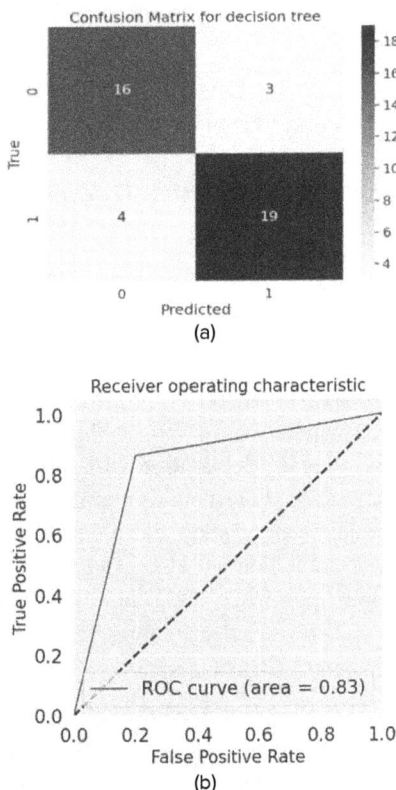

(a)

(b)

Fig. 24.6 (a) Confusion matrix for decision tree, (b) ROC curve for decision tree

curve plots true positive vs. false positive rates, with the AUC indicating model performance. When AUC is zero, the model predictions are 100% wrong and when it is 1, it is 100% correct (Ling and Huang, n.d.). For the decision tree classifier, AUC is observed as 0.83, which is acceptable considering the heterogeneity in the data. The values of AUC are critical as it is very significant in minimizing the false outputs.

3.2 Random Forest

Figure 24.7(a) shows that the Random Forest classification includes 18 samples under true positive and 18 under true negative. Also, 5 samples are under false negative and 1 is under false positive. The AUC (**Error! Reference source not found.** 24.7(b)) is observed to be 0.86, hence Random Forest outperformed than the decision tree.

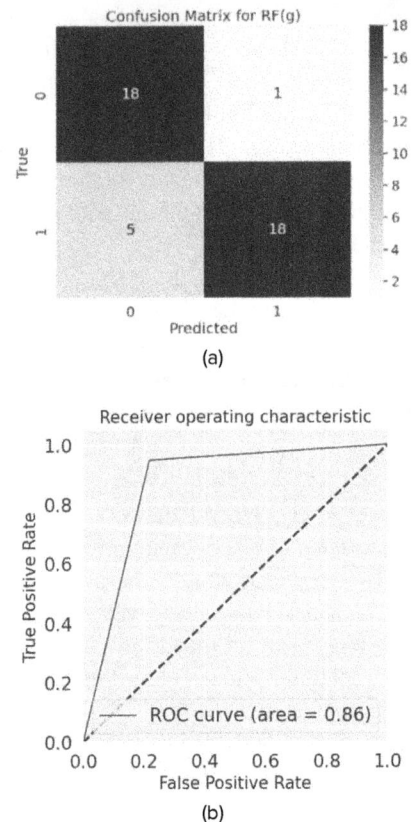

(a)

(b)

Fig. 24.7 (a) Confusion matrix for random forest, (b) ROC curve for random forest

3.3 Neural Network

The confusion matrix of Neural network in Fig. 24.8(a) shows 16 samples each under true positive and true negative, 7 under false negative and 3 under false positive. The AUC in Fig. 24.8(b) gives a value of 0.73 which is least compared to the performance of other two models.

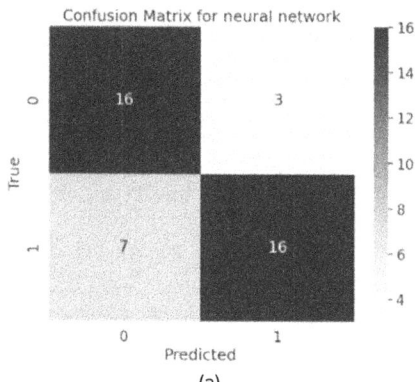

(a)

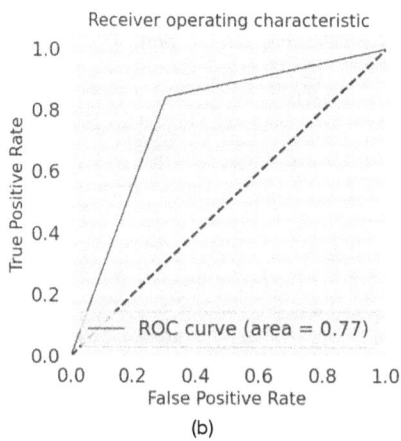

(b)

Fig. 24.8 (a) Confusion matrix for neural network, (b) ROC curve for neural network

The evaluation parameters proposed for the current study is tabulated in Table 24.2.

Table 24.2 Evaluation parameters

Model	Label	Precision	Recall	F1-score	Accuracy	AUC
Decision tree	0	0.80	0.84	0.82	0.83	0.83
	1	0.86	0.83	0.84		
Random forest	0	0.82	0.95	0.88	0.88	0.86
	1	0.95	0.83	0.88		
	1	0.86	0.83	0.84		
Neural network	0	0.70	0.84	0.76	0.76	0.77
	1	0.84	0.70	0.76		

Comparing the evaluation parameters, it can be interpreted that Random Forest classifier performs best for the current dataset for the cases of liquefaction with a precision of 0.95, recall of 0.83, F1-score of 0.88, Accuracy of 0.88 and AUC 0.86. It is because, Random Forest takes considerations of multiple decision trees outcomes before the final prediction. For the same reason Random Forest, well known for its robustness is always preferred for data which is diverse and noisy (errors). Figure depicts the feature importance for the prediction in case of Random Forest. It is seen that N1_60 having the highest importance in the prediction and unit weight of soil has the least. In case of decision tree, CSR, Magnitude, PGA, and N1_60 contributes the most.

For this dataset, performance of neural network is found to be less compared to tree-based methods. It is observed that compared to tree-based methods, significance of feature importance cannot be captured in neural network. This method is suitable when sufficiently large amount of data is used for training. The accuracy of outcome can

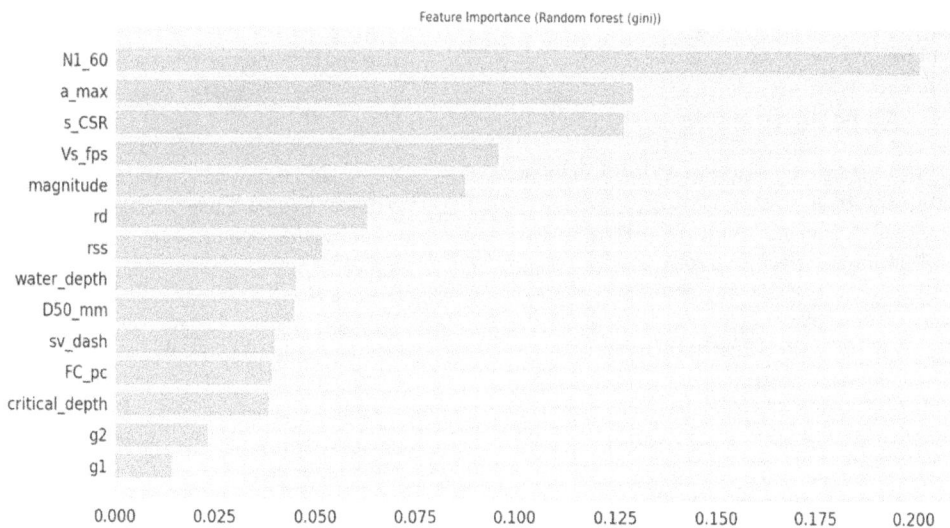

Fig. 24.9 Feature importance of random forest

be improved by deep architectures for getting a better performance.

4. Conclusion

The current research focuses on the application of AI in predicting the liquefaction phenomena of soil. SPT dataset having information of 202 seismic prone sites have taken for the current study. Tree-based algorithms and neural network is used for prediction of liquefaction. The trained models are evaluated with the parameters- Accuracy, Precision, Recall, F1 score and AUC. The tree based Random Forest classifier is evaluated best with the values 0.88, 0.95, 0.83, 0.88 0.86 for Accuracy, precision, recall, F1-score, and AUC respectively. This can be attributed to the fact that Random Forest uses multiple decision trees which help to achieve a better balance between bias and variance. In addition, its higher dimensionality provides estimates of feature importance, which help in understanding the influence of different features on the prediction. Also, they are resilient to noisy data and outliers. The neural network performed the least with the values 0.76, 0.84, 0.70, 0.76 and 0.77 as Accuracy, precision, Recall, F1 score and AUC respectively. It is interpreted that neural networks are less proficient for limited amount of data in a highly heterogeneous dataset. In future, the data corresponding to the significant parameters identified from this study can be collected for a variety of soils and can be used for the improvement of the model performance.

Acknowledgments

The authors express their sincere gratitude to M S Ramaiah University of Applied Sciences, Bangalore for giving the facilities to conduct research and also for the SRF/JRF offered to first and second authors.

References

1. Ali, J., Khan, R., Ahmad, N., Maqsood, I., 2012. Random Forests and Decision Trees.
2. Andrianopoulos, K.I., Papadimitriou, A.G., Bouckovalas, G.D., 2010. Bounding surface plasticity model for the seismic liquefaction analysis of geostructures. Soil Dynamics and Earthquake Engineering 30, 895–911. https://doi.org/10.1016/j.soildyn.2010.04.001
3. Andrus, R.D., Stokoe, K.H., Juang, C.H., 2004. Guide for shear-wave-based liquefaction potential evaluation. Earthquake Spectra 20, 285–308. https://doi.org/10.1193/1.1715106
4. Baziar, M.H., Alibolandi, M., 2024. Assessment of lique-faction potential based on shear wave velocity: Strain ener-gy approach. Journal of Rock Mechanics and Geotechnical Engineering. https://doi.org/10.1016/j.jrmge.2023.10.021
5. Boukpeti, N.;, Mroz, Z.;, Drescher, 2002. A model for static liquefaction in triaxial compression and extension. Canadian Geotechnical Journal 39, 1243–1253. https://doi.org/10.1139/T02-066
6. Cetin, K.O., Seed, R.B., Kayen, R.E., Moss, R.E.S., Bilge, H.T., Ilgac, M., Chowdhury, K., 2018. Dataset on SPT-based seismic soil liquefaction. Data Brief 20, 544–548. https://doi.org/10.1016/j.dib.2018.08.043
7. Chala, A.T., Ray, R., 2023. Assessing the Performance of Machine Learning Algorithms for Soil Classification Using Cone Penetration Test Data. Applied Sciences (Switzerland) 13. https://doi.org/10.3390/app13095758
8. Dafalias, Y.F., Taiebat, M., 2016. Sanisand-Z: Zero elastic range sand plasticity model. Geotechnique 66, 999–1013. https://doi.org/10.1680/jgeot.15.P.271
9. Gao, W., 2024. The Application of Machine Learning in Geotechnical Engineering. Applied Sciences (Switzerland). https://doi.org/10.3390/app14114712
10. Ghani, S., Kumari, S., Ahmad, S., 2022. Prediction of the Seismic Effect on Liquefaction Behavior of Fine-Grained Soils Using Artificial Intelligence-Based Hybridized Modeling. Arab J Sci Eng 47, 5411–5441. https://doi.org/10.1007/s13369-022-06697-6
11. Goren, S, Gelisli, K., Goren, Sevda, 2017. Determination of the liquefaction potential of soils of the northern sea command site (Istanbul, turkey) based on SPT data, Advances in Biology & Earth Sciences.
12. Hateffard, F., Dolati, P., Heidari, A., Zolfaghari, A.A., 2019. Assessing the performance of decision tree and neural network models in mapping soil properties. J Mt Sci 16, 1833–1847. https://doi.org/10.1007/s11629-019-5409-8
13. Jas, K., Dodagoudar, G.R., 2023. Liquefaction Potential Assessment of Soils Using Machine Learning Techniques: A State-of-the-Art Review from 1994–2021. International Journal of Geomechanics 23. https://doi.org/10.1061/ijg-nai.gmeng-7788
14. Kanth, A., Maheshwari, B.K., 2021. Monotonic and Cyclic Loading Behaviour of Solani Sand: Experiments and Finite Element Simulations. International Journal of Geotechnical Engineering. https://doi.org/10.1080/19386362.2021.1966225
15. Kumar, D.R., Samui, P., Burman, A., Kumar, S., 2024. Seismically Induced Liquefaction Potential Assessment by Different Artificial Intelligence Procedures. Transportation Infrastructure Geotechnology 11, 1272–1293. https://doi.org/10.1007/s40515-023-00327-w
16. Kumari, S., Sawant, V.A., 2021. Numerical simulation of liquefaction phenomenon considering infinite boundary. Soil Dynamics and Earthquake Engineering 142. https://doi.org/10.1016/j.soildyn.2020.106556
17. Ling, C.X., Huang, J., n.d. AUC: a Statistically Consistent and more Discriminating Measure than Accuracy.
18. Patel, P., Kumar, V., 2017. Liquefaction potential of soil by empirical and computational method based on SPT-N value-A case study of lucknow city.
19. Pham, B.T., Nguyen, M.D., Nguyen-Thoi, T., Ho, L.S., Koopialipoor, M., Kim Quoc, N., Armaghani, D.J., Le,

H. Van, 2021. A novel approach for classification of soils based on laboratory tests using Adaboost, Tree and ANN modeling. Transportation Geotechnics 27. https://doi.org/10.1016/j.trgeo.2020.100508

20. Robertson, P.K., Campanella, R.G., 1985. Liquefaction potential of sands using the CPT. Journal of Geotechnical Engineering 111, 384–403. https://doi.org/10.1061/(ASCE)0733-9410(1985)111:3(384)

21. Samui, P., Sitharam, T.G., 2011. Machine learning modelling for predicting soil liquefaction susceptibility. Natural Hazards and Earth System Science 11, 1–9. https://doi.org/10.5194/nhess-11-1-2011

22. Sokolova, M., Japkowicz, N., Szpakowicz, S., n.d. Beyond Accuracy, F-Score and ROC: A Family of Discriminant Measures for Performance Evaluation.

23. Tokimatsu, K., Yoshimi, Y., 1981. Field Correlation of Soil Liquefaction with SPT and Grain size, in: Proc of First Internationaal Conference on Recent Advances in Geotechnical Earthquake Engineering and Soil Dynamics. Louis,Missouri, pp. 203–208.

24. Yang, Y., Kavazanjian, E., 2021. Numerical evaluation of liquefaction-induced lateral spreading with an advanced plasticity model for liquefiable sand. Soil Dynamics and Earthquake Engineering 149. https://doi.org/10.1016/j.soildyn.2021.106871

25. Zhang, G., Robertson, P.K., Brachman, R.W.I., 2002. Estimating liquefaction-induced ground settlements from CPT for level ground. Canadian Geotechnical Journal 39, 1168–1180. https://doi.org/10.1139/t02-047

Note: All the figures and tables in this chapter were made by the author.

25

Thermal Stability of the Magnetic Tunnel Junction Stack and their Magnetic Properties

Apoorva Kaul, Gopika C. T.
Nanomaterials Research Laboratory,
Surface Engineering Division, CSIR - National Aerospace Laboratories,
Bangalore, India
Academy of Scientific and Innovative Research (AcSIR),
Ghaziabad, India

Jakeer Khan, Bhagaban Behera
Nanomaterials Research Laboratory,
Surface Engineering Division, CSIR - National Aerospace Laboratories,
Bangalore, India

Sabyasachi Saha, Partha Ghosal
Electron Microscopy Group,
Defence Metallurgical Research Laboratory, PO Kanchanbagh,
Hyderabad, India

P. Chowdhury*
Nanomaterials Research Laboratory,
Surface Engineering Division, CSIR - National Aerospace Laboratories,
Bangalore, India
Academy of Scientific and Innovative Research (AcSIR),
Ghaziabad, India

ABSTRACT: Achieving a high TMR ratio in a Magnetic Tunnel Junction stack requires the crystallization of CoFeB layers along the 001 plane of MgO, which is possible only to soak the TMR stack at an elevated temperature as high as 300°C to 400°C. In this report, we have investigated the thermal stability of the MTJ stack by incorporating a Synthetic Antiferromagnetic (SAF) structure with varying Ru layer thickness, t_{Ru}, from 0.8 nm to 1.1 nm. With t_{Ru} = 0.8 nm, the stack annealed at 280°C shows a stable quasi-antiparallel state due to strong antiferromagnetic (AFM) coupling between the reference and the pinned layer, and above this temperature, it gradually disappears and vanishes at 350 °C. Meanwhile, with t_{Ru} = 1.1 nm, coupling changes to ferromagnetic (FM) in nature, and the thermal stability of the stack was found to be stable up to 400°C. The transport measurements show an enhanced TMR ratio of 16% with t_{Ru} = 1.1 nm with a low exchange bias field; however, lower t_{Ru} shows a lower TMR ratio of 12% at the cost of a higher exchange bias field of 1100 G. It is essential to have a stable anti-parallel state for the fabrication of magnetic sensor for linear sensing application.

KEYWORDS: Annealing, Antiferromagnetic coupling, Exchange bias, Magnetic tunnel junction (MTJ), Ruthenium (Ru), Thermal stability, Tunneling magnetoresistance (TMR)

*Corresponding author: pchowdhury@nal.res.in

DOI: 10.1201/9781003664277-25

1. Introduction

The investigation of epitaxially grown 001-oriented MgO-based MTJs without AFM coupling was reported to have a TMR ratio of 631% for CoFe/MgO/CoFe[1] and 361% for CoFeB/MgO/CoFeB MTJs with AFM coupling measured at room temperature.[2] To achieve such high values of TMR, one of the major requirements is the perfect crystallization of MgO and CoFeB layers sandwiching the MgO spacer layer along (001) planar direction with an increase in annealing temperatures. In practical devices, a SAF structure is incorporated in the MTJ stack, including an AFM, FM, and a Ru spacer layer, creating a stable quasi-antiparallel (AP) state between the FM layers. The Ru spacer layer mediates the interlayer exchange bias coupling between the FM layers, essential for achieving the AP alignment with a stability range of ± 1300 G.[3] It also prevents seeding from the bottom CoFe layer and crystallizes the amorphous CoFeB layer into a bcc texture from the MgO interface side as well as makes the stack withstand high working temperatures as high as 300°C to 400°C.[2]

Annealing can either enhance the TMR% by sharpening the interface between FM and barrier layer and reducing defects, or it can also reduce TMR by causing unwanted diffusion, altering magnetic properties and thermal stability issues.[4,5] Finding optimal annealing conditions is important for maximizing positive effects and involves a delicate balance between specific materials and desired characteristics. According to Shin et al., 2009 [6] when the annealing temperature is increased to a temperature above 330°C, the TMR should increase due to the enhancement in grain-to-grain epitaxy in CoFeB/MgO/CoFeB[7] but leads to a decrease in exchange bias due to the diffusion occurring in CoFe/Ru/CoFeB layers.[2] Only after annealing at a certain temperature does the SAF structure come into effect, leading to an increase in TMR% and reduction of exchange bias for high thermal stability.[8] But annealing these structures above 420°C leads to diffusion of Mn and Ru and disrupts the layers' ideal structure, thereby affecting spin-dependent properties for higher TMR percentages.[5,6] Overall, Ru plays a crucial role in MTJs and can control thermal stability, performance, and functionality by influencing TMR enhancement, magnetic coupling and interfacial properties. Therefore, further research on the role of Ru thickness in MTJs could yield valuable advancements in TMR research, including improvements in high-temperature MTJ performance.

2. Materials and Methods

The multilayer stack with a structure of Ta (5)/Ru (15)/Ta (5)/Ru (15)/Ta (5)/NiFe (2)/IrMn (12.5)/CoFe (3)/Ru (t_{Ru})/CoFe (2)/CoFeB (2)/MgO (1.5)/CoFeB (4)/CoFe (4)/Ta (5)/Ru (10) (thicknesses in nm) with thickness of Ru (t_{Ru}) spacer layer in the SAF structure varied from 0.8-1.1 nm were deposited on a 2-inch Si/SiO$_2$ substrate using an indigenously developed Ultra High Vacuum (UHV) magnetron sputtering system. The system has two processing chambers, each with four sputter targets capable of DC and RF sputtering and a load lock chamber interconnecting the chambers. The metallic layers were deposited using DC sputtering, and the dielectric layer was deposited using the RF sputtering technique at room temperature with a vacuum base pressure of nearly 2×10^{-8} Torr. Argon was used as the working gas at a pressure of 3.5 mTorr during the deposition.

The samples deposited were divided into two categories: small samples for magnetic measurements and the wafers for TMR measurement. The magnetic measurement samples were characterized, ex-situ annealed from 250 - 400°C in a UHV system with a base pressure of 10^{-8} Torr in a magnetic field of 500 Oe. A Vibrating Sample Magnetometer (VSM) (Lakeshore USA) was used to measure the $M(H)$ hysteresis loops at room temperature. Meanwhile, in the case of wafers, the current perpendicular to plane (CPP) geometry was fabricated to measure transport characteristics. The perpendicular pillar formation is an integral part of the CPP geometry, which is fabricated by following a four-step UV-photolithography, Argon ion milling, insulation layer, and top contact, as shown schematically in Fig. 25.1(a). Each wafer consists of a series of pattern devices, each of size 6×3 mm, having a square sensing element of area 0.25 mm^2 consisting of 64 circular pillars connected in series, as shown in Fig. 25.1(b). Each pillar has two steps, with the top step having a diameter of 10 - 30 μm and a height of nearly 20 nm and the middle step having a diameter of 30 - 60 μm and a height of around 40 nm. The devices were thermally annealed in a magnetic field from 250°C to 350°C for 1 hour in the same vacuum conditions mentioned above. Each device was then measured for magneto-transport characteristics by a DC two- and four-probe method (at room temperature).

3. Results and Discussion

3.1 Structural Characterization

Structural study is integral to MTJ characteristics as CoFeB/MgO/CoFeB crystallization is essential for achieving a higher TMR percentage. A multilayer stack of Ta/Ru/Ta/CoFeB /MgO/ CoFeB/Ta was deposited to verify the crystallinity of individual layers. High-resolution transmission electron microscopy (HR-TEM), grazing incidence X-ray diffraction (GI-XRD), and Energy

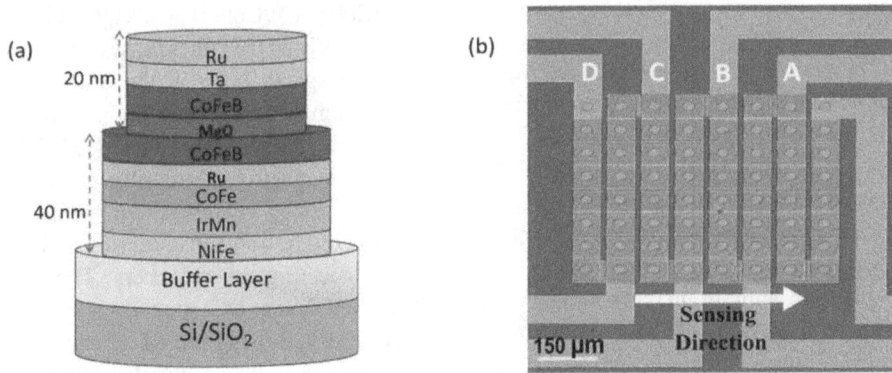

Fig. 25.1 **(a)** Schematic representation of an MTJ pillar with two steps; **(b)** Optical image of the sensing area (0.5 mm X 0.5 mm) of a single device with 64 pillars connected in series.

Dispersive X-ray (EDX) line scan along the cross-section of the sample was conducted for a sample post-annealed at 300°C, as shown in Fig. 25.2. Bright-field image of HR-TEM confirms the textured growth of MgO in (001) bcc structure on top of the CoFeB layer. The presence of Ta atoms in its neighbouring layer confirms a higher degree of diffusion in both Ru and CoFeB layers, though the line profile shows a dip in the diffusion of Ta atoms within the MgO layer.[10,11] GI-XRD confirms the MgO crystallization in the (002) bcc structure with the peak at 42.2° and correlates well with the Selected Area Diffraction (SAD) pattern as shown in the inset of Fig. 25.2(c).

Magnetic Characterization

Figure 25.3 presents the $M(H)$ loops of two MTJ stacks measured at room temperature for t_{Ru} = 0.8 nm and 1.1 nm with an increase in temperature from 280 to 350°C, respectively. The strength of exchange coupling between the reference layer (RL) and pinned layer (PL) depends on t_{Ru}, based on which either FM or AFM coupling occurs as predicted by the Ruderman–Kittel–Kasuya–Yosida (RKKY) interaction.[12] The measured $M(H)$ loops shown in Fig. 25.3(b) and (d) concur that the two samples of t_{Ru} = 0.8 nm and 1.1 nm exhibit AFM and FM coupling

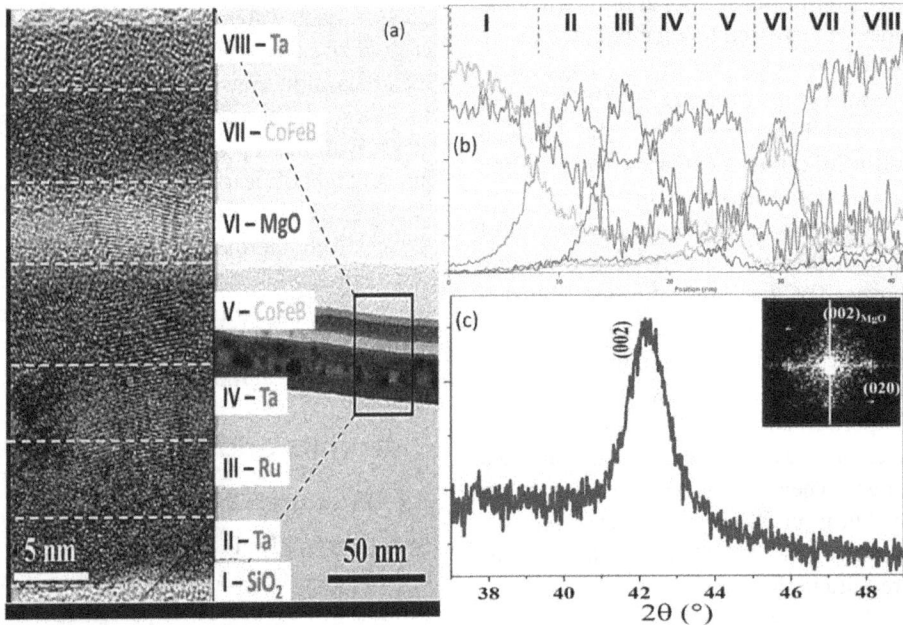

Fig. 25.2 **(a)** Bright-field TEM image revealing the crystallization structure of different layers of the stack at 300°C post-annealing temperature; **(b)** EDX profiles scan of one of the same samples as in exhibiting line profiles of different atoms across the layers; **(c)** GI-XRD scan in agreement with (a) and (b) shows a crystalline structure of MgO; **Inset:** The Selected Area Diffraction (SAD) pattern indicates MgO crystallization in (002) planar direction

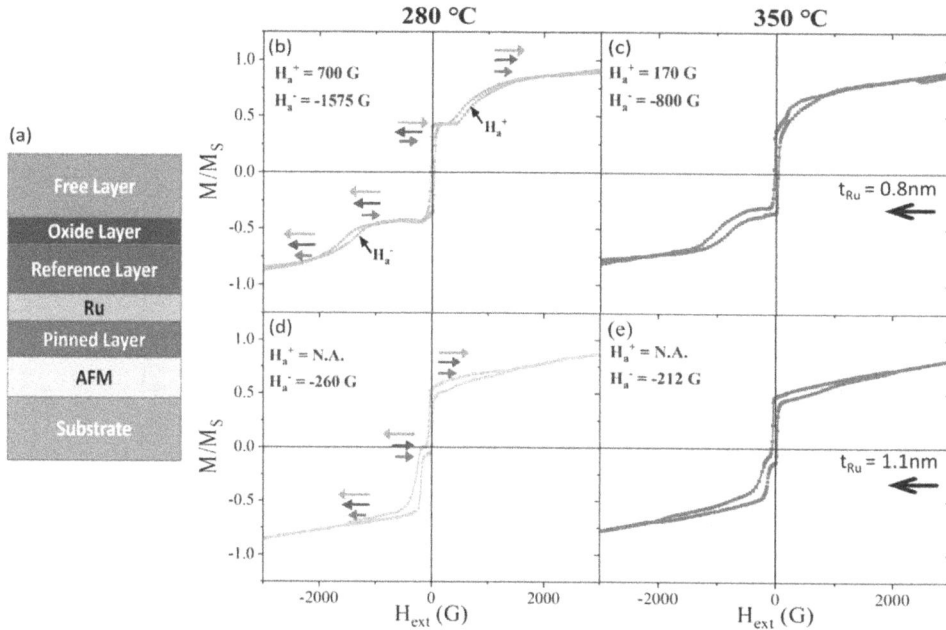

Fig. 25.3 **(a)** Illustration of labels given to different layers in the multilayer stack and indicating the colour reference for the magnetic moments' rotation of a particular layer represented by that specific-coloured arrow; Magnetic characterization of two samples at room temperature for t_{Ru} = 0.8 nm with an increase in temperature from **(b)** 280°C to **(c)** 350°C and for t_{Ru} = 1.1 nm at **(d)** 280°C to **(e)** 350°C

between the reference and pinned layer, respectively. The arrows shown in Fig. 25.3 convey the magnetic moment orientation of each colour-coded layer (Fig. 25.3(a)) concerning the change in an applied magnetic field. After the addition of SAF structure to the AFM layer coupled with PL, two types of exchange biasing occur in the stack: (i) the exchange bias, H_{ex}, due to the AFM layer being coupled with the PL[13] and (ii) the AFM or FM coupling due to RKKY interaction between PL and RL through Ru layer.[14,15] As observed from Fig. 25.3(b) and (d), though the alignment of moments of all three FM layers is in a parallel state at saturation magnetization, the transition of individual FM layers occurs during field sweep depending on the coupling constant value of the dominating one. For the case of t_{Ru} = 0.8 nm (Fig. 25.3(b)), due to the lower thickness of Ru, the AFM coupling between the PL and RL dominates over the exchange bias coupling, and when the magnetic field changes, switching begins with RL layer and rotates at H^+ = 700 G, hence, forming a stable anti-parallel state between the three layers. Further decrease in magnetic field, the free layer transition occurs, and then at H^- = -1575 G, PL switches. In the case of t_{Ru} = 1.1 nm (Fig. 25.3(d)), the FM coupling between RL and PL layers sums up with H_{ex} and hence synchronous switching of the PL with RL can be detected at H^- = -260 G.

The samples were further annealed at 350°C to evaluate the thermal stability of the stack. Figure 25.3(c) and (e)

show the *M(H)* curves for samples with t_{Ru} = 0.8 and 1.1 nm, respectively. As can be seen from Fig. 25.3(c), the stable anti-parallel state, defined by ΔH (= H^+-(-H^-)) = 2275G, at 280°C annealing reduces to 970G.[4] These results correspond to the Ta diffusion occurring across the stack, including the CoFeB electrodes (Fig. 25.2(b)) at higher annealing temperatures. However, in Fig. 25.3(e), for t_{Ru} = 1.1 nm, the H– value remains constant, confirming an increase in the thermal stability of the MTJ stack.

Transport Characterization

Transport measurements for the two samples were carried out and are presented in Fig. 25.4 as a function of the magnetic field. From Fig. 25.4(a) and (c), it was observed that the H_a values of both curves are nearly the same as those of their magnetic counterparts and accordingly correspond to spin alignments, as shown in the figure. However, the ratio percentage for a device with Ru thickness of 0.8 nm and 1.1 nm, annealed at 280°C, was 12% and 16%, respectively. This difference in the percentage of the two devices can be attributed to the principle of spin conservation between the FM layers, which further depends on the magnetic alignment of the three FM layers (FL, RL and PL). For the lower thickness of Ru, a perfect AP state was observed, which is why when the switching occurs, the amount of spin tunneling probability gets reduced between the PL and the RL as

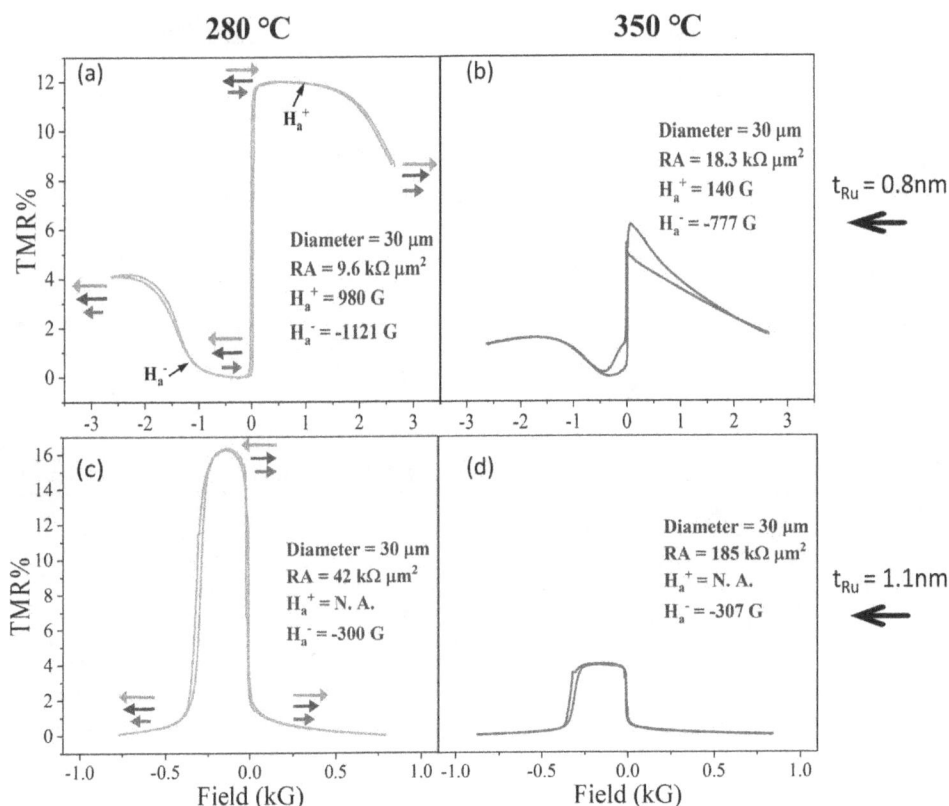

Fig. 25.4   TMR curves for transport characterization of a sample with t_{Ru} = 0.8 nm from annealing temperature **(a)** 280°C to **(b)** 350°C; and for t_{Ru} = 1.1 nm with increase in annealing temperature from **(c)** 280°C to **(d)** 350°C.

well as the RL and the FL. Only when the spins of the two FM layers are aligned in the parallel state the higher probability of spin tunneling between the layers can occur, as seen in Fig. 25.4(c). Here, the spin alignment between the PL and RL is parallel, increasing the spin tunneling probability and a higher TMR% in a lower exchange bias field. Moreover, on annealing the same samples at 350°C, it was observed that even though the TMR percentage was reduced for both samples, the t_{Ru} of 1.1 nm retained the H_a^- field value of 300G. In comparison, the 0.8 nm sample reduced significantly from 980G to 140G, defining the stable FM coupling structure even at high temperature, as shown in Fig. 25.4(b) and (d). However, the deterioration of the stack at higher temperatures can be attributed to the degradation in interfacial quality caused by scattering events for spin-polarized electrons and Ta diffusion across the stack.[19,20]

4. Conclusion

In this report, we investigated the variation in magnetic and transport properties of the MTJ stack with the variation in Ru spacer layer thickness in the SAF structure, which

indicated a correlation depending on the spin alignments between the PL and RL affected by the FM and AFM coupling originating from higher and lower thickness of Ru layer respectively. On comparing the thermal stability of the two stacks, magnetic characterization exhibited better annealing stability than the TMR-characterized devices. The detailed structural analysis of different layers of the stack revealed a textured growth of MgO but with the diffusion of Ta across the whole stack except MgO at 300°C annealing temperature, suggesting the interlayer diffusion of Ta destroying the thermal stability of the stack. However, the high ΔH of the antiparallel state, even with 50% to 100% of TMR, provides a comparable output reported by the commercial sensors. Hence, rather than a higher percentage, the higher stability of the anti-parallel state, which gives a higher field range for the sensitivity measurement, seems more promising for practical operation.

Acknowledgements

The authors thank the Director, CSIR-NAL, for supporting this activity and SERB DST (Ministry of Science and

Technology) for the project. Apoorva Kaul and Gopika C T express their sincere gratitude to AcSIR for allowing them to pursue their doctoral degree at the lab. Apoorva Kaul acknowledges UGC for JRF.

References

1. Scheike, T., Wen, Z., Sukegawa, H., Mitani, S.: 631% room temperature tunnel magnetoresistance with large oscillation effect in CoFe/MgO/CoFe (001) junctions. Applied Physics Letters 122, 112404 (2023)

2. Lee, Y.M., Hayakawa, J., Ikeda, S., Matsukura, F., Ohno, H.: Giant tunnel magnetoresistance and high annealing stability in CoFeB/MgO/CoFeB magnetic tunnel junctions with synthetic pinned layer. Applied Physics Letters 89 (2006)

3. Cuchet, R.B.A.S. L.: Perpendicular magnetic tunnel junctions with a synthetic storage or reference layer: A new route towards pt- and pd-free junctions. Sci Rep 6, 21246 (2016)

4. Okamoto, K., Fuji, Y., Higashi, Y., Kaji, S., Nagata, T., Baba, S., Yuzawa, A., Hara, M.: Enhanced annealing stability of exchange-biased pinned layer in magnetic tunnel junction using Ta/Ru/Ta/Ru underlayer. IEEE Transactions on Magnetics 54(11), 1–4 (2018)

5. Hayakawa, J., Ikeda, S., Lee, Y.M., Matsukura, F., Ohno, H.: Effect of high annealing temperature on giant tunnel magnetoresistance ratio of CoFeB/MgO/CoFeB magnetic tunnel junctions. Applied Physics Letters 89(23) (2006)

6. Shin, I.-J., Min, B.-C., Hong, J., Shin, K.-H.: Effects of Ru diffusion in exchange- biased MgO magnetic tunnel junctions prepared by in situ annealing. Applied Physics Letters - APPL PHYS LETT 95 (2009)

7. Li, Z.-P., Li, S., Zheng, Y., Fang, J., Chen, L., Hong, L., Wang, H.: The study of origin of interfacial perpendicular magnetic anisotropy in ultra-thin CoFeB layer on the top of MgO based magnetic tunnel junction. Applied Physics Letters 109, 182403 (2016)

8. Krizakova, V., Hoffmann, M., Kateel, V., Rao, S., Couet, S., Kar, G., Garello, K., Gambardella, P.: Tailoring the switching efficiency of magnetic tunnel junctions by the field like spin-orbit torque. Physical Review Applied 18 (2022)

9. Ikeda, S., Hayakawa, J., Ashizawa, Y., Lee, Y.M., Miura, K., Hasegawa, H., Tsunoda, M., Matsukura, F., Ohno, H.: Tunnel magnetoresistance of 604% at 300 k by suppression of Ta diffusion in CoFeB/MgO/CoFeB pseudo-spin-valves annealed at high temperature. Applied Physics Letters 93, 082508–082508 (2008)

10. Ibusuki, T., Miyajima, T., Umehara, S., Eguchi, S., Sato, M.: Lower-temperature crystallization of CoFeB in MgO magnetic tunnel junctions by using Ti capping layer. Applied Physics Letters 94, 062509–062509 (2009)

11. Liu, E., Wu, Y.-C., Couet, S., Mertens, S., Rao, S., Kim, W., Garello, K., Crotti, D., Van Elshocht, S., De Boeck, J., Kar, G., Swerts, J.: Synthetic-ferromagnet pinning layers enabling top-pinned magnetic tunnel junctions for high-density embedded magnetic random-access memory. Physical Review Applied 10 (2018)

12. Liu, Y., Zhou, B., Zhu, J.-G.: field-free magnetization switching by utilizing the spin hall effect and interlayer exchange coupling of iridium. Sci Rep 9, 325 (2019)

13. Belokon, V., Nefedev, K., Kapitan, V., Dyachenko, O.: Magnetic states of nanoparticles with RKKY interaction. Advanced Materials Research 774-776, 523–527 (2013)

14. Mouhoub, A., Millo, F., Chappert, C., Kim, J., L'etang, J., Solignac, A., Devolder, T.: Exchange energies in CoFeB/Ru/CoFeB synthetic antiferromagnets. Physical Review Materials 7 (2022)

15. Gawade, T., Borole, U., Behera, B., Khan, J., Barshilia, H., Chowdhury, P.: On-chip full bridge bipolar linear spin valve sensors through modified synthetic antiferromagnetic layers. Journal of Magnetism and Magnetic Materials 587, 171234 (2023)

16. Hong, J., Liang, P., Safonov, V., Khizroev, S.: Energy-efficient spin-transfer torque magnetization reversal in sub-10-nm magnetic tunneling junction point contacts. Journal of Nanoparticle Research 15 (2013)

17. Hong, J., Hadjikhani, A., Stone, M., Liang, P., Allen, F., Safonov, V., Bokor, J., Khizroev, S.: The physics of spin-transfer torque switching in magnetic tunneling junctions in sub-10 nm size range. IEEE Transactions on Magnetics 52, 1–1 (2016)

18. Yang, L.C. CL.: High thermal durability of Ru-based synthetic antiferromagnet by interfacial engineering with re insertion. Sci Rep 11, 15214 (2021)

19. Kim, G., Lee, S., Lee, S., Song, B., Lee, B.-K., Lee, D., Lee, J.S., Lee, M.H., Kim, Y.K., Park, B.-G.:" the influence of capping layers on tunneling magnetoresistance and microstructure in CoFeB/MgO/CoFeB magnetic tunnel junctions upon annealing". Nanomaterials 13(18) (2023)

Note: All the figures in this chapter were made by the author.

Advances in Materials Science and Technology – Dr. Srikari Srinivasan et al. (eds)
© 2025 Taylor & Francis Group, London, ISBN 978-1-041-12342-2

26

Comparative Study of Synthesis Methods of All Inorganic Lead-Free Bromide Perovskite

Suma Jebin*,
Amrita Mandal Bera, Swapna Gijare
Basic Sciences and Humanities Department,
Agnel Charities' Fr. C. Rodrigues Institute of Technology,
Navi Mumbai, India

Dhananjay Panchagade
Mechanical Engineering Department,
Agnel Charities' Fr. C. Rodrigues Institute of Technology,
Navi Mumbai, India

Mini Rajeev
Department of Electrical Engineering,
Agnel Charities' Fr. C. Rodrigues Institute of Technology,
Navi Mumbai, India

ABSTRACT: The recent trend in research focuses on developing novel photocatalytic materials to overcome the issues such as worldwide environmental crisis of growing water and air pollution. Recently, lead halide perovskites have emerged as very efficient absorbers in the field of photovoltaics due to their fascinating optoelectronic properties which make them potential candidates for photocatalytic application too. However, toxic lead is a main barrier for such materials to be implemented in commercial applications. In this context, all inorganic lead free perovskite materials can be very promising materials for photocatalytic devices not only from technical but also from environmental aspects. The properties of the perovskite materials are strongly synthesis route dependent. In this work, the synthesis of all inorganic lead-free bromide perovskites by co-precipitation, direct precipitation and solid state reaction methods is reported and compared. Their structural and optical properties has been analyzed by X-ray diffraction (XRD) and UV-Visible (UV-Vis) absorption techniques. Then the suitable materials can be explored for photocatalytic activity.

KEYWORDS: Absorption spectrum, All–inorganic, Lead free, Synthesis methods, Photocatalysis

1. Introduction

In the last decade, metal halide perovskite materials have made remarkable achievements in the field of photovoltaics due to their unique optoelectronic properties [1-3]. However,

these materials are not being explored much in photo catalytic domain although having potential applications in that field [4]. Among perovskite compounds, organometal lead halide perovskite with $APbX_3$ composition (where, 'A'= methylammonium/formamidinium, and 'X'= I^-, Br^-,

*Corresponding author: suma.jebin@fcrit.ac.in

DOI: 10.1201/9781003664277-26

Cl⁻ or mix of them) have been researched most due to better device performance [5-8]. However, toxic lead and stability issues are the major barriers for commercialization. In this context, proper replacement of toxic lead and volatile organic component by suitable less toxic inorganic materials showed promising path towards new generation of sustainable, eco-friendly perovskites.

Various all-inorganic perovskite materials have been reported till date, however, their stability and applicability are not yet up to the mark compared to lead based perovskite [3]. As the physical, chemical, and optical properties of the perovskite materials are strongly dependent on synthesis procedure, a suitable synthesis method is very much required to obtain the desired properties from the prepared materials.

There are various routes for the synthesis of perovskites material such as gas phase synthesis, liquid phase synthesis and solid phase synthesis which include solution method, solid state reaction, vapour deposition and many more [9]. In the early stage, solution method was adopted by most of the research groups because of simplicity and low power consumption. In 2012, Lee *et al.* reported synthesis of organometal halide perovskites by solution method and achieved 10.9% power conversion efficiency in a single-junction solar cell device [1]. Later on other methods such as vapour deposition, solid state reaction are also explored. In 2019, Chen *et al.* reported a facile vapour deposition method to prepare cesium tin germanium triiodide perovskite which was highly stable [13]. Synthesis techniques affect crystal structure and morphology of the samples resulting further change in properties.

In this paper, an effort has been taken to explore various synthesis techniques to prepare a lead free perovskite by replacing both lead and organic part by suitable inorganic materials (Cs for 'A' site and Sn for Pb). The synthesis of lead free all inorganic bromide perovskite ($CsSnBr_3$) by using three different methods (direct precipitation, co-precipitation and solid state reaction) is addressed. The materials have been characterized by X-ray diffraction studies and UV visible absorption spectra to analyze their structural and optical properties.

2. Experimental

2.1 Materials

Stannous Bromide (SnBr2) (96%) was purchased from Sigma Aldrich. Cesium Bromide (CsBr, 99%), Dimethyl Fluoride (DMF, 99.5%), Dimethylsulfoxide (DMSO) were purchased from Chemco. Apart from these hydrobromic acid which was purchased from SDFCL (a local chem-limited in Mumbai) has been used. All reagents were used without any purification.

2.2 Preperation of $CsSnBr_3$ by Co-Precipitation Method

At first, $CsSnBr_3$ was synthesized by co-precipitation method which has been reported in our previous paper [10]. In this method, DMF was used as a solvent and toluene as antisolvent as shown in Fig. 26.1(a). The method was previously reported for lead perovskite by Liu *et al.*[11]. Precursor solutions in stoichiometric ratio were added to

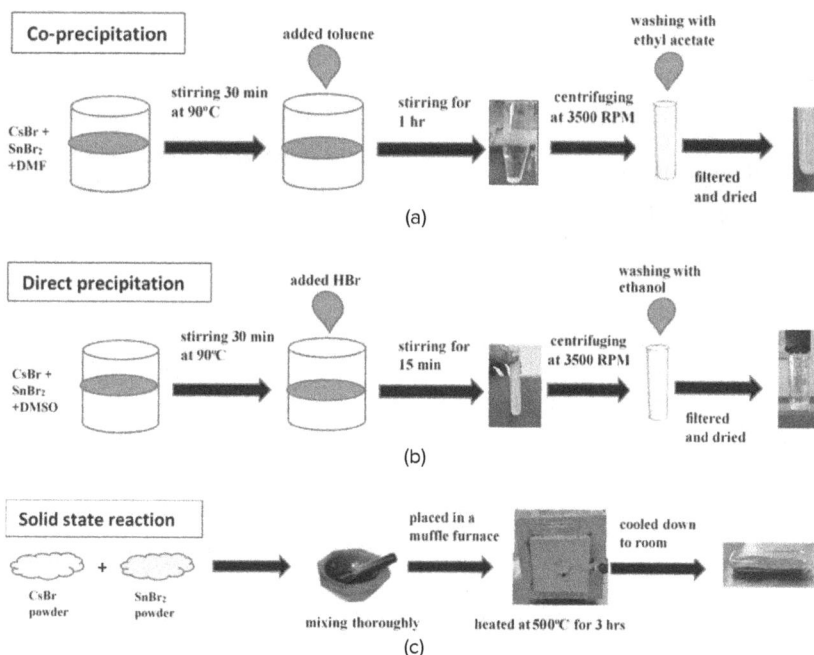

Fig. 26.1 Schematic diagrams of synthesis methods

the solvent DMF and stirred. After this, the mixture was heated at 90°C with continuous stirring for half an hour. In the next step, the antisolvent agent toluene was added to the solution and again stirred for 1 hour. White precipitate was obtained which was collected by centrifugation at 3500 RPM. The final product was washed with ethyl acetate. After filtration, it was kept in hot air oven at 60°C to be completely dried.

2.3 Preperation of CsSnBr$_3$ by Direct Precipitaion Method

Another precipitation method was explored to prepare CsSnBr$_3$ in acidic medium which is named as direct precipitation. In this method, CsSnBr$_3$ was synthesized by using DMSO as a solvent as shown in Fig. 26.1(b) [12]. Precursor solution was made by mixing CsBr and SnBr$_2$ salts proportionally in DMSO and stirred for 5 minutes. After this, the mixture was stirred for another 30 min. 3ml of HBr is dropped in the solutuon and continued stirring for another 15min. After centrifugation, the precipitate was washed with ethanol, filtered and completetely dried in oven at 130 °C.

2.4 Preperation of CsSnBr$_3$ by Solid State Reaction Method

In both the precipitation methods, the yield of final product was very less. Therefore, a solvent free method (solid state reaction) was attempted to synthesize the same material CsSnBr$_3$ as shown in Fig. 26.1(c). The same method was reported earlier for iodide perovskite by Chen *et al.* [13]. The precursor salts, CsBr and SnBr$_2$ are mixed according to stoichiometric proportion thoroughly and placed in a muffle furnace. The precursor heated at 500 °C for 3 hrs. Then the sample is cooled down slowly to room temperature to get the grey coloured final product.

2.5 Characterization of Materials

X-ray diffraction study of the prepared materials have been performed by PANalytical EMPYREAN diffractometer using CuKα radiation of 1.54184 Å in order to obtain their crystallographic information. Additionally, UV visible absorption behavior of the materials was investigated using a UV Visible spectrophotometer (Elico SL159).

3. Result and Discussion

The sample CsSnBr$_3$ has been synthesized by three different methods. The materials prepared are firstly characterized by X-ray diffraction study to identify the phases formed. Figure 26.2 shows the XRD patterns for the samples prepared by co-precipitation, direct precipitation and solid state reaction altogether in one figure for

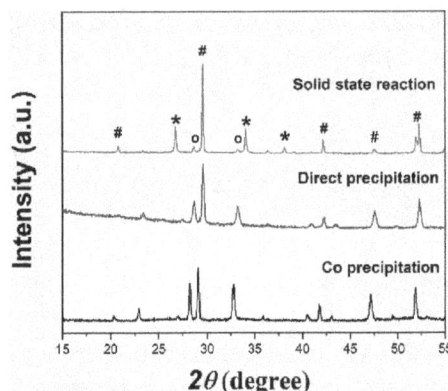

Fig. 26.2 X-ray diffractogram of samples prepared by Co-precipitation, direct precipitation and solid state reaction. The peaks marked by asterisks (*) are corresponding to CsSnBr$_3$. The peaks marked by (#) and (o) are corresponding to CsBr and SnBr$_2$, respectively

comparison. The peaks marked with (*) correspond to CsSnBr$_3$ phase whereas the peaks marked with (#) and (o) correspond to CsBr and SnBr$_2$, respectively [14]. Samples prepared by both the precipitation methods exhibit almost identical XRD patterns indicating the presence of CsBr and SnBr$_2$ which are the precursor salts. It was reported earlier for the sample prepared by co-precipitation method that incomplete reaction or immediate decomposition of the materials on exposure to air might be the reason for the presence of precursor materials. [10] In the case of direct deposition, it was expected that the presence of HBr acid could facilitate the reaction but no such improvement noticed in XRD result [12]. Optimization of precipitation method is more challenging as there are several factors such as precipitating agent, temperature, pH and concentration of reactants which play crucial role in precipitation [15]. On the other hand, the solid state reaction method exhibits better result compared the other two methods. The sample has peaks at 26.78°, 34.07°, 36.12° and 38.14° (marked with '*') which have been identified with cubic phase of CsSnBr$_3$ [14]. Although, the peaks corresponding to precursor salts are still present but intensity of SnBr$_2$ peaks have been reduced significantly. It is evident from XRD results that solid state reaction exhibits better yield and may produce pure phase of CsSnBr$_3$ by complete reaction with further optimization of reaction parameters (e.g. temperature and duration of reaction). As both the parameters play a major role in solid state reaction, further investigations will be done with variable temperature and duration.

UV-Vis absorption behaviour of all the materials were studied to explore their optical properties. Figure 26.3

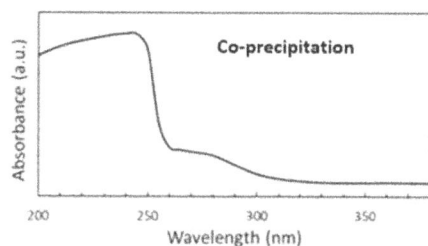

Fig. 26.3 UV-Vis absorption spectrum of the sample prepared by co-precipitation method

shows the absorption spectrum for the sample prepared by co-precipitation method. A sharp band edge around 250 nm along with a small hump around 275 nm were identified. Both correspond to CsBr phase only with different space groups [10]. Figure 26.4 shows the spectrum for the sample synthesized by direct precipitation method. Two sharp peaks at 225 nm and 240 nm were observed. The first one corresponds to CsO_2 phase [16] which might appear due to oxidation of CsBr salt on exposure to air. The other peak at 240 nm represents the presence of CsBr phase. Even for the sample prepared by solid state reaction, two peaks around 220 nm and 245 nm were found which corresponds to CsO_2 and CsBr phases, respectively [Fig. 26.5]. In all the results, there is no peak corresponding to $CsSnBr_3$ phase which may be ascribed due to fast decomposition and oxidation of the samples. However, the spectrum for

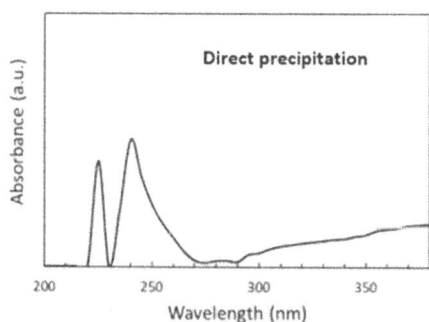

Fig. 26.4 UV-vis absorption spectrum of the sample prepared by direct precipitation method

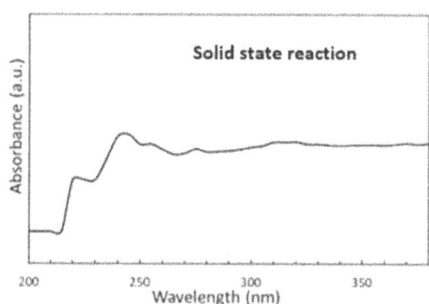

Fig. 26.5 UV-Vis absorption spectrum of the sample prepared by solid state reaction

the sample prepared by solid state reaction shows relatively less intense peaks for CsO_2 and CsBr phases and a wide absorption 250 nm onwards which can be promising for photocatalytic application. This indicates that solid state reaction exhibits comparatively better results however, further optimization is required in synthesis method.

4. Conclusions

In summary, all inorganic lead free perovskite $CsSnBr_3$ have been prepared by three different methods (co-precipitation, direct precipitation and solid state reaction). To compare their structural and optical properties, XRD measurements and UV-Vis absorption studies were performed. XRD results reveal that solid state reaction exhibits better phase formation compared to other two methods. However, further optimization of reaction parameters (e.g. temperature and duration of reaction) is required to achieve pure phase. Absorption spectra showed the peaks corresponding to precursor salt (CsBr) and CsO_2 (a by product due to oxidation). It indicates that the materials undergo either incomplete reaction or immediate decomposition. However, solid state reactions are preferred because they increase yield by reducing waste products. Therefore, solid state reaction method with proper optimization could be a better option to prepare lead free perovskite for photocatalytic applications and can be a promising method for large scale synthesis of such materials.

References

1. Lee, M. M., Teuscher, J., Miyasaka, T., Murakami, T. N., Snaith, H. J. (2012). Efficient Hybrid Solar Cells Based on Meso-Superstructured Organometal Halide Perovskites. Science 338, 643–647.
2. Liu, M., Johnston, M. B., Snaith, H. J. (2013). Efficient planar heterojunction perovskites solar cells by vapour deposition. Nature 501, 395–398.
3. Ke, W., Kanatzidis, M. G. (2019). Prospects for low-toxicity lead-free perovskite solar cells. Nature Communications 10, 965.
4. Huang, H., Pradhan, B., Hofkens, J., M. B. J. Reoffers, J. A. Steele. (2020). Solar-Driven Metal Halide Perovskite Photocatalysis: Design, Stability, and Performance. ACS Energy Letters 5,1107–1123.
5. Park, N.-G. (2013). Organometal Perovskite Light Absorbers Toward a 20% Efficiency Low-Cost Solid-State Mesoscopic Solar Cell. J. Phys. Chem. Letters,4,2423–2429.
6. Eperon, G. E., Stranks, S.D., Menelaou, C., Johnston, M. B., Herz, L. M., Snaith, H. J. (2014). Formamidinium lead trihalide: a broadly tunable perovskite for efficient planar heterojunction solar cells. Energy & Environmental Science 7(3), 982–988.

7. Grätzel, M. (2014). The light and shade of perovskite solar cells. Nature materials13, 838–842.

8. Kojima, A., Teshima, K., Shirai, Y., Miyasaka, T. (2009). Organometal Halide Perovskites as Visible-Light Sensitizers for Photovoltaic Cells. Journal of the American Chemical Society 131, 6050–6051.

9. Kumar, D., Yadav, R.S., Monika, Singh, A. K., Rai, S. B. (1999). Perovskite Materials, Devices and Integration. 1st edition, Intech Open, London.

10. Suma, J., Gijare, S., Mandal Bera, A. (2023). All Inorganic Lead-free Perovskites for Photocatalysis: Preparation and Absorption Study. In: 2023 5th Biennial International Conference on Nascent Technologies in Engineering (ICNTE) pp.1–4. IEEE, Navi Mumbai, India.

11. Liu, S., Zhang, K., Tan, L., Qi, S., Liu, G., Chen, J., Lou, Y. (2021). All-inorganic halide perovskite CsPbBr$_3$@CNTs composite enabling superior lithium storage performance with pseudocapacitive contribution. Electrochimica Acta 367, 1–7.

12. Abib, M.H., Li, J., Yang, H., Wang, M., Chen, T., Jiang, Y. (2021). Direct deposition of Sn-doped CsPbBr$_3$ perovskite for efficient solar cell application. Royal Society of Chemistry 11, 3380–3389 .

13. Chen, M., Ju, M.G., Garces, H.F., Carl, A.D., Ono, L.K., Hawash, Z., Zhang, Y., Shen,T., Qi, Y., Grimm, R.L., Pacifici, D., Zeng, X.C., Zhou., Y., Padture, N.P. (2019). Highly stable and efficient all-inorganic lead-free perovskite solar cells with native-oxide passivation. Nature Communications 101(6),1–8.

14. Ke, W., Stoumpos, C. C., Kanatzidis, M.G. (2018). "Unleaded" Perovskites: Status Quo and Future Prospects of Tin-Based Perovskite Solar Cells. Advanced Materials 31(47), 1–31.

15. https://beingchemist.com/precipitation-gravimetry, last accessed 2024/7/15.

16. Nemade, K. R., Waghuley, S. A. (2017). Synthesis of stable cesium superoxide nanoparticles for gas sensing application by solution-processed spray pyrolysis method. Appl Nanosci. 7,753–758.

Note: All the figures in this chapter were made by the author.

Advances in Materials Science and Technology – Dr. Srikari Srinivasan et al. (eds)
© 2025 Taylor & Francis Group, London, ISBN 978-1-041-12342-2

27

Optimization for Soft Magnetic Properties of [NiFe/Ta]$_n$ Multilayer for Magnetic Flux Concentrator

Gopika C. T.,
Apoorva Kaul, Tejaswini Gawade
[1]Nanomaterials Research Laboratory,
Surface Engineering Division, CSIR - National Aerospace Laboratories,
Bangalore, India
Academy of Scientific and Innovative Research (AcSIR),
Ghaziabad, India

Jakeer Khan, Bhagaban Behera
[1]Nanomaterials Research Laboratory,
Surface Engineering Division, CSIR - National Aerospace Laboratories,
Bangalore, India

P. Chowdhury*
Nanomaterials Research Laboratory,
Surface Engineering Division, CSIR - National Aerospace Laboratories,
Bangalore, India
Academy of Scientific and Innovative Research (AcSIR),
Ghaziabad, India

ABSTRACT: On-chip micro-magnetic flux concentrators (MFC) became a popular choice to enhance the magnetic field sensitivity of a magnetic sensor. However, the magnetic film thickness of several micrometers with significantly reduced coercivity and low saturation field was demanded to achieve the higher gain factor with improved reversibility. In this report, the thick permalloy film of NiFe was replaced by a laminated stack of Ta/Ru/ [NiFe/Ta(t_{Ta})] $_n$/Ta, where Ta layer thickness was varied from 0.1 to 5nm range keeping NiFe thickness constant at 10 nm. Structural investigations revealed that the spray layer of Ta suppresses the grain growth of the NiFe phase and improves the textured growth of the *fcc* (111) phase of NiFe alloy. Magnetic hysteresis loops confirmed the dramatic improvement in the coercivity field in the range of 0.3 ± 0.1Oe, high linearity, and low saturation field (< 10G) while comparing them with the single-layer film of NiFe of similar thickness. Therefore, the on-chip magnetic flux concentrator based on microlaminated thick film is useful In enhancing the sensitivity of a magnetic sensor without affecting linearity performance.

KEYWORDS: Coercivity, Magnetic flux concentrators, Magnetic hysteresis, Permalloy, Saturation field

*Corresponding author: pchowdhury@nal.res.in

DOI: 10.1201/9781003664277-27

1. Introduction

The highest sensitive magnetoresistive sensors are now commercially available based on magnetic tunnel junctions (MTJs) and show great potential for low dynamic field range of applications. To enhance the further sensitivity in the sub-pT range[1, 2] towards health care application[3, 4]. the concept of magnetic flux concentrator (MFC)[5] across the MTJ-based sensing elements was introduced[1]. Two approaches have been made so far. In the first approach, a thick foil (0.5mm) of MFC made of superpermalloy is attached externally across the IC, whereas, in the second one, on-chip integration of MFC was carried out through the deposition of a thick film[6-10]. In the second approach, the simulated footprint with various sizes and shapes can be reproduced to optimize the gain factor[11-13]. However, the need for MFC film with high permeability and low coercivity is significant to maintain the sensor performance in reversibility, linearity, and enhanced sensitivity[11, 14].

Considering the optimized shape and size of the micromagnetic flux concentrator, the thickness of MFC plays an important role in having a higher gain factor[5]. Thicker MFC of NiFe alloy with low coercivity was explored through an electrodeposition process, however, the incorporation of a chemical process was always not viable for on-chip integration. Therefore, physical vapor deposition, such as the sputtering technique, is the alternative approach, though residual stress and magnetostriction may reduce the softness properties of the film[15, 16]. In most of the previous studies for micromagnetic flux concentrators, the thickness of the sputter-deposited film of various alloys was limited to 0.1 microns to 1 micron. The need for post-deposition annealing may relieve the stress[16], however, an on-chip micromagnetic flux concentrator integrated with magnetic sensing elements will restrict the annealing temperature beyond which the sensor elements may damage. More recently, single layer film with micron thick- ness was replaced by a laminated film of conetic alloy ($Ni_{77}Fe_{14}Cu_5Mo_4$) separated by a spacer layer of Cu, where magnetic and the space layer thicknesses were kept at 100 nm and 4 nm respectively[17]. It was reported that soft magnetic properties remain unchanged even with the total film thickness of 7.2 μm while building up the laminated stack[17]. In another report by Luong et al[18], sputtering techniques with a total laminated film of Ta/NiFe of the thickness of 0.8 μm was deposited and the measured coercivity was 0.15 Oe with Ta layer thickness of 3 nm and NiFe thickness of 200 nm. However, no report on a thorough investigation of the structural correlation with magnetic properties exists.

In this work, we have investigated the soft magnetic properties of NiFe/Ta multilayer films deposited using UHV magnetron sputtering. Micro-laminated stacks with thicknesses up to several micrometers were successfully fabricated and the magnetic data confirms the use of these films for on-chip micromagnetic flux concentrator applications.

2. Experimental

In this study, several multilayer stacks, as shown schematically Fig. 27.1 (a and b), were deposited on a SiO_2 coated Si substrate using an Ultra-High Vacuum (UHV) DC magnetron sputtering at room temperature. A base pressure of 4×10^{-8} Torr or less was maintained before the deposition. Deposition rates for NiFe and Ta have been estimated through depth measurements on a thin film (100 nm) deposited over a lithographically patterned wafer using atomic force microscopy (AFM) and found to be 0.2 nm/s and 0.1 nm/s respectively. The stack deposition was carried out under 3.5 mTorr Argon gas pressure. As shown schematically in Fig. 27.1 (b), in the multilayer stack, the bilayer structure of NiFe/Ta was repeated 50 times. For different samples under investigation, the thickness of individual NiFe layers in the bilayer stricture was kept constant at 10 nm, while the Ta lamination layer thickness was varied from 0.1-1nm. For comparison, a single layer of NiFe of thickness around 500nm was deposited under identical conditions, as shown schematically in Fig. 27.1(c).

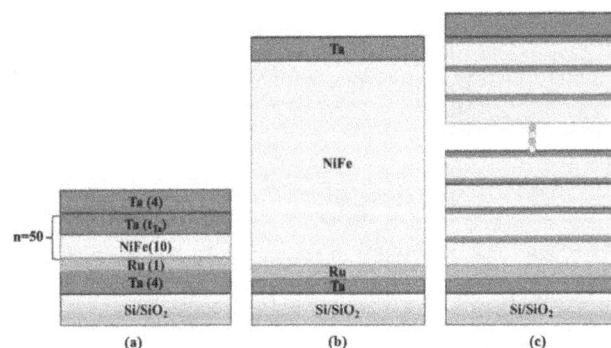

Fig. 27.1 Schematic representation of (a) [Nife/Ta]$_n$ multilayer stack structure deposited using DC magnetron sputtering, (b) Laminated stack with different t_{Ta} values, and (c) 500 nm single layer NiFe film

For comparison, micrometer-thick laminated magnetic films that contain 200 and 400 bi-layers of NiFe (10nm)/Ta (0.1nm) and have a total the magnetic layer thickness of about 2 and 4μm respectively and photolithographically patterned with 1mm ×1 mm square shape was deposited to understand the softness in an extended range of total magnetic thickness.

The magnetic properties of different samples were measured by a Vibrating Sample Magnetometer (VSM) (Lakeshore

USA) at room temperature. The crystallography was investigated using an X-ray diffractometer (XRD) with Cu-Kα radiation.

3. Results and Discussion

3.1 Crystal Structure and Morphology of NiFe Film

Figure 27.2 shows the θ-2θ scan of the X-ray Diffraction (XRD) pattern of different films as a function of Ta layer thickness, and here t_{Ta} was kept at a fixed value in the range from 0.1 to 5nm in a laminated stack. As the number of bilayer stacks was kept at a constant value of 50, the film's total thickness varied from 500nm to 750nm including both magnetic and spacer layer thicknesses. For comparison, we have also shown XRD data generated from a continuous film of thickness 500nm. As shown in Fig. 27.2, all the films show the most prominent peak corresponding to *fcc* (111) phase of NiFe alloy at an angle of $2\theta = 43.83 \pm 0.03°$. The laminated films show the peak broadening with higher intensity, indicating the grain size reduction of the NiFe phase. The inset of Fig. 27.2, presents the variation of grain size with t_{Ta} and it indicates a sharp decrease in the grain even with $t_{Ta}= 0.3$ nm and remains a constant value in the range of 8.17 ± 0.2nm. The higher intensity may relate to the textured growth of the *fcc* (111) phase as the Ta layer acts as a seed layer to allow the growth of the crystalline *fcc* (111) phase of NiFe [19].

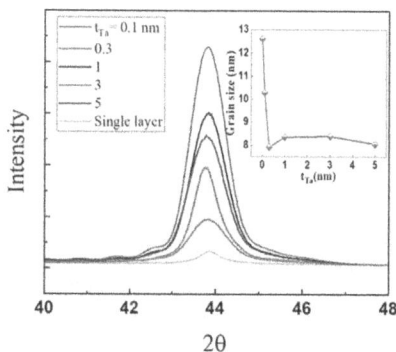

Fig. 27.2 X-Ray diffraction pattern of [Nife/Ta]ₙ multilayer with variation in Ta thickness at room temperature

3.2 Magnetic Properties

Figure 27.3 (a-c) present the M (H) hysteresis loops of the films with a total magnetic layer thickness of 500 µm where 10 nm NiFe layer was interleaved with ultrathin Ta layers in the multilayer structure and the inset of each plot shows an expanded view for better clarity in the low field region. M (H) curve for the single-layer NiFe film with 500 µm layer thickness was shown in Fig. 27.3(d). The extracted

magnetic parameters from these M (H) such as coercive field (H_c), remanence ratio (M_r/M_s), and the saturation magnetization field (H_s) are provided in Fig. 27.3. From these data, it is clear that 500 nm single-layer film shows a higher coercivity of 16.3 G and a high saturation field of 160 G. In contrast, 500 nm laminated films exhibit lower coercivity in the range of 0.3 ± 0.1 G and a low saturation field of 8 ± 3 G. In our further investigations, as the thickness of single-layer NiFe film increases from 10 to 100 nm, the coercivity varies from 0.5 G to 1.7 G (not shown here), however, it further shoots up to 16 G with increasing the film thickness of 500 nm. which is in agreement with previous report on conetic alloy which shows there is drastic increment of the H_c values from 0.5 G to 15 G with increasing the film thickness from 100 to 1000 nm [17]. The origin of this increment was attributed to the increase in stress and magnetostriction in the thicker film. On the other hand, a low value of coercivity in the laminated film indicates the reduction of stress in the film. As shown the Fig. 27.3 (a-c), the thickness of the Ta spacer layer till 1 nm does not impact the overall M (H) characteristics, however with increasing thickness above 1 nm, the coercivity was found to increase to 0.7 0.1 G. Overall, enhancement of interlayer magnetostatic coupling, reduction of stress and magnetostriction play an important role to reduce the coercivity in the laminated film [17].

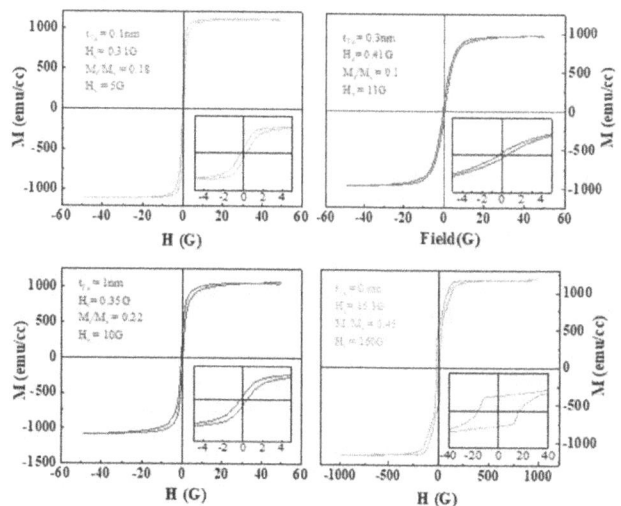

Fig. 27.3 (a-c) Room temperature M-H loops measured along in-plane of film for [NiFe (10 nm)/Ta ($t_{T\,o}$)]₅₀ multilayer with variation in t_{Ta} from 0.1 to 1nm, and (d) same for single layer NiFe film of thickness 500nm

Furthermore, Fig. 27.4 presents comparative M (H) characteristics of the laminated films fabricated with n bilayers in the [NiFe (10 nm)/Ta (0.3 nm)]ₙ stack. Here n was kept at 50, 200, and 400 which correspond to a

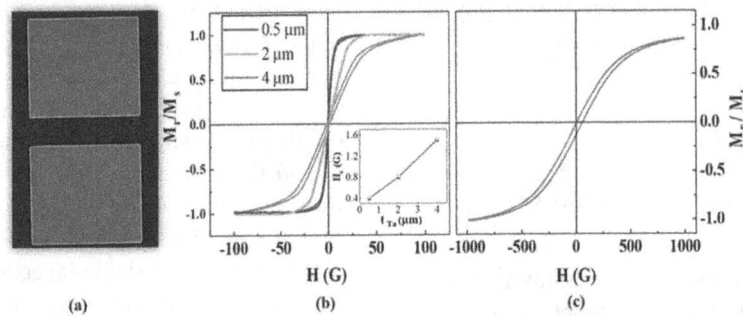

Fig. 27.4 (a) Optical microscope image of 4μm thick [NiFe(10nm)/Ta(0.3nm)]$_n$ multilayer structure successfully patterned to 1 mm 1 mm square shape on SiO$_2$ wafer. (b) M-H loops measurement done on 0.5μm, 2μm, and 4μm thick films (c) M-H loop of commercially used magnetic core measured up to 1000G field

total magnetic layer thickness of 0.5 μm, 2 μm, and 4 μm respectively. It is to be noted that the data was generated from a photolithographically patterned square-shaped film with an area of 1 mm^2 as shown in Fig. 27.4(a), keeping in mind of similar dimension of MFC structure. Enhancement of the coercivity with thickness from 0.3 G to 1.5 G indicates the development stress within the film with an increase in the thickness of the film. However, H$_c$ values even for a 4-micron laminated film comparable to 100 nm single-layer film, demonstrate that the laminated structure substantially reduces stress build up in the thick film and maintains the softness even for few microns thick film in agreement with earlier report[17]. Figure 27.1(c) presents the M(H) curves for a bulk foil of super permalloy (NiFeMoMnSi) which is being used for bulk MFC and EMC-shielded material. This data indicates the laminated film has equivalent properties as the bulk film.

4. Conclusion

In this report, the magnetic properties of several micrometer-thick sputter-deposited laminated structures of permalloy were investigated and compared with single-layer submicron thick film. It was demonstrated that the micrometer-thick laminated film can exhibit significantly lower coercivity and low saturation field in comparison to single-layer film. Along with magnetostatic coupling between successive layers, the reduction of stress and magnetostriction may play an important role in reducing the coercivity in laminated films. The softness properties of laminated film will help to build up the on-chip micromagnetic flux concentrator with enhanced sensitivity for a magnetic sensor based on either MTJ or GMR-based magnetic sensors.

Acknowledgments

The authors would like to thank SERB DST and The Director, CSIR-NAL for supporting this activity. Gopika C T, Apoorva Kaul, and Tejaswini Gawade express their sincere gratitude to AcSIR for allowing them to pursue their doctoral degree. This work is a part of current projects funded by SERB DST, Ministry of Science and Technology India.

References

1. R. C. Chaves, B.O.W.M. P. P. Freitas: Low frequency picoTesla field detection using hybrid MgO based tunnel sensors. Applied Physics Letters 91, 102504 (2007)
2. Chaves, R.C., Freitas, P.P., Ocker, B., Maass, W.: MgO based picoTesla field sensors. Journal of Applied Physics 103(7), 07–931 (2008)
3. Oogane, M., Fujiwara, K., Kanno, A., Nakano, T., Wagatsuma, H., Arimoto, T., Mizukami, S., Kumagai, S., Matsuzaki, H., Nakasato, N., Ando, Y.: Sub- pt magnetic field detection by tunnel magneto-resistive sensors. Applied Physics Express 14(12), 123002 (2021)
4. Ishikawa, K., Oogane, M., Fujiwara, K., Jono, J., Tsuchida, M., Ando, Y.: Investigation of magnetic sensor properties of magnetic tunnel junctions with superparamagnetic free layer at low frequencies for biomedical imaging applications. Japanese Journal of Applied Physics 55(12), 123001 (2016)
5. Sun, X., Jiang, L., Pong, P.W.T.: Magnetic flux concentration at micrometer scale. Microelectronic Engineering 111, 77–81 (2013)
6. Leitao, D.C., Gameiro, L., Silva, A.V., Cardoso, S., Freitas, P.P.: Field detection in spin valve sensors using CoFeB/Ru synthetic-antiferromagnetic multilayers as magnetic flux concentrators. IEEE Transactions on Magnetics 48(11), 3847–3850 (2012)
7. Leitao, D.C., Coelho, P., Borme, J., Knudde, S., Cardoso, S., Freitas, P.P.: Ultra- compact 100 × 100 m2 footprint hybrid device with spin-valve nanosensors. Sensors 15(12), 30311–30318 (2015)
8. Marinho, Z., Cardoso, S., Chaves, R., Ferreira, R., Melo, L.V., Freitas, P.P.: Three-dimensional magnetic flux concentrators with improved efficiency for magnetoresistive sensors. Journal of Applied Physics 109(7), 07–521 (2011)

9. Kulkarni, P.D., Iwasaki, H., Nakatani, T.: The effect of geometrical overlap between giant magnetoresistance sensor and magnetic flux concentrators: A novel comb-shaped sensor for improved sensitivity. Sensors 22(23) (2022)

10. Coelho, P.R.V.: Sub-2μm gap magnetic flux concentrators coupled to nano spin valve sensors. Master's thesis, Instituto Superior T'ecnico, Universidade de Lisboa, Lisboa, Portugal (September 2014)

11. Guedes, A., Almeida, J.M., Cardoso, S., Ferreira, R., Freitas, P.P.: Improving magnetic field detection limits of spin valve sensors using magnetic flux guide concentrators. IEEE Transactions on Magnetics 43(6), 2376–2378 (2007)

12. Silva, M., Silva, J.F., Leitao, D.C., Cardoso, S., Freitas, P.P.: Optimization of the gap size of flux concentrators: Pushing further on low noise levels and high sensitivities in spin-valve sensors. IEEE Transactions on Magnetics 55(7), 1–5 (2019)

13. Zhang, X., Bi, Y., Chen, G., Liu, J., Li, J., Feng, K., LV, C., Wang, W.: Influence of size parameters and magnetic field intensity upon the amplification characteristics of magnetic flux concentrators. AIP Advances 8(12), 125222 (2018)

14. Trindade, I.G., Teixeira, J., Fermento, R., Sousa, J.B., Cardoso, S., Chaves, R.C., Freitas, P.P.: High sensitivity spin valve sensors with af coupled flux guides. IEEE Transactions on Magnetics 44(11), 2472–2474 (2008)

15. Wang, X., Zhu, Z., Ma, L., Feng, H., Xie, H., Wang, J., Liu, Q., Han, G.: Influence of deposition cycle and magnetic annealing on high-frequency magnetic proper- ties of the [Co$_{90}$Fe$_{10}$/Ta]$_n$ multilayer thin films. IEEE Transactions on Magnetics 54(8), 1–7 (2018)

16. Tedzhetov, V., Sheftel, E., Harin, E., Kiryukhantsev-Korneev, P.: Residual stresses in soft magnetic fetib and fezrn films obtained by magnetron deposition. Coatings 11, 34 (2020)

17. Yang, Y., Liou, S.-H.: Laminated magnetic film for micro magnetic flux concen- trators. AIP Advances 11(3), 035004 (2021)

18. Luong, V.S., Nguyen, A.T., Nguyen, T.L., Nguyen, A.T., Hoang, Q.K.: Enhanced soft magnetic properties of [NiFe/Ta]$_n$ laminated films for flux amplification in magnetic sensors. IEEE Transactions on Magnetics 54(6), 1–4 (2018)

19. Padmanapan, S., Hsu, J.-H., Tsai, C.-L., Singh, A., Alagarsamy, P.: Effect of ta-underlayer on thickness dependent magnetic properties of NiFe films. IEEE Transactions on Magnetics 51 (2015)

Note: All the figures in this chapter were made by the author.

Advances in Materials Science and Technology – Dr. Srikari Srinivasan et al. (eds)
© 2025 Taylor & Francis Group, London, ISBN 978-1-041-12342-2

28

Finite Element Analysis of CNC Dry Machining for Alloy Workpiece using Ansys Workbench

Anamika Tiwari*,
Sanjay Mishra, D. K. Singh
Mechanical Engineering Department,
Madan Mohan Malaviya University of Technology,
Gorakhpur, Uttar Pradesh, India

ABSTRACT: In this new advance era of metal cutting, CNC turning process plays a significant role in designing and manufacturing industries and before staring the manufacturing processes, analysis is a crucial part of the manufacturing processes. So, development of simulation model has been done to understand the machining processes and also that can facilitate the understanding of these operations giving results as close as possible to the real time values. This study involves Finite Element Analysis of the CNC turning process of a cylindrical SKD 11 JIS G4404 workpiece. The tool used in this study is Tungsten Carbide tip tool. The combination of Tungsten Carbide cutting tools and SKD 11 Alloy Steel workpieces is ideal for high-precision machining operations requiring durability and resistance to high temperatures and wear. Model of the Single point Tungsten Carbide tip cutting tool was generated using Solid Works software and the analysis was done in ANSYS Workbench 2024 R1 software and the material properties of tool and workpiece are input individually. The Ansys software analysed the Single Point Tungsten Carbide tip cutting tool model under dry cutting by FEA at angular cutting velocity (81.63 rad/sec, 122.48 rad/sec and 163.27 rad/sec) and results obtained are discussed to provide total deformation, equivalent Von-Mises stress. FEA result shows that the boundary conditions for loading of cutting forces are accurate and the prediction of cutting tool stresses statically by finite element analysis can be determined in hard tuning processes. The FE analysis contributes towards the determination of the optimized cutting process variables and parameters.

KEYWORDS: Finite element analysis (FEA), SKD 11 JIS G4404 alloy steel, CNC turning

1. Introduction

Machining is a fundamental metal processing operation performed daily in many manufacturing industries. The quality of metal cutting, which is crucial to the overall product quality, is influenced by numerous processing parameters[1]. These parameters affect the output response and performance characteristics of the final product. Consequently, selecting the appropriate process and parameters is a critical step in the product life cycle. Ensuring optimal machine running conditions involves a careful analysis of machining parameters such as cutting speed, feed rate, depth of cut, and tool geometry. These parameters must be finely tuned to maintain the balance between productivity and quality. Advanced modelling techniques, such as finite element analysis (FEA) and computer-aided manufacturing (CAM), can be employed to simulate and optimize these machining processes [2].

*Corresponding author: arpita10194@gmail.com

DOI: 10.1201/9781003664277-28

The world and the processes within manufacturing industries are changing very rapidly. To keep pace with these changes, the optimization of metal cutting processes is conducted at various levels to meet the new demands faced by manufacturing units [2]. In conclusion, the proper selection and optimization of machining processes and parameters are essential for achieving high-quality products in manufacturing industries. Through rigorous modelling, analysis, and experimental validation, manufacturers can enhance tool life, reduce production costs, and ensure superior product quality, thereby meeting consumer expectations and maintaining a competitive edge in the market[3].

1.1 CNC Turning Process

Turning, a fundamental machining operation, involves the linear movement of a cutting tool against a revolving workpiece to modify its exterior surfaces [4]. While it can be performed manually, the beginning of CNC technology has revolutionized the process, offering unparalleled precision, efficiency, and consistency. CNC technology is not limited to turning but extends to other machining operations like drilling and milling, making it an essential part of modern manufacturing [5]. By leveraging CNC technology, manufacturers can meet the demands of today's dynamic and high-quality production requirements [6]. In this study, the tool used is the Tungsten Carbide cutting tool. Tungsten carbide cutting tools are manufactured using the powder metallurgy method, which involves compacting powdered materials and sintering them at high temperatures to form a solid piece. These tools are renowned for their exceptional hardness and wear resistance, maintaining their hardness up to 1000°C. The following table shows us the tool characteristics. SKD 11 is a high-carbon, high-chromium alloy steel commonly used in the manufacture of dies and moulds due to its excellent wear resistance and toughness [7]. The properties of SKD 11 Alloy Steel are given in the following table. The combination of Tungsten Carbide cutting tools and SKD 11 Alloy Steel workpieces is ideal for high-precision machining operations requiring durability and resistance to high temperatures and wear. This ensures optimal performance and longevity of both the tools and the workpieces in demanding manufacturing environments.

1.2 Software Used

In this research work, two types of software are employed: one for modelling and the other for analysis. The modelling software used is SolidWorks, and the analysis software is Ansys Workbench 2024 R1. The combination of SolidWorks for modelling and Ansys Workbench 2024 R1 for analysis provides a comprehensive approach to the research. SolidWorks facilitates the creation of detailed and precise models, while Ansys Workbench enables thorough analysis of these models under simulated conditions [8]. This integrated approach allows for a deeper understanding of the behaviour and performance of the Tungsten Carbide cutting tool and SKD 11 Alloy Steel workpiece, leading to better optimization and improvement in manufacturing processes.

1.3 Properties

The alloy steel workpiece SKD 11 and a tungsten carbide tip tool were used in this study. The effectiveness of conventional machining operations enhances by the combination of these tool and workpiece combination. The high machinability of SKD 11 and the advanced cutting properties of tungsten carbide cutting tool gives outcome in improved performance, faster machining times, and better overall efficiency in manufacturing processes. The combination of workpiece SKD 11 and a tungsten carbide tip too is highly suitable for research pointed at optimizing machining parameters and improving product quality.

Tool Properties

The tool used in this process is a single point cutting tool made up of tungsten carbide material of which mechanical properties ae listed in Table 28.1. Some of the parameters like shear modulus and bulk modulus in the table has been derived from the young's modulus and poison's ratio, respectively.

Table 28.1 Mechanical properties of the tool

Property	Value (Units)
Density	15700 (kg/m^3)
Young's Modulus	705 GPa
Poisson ratio	0.23
Thermal Conductivity	24 (W/m°-C)
Specific Heat	178 (J/kg-°C)

Workpiece Properties

The workpiece used in this process is a SKD 11 alloy steel comprises of 150mm length and 25.5mm diameter. SKD 11 is a material having content as high-carbon, high-chromium alloy tool steel known for its outstanding wear resistance and toughness. It is commonly used in the manufacturing of dies and moulds. Here are the key properties of SKD 11 alloy steel along with its chemical compositions are shown in Table 28.2. The Johnson-Cook failure model has been used here for the failure model, having a linear connection between the shock and the particle velocity. The mechanical properties of the workpiece have been shown in Table 28.3.

Table 28.2 Chemical composition of SKD 11 workpiece

C	Cr	Cu	Fe	Mn	Mo	Ni	P	Si	S	V
1.4-1.6%	11-13%	0.25%	81.89-86.6%	0.6%	0.8-1.2%	0.5%	0-0.03%	0.4%	0.03%	0.2-0.5%

Table 28.3 Mechanical properties of the SKD 11 workpiece

Properties	Values (Unit)
Density	7.72-8 g/cm^3
Young's Modulus	210 GPa
Shear Modulus	79 GPa
Poisson Ratio	0.3
Specific Heat	420-460 (J/kg-K)
Melting Temperature	1370-1425 °C
Thermal Conductivity	42 (W/m-K)
Thermal Coefficient of expansion	1 e^{-5} – 1.3 e^{-5} (1//K)

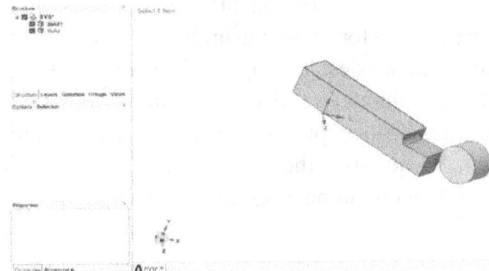

Fig. 28.1 Design of tool and workpiece of given dimension

2. Methodology

2.1 Modelling

Modelling of the workpiece and tool done with the help of the SOLIDWORKS and simulation is done with the help of the ANSYS workbench 2024 R1. The dimension of workpiece is given as 25.5mm diameter and 150mm length with tool as dimension of rake angle and clearance angle as -6° and 6° respectively along with cutting edge length of 15mm and tool corner radius as 1.2mm.the design of tool and workpiece exported to ANSYS workbench is shown in below Fig. 28.1.

2.2 Mesh Generation

For generating mesh, the selection of the geometry has been done and clicked the show mesh icon on left side of the task bar. It generates the mesh of the geometry selected where editing of the meshing parameters has been performed to get the required results. The simulation results get closer to real, finer the mesh gets. Thus, the nodes obtained after meshing were 29245 with element size of 1.5mm for each of the simulation as shown in Fig. 28.2.

2.3 Boundary Conditions

After assigning the proper material properties to the tool and workpiece of prepared model, coordinate systems are also added accordingly to the geometries as one being Cartesian and other being the cylindrical coordinate system. To get a proper cutting action the tool has been align on the workpiece in such accurate manner to get a relative motion between them with coefficient of friction (μ=0.3). The motion is defined in explicit dynamics by displacement command which can be inserted in the modelling by a right click menu on the analysis settings. Displacement 1 is

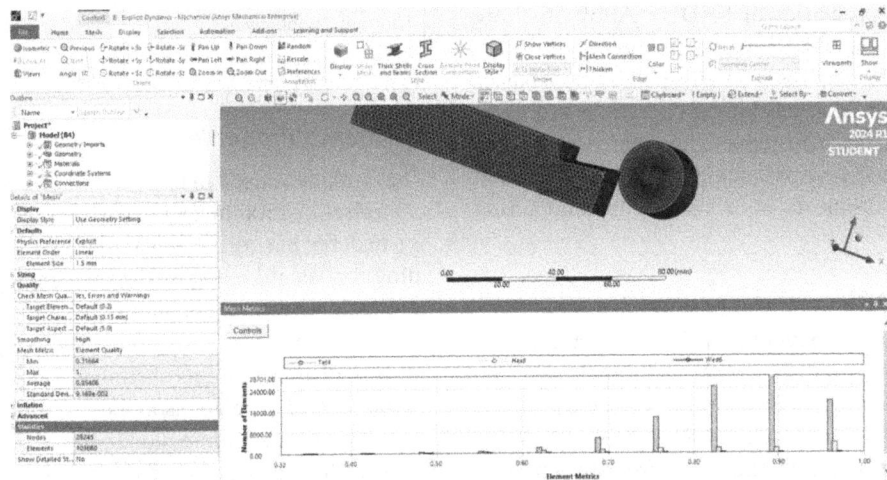

Fig. 28.2 Fine mesh of the model

given to the tool to simulate the feed motion (in direction -Z of the respective body) and displacement 2 is given to the workpiece to simulate the revolution of the workpiece on lathe machine to enable turning operation on the workpiece (rotation in Y direction of the respective body).

3. FEA Results

After the analysis of CNC turning operation of SKD 11 alloy steel with Tungsten Carbide tool, the result has been obtained for Total deformation and Equivalent (Von-Mises) Stress having angular cutting velocity of 81.63 rad/sec, 122.48 rad/sec and 163.27 rad/sec given to workpiece. The result shows us the final stage of the process of the CNC turning process with the desired results to be calculated in the explicit dynamics of the Ansys.

3.1 Total Deformation

The variation of total deformation with time at angular cutting velocity 81.63 rad/sec, 122.48 rad/sec and 163.27 rad/sec are shown in the graph based on same entries depicts the nature of variable with respect as shown in Fig. 28.3 (a), (b) and (c). From the analysis it is obtained that the maximum deformation obtained as 22.604mm obtained at 81.63 rad/sec and 163.27 rad/sec and 22.603mm at 122.48 rad/sec as shown in Fig. 28.3 graph which is drawn between the total deformation and time. And from the Fig. 28.3 (d) it depicted as the time passes along with tool progression the deformation value goes on increasing and after a certain interval of time, maximum stress is obtained at outer surface and minimum deformation analysed at inner surface of workpiece which is also in the favour of machining criteria.

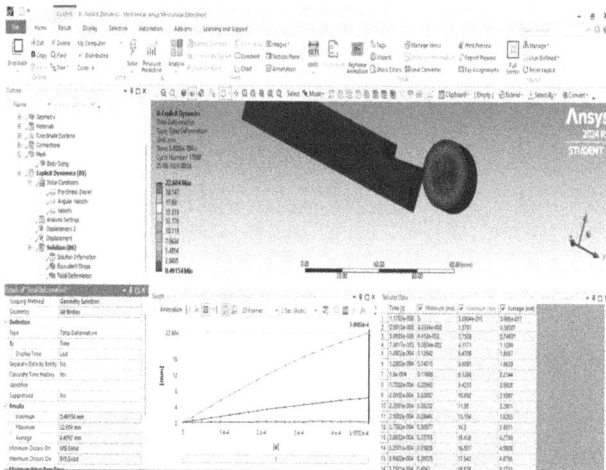

(a)

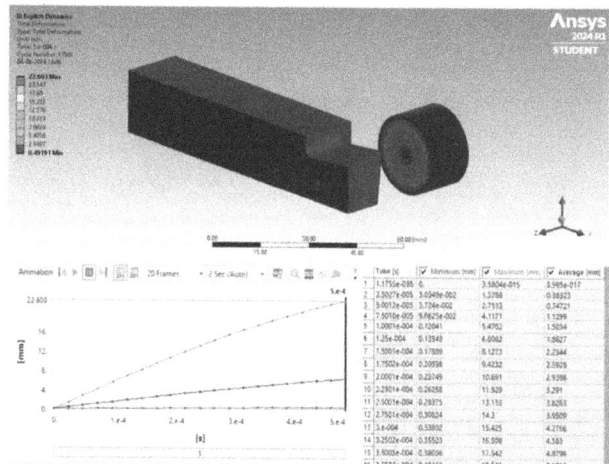

(b)

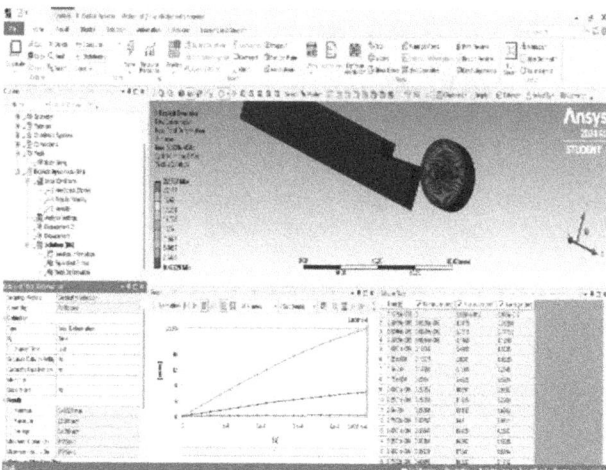

(c)

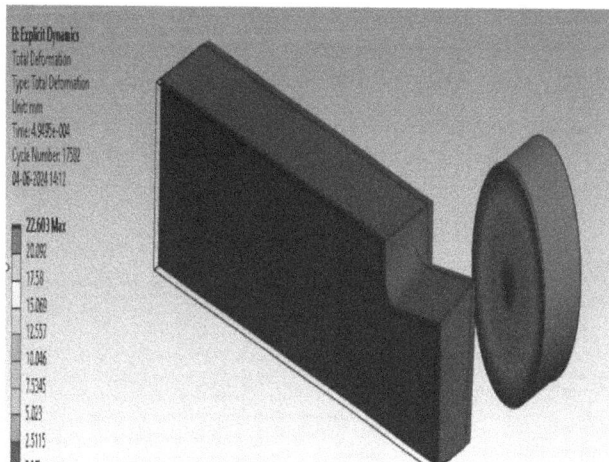

(d)

Fig. 28.3 Total deformation analysis for different radial velocity (a) 81.63 rad/sec, (b) 122.48 rad/sec (c) 163.27 rad/sec (d) Pictorial view of maximum and minimum deformation in workpiece

3.2 Equivalent (VON-Mises) Stress

Wear zone on cutting tool may be forecasted according to Von Mises stress distributions. The variation of equivalent (Von-Mises) stress with time at angular velocity 81.63 rad/sec, 122.48 rad/sec and 163.27 rad/sec are shown in the graph based on same entries depicts the nature of variable with respect as shown in Fig. 28.4 (a), (b) and (c). From the analysis it is obtained that the equivalent (Von-Mises) stress obtained as 251.24MPa, 254.83MPa and 277.9 MPa obtained at 81.63 rad/sec, 122.48 rad/sec and 163.27 rad/sec122.48 rad/sec respectively as shown in Fig. 28.4 graph which is drawn between the equivalent (Von-Mises) stress and time. And from the Fig. 28.4 (d) it depicted as the time passes along with tool progression the equivalent (Von-Mises) stress value goes on increasing from the initial tool-tip interface and after a certain interval of time, maximum stress is obtained at mid of workpiece length and minimum stress analyzed at inner core surface of workpiece.

4. Conclusion

In this study, total deformation and equivalent Von-Mises stress have been analyzed for hard turning of SKD 11 JIS G4404 with Tungsten Carbide tool. The amount of deformation and stress acting over the materials are very difficult ton obtained using the setup within the machine at different angular cutting speed. To compensate this complexity, ANSYS software is used to determine the total deformation and equivalent Von-Mises stress at different angular velocity 81.63 rad/sec, 122.48 rad/sec and 163.27 rad/sec. The effect of dry machining with different angular velocity is analysed using ANSYS FEA explicit dynamics analysis and following are the outcomes:

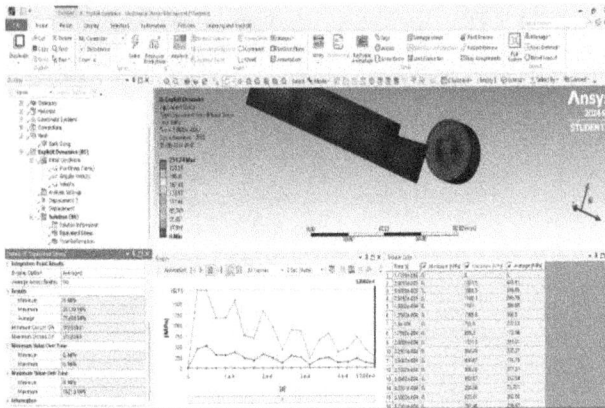
(a)

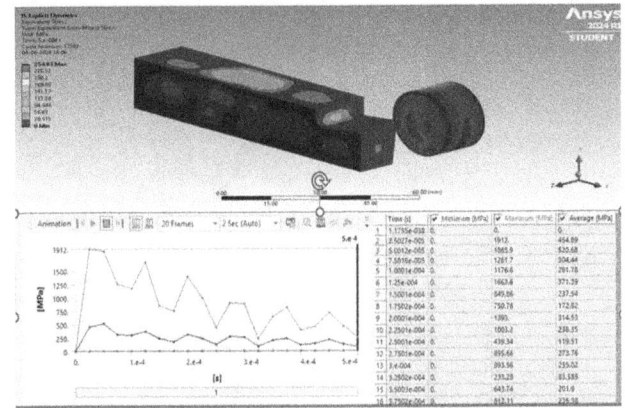
(b)

(c)

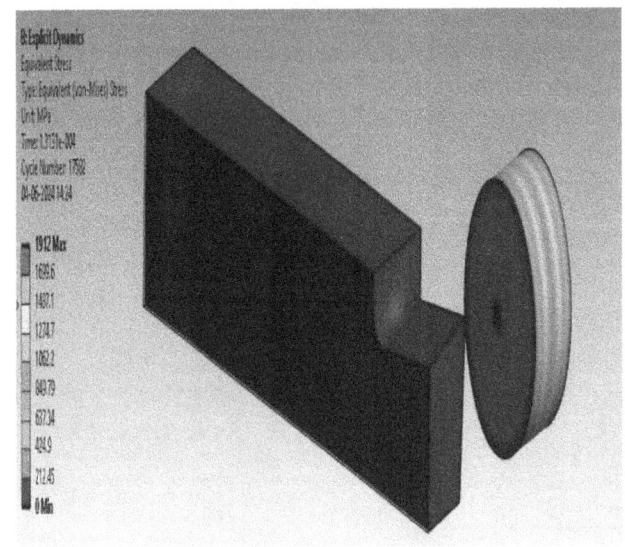
(d)

Fig. 28.4 Equivalent (von-mises) stress for different radial velocity (a) 81.63 rad/sec, (b) 122.48 rad/sec (c) 163.27 rad/sec (d) Pictorial view of maximum and minimum equivalent (Von-Mises) stress in workpiece

The maximum deformation obtained as 22.604mm obtained at 81.63 rad/sec and 163.27 rad/sec and 22.603mm at 122.48 rad/sec and it is depicted that as the time passes along with tool progression the deformation value goes on increasing and after a certain interval of time, maximum stress is obtained at outer surface and minimum deformation analysed at inner surface of workpiece which is also in the favour of machining criteria. From the analysis it is obtained that the equivalent (Von-Mises) stress obtained as 251.24MPa, 254.83MPa and 277.9 MPa obtained at 81.63 rad/sec, 122.48 rad/sec and 163.27 rad/sec122.48 rad/sec respectively as shown in Fig. 28.4 graph which is drawn between the equivalent (Von-Mises) stress and time. And from the Fig. 28.4 (d) it depicted as the time passes along with tool progression the equivalent (Von-Mises) stress value goes on increasing from the initial tool-tip interface and after a certain interval of time, maximum stress is obtained at mid of workpiece length and minimum stress analysed at inner core surface of workpiece.

Acknowledgements

The authors would like to acknowledge the financial support provided by DST INSPIRE Fellowship [sanction order no. DST/INSPIRE Fellowship/2020/IF200127].

Author contributions All authors have contributed to this work. Anamika Tiwari was involved in conceptualization and analysis. In manuscript draft writing and supervision were done with the assistance of Prof. Sanjay Mishra and Prof. D.K. Singh. All authors read and approved the final manuscript.

Conflict of interest The authors declare no conflict of interest.

References

1. Gupta, M. K.; Mia, M.; Pruncu, C. I.; Khan, A. M.; Rahman, M. A.; Jamil, M.; Sharma, V. S. Modeling and Performance Evaluation of Al2O3, MoS2 and Graphite Nanoparticle-Assisted MQL in Turning Titanium Alloy: An Intelligent Approach. *J. Brazilian Soc. Mech. Sci. Eng.*, **2020**, *42* (4), 1–21. https://doi.org/10.1007/s40430-020-2256-z.
2. Abas, M.; Sayd, L.; Akhtar, R.; Khalid, Q. S.; Khan, A. M.; Pruncu, C. I. Optimization of Machining Parameters of Aluminum Alloy 6026-T9 under MQL-Assisted Turning Process. *J. Mater. Res. Technol.*, **2020**, *9* (5), 10916–10940. https://doi.org/10.1016/j.jmrt.2020.07.071.
3. Fernández-Valdivielso, A.; López De Lacalle, L. N.; Urbikain, G.; Rodriguez, A. Detecting the Key Geometrical Features and Grades of Carbide Inserts for the Turning of Nickel-Based Alloys Concerning Surface Integrity. *Proc. Inst. Mech. Eng. Part C J. Mech. Eng. Sci.*, **2016**, *230* (20), 3725–3742. https://doi.org/10.1177/0954406215616145.
4. Tiwari, A.; Singh, D. K.; Mishra, S. A Review on Minimum Quantity Lubrication in Machining of Different Alloys and Superalloys Using Nanofluids. *J. Brazilian Soc. Mech. Sci. Eng.*, **2024**, *46* (3). https://doi.org/10.1007/s40430-024-04676-6.
5. Valiavalappil, A. STRESS, EQUIVALENT STRAIN AND DEFORMATION FOR MACHINING OPERATION UTILIZING EXPLICIT DYNAMIC. **2024**. https://doi.org/10.53370/001c.94740.
6. Camposeco-Negrete, C. Optimization of Cutting Parameters Using Response Surface Method for Minimizing Energy Consumption and Maximizing Cutting Quality in Turning of AISI 6061 T6 Aluminum. *J. Clean. Prod.*, **2015**, *91*, 109–117. https://doi.org/10.1016/j.jclepro.2014.12.017.
7. Nipu, S. A.; Karim, R.; Rahman, A.; Moon, M.; Choudhury, I. A.; Omar, J. Bin; Khushbu, M. S. *Turning SKD 11 Hardened Steel: An Experimental Study of Surface Roughness and Material Removal Rate Using Taguchi Method*; 2023; Vol. 2023. https://doi.org/10.1155/2023/6421918.
8. Wasif, A.; Tousif, K. N.; Nipun, R. S.; Rahman, M. A.; Kharshiduzzaman, M.; Khan, M. E.; Bhuiyan, M. S. FEM Simulation of Dry Machining Using Ansys Workbench. *SSRN Electron. J.*, **2024**, No. June. https://doi.org/10.2139/ssrn.4862415.

Note: All the figures and tables in this chapter were made by the author.

Advances in Materials Science and Technology – Dr. Srikari Srinivasan et al. (eds)
© *2025 Taylor & Francis Group, London, ISBN 978-1-041-12342-2*

29

Efficient CO$_2$ Capture: Activated Porous Carbon from D-Glucose with Non-Corrosive Activating Agent

Jayalekshmi Babu*,
Abhilash K.S., Deepa Devapal

Analytical and Spectroscopy Division,
Analytical, Spectroscopy and Ceramics Group, Propellants,
Polymers, Chemicals and Materials Entity, Vikram Sarabhai Space Centre,
Thiruvananthapuram, Kerala, India

ABSTRACT: Recently, activated porous carbons have garnered considerable attention as potential materials for CO$_2$ capture. These materials have emerged as promising candidates due to two primary advantages: firstly, the physical adsorption of CO$_2$ on their surfaces reduces the energy needed for material regeneration, and secondly, their porous structure allows for high CO$_2$ adsorption capacities. Furthermore, these porous carbon adsorbents can be easily tailored to introduce enhanced surface characteristics and basic functional groups, thereby improving interactions with CO$_2$, which is crucial for achieving high adsorption capacities. A novel all in one-pot approach for the preparation of activated porous carbon through direct carbonization of D-glucose after activation with a non-corrosive chemical, potassium sodium tartrate for carbon dioxide capture is presented here. Carbon dioxide adsorbents were prepared using different ratios of glucose to potassium sodium tartrate (1:0.5, 1:1, 1:2) were chosen for the work. Of which, activated porous carbon prepared using the ratio 1:1 exhibited the best carbon dioxide adsorption capacity of 11.7 mmol/g and 6.2 mmol g−1 at 0 °C/1 bar and 25 °C/1 bar respectively and high specific surface area (1499 m2 g−1) compared to the existing carbon sorbents.

KEYWORDS: Carbonization, D-glucose, Potassium sodium tartrate, Adsorbents

1. Introduction

Porous carbons have emerged as a pivotal class of materials with extensive applications in environmental, energy, and industrial sectors. Characterized by their high surface area, tunable pore size distribution, and versatile surface chemistry, porous carbons are particularly advantageous in adsorption processes, catalysis[1], energy storage[2], and gas separation[3]. Their unique properties arise from a highly interconnected network of varying pore sizes that can be classified into micropores (<2 nm), mesopores (2-50 nm), and macropores (>50 nm). The development of

porous carbon materials has been driven by the need for efficient and sustainable solutions to address the global challenges such as air pollution, energy shortages, and climate change[4]. Their ability to adsorb gases, including carbon dioxide (CO$_2$), methane (CH$_4$), and hydrogen (H$_2$), is of particular interest for applications in environmental remediation and energy storage. Their capacity to adsorb greenhouse gases has also drawn considerable attention. Consequently, the development of sorbents specifically for carbon dioxide has become a significant and burgeoning area of research.

*Corresponding author: jayalekshmi305@gmail.com

DOI: 10.1201/9781003664277-29

The escalating levels of carbon dioxide (CO_2) in the atmosphere have intensified the search for effective and sustainable methods for CO_2 capture and sequestration. Among the myriad of approaches, the development of efficient sorbents for CO_2 capture has garnered significant attention due to their potential to mitigate climate change by reducing greenhouse gas emissions[5]. Sorbents are materials that can selectively adsorb CO_2 from gas mixtures, making them essential for applications in industrial emissions control, air purification, and carbon capture and storage (CCS) technologies. While numerous porous carbon-based materials exist, those derived from biomass are particularly appealing as CO_2 adsorbents due to their superior textural properties, straightforward and efficient synthesis methods, high CO_2 adsorption capacities, and cost-effectiveness and renewability of biomass[6]. Biomass is traditionally converted into biochar through pyrolysis, a heat treatment conducted in the absence of oxygen. This process can be adapted into various forms, such as slow or fast pyrolysis, low or high-temperature pyrolysis, and dry or wet pyrolysis[7]. A wide range of biomass precursors, including wood, non-wood materials, and animal and fruit wastes, have been used to produce biochar. Activated porous carbons are similarly produced but involve the addition of an activating agent either before or after the biochar preparation step.

Activated porous carbons outperform biochars in terms of higher specific surface areas and more advanced porosity, though biochars have the distinct advantage of abundant functional groups present on their surfaces[8]. The overall process for producing activated porous carbons is simpler and has a lower environmental impact compared to the production of other carbon-based adsorbents[9]. Recently, activated porous carbons have rapidly gained considerable attention as promising materials for CO_2 capture due to their cost-effective synthesis, high thermal and chemical stability, easy manipulation of porous properties such as surface area and pore volume, and importantly, lower energy requirements for material regeneration compared to traditional adsorbents.

This paper explores the synthesis of activated porous carbon from D-glucose, utilizing a non-corrosive activating agent. D-glucose, a widely available and renewable biomass, serves as an ideal precursor for carbon materials. The use of a non-corrosive activating agent presents a safer and more environmentally friendly approach compared to traditional methods involving hazardous chemicals such as KOH or H_3PO_4. The porous structure of the synthesized carbon allows for efficient CO_2 adsorption, attributed to its high surface area and the presence of micropores. Moreover, the ability to modify the surface chemistry of these materials enhances CO_2 interactions, further improving adsorption

capacity. This study aims to investigate the preparation, characterization, and CO_2 capture performance of the activated porous carbon produced from D-glucose, highlighting its potential as an effective material for CO_2 capture applications. By focusing on a renewable precursor and a non-corrosive activation process, this research contributes to developing sustainable and efficient CO_2 capture technologies, addressing both environmental and industrial challenges associated with carbon emissions.

2. Experimental

2.1 Materials and Synthesis

The chemicals D-glucose and sodium potassium tartrate utilized in this study were sourced from Sigma Aldrich and employed without additional purification. The activated porous carbon was synthesized in a single step using an all-in-one-pot approach. This method involves heating an aqueous solution of the D-glucose and sodium potassium tartrate to ensure uniform mixing, followed by direct carbonization and activation in a tubular furnace at a high temperature. Typically, 1 g of D-glucose was mixed with varying quantities of sodium potassium tartrate in molar ratios of 1:0.5, 1:1, and 1:2. These composites were then carbonized and activated at 800 °C for 2 hours in the tubular furnace, with a heating rate of 2 °C/min to reach the target temperature. The elevated carbonization temperature was chosen to enhance the activation's influence on the textural characteristics of the synthesized materials. Post-carbonization, the resulting black powders were washed with 2% HCl and distilled water to remove any trapped constituents from the reaction between carbon and sodium potassium tartrate, thereby creating porosity[10]. The final powders were then washed with methanol and given a drying at 100°C in an air oven for 2 h. The resulting black materials are designated as AC n, where 'AC' denotes activated carbon derived from D-glucose and 'n' represents the amount of sodium potassium tartrate used for activation.

2.2 Characterization

The nitrogen sorption isotherms of the porous carbon were analysed at -196°C using a surface area analyzer (Micromeritics Tristar II, USA). The specific surface area was measured from the isotherm in the relative pressure range of 0.05-0.2, according to the Brunauere EmmetteTeller (BET) method. The pore size distribution (PSD) was obtained from density functional theory (DFT). The Barrette Joynere Halenda (BJH) method was used to determine the mesopores from the adsorption curve of the isotherm. The micropore volume (Vm) was calculated from the t-plot. The total pore volume was estimated from the amount of N_2 adsorbed at a relative pressure of 0.99.

The CO_2 adsorption capacity of the samples was measured at 0°C and 25°C by volumetric gas adsorption studies using the surface area analyzer. The PSD obtained by DFT calculation on CO^2 adsorption at 0° C was analyzed for ultra micropore evaluation. The regeneration capacity was studied by a gravimetric method using a thermogravimetric analyzer (Perkin Elmer Pyris 1, Singapore)

X-ray photoelectron spectroscopy (XPS) analysis was carried out with Multilab 2000 (Thermo Fisher Scientific) using Al-Ka radiation. Bruker D8 Discover Small Angle X-ray Scattering Spectrometer was used to study the X-ray diffraction (XRD) pattern of the samples. Elemental analysis was carried out using Perkin Elmer 2400 Series II CHNS analyzer. The morphology of the samples was analyzed using a field emission scanning electron microscope (FESEM, JSM 6060, JEOL, USA). The microstructure of the carbon samples was evaluated using a high resolution transmission electron microscope (HRTEM Tecnai G2 30 S-TWIN, USA) with an accelerating voltage of 300 kV. For TEM measurements, the sample dispersed in acetone was drop-casted on the carbon-coated copper grid and dried under vacuum at room temperature before observation.

3. Results and Discussion

3.1 Surface Area and Porosity

The different pore structure parameters of the samples are given in Table 29.1. The total pore volumes (Vtotal) of the activated porous carbons prepared with sodium potassium tartrate at ratios of 1:0.5, 1:1, and 1:2 were measured to be 0.50 cm³/g, 0.68 cm³/g, and 0.77 cm³/g, respectively. The sample with the 2:1 ratio exhibited the highest total pore volume, indicating a more extensive pore structure likely due to the higher amount of activating agent promoting greater pore formation. The 1:1 ratio sample had the highest micropore surface area and micropore volume, which is critical for CO_2 adsorption as it provides a greater number of adsorption sites.

Table 29.1 Different pore structure parameters of the AC in 3 different ratios

Sample	S_{BET} (m².g⁻¹)	V_{total} (cm³. g⁻¹)	S_{micro} (m².g⁻¹)	V_{micro} (cm³. g⁻¹)
AC 0.5	1459	0.50	670	0.25
AC 1	1499	0.68	815	0.35
AC 2	1479	0.77	730	0.30

where S_{BET} = Surface area, V_{total} = Total volume, S_{micro} = micropore surface area, V_{micro} = micropore volume

The pore size distribution (PSD) of the activated porous carbons was analyzed using BJH and DFT methods. The results (Fig. 29.1) revealed distinct differences among the three samples. The PSD of AC 0.5, 1 were somewhat similar whereas that of AC 2 indicated a wider distribution with an increased presence of mesopores and some macropores. The higher activation ratio led to the merging of smaller pores into larger ones, reducing the overall micropore volume and slightly diminishing the CO_2 adsorption capacity compared to the 1:1 sample.

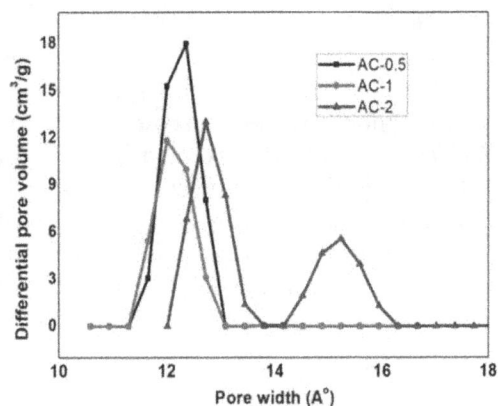

Fig. 29.1 Pore size distribution of AC in 3 different ratios

3.2 CO_2 Adsorption Capacity

The CO_2 adsorption capacity of the samples was evaluated at 0° C and 25° C by volumetric gas adsorption studies using the surface area analyzer. The 1:0.5 ratio sample, with the lowest CO_2 adsorption capacity of 7.8 mmol/g and 4.3 mmol/g @ 0°C and 25°C respectively, due to limited activation and reduced micropore volume, contrasts with the 1:1 ratio sample, which achieved the highest CO_2 adsorption capacity of 11.7 mmol/g and 6.2 mmol/g @ 0°C and 25°C respectively, due to an optimal pore size distribution, while the 1:2 ratio sample, with a moderate capacity of 9.9 mmol/g and 5.8 mmol/g @ 0°C and 25°C respectively, had increased total pore volume but less effective micropore presence, leading to slightly lower CO_2 adsorption compared to the 1:1 sample. The adsorption values are given in Table 29.2 and the plots are presented in Fig. 29.2. The reusability of the activated porous carbons was tested over 10-13 adsorption-desorption cycles. The 1:1 ratio sample retained above 95% of its initial adsorption capacity after thirteen cycles, demonstrating excellent stability and reusability. This suggests that the material can be effectively regenerated with minimal loss in performance, making it a viable option for continuous CO_2 capture applications. The regeneration pattern is given in Fig. 29.3.

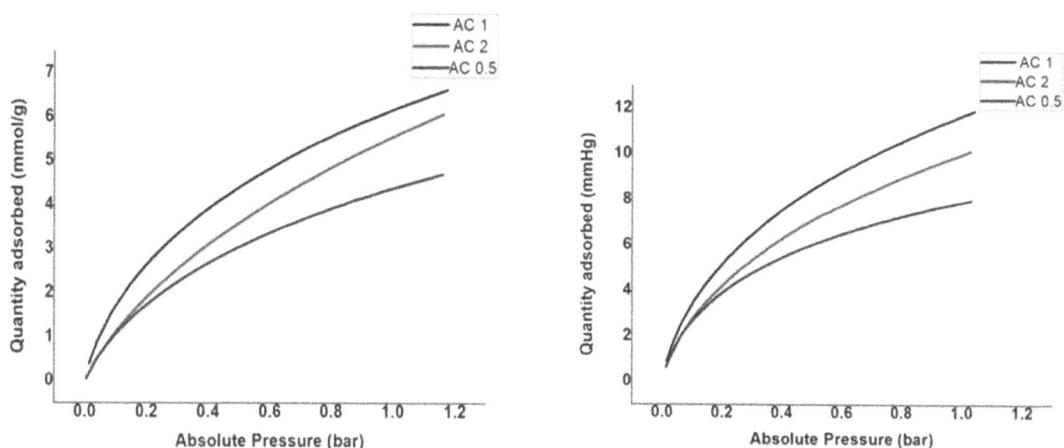

Fig. 29.2 CO_2 adsorption isotherm of AC (0.5,1,2) (a) RT (b) 0°C

Table 29.2 CO_2 adsorption values of AC in 3 different ratios

Samples	CO_2 (mmol/g) @0°C	CO_2 (mmol/g) @25°C
AC 0.5	7.8	4.3
AC 1	11.7	6.2
AC 2	9.9	5.8

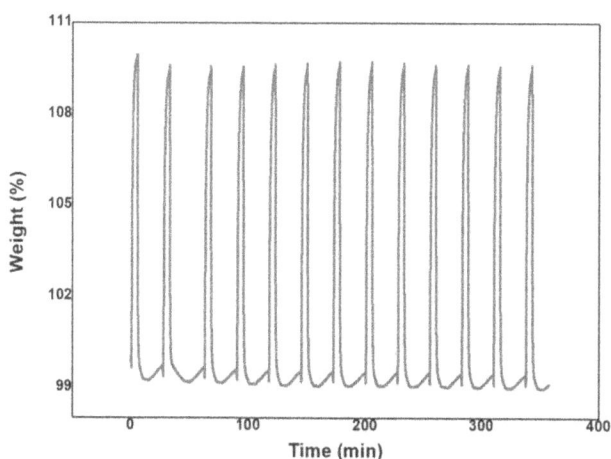

Fig. 29.3 Cyclic studies with AC

3.3 Morphological and Structural Analysis

The Scanning Electron Microscopy (SEM) images (Fig. 29.4) of the 1:1 ratio sample show a porous structure with a uniform distribution of particles. This morphology indicates that the activation process was successful in creating a high surface area with well-defined pore structures, which is essential for maximizing CO_2 adsorption capacity. The observed pores are of appropriate size and distribution, which enhances them accessibility and efficiency of the adsorption sites.

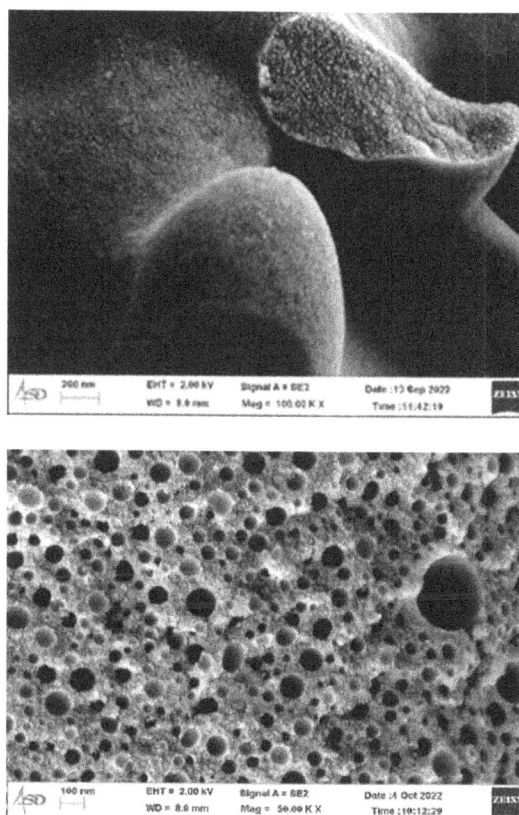

Fig. 29.4 FE-SEM images of AC-1

The microstructure of the optimized carbon material was examined in greater detail using TEM imaging, with the results illustrated in Fig. 29.5. It shows well-defined nanostructures with a uniform particle size. This uniformity and size distribution suggests that the internal structure of the sample is highly ordered. Such characteristics are likely to enhance CO_2 diffusion and adsorption efficiency, contributing positively to the sample's overall performance.

Fig. 29.5 HR TEM image of AC-1

XRD patterns showed broad peaks at 2θ values of 25° and 40° corresponding to amorphous carbon, with no significant difference in crystallinity among the samples. This suggests that the activation process primarily affected the surface area and porosity rather than the crystalline structure. The CHN analysis of the 1:1 ratio sample shows carbon and hydrogen of 93.2% and 1.2%, respectively. The carbon content is consistent with the expected level, indicating a well-developed carbon matrix. This composition is likely to contribute to the effective adsorption of CO_2, as the carbon-rich structure provides a substantial surface for interaction with the gas.

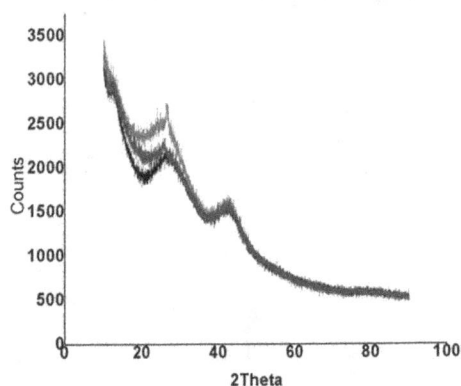

Fig. 29.6 XRD pattern of AC in three different ratios

4. Conclusions

In summary, this study demonstrates the successful synthesis of activated porous carbon from D-glucose using a non-corrosive activating agent, potassium sodium tartrate, via a novel one-pot approach. The activated porous carbon produced with a glucose to potassium sodium tartrate ratio of 1:1 exhibited the highest CO_2 adsorption capacities of 11.7 mmol g^{-1} at 0 °C/1 bar and 6.2 mmol g^{-1} at 25 °C/1 bar, along with a high specific surface area of

1499 m^2 g^{-1}. These results underscore the potential of using a non-corrosive activating agent to enhance the surface characteristics and porosity of carbon materials for efficient CO_2 capture. The physical adsorption mechanism and the ease of regeneration further emphasize the practicality of these materials for sustainable CO_2 capture applications. Future research can explore the scalability of this method and the long-term stability of the adsorbents to further validate their commercial viability.

Acknowledgments

The authors acknowledge Director, VSSC for granting permission to present this paper.

References

1. Hou, Yang, et al. "An advanced nitrogen-doped graphene/cobalt-embedded porous carbon polyhedron hybrid for efficient catalysis of oxygen reduction and water splitting." Advanced Functional Materials 25.6 (2015): 872–882.
2. Sevilla, Marta, and Robert Mokaya. "Energy storage applications of activated carbons: supercapacitors and hydrogen storage." Energy & Environmental Science 7.4 (2014): 1250–1280.
3. Sosa, Julio E., et al. "Adsorption of fluorinated greenhouse gases on activated carbons: evaluation of their potential for gas separation." Journal of Chemical Technology & Biotechnology 95.7 (2020): 1892–1905.
4. Singh, Jasminder, Soumen Basu, and Haripada Bhunia. "CO2 capture by modified porous carbon adsorbents: Effect of various activating agents." Journal of the Taiwan Institute of Chemical Engineers 102 (2019): 438–447.
5. Zhang, Zhen, et al. "Rational design of tailored porous carbon-based materials for CO₂ capture." (2019).
6. Ashourirad, Babak, et al. "A cost-effective synthesis of heteroatom-doped porous carbons as efficient CO 2 sorbents." Journal of Materials Chemistry A 4.38 (2016): 14693–14702.
7. Estevez, Luis, et al. "Hierarchically porous carbon materials for CO2 capture: the role of pore structure." Industrial & Engineering Chemistry Research 57.4 (2018): 1262–1268.
8. Oschatz, Martin, and Markus Antonietti. "A search for selectivity to enable CO 2 capture with porous adsorbents." Energy & Environmental Science 11.1 (2018): 57–70.
9. Singh, Gurwinder, et al. "A combined strategy of acid-assisted polymerization and solid state activation to synthesize functionalized nanoporous activated biocarbons from biomass for CO2 capture." Microporous and Mesoporous Materials 271 (2018): 23–32.
10. Singh, Gurwinder, et al. "A facile synthesis of activated porous carbon spheres from d-glucose using a non-corrosive activating agent for efficient carbon dioxide capture." Applied Energy 255 (2019): 113831.

Note: All the figures and tables in this chapter were made by the author.

Advances in Materials Science and Technology – Dr. Srikari Srinivasan et al. (eds)
© 2025 Taylor & Francis Group, London, ISBN 978-1-041-12342-2

30

Urea-Formaldehyde Microcapsules for the Detection of Microcracks in Polymer Composites

Nisha Balachandran*,
Athira Ajayan, Vishnu S, Deepthi Thomas, Deepa Devapal
Analytical and Spectroscopy Division,
Analytical, Spectroscopy and Ceramics Group, Propellants, Polymers,
Chemicals and Materials Entity, Vikram Sarabhai Space Centre,
Thiruvananthapuram, Kerala, India

ABSTRACT: Polymer materials used for different structural applications viz., automotive, aerospace and space industries are subjected to harsh environmental conditions and may lead to severe mechanical, thermal and chemical damages. Micro cracks and structural flaws plays a major role in the damage mechanisms of polymeric materials. Generally damages in materials initiates as a local phenomenon which gets deteriorated over time leading to the loss of structural rigidity and subsequent failure of the complete system. Structural health assessment of the polymer composites are done by various non-destructive detection techniques. But the challenge is in the detection of small cracks, which is yet to be addressed adequately by the available damage detection techniques. Micro crack detection using fluorescence molecules offer advantages of high sensitivity and ease of detection. Here, we introduce a novel approach of self-sensing mechanism of microcracks in polymer composites focused on microcapsules filled with fluorescent dye (crack-indicating agent). Fluorescent dye is microencapsulated into urea-formaldehyde polymer shell. The formation of microcapsules was confirmed by scanning electron microscopy (SEM), Fourier transform infrared spectroscopy (FT-IR) and thermogravimetric analysis (TGA). The functional properties of the microcapsules are evaluated using fluorescent imaging technique. The microcapsules of different weight loading are taken for composite preparation. The cracks are simulated and are effectively evaluated under 365 nm light source for the cured composites.

KEYWORDS: Crack detection, Microcapsules, Urea-formaldehyde

1. Introduction

Polymeric materials are extensively utilized for the applications viz., aerospace, construction industry, automotive, energy etc. [1, 2]. These polymer materials are subjected to various harsh environmental conditions that can lead to severe mechanical, thermal and chemical damages. The environmental factors can lead to micro cracks and structural flaws of the polymeric materials [3-5]. Generally damages in materials initiates as a local phenomenon which gets deteriorated over time and can lead to the loss of structural rigidity and subsequent failure of the complete system. Structural health assessment of the polymer composites is carried out by various non-destructive detection techniques [6-7].

*Corresponding author: nishapzm@gmail.com

DOI: 10.1201/9781003664277-30

A wide variety of approaches have been reported for sensing microcracks in the polymer materials, but all these systems are expensive since special detection equipment/ sensors are required. Simple and fast approaches for the initial detection of structural cracks on the polymer composites are quite challenging.

Visual detection method to detect damages is a modest, easy approach based on fluorescent dye that does not require any sensor in contact with the materials during the inspection [8, 9]. A novel approach to incorporate these fluorescent molecules into the polymer matrix using a new technique called 'Microencapsulation'. 'Microencapsulation' is a process that permits encapsulation of active materials inside a shell and controlled release of the functional molecules on demand. Theses microcapsules can be interleaved as additives to the desired materials which can impart many functions in various fields of technology [10-13]. Synthesis of microcapsules involves many challenges which include the shell formation conditions, the effective incorporation of the active compounds and its controlled release when required. Figure 30.1 shows the representation of a microcapsule.

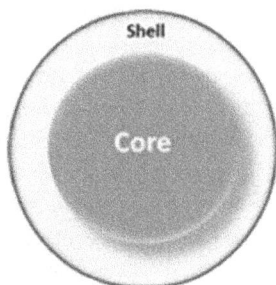

Fig. 30.1 Representation of a core shell microcapsule

A preferred polymer shell material can be Urea–formaldehyde (UF), a type of thermosetting polymer and after curing, it forms insoluble three-dimensional networks and cannot be melted again. Another basic feature of UF resins include their fast curing, high reactivity, good thermal properties, water solubility, non-flammability, hardness and low cost of production[14]. The core material can be a high quantum yield fluorescent dye and should be compatible with the shell polymer.

The first goal in the present work is to optimize the parameters for the formation of microcapsules and to optimize a synthetic protocol for the formation of urea-formaldehyde shell and the effective encapsulation of active agent. In this paper, urea-formaldehyde polymer based microcapsules containing fluorescent dye was synthesized. Dye concentration inside the shell was optimized using various characterization techniques. The functional

testing of the microcapsules was done by UV-Visible Spectroscopy and Fluorescence Microscopy Technique. The microcapsules were loaded to epoxy based polymeric matrix and the functional testing was demonstrated.

2. Experimental

2.1 Materials

Materials used for the synthesis are urea, polyethylene-alt-maleic anhydride (PEMA), resorcinol (Make:M/s Sigma Aldrich), sunflower oil (Food grade), azo dye-orange (M/s Gorilla products). Acetone (AR grade), ammonium chloride (AR grade) and formalin (37% formaldehyde solution) were supplied by S. D. Fine Chem. Ltd., Mumbai. For composite preparation, diglycidyl ether of bisphenol A (LY-556) epoxy resin and tetraethylenepentamine as hardener is used.

2.2 Synthesis and Characterization

Microcapsules are synthesized by the modified procedure. 5 ml (25 wt%) PEMA aqueous solution is taken in round bottom flask. To the above solution, 0.5 g urea, 0.05 g resorcinol and 0.05 g NH_4Cl is added and pH is adjusted to 3.5 by adding 0.1 N NaOH. Azo dye is dispersed in sunflower oil by ultrasonication (~10 minutes). 10 ml of the dye solution is added to the above mentioned RB flask and stirred (350 rpm) for 10 minutes. 2.5 g of formalin (37 wt% aqueous solution) is added to the mixture and heated at 60°C for 4 hours with stirring rate of 400 rpm. After completion of the reaction, the solution is filtered and the precipitate is washed with de-ionized water and acetone to remove the excess PEMA. The precipitate is dried and further characterized. Figure 30.2 shows the scheme of synthesis of microcapsules.

Microcapsules of different weight ratio (5wt%, 10wt %, and 15 wt%) is taken and mixed with 10 gm of LY 556 epoxy. 1.45 gm of the hardner is added to the above mixture and mixed well. Pour the mixture to a flat mould and kept for 24 hours to attain complete curing. The cured epoxy slab was characterized using different tech-niques.

Fourier transform infrared (FTIR) analysis was done using a Nicolet iS50 FTIR spectrometer in the wavelength 4000-550 cm^{-1} at a resolution of 4 cm^{-1}. All the UV-Visible spectra were recorded using a Perkin Elmer lambda 950 UV-Visible spectrometer. Fluorescence measurements were performed using a FluoroMax-4C Spectrofluorometer (Horiba Instruments, USA) at an excitation wavelength of 365 nm with an integration time of 0.1 s. Both excitation and emission slit widths were kept at 1 nm. Thermogravimetry analysis was done using TA instrument at a heating rate 10°C from 50°C to 600°C. Fluorescent images were taken

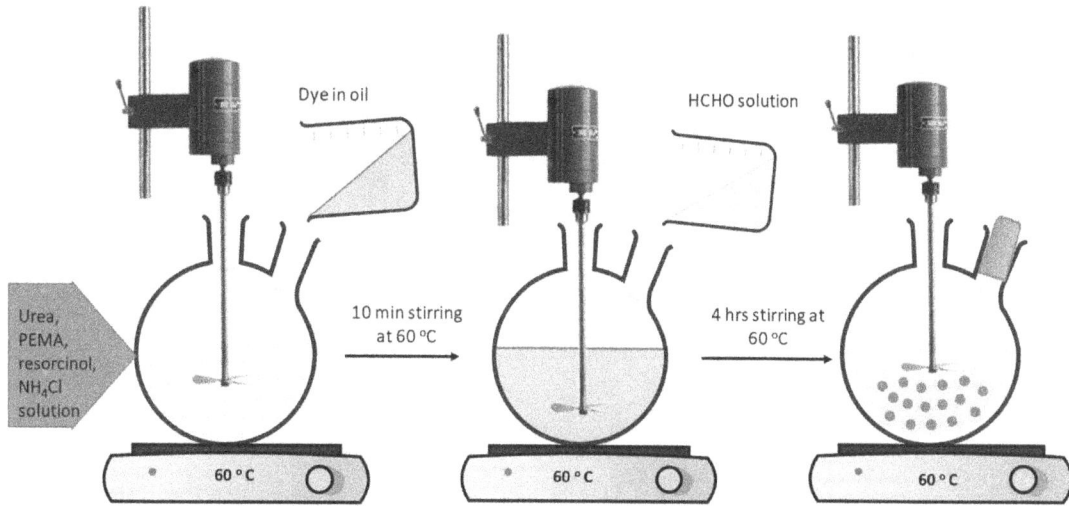

Fig. 30.2 Scheme 1 shows the synthesis of microcapsules

using a Leica DM 1000 LED Microscope by illuminating the samples at 365 nm LED light at various magnifications. Field Emission Scanning Electron Microscopy (FE-SEM) was done using Carl Zeiss Gemini SEM 500 at an operating voltage of 2kV. SEM images were acquired after sputter-coating the samples with gold-palladium to achieve electron conductivity.

3. Results and Discussion

3.1 Microcapsule Characterization

Urea-formaldehyde microcapsules were prepared by in-situ polymerization of urea with formaldehyde. [15]. a Schematic drawing of the overall reaction is shown in Fig. 30.3.

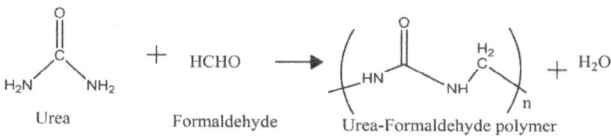

Fig. 30.3 Schematic diagram of the reaction

The UF microcapsules are characterized by FTIR spectroscopy to understand the polymer structure (Fig. 30.4). The major absorption peak at 3100-3500 cm^{-1}, shows the characteristic peaks of the -NH stretching of the bonded -NH group. A distinct absorption band appears around 2995 cm^{-1} which is characteristic of the C-H stretching mode, 1622 cm^{-1} corresponds to carbonyl attached to –NH, 1552 cm^{-1} and 771 cm^{-1} shows the -NH bending vibrations of secondary amines, 1380 cm^{-1} arises from the in-plane bending vibration of CH_2 group and a moderately strong absorption band is appeared from 1300 to 1000 cm^{-1} arises from the C-O linkages [16].

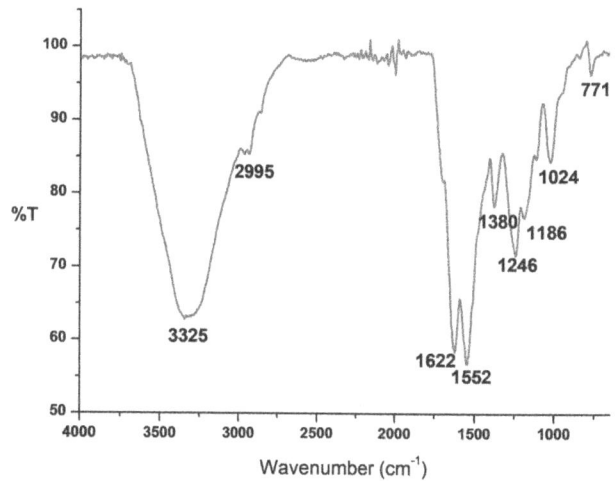

Fig. 30.4 FTIR spectrum of the UF microcapsule with dye as the core

To optimize the concentration of the dye inside the shell, various weight ratios of the dye dispersed in sunflower oil was taken and microcapsules were synthesized by keeping all other parameters constant (see Table 30.1).

Table 30.1 Dye concentration

Sl. No.	Sample designation	Dye concentration
1	UF-D-0	0 Wt%
2	UF-D-0.10	0.10wt%
3	UF-D-0.25	0.25wt%
4	UF-D-0.50	0.50wt%
5	UF-D-0.75	0.75wt%

In order to get the optimized dye concentration fluorescence microscopy (FL) of the samples were attempted. The samples were illuminated with 365 nm LED light and

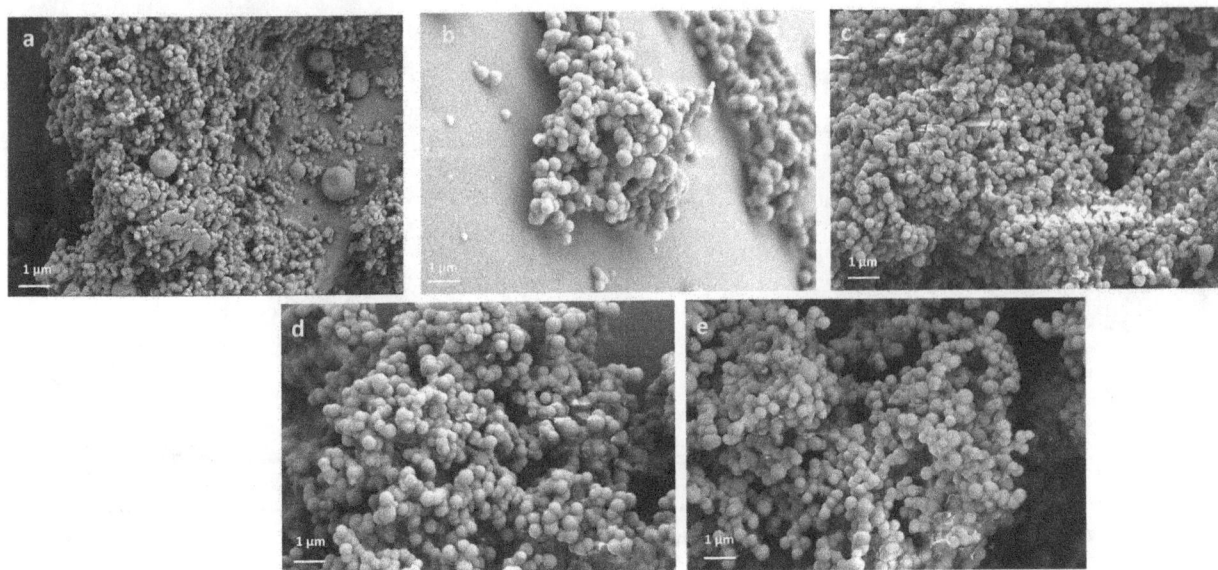

Fig. 30.5 FESEM images of the microcapsules (a) UF-D-0, (b) UF-D-0.10, (c) UF-D-0.25, (d) UF-D-0.50 and (e) UF-D-0.75

images were taken using different objectives (See Fig. 30.6). From the Fig. 30.6 (d), it is clear that the best dye encapsulation is obtained for the Sample UF-D-0.5. To further evaluate the change in fluorescence properties of the UF-D-0.5, it was grinded and FL images were taken Fig. 30.6 (e). After grinding the capsules, the polymer shell layer got broken and the dye inside the capsule was exposed which gives a good fluorescence emission.

The absorption spectrum of the UF-D-0, UF-D-0.50 and dye are also compared (see Fig. 30.7). UF-D-0.50 is not absorbing in the UV-visible region while UF-D-0.50 (after grinding) shows absorption peaks similar to the absorption maxima of the dye. Photoluminescence (PL) spectra also supported the same observation. The peak maximum of the dye at ~580 nm is prominent in the case of grinded capsules.

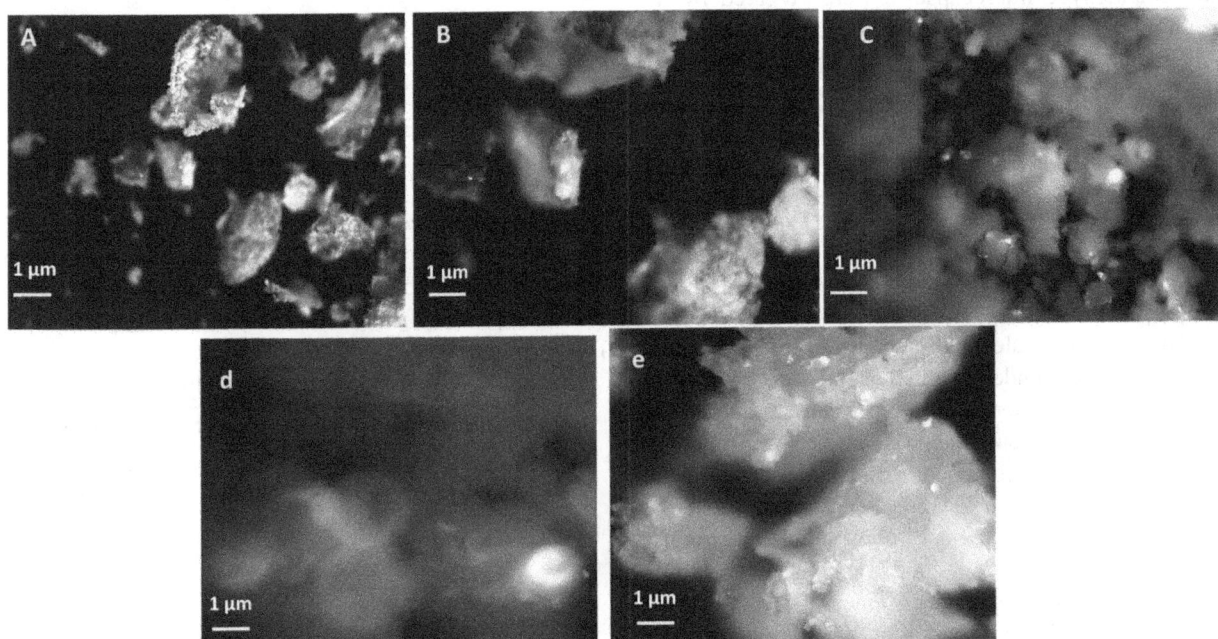

Fig. 30.6 Fluorescent images of the microcapsules (a) UF-D-0.10, (b) UF-D-0.25, (c) UF-D-0.75, (d) UF-D-0.50 and (e) UF-D-0-grinded

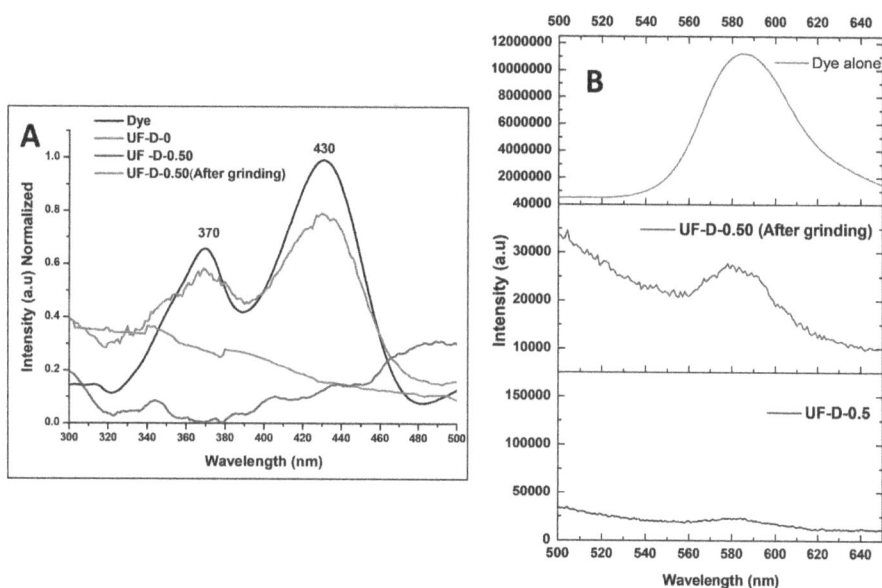

Fig. 30.7 (A) UV-visible spectra and (B) PL spectra of microcapsules

Thermal stability of the microcapsules was evaluated by Thermogravimetry Analysis (TGA). TGA curve shows that the microcapsules are stable upto 200 °C and the decomposition starts above ~200 °C, which can be attributed to the degradation of the UF shell. This proves that the UF capsules can be incorporated into the polymer matrix which can be cured below 200 °C.

3.2 Composite Preparation

UF-D-0.50 microcapsules were chosen for composite preparation due to its better functional properties. Composites of epoxy resin were prepared by loading different weight ratios of UF-D-0.50 microcapsules (5 wt %, 10wt% and 15 wt %).

Figure 30.9 shows representative IR spectrum of Epoxy-UF-D-0.50 cured composite. The complete curing of the

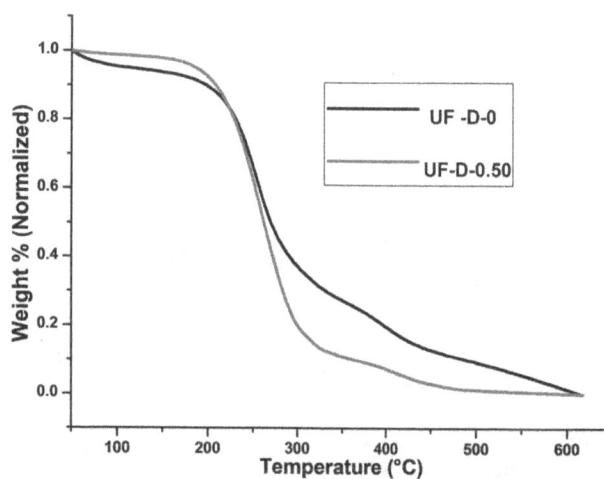

Fig. 30.8 TGA curve of microcapsules

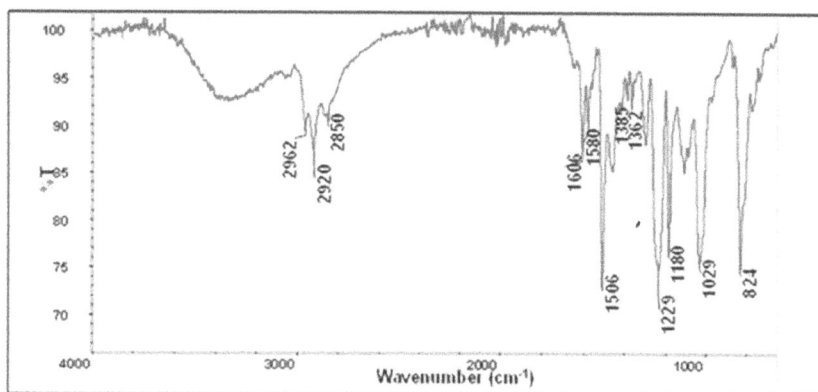

Fig. 30.9 FTIR spectrum of Epoxy-UF-D-0.50 cured composite

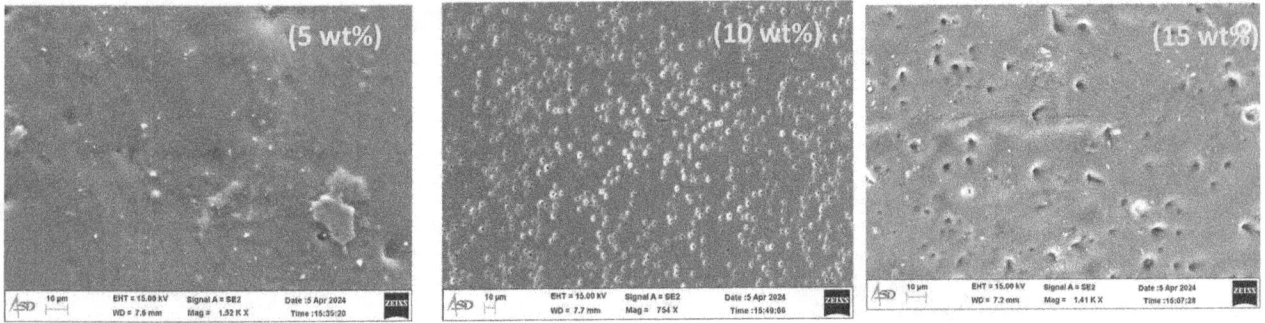

Fig. 30.10   FE-SEM of Epoxy-UF-D-0.50 cured composite

epoxy polymer is assured by the disappearance of peak at 910 cm^{-1}, characteristic of epoxy ring vibration. The peaks in the region 3000-2800 cm^{-1} arises due to the C-H stretching vibrations. Strong peaks at 1606, 1506 and 1580 cm^{-1} are assigned to aromatic C=C vibrations of phenyl groups. Two peaks at 1385 and 1362 cm^{-1} corresponds to bending vibrations of gem dimethyl group. The peak at 1229 cm-1 is due to C-O vibrations. This is followed by Ph-O vibration at 1185 cm^{-1}. The peak at 1029 cm^{-1} corresponds to C-O skeletal and epoxy ring vibrations respectively. A strong peak at 824 cm^{-1} represents benzene ring substitution vibration.

SEM images (see Fig. 30.10) were taken to study the dispersion of UF-D-0.50 microcapsules in the polymer matrix. UF-D-0.50 with 5 Wt% shows less dispersion while 15 Wt % shows particle agglomerations. UF-D-0.50 with 10 Wt % loading shows better dispersion as seen from the SEM images and is further tested for crack detection.

Functional test were carried out by scratching the surface of the composite using a surgical blade and FL images was taken and detection of crack is clearly visible from the FL image (see Fig. 30.11 c). PL spectra of the scratched portion of the sample show the emission peak at ~580 nm. This evidence also supports the FL observations.

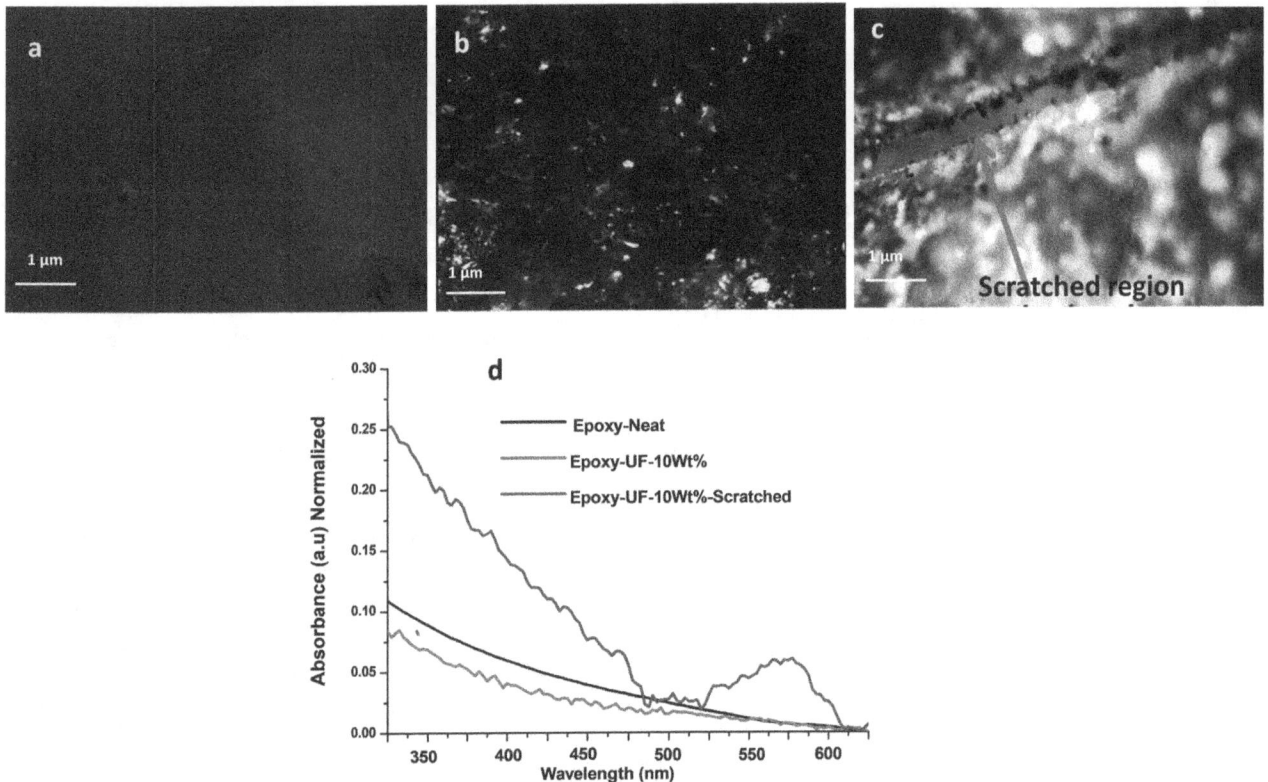

Fig. 30.11   FL images of epoxy-UF cured composite (a) Cured epoxy-neat, (b) Cured epoxy-10wt% UF-D-0.50 and (c) Cured epoxy-10wt% UF-D-0.50-after scratching

Thermal stability of the cured epoxy composite was also evaluated and the Thermogram (see Fig. 30.12) shows that there is no reduction in the thermal stability of the epoxy cured composites by the addition of the microcapsules. This shows the compatibility of the UF microcapsules with the epoxy matrix.

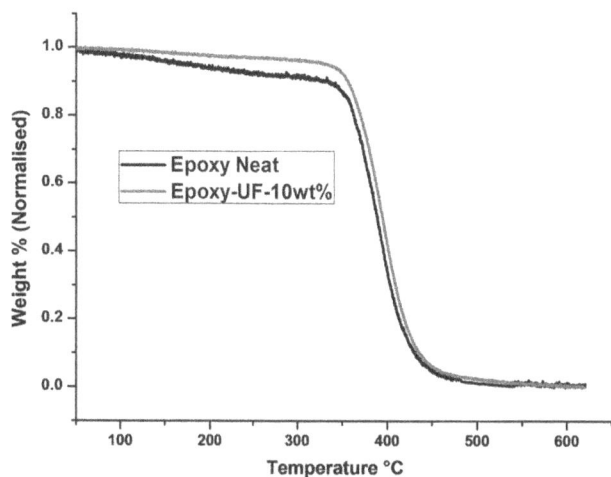

Fig. 30.12   Thermogram of cured epoxy and epoxy systems with microcapsule (UF-D-0.50) of 10 Wt % loading

4. Conclusions

This work has validated the feasibility of using an inexpensive commercial fluores-cent dye for self-sensing of microcracks. In this work, one-stage in-situ polymeriza-tion technique was attempted for fluorescent dye encapsulation in urea-formaldehyde (UF) microcapsules. The dye concentrations for the synthesis of microcapsules were optimized using morphology analysis, fluorescent imaging, UV-Visible spectrum and photoluminescence spectrum. Thermal properties of the microcapsules were also evaluated to study the compatibility of the material. The effect of the addition of 5 Wt %, 10Wt % and 15 Wt % microcapsules on the functional properties of the epoxy polymer was studied, which revealed that the optimum concentration of microcapsules loading as 10 Wt% can give a better dispersion and visual detection of cracks formed on the surface of the cured polymer matrix. Thus optical damage visualization is simple, quick and hence can be used for the detection of microcrack in polymer composites.

Acknowledgments

The authors acknowledge Director, VSSC for granting permission to present this paper.

References

1. Falk, Hille., Damian, Sowietzki., Ruben. Makris. (2020): Luminescence-based early detection of fatigue cracks. Materials Today: Proceedings 2 (32), 78–82.
2. L. Pook., (2007):Metal Fatigue What It Is, Why It Matters, springer.
3. S. Suresh., (1994): Fatigue of Materials, Cambridge University Press.
4. Mihai, Brebu.(2020): Environmental Degradation of Plastic Composites with Natural Fillers-A Review, Polymers (Basel). 12(1): 166.
5. Sudip Ray, Ralph P. Cooney.(2018): Chapter 9 - Thermal Degradation of Polymer and Polymer Composites, Handbook of Environmental Degradation of Materials, (3), 185–206.
6. S, Gholizadeh. (2016): A review of non-destructive testing methods of composite materials, Procedia Structural Integrity, (1), 50–57.
7. Duchene, P., Chaki, S., Ayadi, A. et al. (2018): A review of non-destructive techniques used for mechanical damage assessment in polymer composites. J Mater Sci (53), 7915–7938.
8. Salim, Chaki., Patricia, Krawczak., (2022): on-Destructive Health Monitoring of Structural Polymer Composites: Trends and Perspectives in the Digital Era, Materials, 15 (21), 7838.
9. Naebe, M., Abolhasani, M. M., Khayyam, H., Amini, A., Fox, B., (2016): Crack Damage in Polymers and Composites: A Review. Polymer Reviews, 56(1), 31–69.
10. Nitamani, Choudhury., Murlidhar, Meghwal., Kalyan. Das.,(2021): Microencapsulation: An overview on concepts, methods, properties and applications in foods, Food Frontiers., (2), 426–442.
11. Denis, Poncelet,. (2006): Microencapsulation: Fundamen-tals, methods and applications, In book: Surface Chemistry in Biomedical and Environmental Science, 23–34.
12. M,N, Singh., K,S, Y, Hemant., M, Ram., H,G, Shivakumar, (2010): Microencapsulation: A promising technique for controlled drug delivery, Res Pharm Sci. Jul-Dec; 5(2): 65–77.
13. Orive, Gorka,.et al. (2004): History, challenges and perspectives of cell microencapsulation, TRENDS in Biotechnology 22 (2) 87–92.
14. Fayyad, Eman, M., Mariam, A, Almaadeed,. Alan, Jones,.(2016): Preparation and characterization of urea–formaldehyde microcapsules filled with paraffin oil, Polymer Bulletin (73), 631–646,
15. Elham, Katoueizadeh., Seyed, Mojtaba, Zebarjad,. Kamal, Janghorban, (2019): Investigating the effect of synthesis conditions on the formation of urea–formaldehyde microcapsules, jmaterrestechnol; 8(1):541–552.
16. Myers GE., (1981): Investigation of urea–formaldehyde polymer cure by infrared, J Appl Polymer Sci. 26(3):747–6.

Note: All the figures and table in this chapter were made by the author.

Advances in Materials Science and Technology – Dr. Srikari Srinivasan et al. (eds)
© 2025 Taylor & Francis Group, London, ISBN 978-1-041-12342-2

31

Development of Linear Bipolar and Non-Hysteretic Spin Valve GMR Sensor

Tejaswini C. Gawade[1,2]
[1*]Nanomaterials Research Laboratory,
Surface Engineering Division, CSIR - National Aerospace Laboratories,
Bangalore, India
[2]Academy of Scientific and Innovative Research (AcSIR),
Ghaziabad, India

Bhagaban Behera[1]
[1*]Nanomaterials Research Laboratory,
Surface Engineering Division, CSIR - National Aerospace Laboratories,
Bangalore, India

Umesh Borole[3]
[3]Sensohm Pvt. Ltd., Bangalore, India

P. Chowdhury[1,2*]
[1*]Nanomaterials Research Laboratory,
Surface Engineering Division, CSIR - National Aerospace Laboratories,
Bangalore, India
[2]Academy of Scientific and Innovative Research (AcSIR),
Ghaziabad, India

ABSTRACT: This study aims to advance spin valve-based magnetic sensors for industrial applications, emphasizing three critical attributes: bipolar operation, linear response, and hysteresis-free performance. Four distinct strategies were explored to achieve sensor linearization: first, employing high aspect ratio patterning of sensing elements to induce a demagnetizing field, resulting in reduced hysteresis to 1-2 Oe compared to bulk film; second, implementing ex-situ coil biasing to modulate the sensor's response; third, utilizing in-situ stack biasing to fine-tune linearity; and fourth, employing in-situ thick film permanent magnet field biasing with strategically placed magnets to influence sensor behavior, achieving hysteresis reductions to 0.2-0.3 Oe.

KEYWORDS: External coil biasing, Linear and bipolar magnetic sensor, PM biasing, Soft biasing, Spin valve, Synthetic antiferromagnetic layer, Wheatstone bridge

1. Introduction

Recently, there has been a growing interest in the development of bipolar sensors with a wide linear operating field range and higher sensitivity[1-3]. Subsequently a concept of push-pull Wheatstone bridge configuration was deployed to develop a magnetoresistive sensor based on both spin valve and tunneling magnetoresistance (TMR) technologies[4-5].

*Corresponding author: pchowdhury@nal.res.in

DOI: 10.1201/9781003664277-31

Longitudinal biasing (LB) in magnetic sensors has become a focal point of research to achieve hysteresis-free and symmetric linear responses in MR systems. The sensing layer of a magnetic multilayer stack typically exhibits a hysteric response due to the formation of a magnetic multidomain structure. Applying an LB field perpendicular to the measurement direction (i.e., the sensitivity direction) suppresses this multidomain structure into a single-domain configuration, resulting in a hysteresis-free, linear response and reducing Barkhausen noise characteristics[6-7]. Two approaches have been explored to generate the LB field across the sensing layer. The first approach involves in-stack biasing by in-situ depositing a ferromagnetic exchange-coupled antiferromagnetic layer on top of the sensing layer[8-9]. This technique requires additional AFM materials with varying blocking temperatures, followed by a post-annealing procedure to create a soft exchange bias in the sensing layer perpendicular to the measuring/pinning layer. Although this process broadens the operating field range, it reduces sensitivity.

The second approach employs external field biasing, either through an on-chip current line loop or permanent magnet (e.g., CoCrPt thick film) placed on top of or across the sensing layer[10–12].

This study focuses on developing high-sensitivity bipolar linear magnetic sensors that are hysteresis-free and suitable for diverse industrial applications. Four key strategies were explored for optimizing linear response: shape anisotropy to introduce non-zero coercivity (2-3 Oe), external coil biasing for dynamic response modulation, in-situ stack biasing for enhanced linearity and reduced hysteresis, and permanent magnet biasing for stable sensor behavior. Each strategy was evaluated for effectiveness in achieving desired characteristics like linearity and sensitivity in linear field applications.

2. Experimental

SV stacks were deposited on SiO_2 coated Si-substrates using the ingeniously developed Ultra-High Vacuum (UHV) magnetron sputtering system equipped with eight sputter targets which are loaded in two separate sputtering chambers having base pressure of $< 1.2 \times 10^{-8}$ Torr. During the deposition, the working Ar pressure was kept at 3.5×10^{-3} Torr. The compositions of NiFe, CoFe, IrMn, and PtMn layers were $Ni_{81}Fe_{19}$(at. %), $Co_{90}Fe_{10}$(at. %), $Ir_{22}Mn_{78}$(at. %), and $Pt_{50}Mn_{50}$(at. % respectively. The growth rates of CoFe, NiFe, Ta, Cu, Ru, IrMn, and PtMn were 0.32, 1, 0.65, 0.44, 1.28, 1.42, and 0.5 ^{o}A/s, respectively. After thickness optimization of each layer, SV stacks were deposited using two different antiferromagnetic layers, i.e., IrMn and

PtMn. Two different types of SV stacks were grown using two different antiferromagnetic layers:

(i) in Stack-A, where IrMn was used as AFM-1 with synthetic antiferromagnetic layer as shown in Fig. 31.1 (a), whereas (ii) in Stack-B, with soft pinning of sensing layer was incorporated using PtMn as AFM-1 and IrMn as AFM-2 layer. After deposition, Stack-A was annealed for 60 min in a vacuum ($\approx 10^{-7}$ Torr) at a temperature of 210°C, followed by a field cool to room temperature in a magnetic field of 1300 Oe, to induce the exchange bias in the multilayer stack. In the case of Stack B, the annealing process was carried out in two steps. To fix the magnetization direction of the pinned layer, at the first step, in-field annealing was carried out at 280°C for an hour along the length direction and the second step annealing at 210°C for 30 min was followed by rotating the sample in 90°, to set sensing layer orthogonal to the pinned layer.

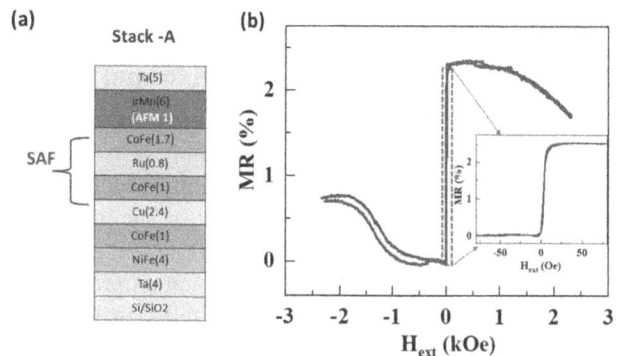

Fig. 31.1 (a) Schematic diagram of GMR SV- stack, and (b) major MR curve of GMR spin-valve stack (inset shows minor curve in low field range)

The MR properties of the stacks are measured using the four-probe method. Mag- netoresistance is determined by the expression MR% = $(R_{max} - R_{min})/R_{min} \times 100$, where R_{max} and R_{min} are the resistance when pinned and free layer are antiparallel, i.e., $\theta = 180^0$ parallel align, i.e., for $\theta = 0^0$ respectively.

During device fabrication, SV stacks were deposited onto a 2-inch Si/SiO_2 wafer. The resistive elements were created using photolithography followed by an ion milling process. The magnetization vectors in the resistive elements were aligned in a cross configuration for each resistor through a stepwise annealing process, as previously described. To achieve a full Wheatstone bridge with push-pull geometry, four identical resistors were configured with positive ($dR/dH > 0$) and negative ($dR/dH < 0$) slopes. This configuration was obtained by mechanically rotating one of the dies containing two resistive elements by 180° and wire bonding it between two dies to form the Wheatstone bridge [13].

3. Results and Discussion

3.1 Spin Valve Sensor Fabrication

Figure 31.1 (a) shows the spin-valve stack used in this work using IrMn as AFM layer. The MR *(H)* curve of Stack-A is shown in Fig. 31.1 (b) with a minor loop. Two approaches are being made to design and develop the linear and bipolar sensors with push-pull Wheatstone bridge configuration. The first approach was known as mechanical assem-bly, where a single-layer SAF structure was used to fabricate the sensor. The second one is termed an on-chip assembly, where both types of single-layer and double-layer SAF structures were used to develop the sensor. In this work, we utilized mechani- cal assembly to fabricate SV sensors with parallel field biasing using an external coil, as well as SV sensors with soft biasing of the sensing layer. However, for SV sensors with PM biasing, we employed an on-chip assembly. The details of the fabrication are published elwhere[13-14].

3.2 Parallel Field Biasing using External Coil

To obtain a full Wheatstone bridge geometry, an approach was made toward the mechanical assembly of two die in a single lead frame with closed proximity and opposite pinned directions as shown in Fig. 31.2(a). The latter was achieved through mechanical rotation of one of the dies by 180° as shown in 31.2(a) and it is a common practice in the semiconductor industry. A batch of five 2-inch wafers was fabricated containing 5000 dies per wafer. Each die of dimension 0.5 mm × 0.5 mm contains two GMR spin-valve resistors and each of them has dimensions: l = 150 μm and w = 3 μm. Two dies were packaged in a plastic package of 3 mm × 3 mm × 0.75 mm as shown in Fig. 31.2 (b). The fabricated chip was labeled as 'SVF 2201' where 'SVF' indicates Spin-Valve Full-bridge. The fabricated chip was mounted individually on the PCB for further characterization without and with parallel field biasing as shown in the inset of Figs. 31.2(c) and (d). Here, the bias field along the easy axis was generated externally by winding a solenoid coil having 150 turns. The estimated field generated by this coil was around 1 Oe with a supply of DC of 1 mA.

Keeping I_{bias} (parallel field biasing using external coil) = 0, the transport charac- teristics of an SV resistor and the measured output voltage from a chip configured in Wheatstone bridge configuration are superimposed and shown in Fig. 31.2(c). Both the curves show a linear operating range of ± 20 Oe, with a H_c > 4 Oe, however, the bridge output has symmetric bipolar characteristics with respect to the zero applied field. The observed high H_c value in the stripped geometry with crossed anisotropy

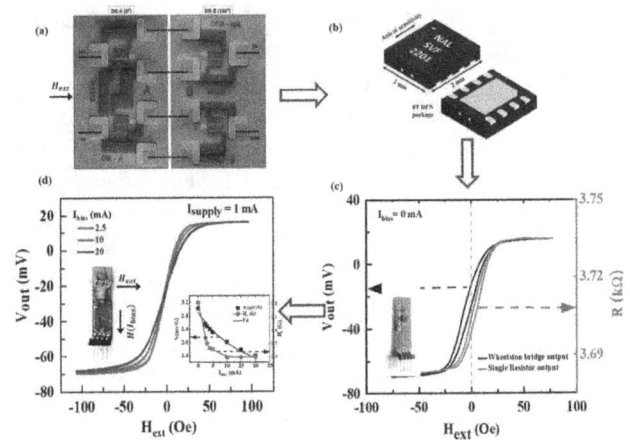

Fig. 31.2 (a) FESEM images of two die placed 180° opposite to each other and connected in full bridge sensor assembly. Each die contains identical spin-valve resistors, (b) a final packaged chip containing spin-valve resistors, (c) transfer curves of a single resistor and the bridge output characteristics with coil current, I_{bias} = 0 mA. Inset shows the spin-valve chip mounted on the evaluation PCB, (d) output characteristics of the full bridge spin-valve sensor with different biasing currents, and inset (down right) is a plot of the sensitivity of the spin-valve sensor and Hc at different bias currents calculated at Hext = 0, inset (left) is a spin-valve chip with the biasing coil

was not expected as the bulk sample denotes H_c < 2 Oe as shown in Fig. 31.1(b) and indicates that the strip geometry introduces the non-uniform magnetization across the free layer with multi-domain characteristics. The origin of multi-domain characteristics is not yet clearly understood, however, it may be at the edge boundaries of the strip. To have a uniaxial anisotropy of the free layer along the length of stripped geometry, it is necessary to apply an external biasing field along the easy axis of the free layer. Hence, with the application of H_{ext}, along with the magnetization direction of the reference layer and the parallel bias field, $H(I_{bias})$ along with the free layer, the condition will result in coherent rotation of free layer magnetization with zero hysteresis.

The typical output characteristics of the Wheatstone bridge formed by the four resistors under excitation of constant DC of 1 mA across the bridge of the sensing device and with different set values of I_{bias} of 2.5 mA, 10 mA, and 20 mA are shown in Fig. 31.2(d). With the increasing I_{bias}, two observations were made as follows: (i) H_c reduces exponentially and becomes zero with a biasing current of 10 mA, and (ii) the sensitivity, $S = (\Delta V/\Delta H)$, decreases with an increase in the biasing current.

3.3 SV Sensor with Soft Exchange Biasing of the Sensing Layer

Micro-fabrication with aspect ratio, $(l/w) > 50$ (i.e., $l = 100$ μm, and $w = 2\mu$m) across the whole wafer is always a challenging task using a UV photolithography process. For a smaller width, edge roughness and other local defects serve as large distributions of shape anisotropy values even within a single device causing dispersion in the transfer curve and deviating from the linearity. Recently, soft biasing of the SL was reported[15] as an alternate route to enhance field dynamic range. For this, the SV stack was modified by adding a new antiferromagnetic (AFM) layer adjacent to the FL(or sensing layer (SL)) with a lower blocking temperature in comparison to the existing AFM layer adjacent to the PL. Cross configuration was introduced through post- deposition field annealing two times at two different temperatures. These annealing steps were conducted with the field directions aligned along and perpendicular to the SL direction. The choice of two different AFM materials is very important to induce the crossed anisotropy. In this context, we utilized AFM-1 as the PtMn layer and AFM-2 as the IrMn layer. IrMn was chosen due to its lower blocking temperature compared to PtMn, and its extensive research background.

Stack-B presents the optimized multilayer stack with soft pinning of the sens-ing layer (Fig. 31.3(a)). The sensing FM layer was modified by a multilayer stack of AFM/FM/ NM/FM layer. In this modified sensing layer stack, Ru was used as the NM layer, and the thickness of Ru thickness was chosen in the ferromagnetic exchange cou-pling region between the SL and SPL layers, with the goal of having a lower saturation field than that of the direct biasing of SL by the AFM2 layer.

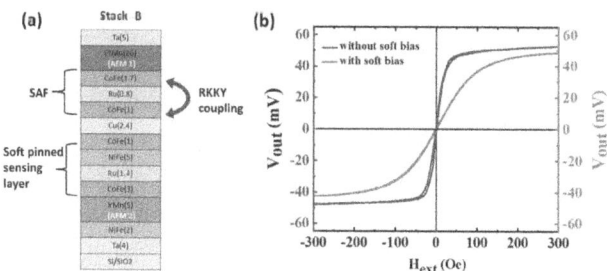

Fig. 31.3 (a) Schematic diagram of the optimized SV-stack with soft pinning of the sensing layer, and (b) the output characteristic of sensors fabricated from Stack-A and Stack-B

The sensor output characteristics from two devices fabricated using Stack-A and Stack-B at room temperature under excitation of 5 V input voltage are shown in Fig. 31.3(b). As expected, the device formed with Stack-C

can be operated in the field range of $\pm$ 15 Oe and have a sensitivity of the order of 0.6 $\pm$ 0.1 mV/V/Oe and the coercivity field, H_c lying in the range of 1-2 Oe. Whereas, the operating linear field range was extended $\pm$ 100 Oe at the cost of reduction of the sensitivity of 0.1 $\pm$ 0.05 mV/V/Oe for the device fabricated from Stack-B. Most importantly, the hysteresis was found to reduce to 0.2 Oe, which ensures the enhancement of the dynamic field for application for developing current sensors with a higher current range.

3.4 Implementation of PM in Spin Valve Sensor

CoCrPt thick film was chosen as a permanent magnet (PM) for this study. The sputter deposited films with varying thicknesses from 40 to 600 nm were characterized for higher coercivity (> 1500 Oe) and higher remanent ratio. Figure 31.4(a) represents the bulk magnetization-field loop with varying $t_{CoCrP\ t}$ at 20 mTorr pressure. Numerical simulation was carried out with varying gaps and thickness of the films and was found that it can generate a magnetic field up to 40 to 50 Oe between two pole caps for a gap of 90 microns with 1 micron thick film. The mapping of flux variation across

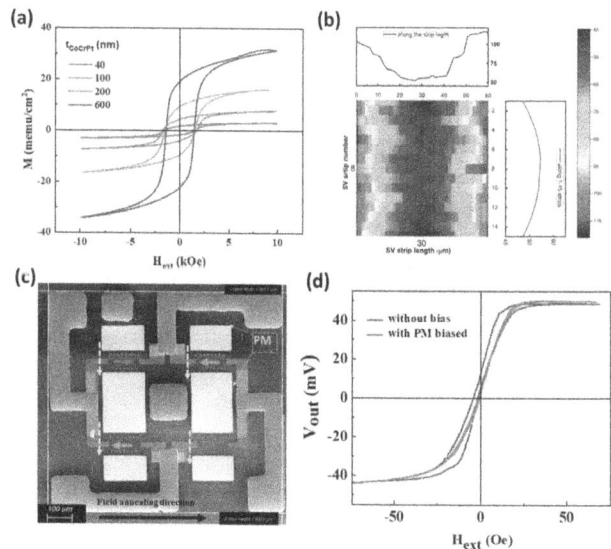

Fig. 31.4 (a) M(H) curves of the trilayer stack Cr/CoCrPt/Ta films with different $t_{CoCrP\ t}$ deposited at 20 mTorr Ar pressure, (b) magnetic flux generated across the SV element within the pole gap. Magnetic flux distribution, (top) along the length of the SV strip and (right) along the 1 to 15 strips of SV element, (c) FESEM image of on-chip integrated PM biased SV sensor, and (d) the output characteristics of sensors fabricated with and without PM bias

A single SV element is displayed in Fig. 31.4(b). The fabrication process involved the creation of an on-chip full Wheatstone bridge configured SV sensor. The PM films were deposited by a lift-off process across the elements. Figure 31.4(c) presents a FESEM image of the PM integration in the SV sensor structure. Figure 31.4(d) illustrates the sensor output obtained with a bridge excitation of 1 mA current. Before biasing through the PM, the transverse curve demonstrates hysteric behavior in the order of 2-3 Oe. This behavior is attributed to the edge boundary effect observed at the submicron SV strip. Upon biasing the SV elements through the PM, oriented orthogonally to the sensing direction, the measured output shows a coercivity of 0.4 Oe (H_c). Furthermore, there is an evident increase in the linear operating field region from ±15 to ±25 Oe. This adjustment signifies an enhancement in the sensor's performance, particularly in terms of its reduced hysteresis and linearity, thereby improving its overall functionality and reliability.

4. Longitudinal Field Biasing Comparison

Figure 31.5 illustrates the transfer curves of the SV sensor, showcasing different biasing techniques. This graph provides a comprehensive comparison of the sensor's performance under different biasing conditions, enabling a detailed analysis of its characteristics and response across a range of operating parameters. Notably, the transfer curve of the SV sensor biased by PM exhibits an asymmetry across zero field, potentially attributed to the different field distributions faced by all four elements. Conversely, in coil biasing, the generated field is deemed uniform along all four elements in the bridge configuration, resulting in a symmetric output curve. Furthermore, in PM biasing, the linear field range may vary depending on the thicknesses of the PM film, similar to the alterations observed with the increase in current in coil biasing. Soft biasing, on the

other hand, offers a higher linear field range compared to other methods, requiring no additional steps, rendering it a promising technique for future applications in MTJ structures. Each of the techniques mentioned is capable of reducing the hysteresis in the sensor transfer curve within the range of 0.2-0.4 Oe. This reduction is achieved by enabling the sensing layer to transition from a multidomain to a single-domain state.

5. Conclusions

In this report Spin valve sensor was successfully developed high-sensitivity bipolar lin-ear magnetic field tailored for industrial applications. By employing shape anisotropy, external coil biasing, in-situ stack biasing, and permanent magnet biasing, we opti- mized sensor performance to achieve hysteresis-free operation and a robust linear response.

Acknowledgment

The authors would like to thank the Director, CSIR-NAL for supporting this activity. Tejaswini C. Gawade acknowledges the Department of Science and Technology (DST), Government of India for providing the Inspire Fellowship.

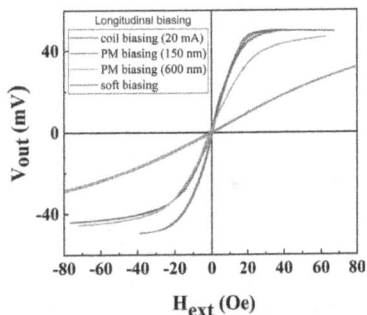

Fig. 31.5 Comparison of the output characteristics of SV sensor linearised using different longitudinal bias techniques

References

1. Freitas, P.P., Ferreira, R., Cardoso, S.: Spintronic sensors. Proceedings of the IEEE 104(10), 1894–1918 (2016).
2. Han, X., Zhang, Y., Wang, Y., Huang, L., Ma, Q., Liu, H., Wan, C., Feng, J., Yin, L., Yu, G., *et al.*: High-sensitivity tunnel magnetoresistance sensors based on double indirect and direct exchange coupling effect. Chinese Physics Letters 38(12), 128501 (2021).
3. Wang, D., Daughton, J., Nordman, C., Eames, P., Fink, J.: Exchange coupling between ferromagnetic and antiferromagnetic layers via ru and application for a linear magnetic field sensor. Journal of applied physics 99(8), 08–703 (2006).
4. Wang, D., Tondra, M., Daughton, J.M.: Push-pull bridge sensor using sdt junctions without shields. In: 2002 IEEE International Magnetics Conference (INTERMAG), p. 5 (2002)
5. TMR2703 - High Sensitivity and Low Hysteresis TMR linear sensor. https:// www.dowaytech.com/en/1398.html. Accessed on 20 October 2022.
6. Jeffers, F., Freeman, J., Toussaint, R., Smith, N., Wachenschwanz, D., Shtrik- man, S., Doyle, W.: Soft-adjacent-layer self-biased magnetoresistive heads in high-density recording. IEEE Transactions on Magnetics 21(5), 1563–1565 (1985).
7. Yamada, K., Maruyama, T., Tatsumi, T., Suzuki, T., Shimabayashi, K., Moto- mura, Y., Aoyama, M., Urai, H.:

Shielded magnetoresistive head for high density recording. IEEE transactions on magnetics 26(6), 3010–3015 (1990).

8. Negulescu, B., Lacour, D., Montaigne, F., Gerken, A., Paul, J., Spetter, V., Marien, J., Duret, C., Hehn, M.: Wide range and tunable linear magnetic tunnel junction sensor using two exchange pinned electrodes. Applied Physics Letters 95(11) (2009).

9. Chen, J., Feng, J., Coey, J.: Tunable linear magnetoresistance in mgo magnetic tunnel junction sensors using two pinned cofeb electrodes. Applied Physics Letters 100(14) (2012).

10. Zhu, J.-G., Zheng, Y., Liao, S.: Patterned exchange stabilized spin valve heads at very narrow track width. IEEE transactions on magnetics 37(4), 1723–1726 (2001).

11. Chaves, R., Cardoso, S., Ferreira, R., Freitas, P.: Low aspect ratio micron size tunnel magnetoresistance sensors with permanent magnet biasing integrated in the top lead. Journal of Applied Physics 109(7) (2011).

12. Xiao, M., Devasahayam, A.J., Kryder, M.H.: Fabrication and characterization of contiguous permanent magnet junctions. IEEE transactions on magnetics 34(4), 1495–1497 (1998).

13. Gawade, T.C., Borole, U.P., Behera, B., Khan, J., Barshilia, H.C., Chowdhury, P.: Giant magnetoresistance (gmr) spin-valve based magnetic sensor with linear and bipolar characteristics for low current detection. Journal of Magnetism and Magnetic Materials 573, 170679 (2023).

14. Gawade, T.C., Borole, U.P., Behera, B., Khan, J., Barshilia, H.C., Chowdhury, P.: On-chip full bridge bipolar linear spin valve sensors through modified syn-thetic antiferromagnetic layers. Journal of Magnetism and Magnetic Materials 587, 171234 (2023).

15. Paz, E., Ferreira, R., Freitas, P.P.: Linearization of magnetic sensors with a weakly pinned free-layer mtj stack using a three-step annealing process. IEEE Transactions on Magnetics 52, 1–4 (2016).

Note: All the figures in this chapter were made by the author.

Advances in Materials Science and Technology – Dr. Srikari Srinivasan et al. (eds)
© *2025 Taylor & Francis Group, London, ISBN 978-1-041-12342-2*

32

Prediction of Equivalent Plastic Strain in Cold Spray Process using Feedforward Artificial Neural Network

Urjit Parab, Nitin Gautam
Department of Mechanical Engineering,
Birla Institute of Science and Technology Pilani, K. K. Birla Goa Campus,
Goa, India

Digvijay G. Bhosale
Department of Mechanical Engineering,
Dr. D. Y. Patil Institute of Technology,
Pune, Maharashtra, India

Biswajit Das,
Manik A. Patil*, Dhananjay M. Kulkarni
Department of Mechanical Engineering,
Birla Institute of Science and Technology Pilani, K. K. Birla Goa Campus,
Goa, India

ABSTRACT: Cold spray is an additive manufacturing process used for fabrication and repair in various applications. The critical velocity in the cold spraying technique decides the possibility of successful deposition of particles on the substrate. The increase in particle's impact velocity leads to a gradual rise in powder deposition efficiency. The higher kinetic energy of powder particles increases the equivalent plastic strain (PEEQ), which helps to improve coating quality. In this work, a feedforward Artificial Neural Network (ANN) model is developed and trained on an extensive dataset obtained from simulation using the finite element method. The PEEQ evolved by the particle striking the surface is predicted using ANN model. The input dataset used for the model has been created using Abaqus software with the Smoothed Particle Hydrodynamics (SPH) approach. The dataset includes values of Von Mises Stress, temperature, and PEEQ obtained for different velocities in progress with time, as well as results of various diameters of particles. The present ANN model shows around 90% accuracy with the results obtained from numerical modelling. The output of the ANN model shows that, compared to traditional numerical modelling, the prediction of equivalent plastic strain can be made with greater accuracy and in a shorter amount of time.

KEYWORDS: Cold spray, PEEQ, ANN model, SPH technique

1. Introduction

Cold spraying technique holds great potential for coating delicate metals and composites. It is extensively employed in various fields, such as surface engineering and repair. In contrast to traditional thermal spraying, cold spraying offers numerous benefits, like reduced heat impact during processing, low operating temperature, absence of

*Corresponding author: p20210062@goa.bits-pilani.ac.in

DOI: 10.1201/9781003664277-32

oxidation, and phase transformation. It is beneficial for forming temperature-sensitive materials [1-2]. The critical velocity is a crucial factor in determining the successful deposition of the particle on the substrate during the cold spraying process.

The powder deposition rate increases gradually with particle velocity, which leads to a higher kinetic energy for the powder. This results in high plastic strain causing superior coating quality [3]. The particular value of velocity at which powder particle can be successfully deposited onto the substrate is the critical velocity [4]. Subsequent studies revealed that the particle velocity variation range stays within a minimal interval, and the deposition efficiency increases rapidly after reaching roughly 50% [5]. In addition, a range rather than a specific quantity of particle size is maintained during the spraying process. Thus, the velocity corresponding to a 50% deposition efficiency is commonly considered the particle's critical velocity. If the particle velocity surpasses a particular value, roughly two or three times the critical velocity, the excessive speed of the powder particle will cause erosion of the previous coating, causing the deposition rate to decrease from 100% to a negative value [6]. As a result, during spray processing, the particle velocity usually falls within an appropriate range. Internal and external factors can impact a particle's critical velocity, but it is primarily the internal characteristics of the material that determine whether a particle can be deposited. Lower melting point particles typically soften and deform more efficiently during spraying, quickly reaching the critical velocity [7-8]. In the meantime, the particle's hardness is also significant. The deformation capacity of a particle decreases with increasing particle hardness. As a result, it is challenging to deposit on the substrate's surface but simple to erode and impact damage the substrate [8-9]. For instance, it is difficult to directly deposit W, Cr and other hard-phase particles using cold spraying. Additionally, the level of particle oxidation directly impacts the critical velocity. Some of the particle's kinetic energy is used to break the oxide layer and allow the metal to come out and mix with the substrate, raising the particle's critical velocity [10]. It is evident from this that the material's characteristics, particularly its capacity for plastic deformation, are the primary factors influencing the critical velocity. When the same material is deposited on various substrates, the harder substrate requires a smaller critical velocity of the particle than the soft substrate. This is because the harder substrate allows for easier formation of self-locking during impact, and the particle experiences high plastic deformation [11]. Increasing the temperature of particle carrying gas allows the in-flight particles to undergo additional softening within the Laval nozzle, leading to increased plastic deformation upon impact with the substrate [12]. Thus, to keep the material from melting during the actual formation process, the critical velocity of the powder particle can be lowered by raising the temperature of carrier gas below the melting point.

Two primary methods are utilized to measure and forecast the critical velocity during cold spraying: experiment and numerical simulation. The pulse signal reflected back is then collected by the CCD detector and utilized for data processing, after that the actual velocity of powder particle is measured [13-14]. Researchers have proposed an empirical formula of critical velocity, which includes tensile stress, density, reference temperature, melting point and other parameters, based on experimental results and theoretical derivation [6, 15]. These formulas also clarify that the physical characteristics of the particles play a major role in determining the critical velocity influence factors.

However, with the advancement of computer finite element technology, it is now possible to accurately observe instantaneous phenomena that are challenging to achieve through experiments and effectively simulate the impact and interaction behaviour of high-speed powder particles. The adiabatic shear instability (ASI) is the standard to predict the successful deposition of powder particles, according to Assadi et al. [15]. They also developed the cold spray model relied on the Lagrangian method. Later, other researchers replicated the deposition of particles at various speeds using the same methodology [16]. The findings demonstrate how closely the ASI velocities at particle temperature, PEEQ, and equivalent normal stress match the critical velocities discovered through experimentation. As a result, ASI is frequently employed as a standard for determining the critical speed for cold spraying deposition technique [17-19]. Furthermore, the Eulerian method, the Arbitrary Lagrange-Euler (ALE) method, and the SPH method has been extensively applied in the numerical modelling of cold spraying [20,21,22] based on ASI to avoid the demerits of excessive distortion in traditional Lagrangian method. Nevertheless, predicting the particle's critical velocity through experimentation or simulation is time-consuming and resource-intensive, and considering the influence law when numerous factors act simultaneously is challenging.

2. Methodology

2.1 Dataset Used

The dataset used in this research was simulated using Abaqus software. At different velocities, the von-mises stress, temperature, and PEEQ (plastic equivalent strain) were simulated for three different diameters of particles (20 mm, 25 mm, and 35 mm) and for different velocities ranging from 400 to 800 m/s with respect to time. The critical velocity of the particle was identified to be 772 m/s by inspecting the graph of PEEQ vs time, where PEEQ remained constant after it came the particle met the object. To provide the input to the machine learning model, a new dataset was created using this dataset, consisting of time, S-mises stress, temperature, and velocity as the inputs. The PEEQ strain was taken as the output. The Artificial Neural Network (ANN) model was chosen since it had been applied to an earlier study based on the analysis of the critical velocity, and it had given good results [23].

2.2 Artificial Neural Network

The Artificial Neural Network in a system comprises these units arranged in a sequence of layers. The number of units in a layer can range from a few dozen to millions, depending on how many complex neural networks are needed to uncover the dataset's hidden patterns. Artificial neural networks typically consist of hidden, output, and input layers, as shown in Fig. 32.1. The input layer feeds external data into the neural network for analysis or education. After that, the data goes through one or more hidden layers, which convert the input into valuable data for the output layer. Lastly, the response of the Artificial Neural Network model to the supplied input data is provided as an output by the output layer.

A hidden layer was added to the ANN model so that the inputs could be converted into valuable outputs to provide the last output. Also, Standard Scaling was used to scale the parameters using the standard normal distribution, which takes in the value, subtracts the mean from it, and then divides the standard deviation of that feature. Standard scaling helps ensure that all the parameters' values lie between -3 and 3 to ensure consistency in the model and improve the model's efficiency at recognizing the parameters that are more likely to cause machine

failure. After the data was split into train set and test set, the train set data was fed into the models to train them with the data. (An 80-20 split was used, which means that 80% of the dataset was used as a train set, and 20% was used as the test set). This helped the model to be trained effectively using enough data, which was then used to test on the test dataset. The Keras library was imported from the TensorFlow library, and the sequential function was used to create the input layer for the ANN model. The hidden layer was created using the Dense function, where the activation function was set to ReLU (rectifier linear unit), as shown in Fig. 32.2, which helps introduce the non-linearity property to a deep learning model.

3. Numerical Modelling for PEEQ

The bonding and interaction between substrate and particle are highly dynamic and non-linear, which makes it extremely difficult and costly to understand through experimental work. Hence, simulating the model using the finite element method can be a simple and less expensive approach. The SPH technique with Abaqus 6.14V software is used to numerically model the cold spraying process. Compared to Lagrangian and ALE, the SPH technique shows an entirely different deformation pattern where the particle penetrates the substrate, and no jet formation occurs. However, there is severe plastic deformation near the interface [24].

A model consisting an assembly of deformable cylindrical substrate with radius R and Height H and a deformable spherical particle with a radius (1/10th) of the substrate is developed (Fig. 32.2b). The particle and substrate are meshed using C3D8R element with number of elements as 57000 and 4779417 respectively as shown in Fig. 32.3c). The particle and substrate are assigned with material properties of Aluminium [25].

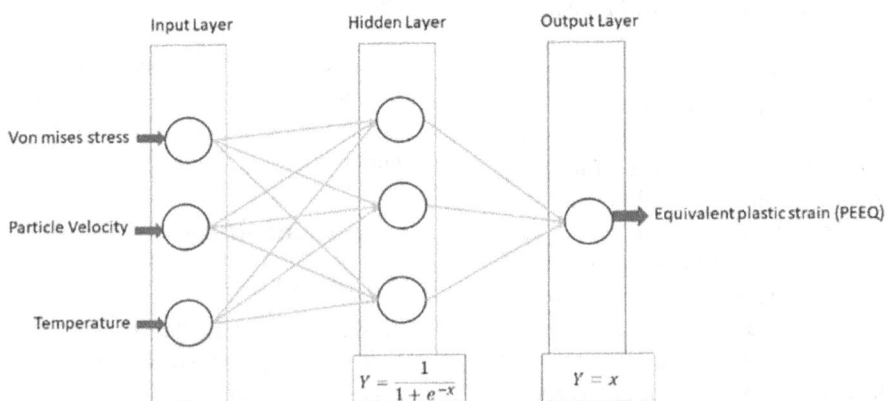

Fig. 32.1 An artificial neural network (ANN) showing the input, hidden and output layers

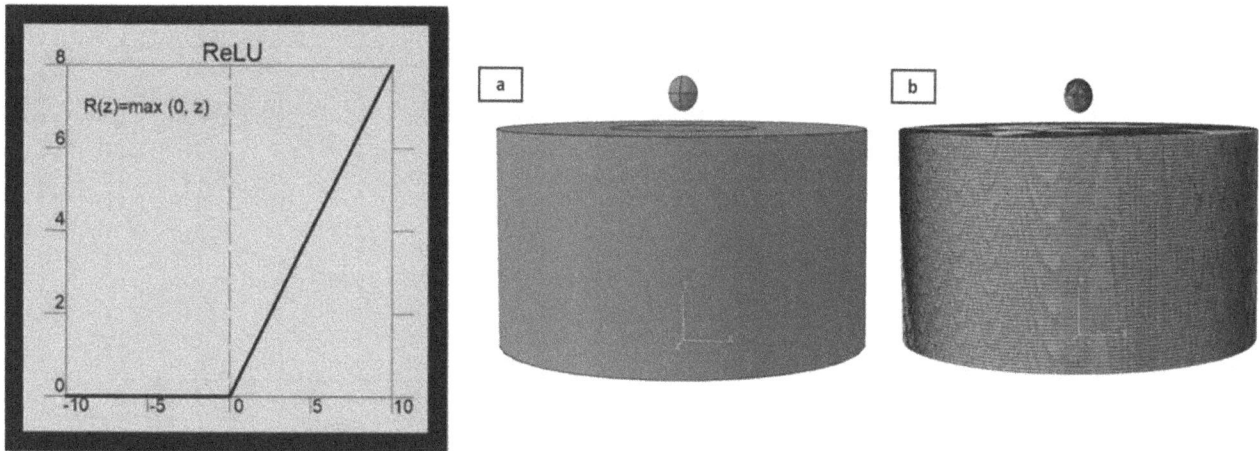

Fig. 32.2   a) Rectifier linear unit **b)** Assembly of substrate and particle **c)** Meshed particle and substrate

4. Result and Discussion

4.1 ANN Model Result

The output layer was also created using the Dense function. The model was compiled using the Adam (Adaptive moment estimation) optimizer function, and the mean squared error for each epoch step was calculated. The batch size for the model was kept at eight throughout the testing of the model, but the number of epochs (an epoch means one complete pass of the training dataset through the algorithm) was varied (from 100 to 500) to decrease the mean squared error and increase the accuracy of predicting the PEEQ strain.

When the number of epochs=100, the mean squared error=0.812, and when the number of epochs=500, the mean squared error =0.2284. Figure 32.3 a) shows the

graph between PEEQ and time for velocity=700 m/s and number of epochs=100. This indicates that when the number of epochs=100, the PEEQ vs time graph is a straight line and increases linearly until 3.50×10^{-08} seconds and then reaches a plateau. This tells us that the PEEQ vs time graph for epochs=100 is incorrect if we compare it to the graph provided in the initial dataset simulated using Abaqus. Also, the low mean squared error found in epochs=500 shows that the epochs=500 model is more accurate than the epochs=100 model. So, we obtained graphs of PEEQ vs time for all the velocities in the initial dataset for epochs=500 as shown in Fig. 32.3b.

In Fig. 32.4a), the 1st column shows the ANN model predictions, and the 2nd column shows the test cases. Here, we can see 21 cases, of which 12 are approximately correct. (A difference of less than 0.4 between the predictions and

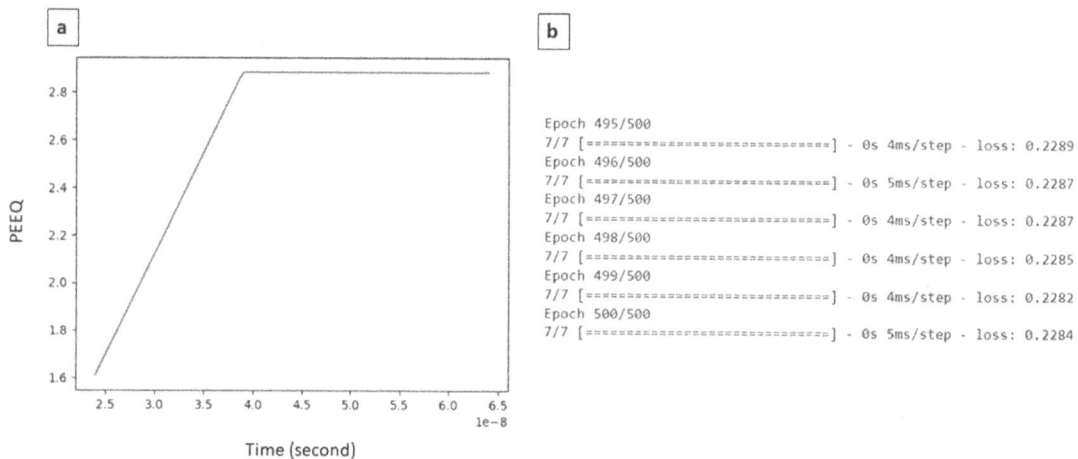

Fig. 32.3   a) PEEQ vs time graph for velocity=700 m/s (epochs=100) **b)** The last six steps of epochs=500. The loss (mean square error) =0.2284 for epoch 500/500

```
y_pred=ann.predict(X_test)
np.set_printoptions(precision=2)
print(np.concatenate((y_pred.reshape(len(y_pred),1), y_test.reshape(len(y_test),1)),1))

1/1 [==============================] - 0s 83ms/step
[[2.    2.63]
 [3.69 3.34]
 [0.68 0.15]
 [2.42 3.27]
 [3.42 3.23]
 [1.23 1.2 ]
 [1.78 1.8 ]
 [2.75 3.05]
 [3.   3.33]
 [1.74 1.79]
 [2.85 3.32]
 [3.08 2.98]
 [3.26 3.  ]
 [1.22 0.33]
 [1.53 1.49]
 [2.49 3.51]
 [0.68 0.01]
 [1.64 1.45]
 [2.99 3.53]
 [1.61 1.51]
 [3.07 3.12]]
```

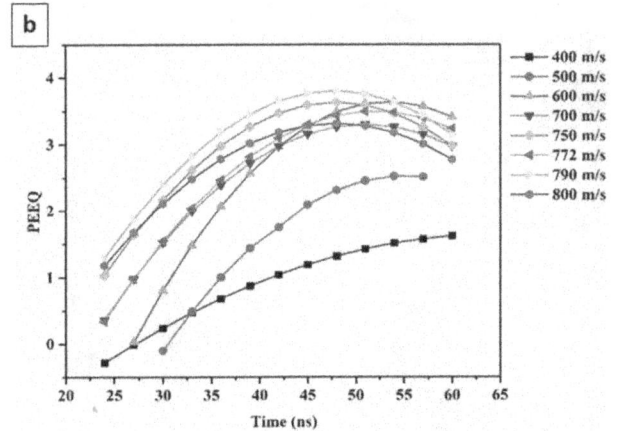

Fig. 32.4 **a)** Comparison of predictions (column 1) and the test cases (column 2) **b)** PEEQ vs time for all velocities (ANN predicted)

the actual value corresponds to accurate predictions, and a difference of greater than 0.4 corresponds to incorrect predictions. Fig. 32.4b) shows the graph of PEEQ against time for all velocities. The critical velocity prediction is close to the actual simulation since the graphs show that the non-linear increase is similar to the original dataset graphs. The peak plastic equivalent strain for the graphs with velocities from 600 to 800 m/s occurs around 50 ns. Also, the equivalent plastic strain decreases as we cross the critical velocity (772 m/s) for velocities 790 m/s and 800 m/s, which tells us that at higher velocities, there are chances of substrate erosion by the particle. These graphs are much more similar to the original dataset graphs in that they increase gradually and non-linearly compared to the graphs with epochs=100, where they increased linearly, which is incorrect. This shows us that the graphs produced when the number of epochs is increased to 500 shows more accurate results of the graphs of PEEQ vs time, which tells

us that Artificial Neural Network (ANN) model can be applied to the simulated cold spray dataset to give almost accurate predictions.

The graphs become finer, and the mean squared error decreases as we increase the number of epochs, which fits well with ANN and machine learning theory. Smaller size of the original dataset produced using simulation may face the problem of overfitting for higher epochs value. Hence presently, 500 epochs show good prediction for velocities corresponding to 600-790 m/s. However, if the dataset has many more parameters that affect the critical velocity of a particle attaching to the substrate, the ANN can promise to provide results that are close to actual.

4.2 Numerical Model Result

The contour of equivalent plastic strain obtained for the particle velocity of 400 m/s from the Abaqus simulation is shown in Fig. 32.5a). The highest PEEQ value is observed

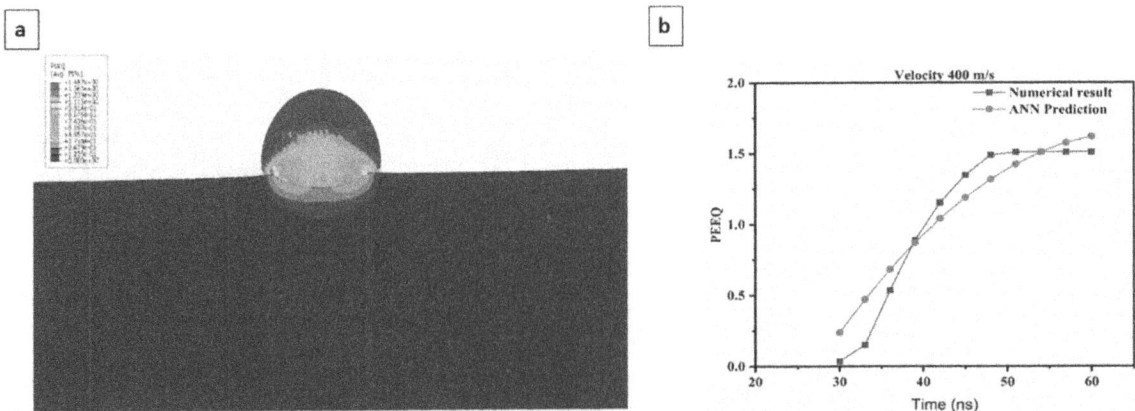

Fig. 32.5 **a)** PEEQ value obtained through numerical result **b)** Comparison between Numerical and ANN prediction of PEEQ

at the interface between substrate and powder particle, where the probability of severe plastic deformation is high. The PEEQ distribution during the impact of powder particle shows the trend of plastic deformation [26]. Fig. 32.5b) shows the comparison of PEEQ value predicted numerically and using ANN model. The higher epoch iterations give an accuracy of around 90% for the PEEQ values. For particle velocities beyond critical value, the PEEQ starts decreasing after the peak value owing to solid erosion.

5. Conclusion

In the current study, a feedforward artificial neural network is used to predict the equivalent plastic strain for different impact velocities. The ANN model that was developed shows high accuracy in data prediction. The graphs obtained from predicted data of PEEQ value against time converge well with the numerical results. The critical velocity predicted by SPH method is around 772 m/s. The powder particle velocity significantly affects the prediction of PEEQ values. The performance of the ANN model is heavily dependent upon the size and quality of input data. For a more robust prediction model, the input dataset should have a high volume. The ANN model gives more accurate results, around 90% as compared to conventional prediction techniques.

References

1. S. Yin, P. Cavaliere, B. Aldwell, R. Jenkins, H. Liao, W. Li and R. Lupoi, Cold Spray Additive Manufacturing and Repair: Fundamentals and Applications, Addit. Manuf., 2018, 21, p 628–650.
2. H. Assadi, H. Kreye, F. Gärtner and T. Klassen, Cold Spraying – a Materials Perspective, Acta Mater., 2016, 116, p 382–407.
3. S. Guetta, M.H. Berger, F. Borit, V. Guipont, M. Jeandin, M. Boustie, Y. Ichikawa, K. Sakaguchi and K. Ogawa, Influence of Particle Velocity on Adhesion of Cold-Sprayed Splats, J Therm Spray Tech, 2009, 18(3), p 331–342.
4. A. Manap, T. Okabe and K. Ogawa, Computer Simulation of Cold Sprayed Deposition Using Smoothed Particle Hydrodynamics, Procedia Engineering, 2011, 10, p 1145–1150.
5. D.L. Gilmore, R.C. Dykhuizen, R.A. Neiser, T.J. Roemer, and M.F. Smith, Particle velocity and deposition efficiency in the cold spray process, J. Therm. Spray Technol, 1999, 8, p 576–582.
6. T. Schmidt, F. Gärtner, H. Assadi and H. Kreye, Development of a Generalized Parameter Window for Cold Spray Deposition, Acta Mater., 2006, 54(3), p 729–742.
7. T. Schmidt, F. Gaertner and H. Kreye, New Developments in Cold Spray Based on Higher Gas and Particle Temperatures, J. Therm. Spray Technol., 2006, 15(4), p 488–494.
8. S. Shin, S. Yoon, Y. Kim and C. Lee, Effect of Particle Parameters on the Deposition Characteristics of a Hard/Soft-Particles Composite in Kinetic Spraying, Surf. Coat. Technol., 2006, 201(6), p 3457–3461.
9. W.-Y. Li, C.-J. Li and H. Liao, Significant Influence of Particle Surface Oxidation on Deposition Efficiency, Interface Microstructure and Adhesive Strength of Cold-Sprayed Copper Coatings, Appl. Surf. Sci., 2010, 256(16), p 4953–4958.
10. S. Yin, X.-F. Wang, W.Y. Li and H J. Vlcek, L. Gimeno, H. Huber and E. Lugscheider, A Systematic Approach to Material Eligibility for the Cold-Spray Process, J. Therm. Spray Technol., 2005, 14(1), p 125–133.
11. E. Jie, Effect of Substrate Hardness on the Deformation Behaviour of Subsequently Incident Particles in Cold Spraying, Appl. Surf. Sci., 2011, 257(17), p 7560–7565.
12. S. Yin, X. Wang, X. Suo, H. Liao, Z. Guo, W. Li and C. Coddet, Deposition Behavior of Thermally Softened Copper Particles in Cold Spraying, Acta Mater., 2013, 61(14), p 5105–5118.
13. H. Fukanuma, N. Ohno, B. Sun and R. Huang, In-Flight Particle Velocity Measurements with Dpv-2000 in Cold Spray, Surf. Coat. Technol., 2006, 201(5), p 1935–1941.
14. B. Jodoin, F. Raletz and M. Vardelle, Cold Spray Modeling and Validation Using an Optical Diagnostic Method, Surf. Coat. Technol., 2006, 200(14–15), p 4424–4432.
15. M. Grujicic, C.L. Zhao, W.S. DeRosset and D. Helfritch, Adiabatic Shear Instability Based Mechanism For Particles/Substrate Bonding in the Cold-Gas Dynamic-Spray Process, Mater. Des., 2004, 25(8), p 681–688.
16. S. Yin, X. Suo, Z. Guo, H. Liao and X. Wang, Deposition Features of Cold Sprayed Copper Particles on Preheated Substrate, Surf. Coat. Technol., 2015, 268, p 252–256.
17. F. Meng, S. Yue and J. Song, Quantitative Prediction of Critical Velocity and Deposition Efficiency in Cold-Spray: A Finite-Element Study, Scripta Mater., 2015, 107, p 83–87.
18. R. Ghelichi, S. Bagherifard, M. Guagliano and M. Verani, Numerical Simulation of Cold Spray Coating, Surf. Coat. Technol., 2011, 205(23–24), p 5294–5301.
19. W.-Y. Li, S. Yin and X.-F. Wang, Numerical Investigations of the Effect of Oblique Impact on Particle Deformation in Cold Spraying by the SPH Method, Appl. Surf. Sci., 2010, 256(12), p 3725–3734.
20. G. Qiu, S. Henke and J. Grabe, Application of a Coupled Eulerian-Lagrangian Approach on Geomechanical Problems Involving Large Deformations, Comput. Geotech., 2011, 38(1), p 30–39.
21. M. Yu, W.-Y. Li, F.F. Wang and H.L. Liao, Finite Element Simulation of Impacting Behavior of Particles in Cold Spraying by Eulerian Approach, J Therm Spray Tech, 2012, 21(3–4), p 745–752.

22. Ziyu Wang, Shun Cai, Wenliang Chen, Raneen Abd Ali & Kai Jin, Analysis of critical velocity of cold spray based on machine learning method with feature selection, Journal of Thermal Spray Technology, 2021, 30, p 1213

23. Li WY, Liao H, Li CJ, Bang HS, Coddet C. Numerical simulation of deformation behavior of Al H. Assadi, F. Gärtner, T. Stoltenhoff and H. Kreye, Bonding Mechanism in Cold Gas Spraying, Acta Mater., 2003, 51(15), p 4379–4394.

24. particles impacting on Al substrate and effect of surface oxide films on interfacial bonding in cold spraying. Applied Surface Science. 2007 Mar 30;253(11):5084–91.

25. Gnanasekaran B, Liu GR, Fu Y, Wang G, Niu W, Lin T. A Smoothed Particle Hydrodynamics (SPH) procedure for simulating cold spray process-A study using particles. Surface and Coatings Technology. 2019 Nov 15;377:124812.

26. Li WY, Yin S, Wang XF. Numerical investigations of the effect of oblique impact on particle deformation in cold spraying by the SPH method. Applied Surface Science. 2010 Apr 1;256(12):3725–34.

Note: All the figures in this chapter were made by the author.

Advances in Materials Science and Technology – Dr. Srikari Srinivasan et al. (eds)
© 2025 Taylor & Francis Group, London, ISBN 978-1-041-12342-2

33

One-pot Hydrothermal Synthesis of Zinc Gallate Nanoparticles: A Study of the Structural, Optical, Morphological and Electrical Properties

Indu Treesa Jochan

Optoelectronic and Nanomaterials' Research Laboratory,
Department of Physics and Electronics, CHRIST University,
Bengaluru, Karnataka, India

Department of Physics, St. Aloysius College,
Edathua, Kerala, India

Chrisma Rose Babu and E. I. Anila*

Optoelectronic and Nanomaterials' Research Laboratory,
Department of Physics and Electronics, CHRIST University,
Bengaluru, Karnataka, India

ABSTRACT: Spinel zinc gallate ($ZnGa_2O_4$) is an important ternary oxide semiconductor that paves the way for researchers to explore the potential applications in electronic devices like display panels, phototransistors and so on due to its wide band gap, high thermal and chemical stability. The present work investigates the structural, optical, morphological and electrical properties of zinc gallate nanoparticles synthesised using a one-pot hydrothermal method. Nanocrystalline zinc gallate powder was obtained with an estimated crystallite size of 14.67 nm using X-ray diffraction studies. The different vibrational bands in the synthesised nanoparticles were confirmed by the Fourier transform infrared spectrum (FTIR). Field emission scanning electron microscopic images assessed the surface morphology of the $ZnGa_2O_4$ nanoparticles. The optical properties of the prepared nanoparticles were analysed using the diffused reflectance spectrum (DRS) and photoluminescence (PL) spectrum). The direct bandgap, calculated from the Kubelka-Munk plot, was 4.8 eV. Blue emission was obtained upon UV excitation of the sample.

KEYWORDS: Conductivity, Hydrothermal, Photoluminescence, Spinel oxide, $ZnGa_2O_4$

1. Introduction

Zinc gallate ($ZnGa_2O_4$), a spinel with Fd3m symmetry and a face-centred cubic structure, features Ga^{3+} ions in octahedral sites, Zn^{2+} ions in tetrahedral sites, and oxygen anions in a cubic close-packed lattice[1,2,3]. Known for its vacuum stability, high UV transmittance, broad bandgap, and blue emission under UV or electron irradiation, zinc gallate is a member of the transparent semiconducting oxides family with a bandgap ranging from 4.4 to 5.0 eV[1,4,5,6]. Zinc gallate can be synthesized via hydrothermal methods[7], solid-state reactions[8], chemical vapour deposition[9], or sol-gel processes[10], with hydrothermal synthesis preferred for low-temperature, controlled particle morphology and reduced agglomeration[5].

In this study, we synthesised $ZnGa_2O_4$ spinel particles using a hydrothermal method at 110°C for 12 hours. After

---

*Corresponding author: anila.ei@christuniversity.in

DOI: 10.1201/9781003664277-33

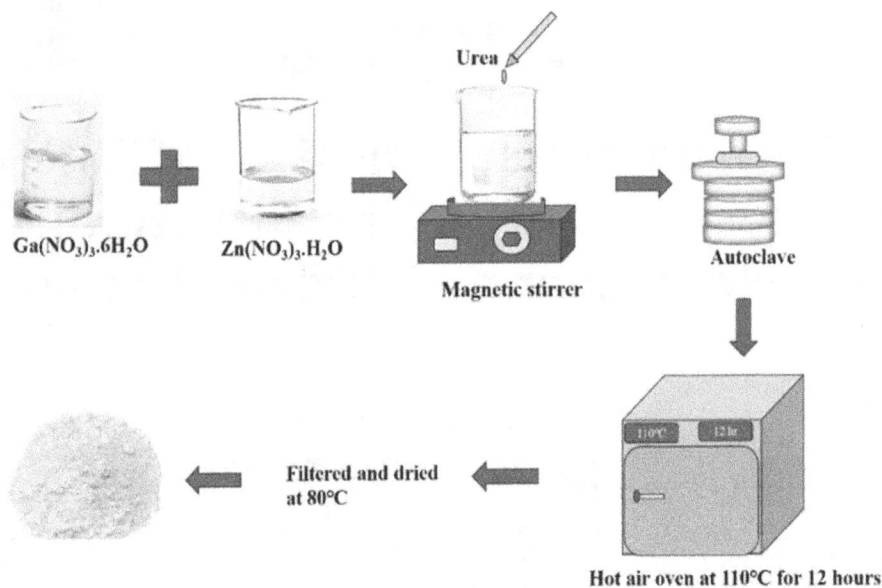

Fig. 33.1 A diagrammatic representation of the hydrothermal process of zinc gallate

synthesis, we performed comprehensive morphological, optical, structural and conductivity analyses to characterise the material. These investigations provide insights into the properties and potential applications of $ZnGa_2O_4$ spinel.

2. Materials and Methods

Zinc gallate was synthesized hydrothermally using $Zn(NO_3)_3.H_2O$ and $Ga(NO_3)_3.6H_2O$ (1:2 molar ratio) in 75 mL deionized water with urea, adjusted to pH 1.95. The solution was stirred, moved to a 100 mL Teflon-lined autoclave, and heated at 110°C for 12 hours. The product was then filtered (Whatman paper), dried at 80°C, and yielded a white powder, as depicted schematically in **Fig. 33.1**. Crystallographic data were obtained via XRD (Rigaku-mini flex diffractometer, 40 kV, Cukα), FTIR using IR-Prestige21 Shimadzu, morphology via FESEM (Apero 2 SEM, ThermoFisher Scientific), diffuse reflectance using UV-2600i Shimadzu UV-VIS spectrophotometer,

photoluminescence using Horiba-Duetta fluorescence and absorbance spectrometer, and DC conductivity using Keithley SMU 2450 Source Meter.

3. Results and Discussions

Figure 33.2(a) presents the XRD pattern of hydrothermally synthesized zinc gallate nanoparticles, confirming the cubic $ZnGa_2O_4$ structure (ICDD card no. 00-038-1240)[11]. The prominent diffraction peaks correspond (1 1 1), (2 2 0), (3 1 1), (2 2 2), (4 00), (4 2 2), (5 1 1) and (4 4 0) planes, with the highest peak at (3 1 1) located at 2θ = 35.7°. Impurity peaks, marked as (*), arise from gallium oxyhydroxide (GaOOH) and zinc nitrate urea hydrate[7]. Zinc nitrate and gallium nitrate react under hydrothermal conditions with urea, creating a basic environment for zinc gallate formation, while gallium nitrate hydrolysis and excess urea contribute to impurity phases. Using the Scherrer equation, the crystallite size (D) can be calculated

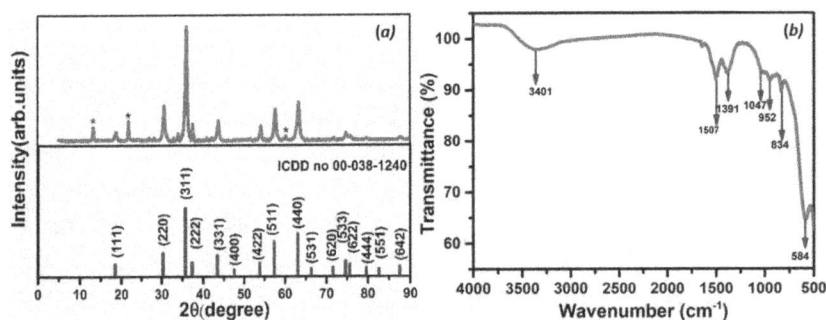

Fig. 33.2 (a) XRD pattern (b) FTIR spectra of zinc gallate

as,

$$D = \frac{0.9\lambda}{\beta \cos\theta} \qquad (1)$$

where θ represents the Bragg's angle, λ is 1.5406 Å (wavelength of CuKα), and β denotes the full-width at half maximum (radians)[12]. From the lattice geometry equation, the lattice constant (a) is given as

$$a = d\sqrt{h^2 + k^2 + l^2} \qquad (2)$$

where d denotes interplanar spacing[13]. The average crystallite size (D) evaluated by considering the planes (1 1 1), (2 2 0), (3 1 1), (5 3 3) and (6 2 2) is 14.67 nm, with a lattice parameter of 8.2954 Å.

FTIR analysis, shown in **Fig. 33.2(b)**, identifies vibrational bands in zinc gallate. Ga-O stretching vibrations appear at 584 cm^{-1}, while Zn-O stretching occurs at 834 cm^{-1} [14]. O-H bending vibrations cause peaks at 1507 cm^{-1} and 3401 cm^{-1}, and H-O-H bending contributes to the 1652 cm^{-1} band, indicating water presence [15,16]. The peak at 1391 cm^{-1} represents adsorbed nitrate ions [17]. Peaks at 1047 cm^{-1} (C-H bending) and 952 cm^{-1} (bending vibration of =C–H) arise from urea [18].

FESEM images in **Fig. 33.3(a)** show that the ZnGa$_2$O$_4$ particles have a cubic-like shape. **Figure 33.3(b)** presents the Energy-Dispersive X-ray Spectroscopy (EDX) analysis, confirming the presence of Ga, Zn, and O with atomic percentages of 22.94%, 17.10%, and 59.97%, respectively[19].

Figure 33.4(a) shows that ZnGa$_2$O$_4$ has low reflectance in the UV range (200-400 nm) and high reflectivity in the visible to near-infrared region (400-1000 nm)[20]. The optical gap, the energy difference between the band edges, is determined by the Kubelka-Munk equation [21,22]. The direct bandgap energy of the sample, determined to be 4.8 eV, was calculated by plotting photon energy (**hν**) vs (($\mathbf{k/s}$)**hν**)^2, where s and k are the scattering and absorption coefficients, respectively [23], and extrapolating linearly to the x-axis, as depicted in **Fig. 33.4(b)**.

The photoluminescence (PL) analysis of ZnGa$_2$O$_4$ nanoparticles (**Fig. 33.5(a)**) shows three emission peaks at 402 nm, 445 nm, and 488 nm for 360 nm excitation. The 360 nm peak is linked to self-activated optical centres from tetrahedral Ga-O groups[4], while the 402 nm emission originates from octahedral Ga-O groups[4,16]. Peaks at 445 nm and 488 nm are attributed to point defects like antisite

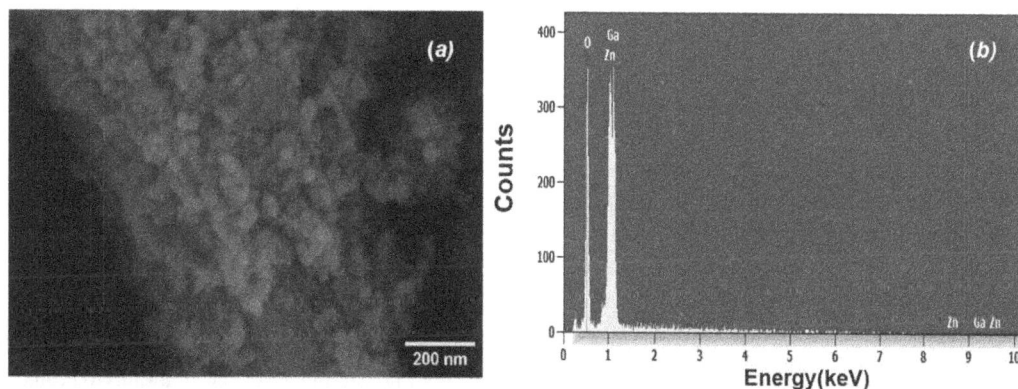

Fig. 33.3   (a) FESEM (b) EDAX of zinc gallate sample

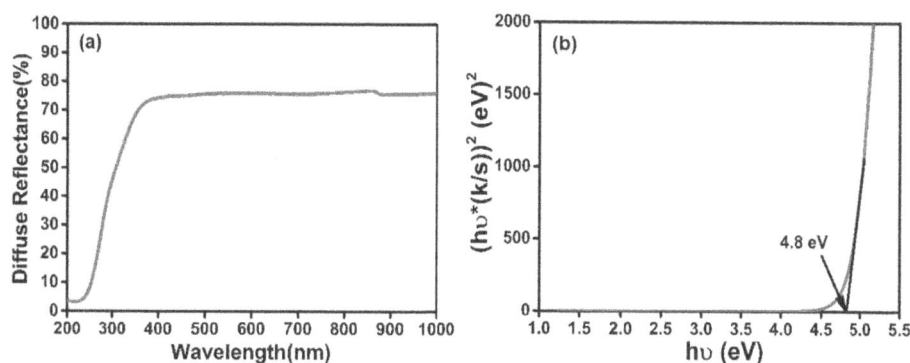

Fig. 33.4   (a) Diffuse Reflectance spectra (b) Kubelka-Munk plot

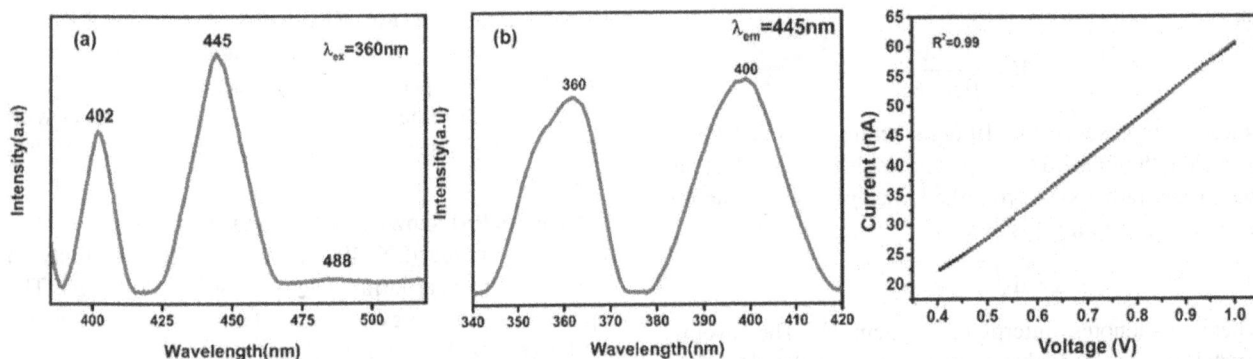

Fig. 33.5 (a) PL emission spectra (b) PL excitation spectra (c) Current(I) vs Voltage(V) graph

defects and oxygen vacancies[24,16]. **Figure 33.5(b)** shows the PL excitation spectrum for the 445 nm emission wavelength.

Point defects, particularly anti-site defects, significantly influence the optical and electronic properties of spinels. In ternary spinels, doping often involves anti-site defects through cross-substitution or self-doping[25]. Conductivity studies of $ZnGa_2O_4$, performed using a Keithley SMU 2450 Source Meter on the pelletized sample, confirm ohmic behavior (**Fig. 33.5(c)**) with a conductivity of 1.4×10^{-7} S·cm^{-1}.

4. Conclusion

Zinc gallate nanoparticles were synthesized successfully through the hydrothermal method at a reaction temperature of 110°C over 12 hours. XRD and FESEM analyses confirmed the cubic crystal structure of the zinc gallate. FTIR identified various vibrational bands in the sample. DRS revealed high reflectivity in the visible to near-infrared region (400-1000 nm) and low reflectivity in the UV region (200-400 nm). A blue shift in bandgap was observed compared to the theoretical bandgap value (4.6 eV) due to the size effect. Photoluminescence (PL) revealed blue emission with peaks at 402 nm, 445 nm, and 488 nm when excited at a wavelength of 360 nm. DC electrical conductivity of the sample was evaluated as 1.4×10^{-7} S·cm^{-1}.

Acknowledgements

We gratefully acknowledge the Centre for Advanced Research and Development (CARD), Christ (Deemed to be University), for their assistance with the characterisation throughout the entire experiment.

CRediT taxonomy: Conceptualisation, Data curation, Formal analysis, Investigation, Methodology, Validation, Visualization, Writing – original draft, Writing – review & editing: [Indu Treesa Jochan]; Formal analysis, Methodology, Validation: [Chrisma Rose Babu]; Supervision, Validation, Formal analysis, Writing – review & editing: [E.I. Anila].

Data Availability

The data that has been used is confidential.

Conflict of Interest

The authors declare that, concerning the work submitted, there is no conflict of interest arising from any commercial or associative interest.

References

1. Sampath SK, D G K, Pandey R. Electronic structure of spinel oxides : zinc aluminate and zinc gallate. J Phys Condens Matter. 1999;3635.
2. Dazai T, Yasui S, Taniyama T, Itoh M. Cation-De fi ciency-Induced Crystal-Site Engineering for ZnGa 2 O 4 :Mn 2+ Thin Film. 2020;0–4.
3. Zhou Y, Sun S, Wei C, Sun Y, Xi P, Feng Z, et al. Significance of Engineering the Octahedral Units to Promote the Oxygen Evolution Reaction of Spinel Oxides. 2019;1902509:1–11.
4. Communications SS, No V. Pergamon. 1998;105(3): 179–83.
5. Chen MI, Singh AK, Chiang JL, Horng RH, Wuu DS. Zinc gallium oxide—a review from synthesis to applications. Nanomaterials. 2020;10(11):1–37.
6. Search H, Journals C, Contact A, Iopscience M. Electronic structure and band gap of zinc spinel oxides beyond LDA: ZnAl2O4, ZnGa2O4 and ZnIn2O4. 063002.
7. Hirano M, Sakaida N. Synthesis of zinc gallate spinel nanoparticles by hydrothermal method in the presence of urea and their sintering. J Ceram Soc Japan. 2003;111(1291):176–80.
8. Safeera TA, Johns N, Krishna KM, Sreenivasan P V., Reshmi R, Anila EI. Zinc gallate and its starting materials in solid state reaction route- A comparative study. Mater

Chem Phys [Internet]. 2016;181:21–5. Available from: http://dx.doi.org/10.1016/j.matchemphys.2016.06.029

9. Knapp CE, Manzi JA, Kafizas A, Parkin IP, Carmalt CJ. Aerosol-assisted chemical vapour deposition of transparent zinc gallate films. Chempluschem. 2014;79(7):1024–9.

10. Zhang W, Zhang J, Li Y, Chen Z, Wang T. Preparation and optical properties of ZnGa 2 O 4 :Cr 3+ thin films derived by sol-gel process. Appl Surf Sci [Internet]. 2010;256(14):4702–7. Available from: http://dx.doi.org/10.1016/j.apsusc.2010.02.077

11. Wang X, Qiao H, Wang X, Xu Y, Liu T, Song F, et al. NIR photoluminescence of ZnGa2O4:Cr nanoparticles synthesized by hydrothermal process. J Mater Sci Mater Electron [Internet]. 2022;33(24):19129–37. Available from: https://doi.org/10.1007/s10854-022-08750-4

12. Rekha S, Martinez AI, Safeera TA, Anila EI. Author's Accepted Manuscript Enhanced Luminescence of Triethanolamine Capped Calcium Sulfide Nanoparticles Synthesized using Wet Chemical Method. J Lumin [Internet]. 2017; Available from: http://dx.doi.org/10.1016/j.jlumin.2017.05.042

13. Babu CR, Avani A V., Xavier TS, Tomy M, Shaji S, Anila EI. Symmetric supercapacitor based on Co3O4 nanoparticles with an improved specific capacitance and energy density. J Energy Storage. 2024;80(August 2023).

14. Valerio TL, Maia GAR, Gonçalves LF, Viomar A, Banczek EP, Rodrigues PRP. Study of the Nb 2 O 5 Insertion in ZnO to Dye-sensitized Solar Cells 2 . Material and Methods. Mater Res. 2019;22:1–5.

15. Mondal A, Manam J. Structural and Luminescent Properties of Si 4 + Co-Doped MgGa 2 O 4 : Cr 3 + Near Infra-Red Long Lasting Phosphor. 2017;6(7):88–95.

16. Safeera TA, Khanal R, Medvedeva JE, Martinez AI, Vinitha G, Anila EI. Low temperature synthesis and characterization of zinc gallate quantum dots for optoelectronic applications. J Alloys Compd [Internet]. 2018;740:567–73. Available from: https://doi.org/10.1016/j.jallcom.2018.01.035

17. Kim JS, Park HL, Chon CM, Moon HS, Kim TW. The origin of emission color of reduced and oxidized ZnGa 2 O 4 phosphors. 2004;129:163–7.

18. Hocart AM. Caste: A comparative study. Caste A Comp Study. 2018;1–159.

19. Chen L, Liu Y, Lu Z, Huang K. Hydrothermal synthesis and characterization of ZnGa2O4 phosphors. Mater Chem Phys. 2006;97(2–3):247–51.

20. Safeera TA, Anila EI. Wet chemical approach for the low temperature synthesis of ZnGa 2 O 4 : Tb 3 þ quantum dots with tunable blue-green emission. J Alloys Compd [Internet]. 2018;764:142–6. Available from: https://doi.org/10.1016/j.jallcom.2018.06.048

21. Anila EI, Jayaraj MK. Effect of dysprosium doping on the optical properties of SrS:Dy,Cl phosphor. J Alloys Compd [Internet]. 2010;504(1):257–60. Available from: http://dx.doi.org/10.1016/j.jallcom.2010.05.104

22. Kubelka P. New Contributions to the Optics of Intensely Light-Scattering Materials Part II: Nonhomogeneous Layers*. J Opt Soc Am. 1954;44(4):330.

23. Krishna KM, Nisha M, Reshmi R, Manoj R, Asha AS, Jayaraj MK. Electrical and optical properties of ZnGa2O4 thin films deposited by pulsed laser deposition. Mater Forum. 2005;29:243–7.

24. Shrivastava N, Guffie J, Moore TL, Guzelturk B, Kumbhar AS, Wen J, et al. Surface-Doped Zinc Gallate Colloidal Nanoparticles Exhibit pH-Dependent Radioluminescence with Enhancement in Acidic Media. Nano Lett. 2023;23(14):6482–8.

25. Shi Y, Ndione PF, Lim LY, Sokaras D, Weng TC, Nagaraja AR, et al. Self-doping and electrical conductivity in spinel oxides: Experimental validation of doping rules. Chem Mater. 2014;26(5):1867–73.

Note: All the figures in this chapter were made by the author.

Advances in Materials Science and Technology – Dr. Srikari Srinivasan et al. (eds)
© 2025 Taylor & Francis Group, London, ISBN 978-1-041-12342-2

34

Graphene Nanoplatelets Based Material for Energy Storage Applications

Vishnu S.*, Nisha Balachandran,
Chithra A., Deepthi Thomas, Deepa Devapal
Analytical and Spectroscopy Division,
Analytical, Spectroscopy & Ceramics Group,
Propellants Polymers Chemicals and Materials, Vikram Sarabhai Space Centre,
Thiruvananthapuram

ABSTRACT: Carbon nanomaterials are potential candidates for energy storage applications. In this work, we are exploring energy storage ability of graphene nanoplatelet based system. Specific capacitance of a material is critically affected by its surface area and electrical conductivity. Even though bare graphene nanoplatelet possess higher surface area, its specific capacitance was not quite significant, mostly due to its comparatively lower electrical conductivity. Hence attempts were made to enhance the charge storage capability by physical mixing with expanded graphite. Grinding method and sonication method were adopted for the components mixing. By taking different weight ratios of the components, a correlative study between the specific capacitance and material properties such as specific area and electrical conductivity could be carried out. Graphene nanoplatelet-expanded graphite mixture showed superior specific capacitance compared to individual components. Detailed analysis of the specific capacitance revealed that the method adopted for physical mixing also critically affects the material functional property.

KEYWORDS: Energy storage, Graphene nanoplatelets, Expanded graphite

1. Introduction

Energy storage technologies such as supercapcitors and batteries have applications ranging from societal to aerospace field. These energy storage systems are always in the lime light of research due to continuous development in the material science. Since, improvement in the charge or energy storage ability of the active material is the fundamental topic of research in energy storage technology, advanced material development is an evergreen theme in material science. Among the various advanced nanomaterials, 2D nanomaterials received enormous attention in energy storage capability. Their properties such as layered structure, high surface area, chemical tunability and flexibility are preferred criteria for advanced energy storage systems. [1-4] Even though class of 2D nanomaterials include various systems such as layered double hydroxides, transition metal dichalcogenides, transition metal oxides etc., graphene-based materials are most studied among them for their stability and feasibility.

Graphene nanoplatelet is a potential candidate for energy storage applications due to its tunability, thermal and chemical stability, high surface area, low toxicity etc. However, most of the graphene nanoplatelet based systems suffers from undesired aggregation and restacking due to π-π interaction. This leads to low surface area and poor charge carrier mobility. Hence proper tuning of these systems is essential to reduce the aggregation tendency and maintain high surface area and charge carrier mobility.

*Corresponding author: vishnusn0007@gmail.com

DOI: 10.1201/9781003664277-34

Charge storage and charge transport are the two important processes in any energy storage system. From a material point of view, surface area and electrical conductivity are the two major factors which affect the above-mentioned processes. [5-6] Herein, we are tuning these properties of graphene nanoplatelets by combining it with expanded graphite and associated change in the energy storage or specific capacitance is evaluated.

2. Experimental

2.1 Materials and Characterization

Commercial graphene nanoplatelets (Sigma Aldrich) was used as received for the work and expanded graphite was prepared from natural graphite as described below. Morphological analyses were carried out using Carl Zeiss GeminiSEM 500 at 2kV. Raman analyses were done using Witec Alpha300R confocal raman microscope with 532 nm as the excitation laser. X-ray powder diffraction (XRD) analysis was carried out using Bruker D8 Discover diffractometer from 10 to 80° with Cu-Kα radiation (λ = 1.5406 A°) at a scanning rate of 4°/min. For Brunauer–Emmett–Teller (BET) specific surface area measurement, N2 adsorption isotherms of the materials (degassed at 200 °C for 3 h) were recorded using Quantachrome Nova 1200e surface area analyser. Electrochemical characterization done using Metohm Autolab M204 in a three electrodes configuration in 1M KCl solution. Glassy carbon, Platinum rod, Ag/AgCl electrode as working electrode, counter electrode and reference electrode respectively. Working electrode was modified as described below. Electrical conductivity of the material was measured in the pelletized form using two electrode setups in Metohm Autolab M204

2.2 Preparation of Expanded Graphite

Initially 5g of $(NH_4)_2S_2O_8$ and 3ml Conc. H_2SO_4 were mixed by sonication for 5 min. Then 1g of natural graphite was added to the above prepared mixture and stirred for 15 min in ambient conditions. The resultant slurry kept under room temperature for 24h. Finally, the sample was washed with deionized water until pH of the filtrate is neutral. [7].

2.3 Preparation of Graphene Nanoplatelets - Expanded Graphite (GNP-EG) Mixture

Different weight ratios of GNP and EG has been mixed by two ways (a) Grinding method: Components mixed together by manual grinding; (b)Sonication method: Components are dispersed in isopropyl alcohol and sonicated for 20 minutes. Further dispersion is filtered, washed with deionized water and dried.

2.4 Preparation of Working Electrode

Glassy carbon electrode was polished using alumina slurry and cleaned in de-ionized water. 20 mg of the active material and polyvinyledene fluoride (binder) are mixed in 90:10 weight ratio and sonicated for five minutes in DMF. 5 ml of the resultant slurry was deposited onto glassy carbon. [8]

3. Results and Discussions

Raman analysis of graphene nanoplatelets (GNP) showed characteristic peaks corresponds to D (1350 cm^{-1}), G (1570 cm^{-1}) and corresponding secondary bands as shown in Fig. 34.1. Well resolved secondary bands indicates the quality of the exfoliated graphene nanoplatelets. Morphological analysis revealed the GNP of size in the range of 20-60 nm. (Fig. 34.1) BET surface area of GNP is about 740 m^2/g, which can be attributed to the well exfoliated nanoplatelets.

Expanded graphite (EG) is prepared from natural graphite as mentioned earlier. Comparing the Raman spectrum, it can be seen that both the natural graphite and the expanded graphite shows characteristic D, G and 2D bands. (Fig. 34.2) Higher ID/IG ratio along with broadness and shift in the (002) peak in the XRD confirms the increase in

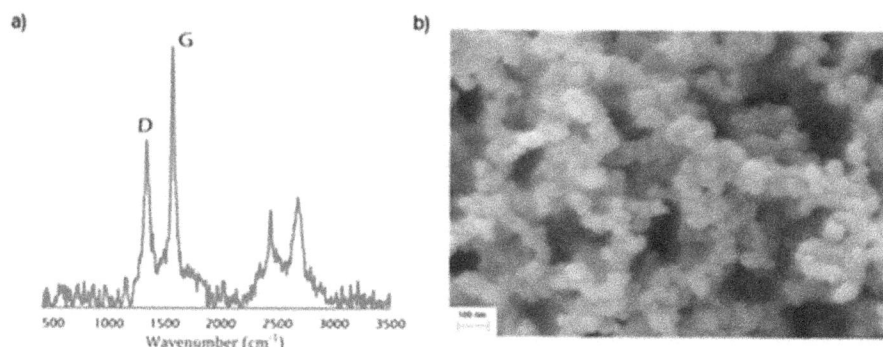

Fig. 34.1 (a) Raman spectrum of GNP; (b) SEM image of GNP

a)

b)

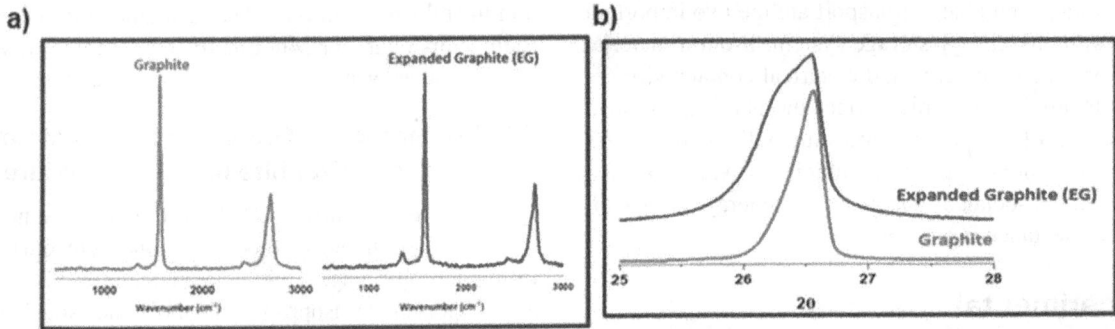

Fig. 34.2 (a) Raman spectrum of natural graphite and expanded graphite; (b) XRD natural graphite and expanded graphite

the interlayer spacing ($d_{(002)}$) between the graphite layers. [5] Change in the interlayer spacing is clear from the SEM images shown in Fig. 34.3. BET analysis showed surface area of about $43 m^2/g$.

Table 34.1 Summary of raman analysis

Sample	D band	G band	2D band	I_G/I_D
Natural graphite	1351 cm^{-1}	1575 cm^{-1}	2715 cm^{-1}	0.05
Expanded graphite	1340 cm^{-1}	1564 cm^{-1}	2704 cm^{-1}	0.17

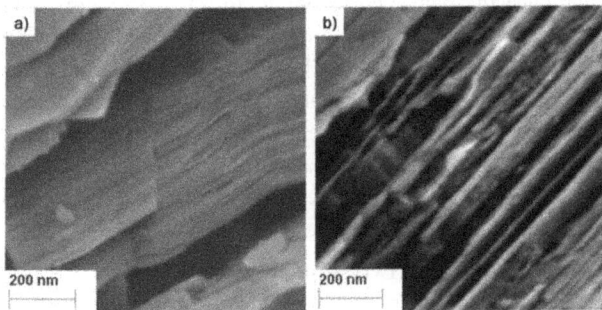

Fig. 34.3 SEM images of (a) natural graphite; (b) expanded graphite

Specific capacitance (Cp) of GNP and EG was evaluated from galvanostatic charge discharge (GCD) curve using

equation 1. Galvanostatic charge discharge analysis was carried out at a current density of 1 A/g.

$$Cp = \frac{I\Delta t}{m\Delta V} \quad (1)$$

Where, Cp is specific capacitance (F/g), I- current, m-mass of the material coated over electrode, ΔV & Δt- change in potential and time (from the slop of the GCD curve) respectively. GCD curves of GNP and EG are shown in Fig. 34.4 and the linear nature of the GCD curves indicates capacitive behavior of the materials. Cp calculated for GNP and EG are 350 F/g and 30 F/g respectively. Difference in their specific capacitance can be correlated to their surface area and electrical conductivity. Electrical conductivity of the GNP and EG are 12 S/cm and 88 S/cm respectively. Even though EG has higher conductivity than GNP, its low surface area may be the reason for the inferior specific capacitance. Hence for improving the performance of the GNP, mixtures of GNP-EG having different weight ratios of components by grinding and sonication method as mentioned in the experimental section.

Chronopotentiometric analysis of materials prepared by grinding is summarized in Fig. 34.5. It can be seen that, increase in the wt% of EG specific capacitance is initially increasing and then decreasing. It was observed that

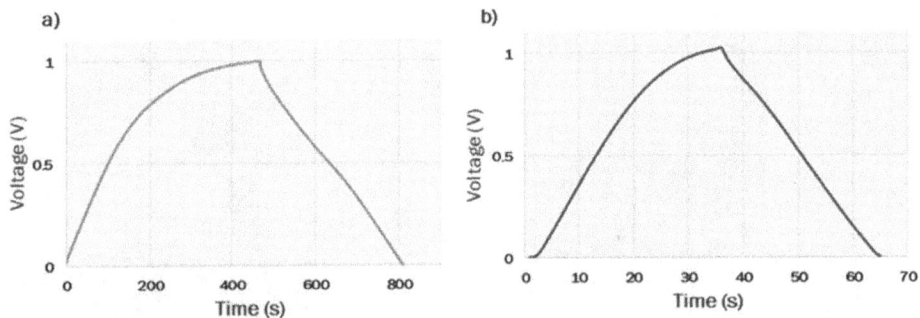

a)

b)

Fig. 34.4 Galvanostatic charge/discharge curve of (a) GNP & (b) EG using 1M KCl electrolyte at constant current density of 1A/g

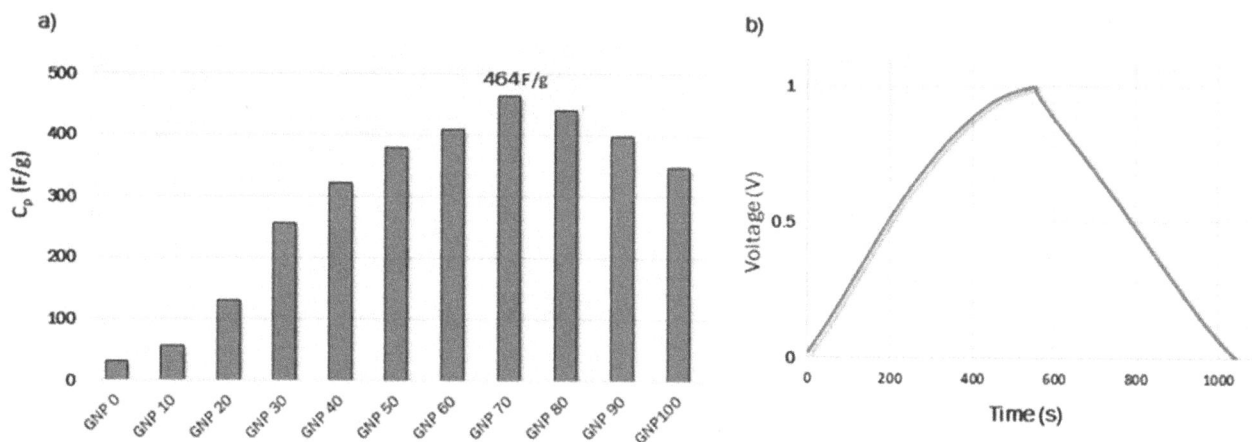

Fig. 34.5 (a) Specific capacitance of the GNP-EG mixture prepared by grinding method (Note: Number in the naming indicates the wt% of GNP in the mixture); (b) GCD curve of GNP-70

GNP-70 (GNP-70 wt% and EG-30 wt%) shows higher performance (464 F/g) among the materials.

Like grinding, sonication also showed similar trend in the specific capacitance with the addition of EG and GNP-70 showed highest specific capacitance of about 526 F/g. (Fig. 34.6) As shown in the SEM images of GNP-70 samples,

it is clear that the grinding induces undesired aggregation of the material which in turn reduces the surface area (677 m^2/g) compared to that of sonication (720 m^2/g). (Figure 34.7) It is observed that the mixture containing 70 wt% of GNP shows fitting surface area and electrical conductivity for higher energy storage. (Table 34.2)

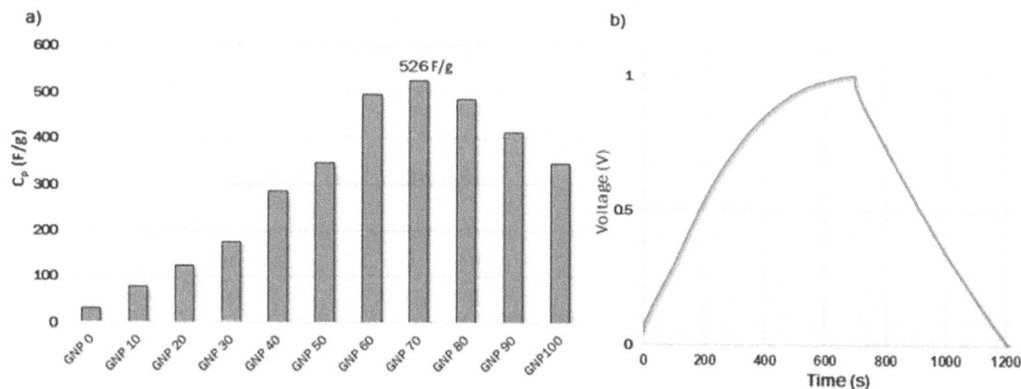

Fig. 34.6 (a) Specific capacitance of the GNP-EG mixture prepared by sonication method (Note: Number in the naming indicates the wt% of GNP in the mixture); (b) GCD curve of GNP-70

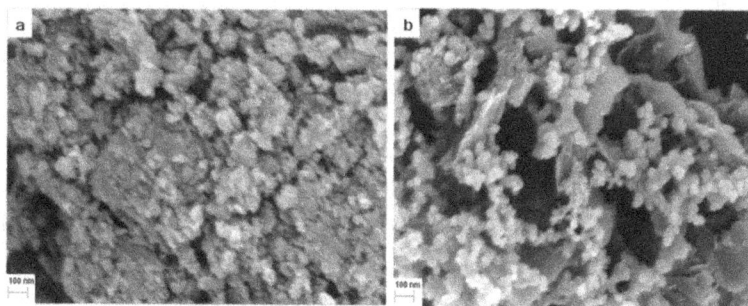

Fig. 34.7 SEM images of GNP-70 (a) grinding method; (b) sonication method

Table 34.2 Summary of raman analysis

Sample	Surface area (m²/g)	Electrical conductivity (S/cm)	Specific capacitance (F/g)
Expanded graphite	43	88	30
Graphene nanoplatelets	740	12	350
GNP-70 (Grinding)	677	31	464
GNP-70 (Sonication)	720	36	526

Figure 34.8 shows comparative changes in the surface area and electrical conductivity of GNP-EG mixture prepared by sonication. It can be seen that when GNP wt% reaches 70-80% surface area comes to plateau region with comparable conductivity to expanded graphite. From the morphological analysis it can be seen that GNP-70 showed well dispersed components compared to other mixtures. (Figure 34.9) From the SEM images it was clear that the self-sorted aggregation (preferential aggregation of GNP alone or EG) tendency was lesser in GNP-70. These peculiar molecular level interactions in GNP-70, facilitates its superior energy storage ability compared to other mixtures.

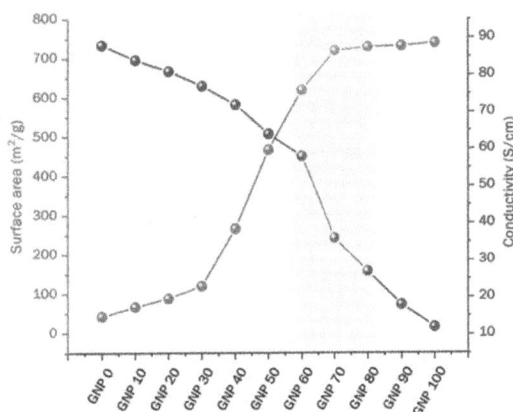

Fig. 34.8 Surface area and electrical conductivity of GNP-EG mixture prepared by Sonication method. (Note: Highlighted portion shows mixtures with better specific capacitance)

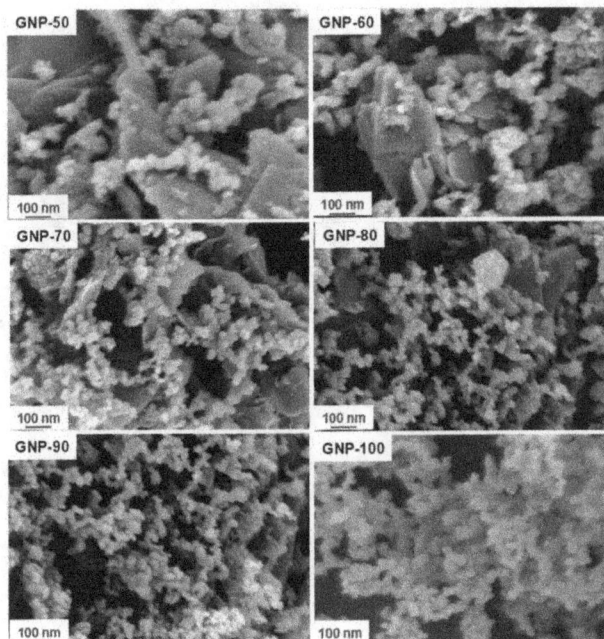

Fig. 34.9 SEM images of GNP-EG mixtures prepared by sonication method

unwanted aggregation tendency of the materials in the latter method. A correlative study of the dependence of specific capacitance and physical properties of the materials such as surface area and electrical conductivity has been done, which might be crucial in designing and development of new materials for energy storage applications.

Acknowledgement

The authors acknowledge Director, VSSC, Deputy Director VSSC (PCM) and colleagues in Analytical and Spectroscopy Division, VSSC for their support.

4. Conclusions

In summary a highly efficient electrode material for super capacitor applications from graphene nanoplatelets and expanded graphite was developed. GNP-EG mixture containing 70 wt% of GNP and 30 wt% of EG showed best performance among the prepared materials. It has been proven that the method of preparation like grinding and sonication is playing a vital role in the material functional performance. In this work, it has been observed that sonication has advantage over grinding method due to the

References

1. H. N Alshareef, F. Zhang, N. Kurra, C. Xia., J. Material Sci. Eng. 6, 10 (2017).
2. S. B. Mujib, Z. Ren, S. Mukherjee, D. M. Soares, G. Singh., Mater. Adv. 1, 2562 (2020).
3. Y. Gogotsi ACS Nano 8, 5369–5371 (2014).
4. E. Pomerantseva, F. Bonaccorso, X. Feng, Y. Cui, Y. Gogotsi., Science 366, 969 (2019).
5. E. A. Worsley, S. Margadonna, P. Bertoncello., Nanomaterials 12, 3600 (2022).
6. B. A. Ali, A. H. Biby, N. K. Allam., Energy Fuels 35, 13426–13437 (2021)
7. T. Liu, R. Zhang, X. Zhang, K. Liu, Y. Liu, P. Yan., Carbon 119, 544-547 (2017).
8. A. Yu, A. Sy, A. Davies., Synthetic Metals 161, 2049–2054 (2011).

Note: All the figures and tables in this chapter were made by the author.

Advances in Materials Science and Technology – Dr. Srikari Srinivasan et al. (eds)
© 2025 Taylor & Francis Group, London, ISBN 978-1-041-12342-2

35

Biopolymer-Based Plant Extracts Mediated-Silver Nanoparticles as Antibacterial Dressing

Rebika Baruah*,
Archana Moni Das
Natural Product Chemistry Group,
Chemical Science and Technology Division,
CSIR-North East Institute of Science and Technology,
Jorhat, Assam, India
Academy of Scientific and Innovative Research (AcSIR),
Ghaziabad, India

ABSTRACT: Stable multifunctional silver-containing materials are suitable for antibacterial wound dressings. Maintaining of stability of nanoparticles is a big challenge for researchers; therefore cellulose and chitosan were used in the synthesis of biopolymer-based plant extracts mediated-silver nanoparticles [CCPA NPs]. Cellulose and chitosan were served as stabilizing and capping agents and antibacterial activity enhancing agents respectively. Plant extracts was used as reducing agents in the reduction of Ag+ ions to Ag (0) NPs. To confirm the shape, size, optical properties, and involvement of phytochemicals and biopolymer in the synthesis and stabilization of naocompositess Transmission Electron Microscope (TEM), Scanning Electyron Microscope (SEM), X-Ray Diffraction (XRD), UV-Visible, and Fourier Transform Infrared (FTIR) analysis were performed respectively. To study the biomedical property of the CCPA; the antimicrobial and antioxidant activities of the NPs were examined. CCPA showed excellent antibacterial activities against wound infected bacteria like *Staphylococcus aureus* and *Escherichia coli*. Therefore, CCPA will have good application prospects for wound dressings.

KEYWORDS: Biopolymer, Cellulose, Chitosan, Ag nanoparticles, Antibacterial activity

1. Introduction

Nanostructured inorganic-organic biocomposites are the multifunctional materials in the field of nanotechnology. They have tremendous application in various fields such as medicine, water, purification, biosensor, food packaging, etc. (Gebre, 2023 & Baruah et al., 2023). Multifunctional nanohybrid materials are synthesized by different techniques which have a lot of disadvantages including multistep procedures, costly, time consuming, etc. (Baruah et al., 2021).

Chemical reduction involves many hazardous chemicals as reducing and capping agents that causes severe environmental pollutants. Hence, the synthesis of renewable biomaterials by using biocompatible and environmentally benign methods is an emerging area of research (Baruaj et al., 2022). These biomaterials have a potential application

---

*Corresponding author: baruahrebika9@gmail.com

DOI: 10.1201/9781003664277-35

in medicine and food industry. Recently, the synthesis of metal nanoparticles on natural fibre by green in situ method has drawn enormous attention due to their stability and uniform distribution on the supporting agents (Baruah et al., 2024). Plant extracts derived synthesis of nanoparticles such as Ag, Au, Pt, Pd, etc. is an ideal alternative in case of size and shape of the nanoparticles and it also provides superior characteristics of the nanoparticles (Goswami and Das 2018). Plant extracts act both as reducing and capping agents due to the presence of various phytochemicals; phenols, terpenoids, flavonoids, steroids, tannins, alkaloids, etc (Goswami et al, 2028).

Silver nanoparticles are the most applicable NPs due to their outstanding physiochemical properties and potent antimicrobial activities which make them a potential applicant in biomedical and food packaging industry (Baruah et al., 2029). In situ synthesis of Ag NPs makes them as safe candidates in food and water contact application due to their controlled release (Nariya et al., 2020).

Cellulose is the most prominent natural fibre for the impregnation of Ag NPs due to its biocompatibility, renewability and abundance. The fibre is capable of binding to positively charged ions through electrostatic interaction (i. e. ion dipole) (Shkir et al., 2024). Chitosan is also a suitable fibre for the stabilization of the NPs due to their unique polycationic and chelating characters. The fibre can also be use as moderate reducing agents and ion capping agents to prevent the agglomeration of the NPs (Sharifi-Rad et al., 2024).

We herein reported the in-situ synthesis of Ag NPs by using the *Livistona jekinsiana* plant extracts as reducing agents on cellulose and chitosan film as support. The plant extracts of *Livistona jekinsiana* was used for the first time to synthesize the Ag NPs which is an evergreen, unbranched and thin plants and found in Africa, South Arabia, South East, and Eastern Asia, Malaysia, Australia, China, and Thailand. Antibacterial and antioxidant activity of the synthesized nanocomposites was studied.

2. Materials and Methods

2.1 Materials

All the chemicals used in the synthesis of bionanocompsitesand its application were purchased from Merk chemicals and used as received without purification. Double distilled H_2O was used for the preparation of solutions. Cellulose was extracted from *Livistona jekinsiana* stem.The *Livistona jekinsiana* leaves were collected from North-West Jorhat, Assam and washed thoroughly with distilled water before the extraction of the leaves.

2.2 Analytical Methods

The characteristics UV-Vis absorption peak of the Ag NPs was recorded using a UV-Vis spectrophotometer (Hitachi Model No –U-3900). The involvement of phytochemicals in the synthesis of nanoparticles was evaluated by FTIR spectroscopy (Perkin-Elmer FTIR-2000 spectrometer). The phase, crystalline structure and grain size were determined by X-ray diffractometer (Rigaku Ultima IV diffractometer). The phase, shape, and size of the synthesized ZnO NPs were analysed by the transmission electron microscope (JEM-2100 Plus). The zone of inhibition of every test organism in antibacterial activity was measured using an Antibiotic Zone Scale.

2.3 *Livistona Jekinsiana* Leaf Extraction

The cleaned *Livistona jekinsiana* leaves were cut into small pieces and 10 g leaves were added to distilled water and heated to 80°C for 20 min. The decant was separated, filtered and stored till further use.

2.4 Preparation of Cellulose/Chitosan Wet Films with Diffused Leaf Extract

The mixture of cellulose and chitosan solution was cast on glass plates and regenerated with alcohol bath. There generated wet films were washed thoroughly in distilled water to remove the remaining alcohol and salts. These wet films were then kept in leaf extract in a beaker for about two hours to ensure its uniform diffusion into the films.

2.5 Preparation of Cellulose/Chitosan/AgNP Composite Films

1 mM aqueous silver nitrate solution was prepared in a beaker and to each; the leaf extract diffused wet films were added and stirred with the help of a magnetic stirrer at 100 rpm for about 2 h. The colour change of the film indicated the *in-situ* generation of Ag NPs in the matrix. These films were washed thoroughly with water and dried at room temperature.

3. Result and Discussion

3.1 UV, XRD and FTIR Analysis

The characteristics UV-vis spectra of the Ag NPs formed outside the matrix was appeared at 417 nm due to the surface plasmon resonance (SPR) of the Ag NPs (Fig. 35.1. [a]).

In the XRD pattern of Ag/ cellulose/chitosan bionanocomposites films, the common peaks in the matrix and the nanocomposites were observed at 2θ = 12.2°, 20.4° and 22.1° which were due to the crystal planes (1 -1 0),

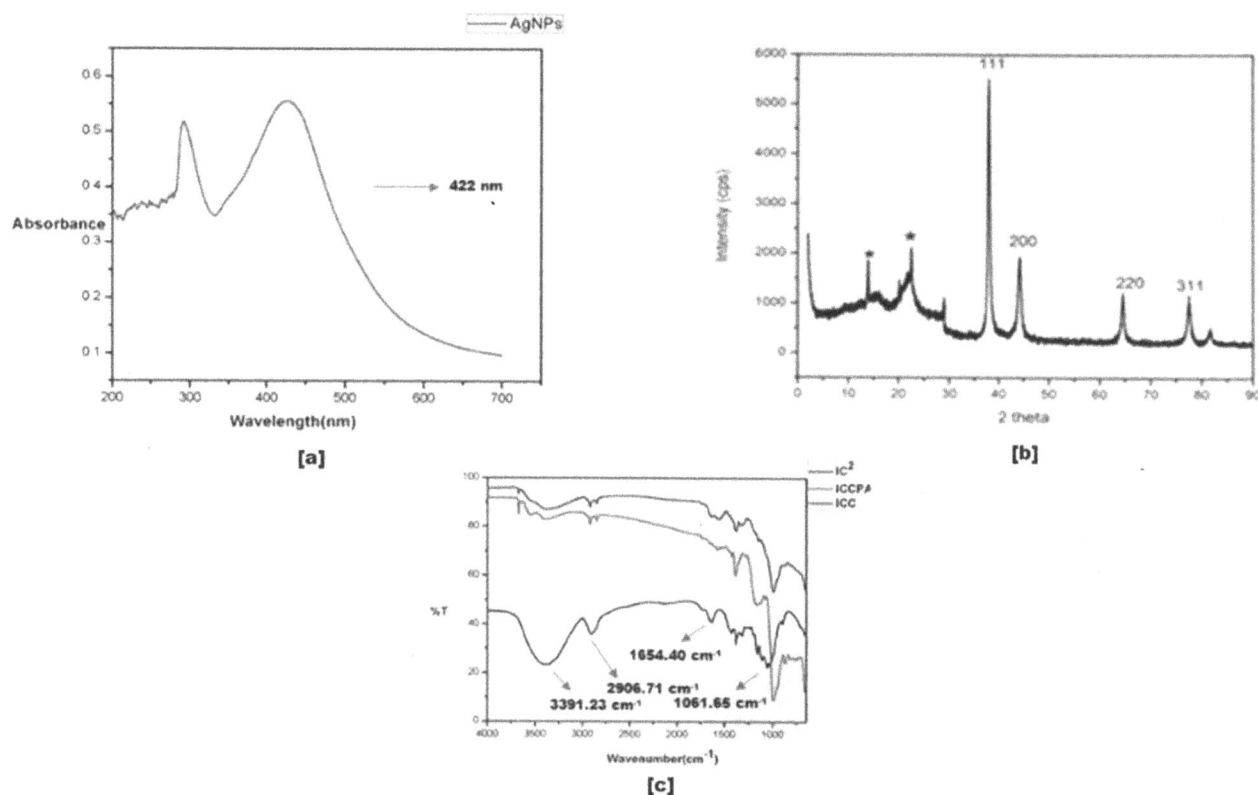

Fig. 35.1 UV-Vis spectrum of Ag NPs outside the matrix [a], XRD pattern of Ag/C/C bionanocomponsites film and FTIR spectra of cellulose (IC^2), cellulose-chitosan (ICC) and Ag/C/C (ICCPA) film

(1 1 0) and (2 0 0) of cellulose-II (Maruthai et al., 2022 & Gulati et al., 2022). The peaks observed at $2\theta =28°$, $32.2°$, $38.5°$, $46.5°$, $64.4°$ and $77.4°$ of which $38.5°$, $64.4°$ and $77.4°$ were attributed to (111), (220) and (311) crystal planes of FCC structure of silver (Shen et al., 2021). This further confirms the *in-situ* generation of AgNPs in the matrix (Fig. 35.1. [b]).

The chemical composition of the matrix and nanocomposite films was studied by FTIR analysis. The broad band at 3320 cm^{-1} was attributed to the OH groups of cellulose and the other main bands at 2881 cm^{-1} and 1010 cm^{-1} arose due to the stretching vibrations of CH and C-O-H groups of cellulose. The other prominent peak at 1636 cm^{-1}was attributed to the crystallization of water. The presence of common groups both in the matrix and the nanocomposite films under study indicate no chemical interactions between them except possible physical interactions such as electrostatic forces (Fig. 35.1. [c]).

3.2 SEM and TEM

The TEM and SEM images of the particles formed outside the matrix and on the matrix are shown in (Fig. 35.2. [d] and [e]). From (Fig. 35.2. [e]), it is evident that the Ag

NPs formed were spherical in shape. The particle size distribution of the Ag NPs formed on the film is showed in Fig. 35.2. The average diameter of the nanoparticles produced on the film was 69 nm. EDX pattern confirms elemental composition of the films i. e. Ag, C and O.

3.3 Antibacterial Activity

Ag/C/C behaved as a broad-spectrum antibiotics film. Table 35.1. showed the inhibition diameter of tested bacteria in the presence of ZnO NPs.

The zone of inhibition (mm) against *S. aureus* measured 0, 10, 11, 12 and 13 mm; against *E. coli* measured 0, 10, 12, 13 and 13 mm for cellulose/chitoisan film; against *S. aureus* measured 10, 13, 16, 18 and 19 and against *E. coli* measured 9, 11, 13, 14 and 15 mm for concentrations at 20, 40, 60, 80 and 100 µg/ml, respectively. Based on the results, it can be summarized that Ag/C/C more efficiently inhibited the growth of gram-positive bacteria and gram-negative bacteria than the cellulose/chitosan film.

3.4 Antioxidant Activity

DPPH scavenging assay is utilized to examine the antioxidant activity of the Ag/C/C films. Characteristics

[d]

[e]

[f]

Particle size (nm) [g]

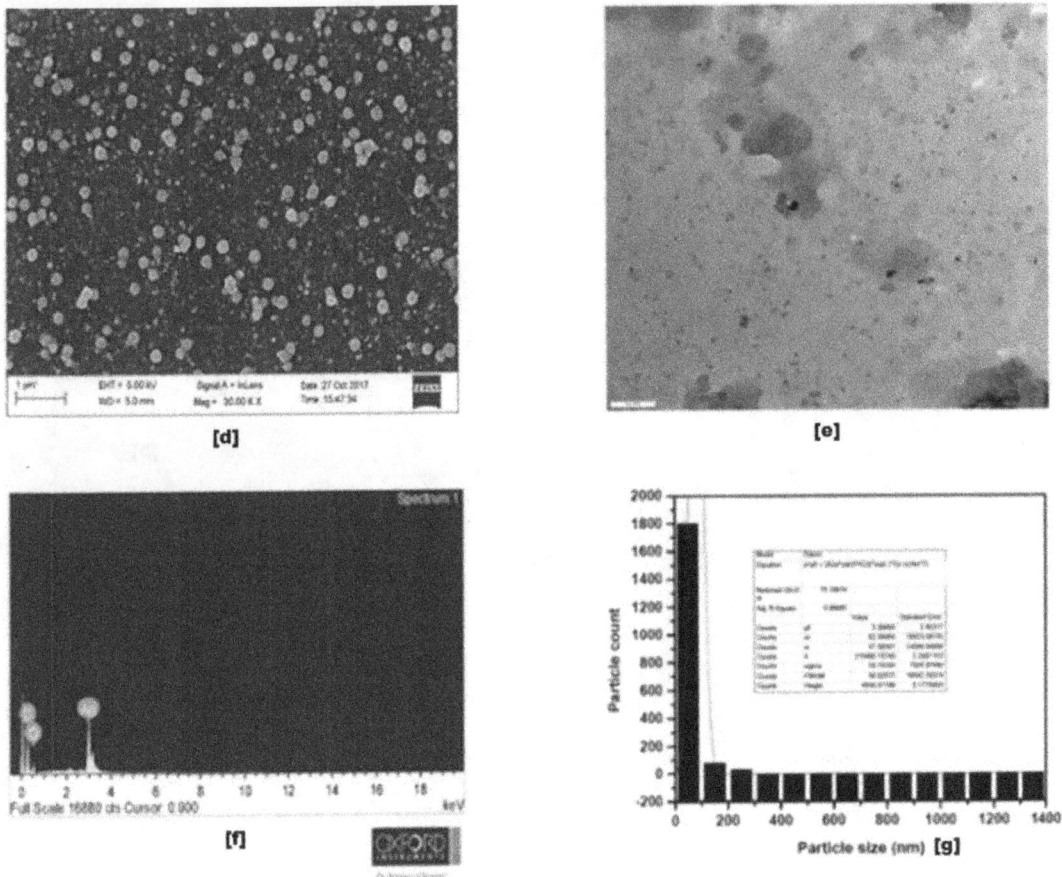

Fig. 35.2 SEM images of Ag/C/C film [d], TEM images of Ag NPs [e], EDX pattern of Ag/C/C film [f] and particle distribution of Ag NPs on the matrix [g]

Table 35.1 Diameter of inhibition zone of the bacteria in the presence of Ag/C/C film

S No.	Test Organisms	Zone of Inhibition (in mm) for different concentrations of Ag NPs					Neomycin (Standard) (ZI)[b]
		1	2	3	4	5	
1	SA	0	10	11	12	13	14
2	EC	0	10	12	13	13	16
3	SA	10	13	16	18	19	21
4	EC	9	11	13	14	15	18

maximum absorption peak of DPPH is 517 nm. DPPH reacts with an antioxidant in the presence of a hydrogen donor and becomes paired off and the reduction of DPPH to the DPPHH (Diphenyl picryl hydrazine) takes place. Along with this, the absorbance of the DPPH also decreases, and the coloured solution becomes decolorized (yellow colour).

The IC50 values of the Ag/C/C film was 43.18 µg/ml. Ascorbic acid is used as a standard with IC50 value 15.22 µg/ml (Fig. 35.3.).

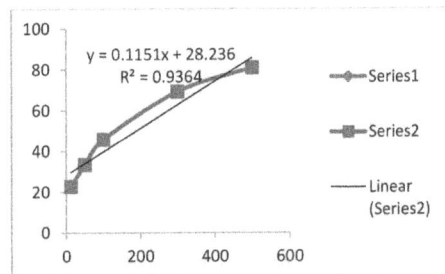

Fig. 35.3 Antioxidant activity of the Ag/C/C bionanocomposites film

4. Conclusion

Ecologically and economically sound approach was employed to synthesize multifunctional cellulose/chitosan/Ag bionanocomposites films. Phytochemical analysis of the plant extract of *Livistona jekinsiana* confirmed the involvement of biomolecules in the synthesis of nanocomposites. SEM and TEM analysis confirmed the well distribution of nanoparticles on the polymer matrix. The average size of the NPs on the matrix is around 69 nm. FTIR analysis confirmed the presence of cellulose and chitosan on the bionanocomposites filmThe film showed a potential antibacterial activity against *Staphylococcus aerus* and *Escherichia coli*. It also possesses notable antioxidant activity in DPPH assay. Therefore, the synthesized bionanocomposite film is a potential applicant in biomedical and water remediation field.

Acknowledgement

The authors like to acknowledge Director of CSIR-NEIST to give the permission and facility to carry out the research in CSIR-NEIST and DST-INSPIRE, for the funding. The authors also acknowledge the VC and organizing committee of ICAMST 2024 to give us the opportunity to present our work. At last.

Conflict of Interest

There is no conflict of interest

References

1. Baruah, D., Yadav, R.N.S., Yadav, A. and Das, A.M. (2019). *Alpinia nigra* fruits mediated synthesis of silver nanoparticles and their antimicrobial and photocatalytic activities. J Photochem Photobiol B. 201:111649.
2. Baruah, R., Goswami, M., Das, A.M., Nath, D. and Talukdar, K. (2023) Multifunctional ZnO Bionanocomposites in the Treatment of Polluted Water and Controlling of Multi-drug Resistant Bacteria. J. Mol. Struct. 1283:135251.
3. Baruah, R., Hazarika, M.P., Das, A.M., Sastry, G.N., Nath, D. and Talukdar, K. (2024) Green Synthesis of Nanocellulose Supported Cu-Bionanocomposites and Their Profound Applicability in the Synthesis of Amide Derivatives and Controlling of Food-Borne Pathogens. Carbohydr. Polym. 330:121786.
4. Baruah, R., Yadav, A. and Das A.M. (2022) Evaluation of the multifunctional activity of silver bionanocomposites in environmental remediation and inhibition of the growth of multidrug-resistant pathogens†. New J chem. 46:10128-10153.
5. Baruah, R., Yadav, A. and Das A.M. (2021) *Livistona jekinsiana* fabricated ZnO nanoparticles and their detrimental effect towards anthropogenic organic pollutants and human pathogenic bacteria. Spectrochim Acta A Mol Biomol Spectrosc. 251:119459.
6. Gebre, S.H. (2023) Recent developments of supported and magnetic nanocatalysts for organic transformations: an up-to-date review. Appl. Nanosci. 13:15–63.
7. Goswami, M., Baruah, D. and Das A.M. (2018) Green synthesis of silver nanoparticles supported on cellulose and their catalytic application in the scavenging of organic dyes. New J chem. 42:10868–10878.
8. Goswami, M. and Das, A.M. (2018) Synthesis of cellulose impregnated copper nanoparticles as an efficient heterogeneous catalyst for C-N coupling reactions under mild conditions. Carbohydr. Polym. 195: 189–198.
9. Gulati, U., Rajesh, U.C., Rawat, D.S. and Zaleski, J.M. (2022) Development of Magnesium Oxide-Silver Hybrid Nanocatalysts for Synergistic Carbon Dioxide Activation to Afford Esters and Heterocycles at Ambient Pressure. Green Chem. 22(10):3170–3177.
10. Maruthai, J., Ramachandran, K., Muthukumarasamy, A., Chidambaram, S., Gaidi, M. and Daoudi, K. (2022) Bio fabrication of 2D MgO/Ag nanocomposite for effective environmental utilization in antibacterial, anti-oxidant and catalytic applications. Surf. Interfaces. 30:101921.
11. Nariya, P., Das, M., Shukla, F. and Thakore, S. (2020) Synthesis of magnetic silver cyclodextrin nanocomposite as catalyst for reduction of nitro aromatics and organic dyes. J. Mol. Liq. 300:112279.
12. Sharifi-Rad, M., Elshafie, H.S. and Pohl, P. (2024) Green synthesis of silver nanoparticles (AgNPs) by Lallemantia royleana leaf Extract: Their Bio-Pharmaceutical and catalytic properties. J. Photochem. Photobiol. A. 448:115318.
13. Shen, A., Luo, R., Liaoa, X., Hea, C. and Li, Y. (2021) Highly dispersed silver nanoparticles confined in a nitrogen-containing covalent organic framework for 4-Nitrophenol reduction. Mater. Chem. Front. 5:6923–6930.
14. Shkir, M., AlAbdulaal, T.H., Manthrammel, M.A. and Khan, F.S. (2024) Novel MgO and Ag/MgO nanoparticles green-synthesis for antibacterial and photocatalytic applications: A kinetics-mechanism & recyclability. J. Photochem. Photobiol. A. 449:115398.

Note: All the figures and table in this chapter were made by the author.

Advances in Materials Science and Technology – Dr. Srikari Srinivasan et al. (eds)
© 2025 Taylor & Francis Group, London, ISBN 978-1-041-12342-2

36

Simulations of Impacts on AA2219-T87 Plates

Mathiazhagan S.
Liquid Propulsion Systems Centre (LPSC),
Valiamala, Trivandrum, Kerala

Narendra H. Wankhede*
Indian Institute of Space Science and Technology (IIST),
Valiamala, Trivandrum, Kerala

V. Viswanath,
R. Vasudevan, and A.K. Asraff
Liquid Propulsion Systems Centre (LPSC),
Valiamala, Trivandrum, Kerala

ABSTRACT: This research work focuses on understanding the behaviour of Plates under impact loads by cylindrical and spherical projectiles using ANSYS Explicit Dynamics simulations. For this purpose, the effects of initial velocities on the resulting residual velocities of projectiles after impact is analysed for the impacts against AA2219-T87 plates. Additionally, the influences of mass and contact lengths of projectiles on the impact response of plates are examined. Through exhaustive simulations, residual velocities are evaluated for different projectile configurations and impact conditions. This study culminates in the determination of the ballistic limit velocity for both cylindrical and spherical projectiles; ballistic limit velocity is an important parameter in characterizing the impact resistance of metals. These findings provide valuable insights into the impact resistance of materials used for propellant tanks, contributing to the enhancement of design strategies and safety measures in aerospace applications.

KEYWORDS: Ballistic limit velocity, Explicit dynamics, Impact

1. Introduction

Impact simulations are computational methods used to predict the behaviour of structures when subjected to various types of impacts, such as collisions, crashes, or blasts. These simulations are employed across a wide range of industries, including automotive, aerospace, defence, and civil engineering, to assess the performance, safety, and structural integrity of designs before physical prototypes are built.

Studying impact phenomena holds immense importance across various fields, ranging from civil aviation to spacecraft design and survivability analysis. The motivation to delve into impact studies stems primarily from the need to understand and mitigate the vulnerabilities of critical systems such as aircraft, vehicles, and spacecraft to a wide array of threats, including projectile impacts and space debris. However, traditional methods of conducting impact experiments have their limitations. They can be costly, time-consuming, and inherently stochastic, making it

*Corresponding author: narendraw262002@gmail.com

DOI: 10.1201/9781003664277-36

challenging to replicate results consistently or apply them comprehensively to vulnerability analyses. Moreover, the complexity of measurements increases with the velocity of the impacting projectile, further adding to the experimental challenges.

To address these limitations, researchers have turned to computational methods such as numerical analysis and computer simulation. These approaches leverage the increasing computational power and advanced finite element (FE) codes to model ballistic impact events accurately. These computational simulations enable the determination of residual velocities of fragments and ballistic limits, providing valuable insights into the potential damage caused by impact events.

In the present research work, we aim to study the effect of impacts of projectiles on Plates made of AA2219-T87 material. The AA2219-T87 material has many applications in various industries due to its very special properties. But Impact study on AA2219-T87 as a material is not carried out extensively till date. Therefore, the following works are carried out and their results are discussed in this article. Initially the impact of Spherical and Cylindrical Projectile on Plate structure made up of AA2219-T87 is studied. Subsequently, the effect of varying Spherical and cylindrical Projectiles' mass on Plate structure is studied. Further, the effect of increasing contact area during impact on plate structure is studied. Finally, ballistic limit velocity of all the projectile with different masses is found.

2. Materials and Methods

Impact simulations are carried out on Plates structures made of AA2219-T87 materials. The 2000-series Aluminum alloys are highly valued in aerospace for their strength-to-weight ratio and toughness. Among them, AA2219-T87 stands out as a key material in the construction of liquid cryogenic rockets, such as NASA's SLS rocket and historic vehicles like the Saturn V and Space Shuttle. Its popularity is owed to several factors: excellent weldability, impressive strength-to-weight ratio, and exceptional resistance to oxidation and corrosion. The T87 designation signifies a specific treatment process: the alloy undergoes solution heat treatment at 535°C, followed by a 7% strain hardening process, and finally aging for 18 hours at temperatures ranging between 170°C and 180°C. This treatment sequence enhances the alloy's mechanical properties, making it ideal for the demanding conditions of rocketry [1]. The T87 temper condition of AA2219 ensures that the alloy possesses the necessary strength and durability to withstand the extreme conditions and pressures encountered during launch and spaceflight [6].

2.1 Explicit Dynamics Method

Ansys Explicit Dynamics simulation tool available in ANSYS Workbench is used to perform all the impact Simulations. In these simulations, explicit time integration method is used and is capable of simulating scenarios involving rapid changes, such as shock wave propagation, significant deformations, nonlinear material behaviour, and intricate contact interactions. It is commonly used in simulations like drop tests, impact studies, and penetration analyses due to its accuracy and efficiency in capturing dynamic behaviour [2].

This simulation method is capable of accurately predicting responses to severe loads, such as material deformations and failures, and interactions between components. It uses advanced mathematical algorithms to handle complex problems, especially those with high strain rates where implicit methods may struggle. While implicit dynamics is suited for scenarios with mild nonlinearity and large time increments, Ansys Explicit Dynamics is ideal for highly complex, nonlinear scenarios involving rapid changes.

2.2 Material Model

Materials have complex response under dynamic loading. Material models define how materials respond to different types of loading, such as compression, tension, shear, and thermal effects. They provide constitutive equations that describe the relationship between stress and strain under various loading conditions. Accurate material models are required to simulate the behaviour of materials under such extreme conditions, accounting for factors like strain rate effects and material non-linearity.

Material models can incorporate criteria for predicting failure and damage initiation in materials. By specifying appropriate failure criteria and damage models, the structural integrity of components and predict failure modes under dynamic loading can be assessed. There are various material models available in Ansys Explicit Dynamics but in this study, we have used two material models.

Johnson-Cook Strength Model

This model describes the strength behaviour of materials, particularly metals, when they experience significant strains, high strain rates, and elevated temperatures, such as those encountered during high-velocity impacts. This model is given by following equation:

$$Y = [A + B\epsilon_p^n][1 + C\ln\epsilon_p^*](1 - T_H^m)$$

In this model, the yield stress (Y) of the material varies based on three key factors: effective plastic strain (ϵ_p), normalized effective plastic strain rate (ϵ_p^*), and homologous temperature (T_H). Homologous temperature is a measure of the material's temperature relative to its melting point [2]. The model comprises five material constants (A, B, C, n, and m), which influence the yield stress. 'A' represents the basic yield stress at low strains, while 'B' and 'n' account for strain hardening effects [2].

Effective Plastic Strain Failure Model

In the plastic strain failure model, ductile failure in materials is simulated based on the accumulation of plastic strain. The user specifies a maximum allowable plastic strain value. When the effective plastic strain in the material exceeds this user-defined maximum, failure initiation is triggered. The element which exceeded the defined effective plastic strain value is removed from the analysis.

2.3 Impact Simulation Setup

Impact on target plates was performed with two different shaped projectiles: spherical and cylindrical projectiles. Simulations are performed by artificially varying masses of projectiles in the range of 0.1kg to 500kg. This analysis was performed in order to check how varying mass and shape can affect the geometry. Then impact analysis with increased contact length i.e. increased surface area was done for cylindrical projectile, in order to study how changing the radius of cylinder can affect the impact or penetration. In this case, the radius of cylindrical projectile was increased but mass and volume are kept constant. Finally, the Ballistic limit velocity was found for both cylindrical and spherical projectile with varying masses. In all simulations material used for target is AA2219-T87 and for projectile is Steel. Moreover, in order to simulate actual impact of foreign objects falling due to gravity during regular operation of propellant tanks, three different impact velocities for projectiles are considered, which are 6264mm/s, 8858mm/s, 14007mm/s.

In the simulation of the perforation process, computational efficiency is prioritized by assuming axisymmetric geometry and loading conditions. This means that the setup is symmetrical around an axis, allowing for simpler calculations and faster processing. The projectile is treated as rigid, i.e. it does not deform during impact. On the other hand, the plate is modelled as deformable, i.e. it can undergo changes in shape and deformation in response to the impact. This approach strikes a balance between accuracy and computational cost, enabling the simulation to effectively capture the perforation process while minimizing computational time.

In the simulations, the finite element method is used to model the perforation process. To account for the material behaviour during impact, the Johnson–Cook thermo-visco-plastic constitutive equation and the Plastic Strain Failure

model are employed. These models consider factors such as yield stress, strain hardening, strain rate effects, thermal softening, and material failure.

Any contact between the projectile and the plate is modelled using a surface-to-surface contact approach based on the penalty method. In this method, contact forces are calculated based on the penetration of one surface into another. Importantly, friction effects between the surfaces are neglected as the friction coefficients are set to zero in all simulations. This setup allows for an accurate representation of the perforation process while simplifying the computational complexity. The following table shows the properties of AA2219-T87 used for target.

Table 36.1 Material model properties of AA2219-T87[5]

	AA2219-T87
Density(g/cm^3)	2.84
Young's Modulus (GPa)	73.1
Poisson's Ratio	0.33
Specific Heat(J/kg.C)	864
A(MPa)	390
B(MPa)	300
n	0.11
C	0.72
m	0.04
$T_m(^0C)$	550
Maximum Equivalent Plastic Strain (EPS)	0.15

As mentioned earlier three different cases are studied, they are as follows: -

Case 1: Impact on Plate due to Spherical Projectile (**Fig. 36.1**)

Case 2: Impact on Plate due to Cylindrical Projectile (**Fig. 36.2**)

Case 3: Impact by Cylindrical Projectile with increased contact length (**Fig. 36.3**)

For each case the test parameters for Projectile and Target are listed in the Table 36.2.

3. Results

For all masses (0.1 to 500 kg) and impact velocities (6264mm/s, 8858mm/s, and 14007mm/s), residual velocities obtained are listed below in Table 36.3 and Table 36.4 for Case 1 and 2 respectively.

For three different radius of projectile and two different initial velocities residual velocity is calculated and shown in Table 36.5 for Case 3. Here for convenience velocity

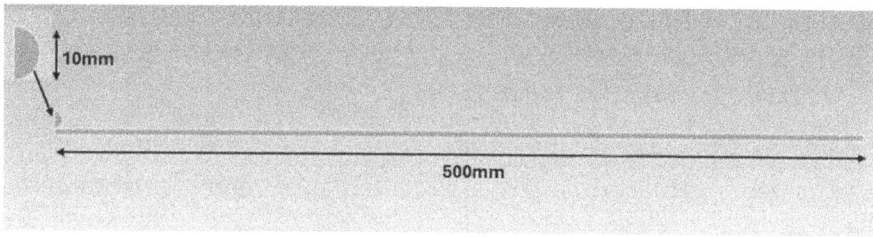

Fig. 36.1 Spherical projectile impact case 1

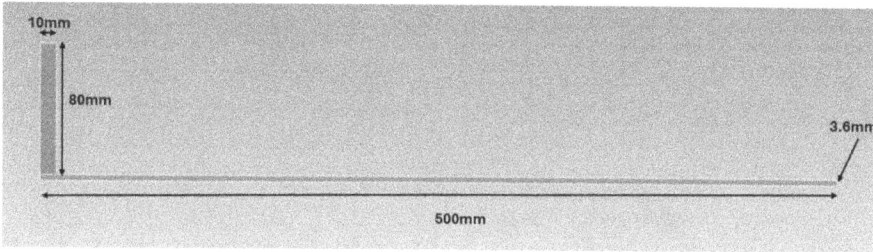

Fig. 36.2 Cylindrical projectile impact Case 2

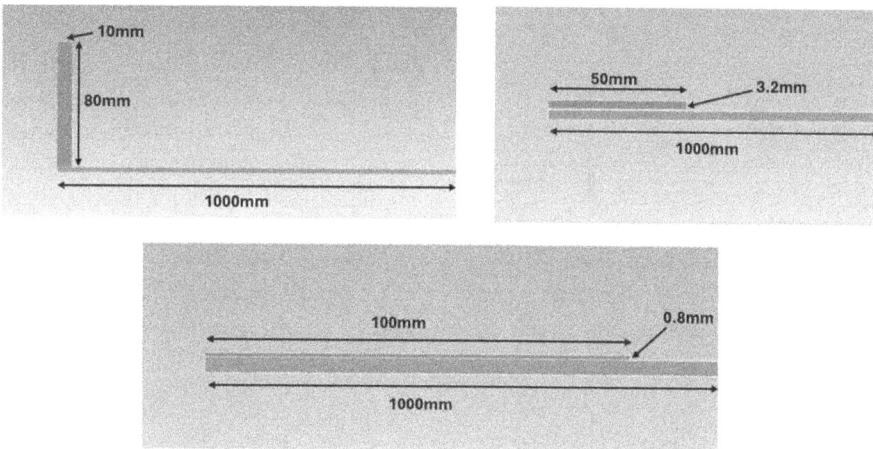

Fig. 36.3 Three projectile with different contact region during impact Case 3

Table 36.2 Test parameters for target and projectile

	Projectile	Target
Case 1	Diameter = 10mm	Diameter = 1000mm
		Thickness = 3.6mm
Case 2	Radius = 10mm	Diameter = 1000mm
	Height = 80mm	Thickness = 3.6mm
Case 3	Projectile 1	
	Radius = 10mm	Radius = 1000mm
	Height = 80mm	Thickness = 3.6mm
	Projectile 2	
	Radius = 50mm	Radius = 1000mm
	Height = 3.2mm	Thickness = 3.6mm
	Projectile 3	
	Radius = 100mm	Radius = 1000mm
	Height = 0.8mm	Thickness = 3.6mm

Table 36.3 Residual velocity(mm/s) for a given mass and initial velocity for spherical projectile

Initial Velocity (mm/s) / Mass of Projectile (kg)	6264	8858	14007
0.1	-369.01	-566.8	-984.08
0.25	-826.4	-1101.2	-1875.6
0.5	-1546	-1682.2	-2860
1	-3281.3	-4312	-4938.8
2	-4282	-5317.6	8380.5
5	1240.4	5773.3	12165
10	2221.8	6161.3	12460
50	5402	8419.1	13701
100	5988.6	8613.5	13857
250	6151	8731.2	13949
500	6204.7	8811.5	13978

Table 36.4 Residual velocity(mm/s) for a given mass and initial velocity for cylindrical projectile

Mass of Projectile (kg) \ Initial Velocity (mm/s)	6264	8858	14007
0.1	-369.3	-661.04	-1051
0.25	-827.61	-1025.9	-1775.4
0.5	-1414.4	-1719.4	-3047.2
1	-3308.8	-4868.1	-5882
2	-4706.5	-6681.1	-10098
5	-5806.6	-7758.7	-2399.8
10	-5091.4	406.13	6503.2
50	4760.7	7071.4	12933
100	5504.7	7964.4	13501
250	5956.5	8508.2	13826
500	6109.8	8730.4	13919

Table 36.5 Residual velocity for particular radius and initial velocity

Projectile Radius(mm) \ Initial Velocity(mm/s)	6264	8858
10	-243.6	-387.93
50	-238.92	-268.22
100	-310.24	-629.66

in negative y direction denotes that the projectile moves away from target and positive sign has assigned for moving towards target.

Ballistic limit velocity, often referred to as BLV, is the maximum velocity at which a projectile can impact a particular material without perforating it. Ballistic limit velocities for both projectiles are estimated for all masses. This is done by performing analysis for a greater number of initial velocities for all cases. BLV for different masses are listed in the Table 36.6.

Table 36.6 Ballistic limit velocity for particular mass and shape of projectile

Mass of Projectile (kg) \ Shape of projectile	Spherical Projectile	Cylindrical Projectile
0.1	45496.3	80841.7
0.25	30893.3	51624.4
0.5	21492.2	46056.3
1	16598.9	31950.5
2	10856.8	23519.1
5	5557.6	15221.1
10	4690.7	8666.3
50	1743.8	4068.2
100	1620.0	2716.7

4. Discussion

4.1 Impact on Plate due to Spherical Projectile

Variation of residual velocity with initial impact velocity of spherical projectile is plotted and shown in Fig. 36.4. It can be seen that residual velocity increases with Initial velocity for higher masses. But for small masses the projectile does not penetrate through target and bounce back which results in change in sign of velocity in graph. For masses above 5kg, residual velocity is increasing in same direction i.e. projectile is penetrated through target. For projectile of mass 2kg some different behaviour is observed, it bounces back for first two initial velocities, and it penetrates for 14007mm/s.

The conservation of energy during the impact simulation is shown in Fig. 36.5. It is observed that reference energy is constant which is sum of Kinetic energy, Internal energy

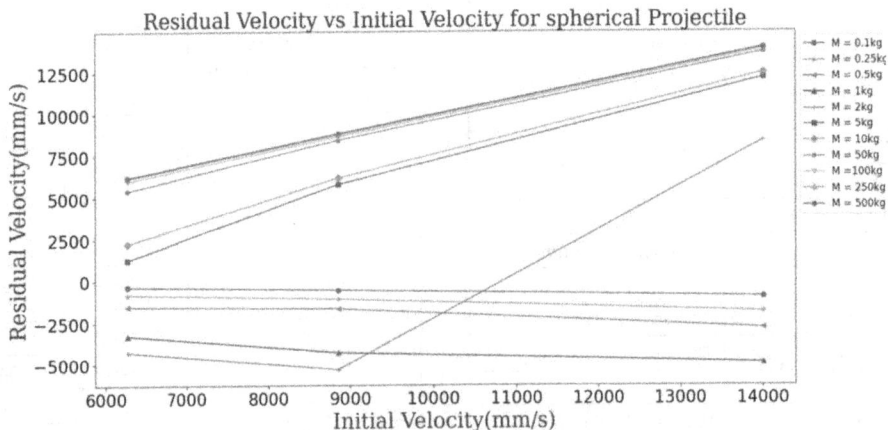

Fig. 36.4 Residual velocity vs initial velocity for spherical projectile

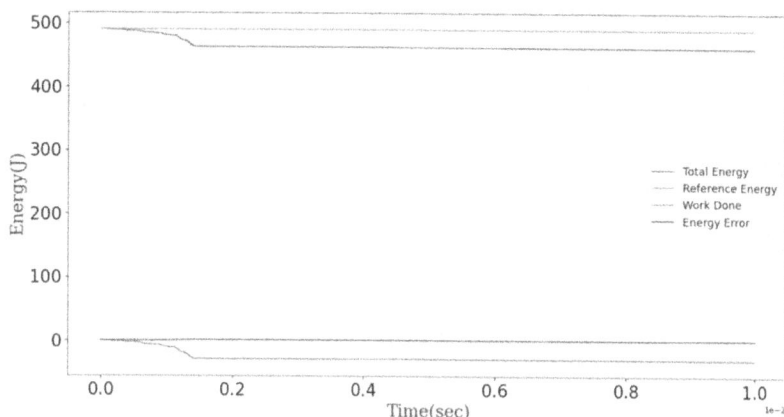

Fig. 36.5 Energy conservation for M = 5kg with initial velocity = 14.007m/s

and hourglass energy. Total energy decreases for some time interval due to impact, as kinetic energy of projectile decreases and elements are getting eroded which results in decrease in energy and remain constant after that. Energy error is zero which tells that there is proper energy conservation taking place during impact.

4.2 Impact on Plate due to Cylindrical Projectile

Variation of residual velocity with initial impact velocity for cylindrical projectile was obtained and shown in Fig. 36.6. It is observed that increasing initial velocities leads to increases in Residual velocities for all the masses. For masses below 5kg projectile do not penetrate the target and bounce back but for masses above 10kg it completely penetrates. Different behaviour is observed for 5kg and 10kg projectile. For 5kg projectile, it does not penetrate through target for all the initial velocities but for initial velocity 14007mm/s its residual velocity decreases as compared to first two initial velocities. For 10kg projectile it does not penetrate for initial velocity 6264mm/s but penetrate for other two cases.

The conservation of energy during the impact simulation is shown in Fig. 36.7. Energy conservation is observed as Energy error is zero, which explains that all energies are balanced.

4.3 Ballistic Limit Velocity

The ballistic limit velocity depends on various factors, including the projectile's shape, size, composition, and the material's properties. Understanding the ballistic limit velocity allows engineers to develop materials and structures that can resist various threats, including bullets, shrapnel, and fragments.

The variation of BLV with different mass of projectile is shown in Fig. 36.8. It can be observed that as mass of the Projectile increases Ballistic Limit Velocity of Projectile decreases. This is due to increase in kinetic energy during impact. For cylindrical projectile, the BLV is higher compared to that BLV of spherical projectile. This could be due to more contact area of cylindrical projectile during impact. On the other hand, small size of spherical projectile could lead to more mass concentration in small volume and high stress concentration and damage. In both cases,

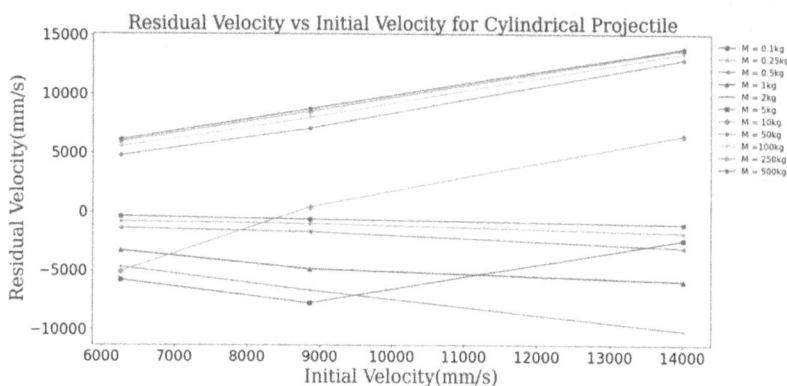

Fig. 36.6 Residual velocity vs initial velocity for cylindrical projectile

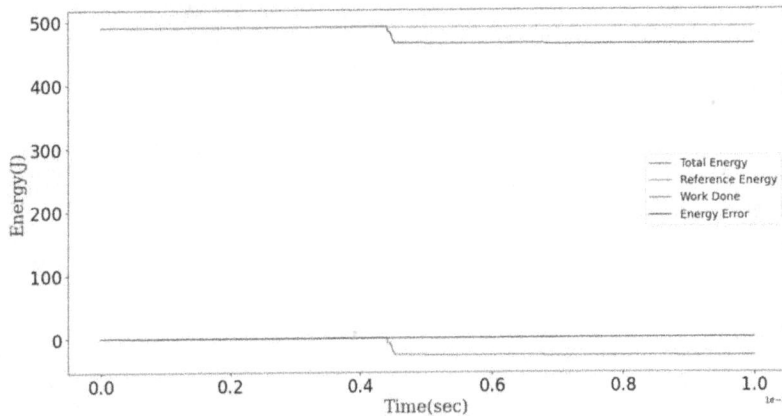

Fig. 36.7 Energy conservation for cylindrical projectile's mass 5kg, initial velocity = 14.007m/s

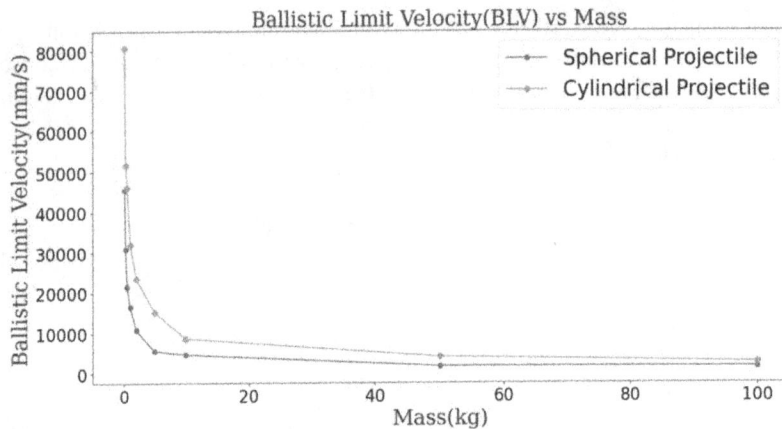

Fig. 36.8 Ballistic limit velocity (BLV) vs mass of projectile

as mass increases BLV decreases as higher mass have high kinetic energy.

5. Conclusion

In this study, we investigated the behaviour of Plate Structure under impact through using ANSYS Explicit Dynamics simulations. We focused on analysing the effects of varying projectile masses and initial velocities, as well as the influence of increased contact length on the structural response. Additionally, we determined the ballistic limit velocity for both cylindrical and spherical projectiles. The main findings from the study are:

- Varying projectile masses and initial velocities significantly affect the resulting residual velocities upon impact. As the initial velocity increases, residual velocity also increases. But this trend was not followed for some masses.
- Increasing the contact length has a noticeable effect on the structural response, highlighting the importance

of the extent of contact between the projectile and the plate structure.

- Energy conservation principles hold true in impact simulations of plates, providing in-sights into energy transfer and dissipation during impact events.
- The determination of ballistic limit velocities establishes critical thresholds for structural integrity under ballistic impacts. BLV is higher for cylindrical projectile compared to spherical projectile. As mass increases ballistic limit velocity decreases.

These findings from this study have significant implications for the design and safety of structure in aerospace applications. By understanding the factors influencing impact behaviour and ballistic limit velocities, engineers can develop more robust and resilient structures, enhancing the overall performance and survivability of aerospace vehicles. Future research could focus on exploring modelling propellant tank with advanced modelling techniques, material properties, and structural configurations to further refine the understanding of impact dynamics and optimize tank design for enhanced safety and reliability.

References

1. Kathryn V Anderson. Characterization of the evolution of 2219-T87 aluminum as a function of the self-reacting friction stir welding process. The University of Alabama, 2019.
2. ANSYS, Inc. ANSYS Explicit Dynamics User's Guide. ANSYS, Inc. 2022. URL: https://www.ansys.com/products/platform/ansys-explicitdynamics.
3. Bareggi, A., Finite Element Analysis of High-Speed Impact on Aluminium Plate. In Proceedings of the 7th International Conference on Mechanics and Materials in Design Albufeira/Portugal, (2017) pp. 1025–1030.
4. T Børvik et al. "Ballistic penetration of steel plates". In: International journal of impact engineering 22.9-10 (1999), pp. 855–886.
5. Andrew Gilmore and Xun Liu. "Mechanical Behavior and Microstructure Evolution during Hot Torsion Deformation of Aluminum Alloy AA2219". In: Advanced Engineering Materials 24.9 (2022), p. 2200048.
6. P Manikandan et al. "Tensile and fracture properties of aluminium alloy AA2219T87 friction stir weld joints for aerospace applications". In: Metallurgical and Materials Transactions A 52.9 (2021), pp. 3759–3776.

Note: All the figures and tables in this chapter were made by the author.

Advances in Materials Science and Technology – Dr. Srikari Srinivasan et al. (eds)
© 2025 Taylor & Francis Group, London, ISBN 978-1-041-12342-2

37

Evaluation of Antimicrobial Property of Natural C-Dots

Kora Ramya Reddy*, Kavitha Prasad
Department of Oral and Maxillofacial Surgery,
Faculty of Dental Sciences,
M.S Ramaiah University of Applied Sciences,
Bangalore, Karnataka, India

Nagaraju Kottam
Department of Chemistry,
M S Ramaiah Institute of Technology,
Bangalore, Karnataka, India

Smrithi S. P.
Department of Applied Sciences,
School of Advanced Studies, S-Vyasa University,
Bengaluru, Karnataka, India

Prashanth G.
Department of Chemical Engineering,
M S Ramaiah Institute of Technology,
Bengaluru, Karnataka, India

ABSTRACT: The high rates of morbidity and mortality associated with pathogenic bacteria infection, coupled with the heightened costs associated with patient management, make it a significant public health concern. Due to the inappropriate use of these drugs emergence of drug-resistant bacteria has increased, rendering them ineffective. The significant potential of nanocarbon science in biomedicine is causing it to gain new ground. Discrete quasi-spherical carbon nanoparticles < 10 nm are referred to as Carbon dots (C-Dots). Due to their suitability for green synthesis techniques, they have attracted a lot of attention. C-Dots possess chemical and photoelectric properties that make them excellent candidates for antibacterial applications. To examine the efficacy of Natural Carbon nanodots as antimicrobial agents. To identify the source of natural carbon nanodots with the highest antimicrobial property. Locally sourced Neem and Moringa leaves were used to extract C-Dots via hydrothermal treatment at 180^0 C for 8 - 12 hrs. Following filtration, they were kept in a vacuum oven, 60^0 C at 400atm pressure to increase the concentration to 5mg/ml. Characterization test and antimicrobial activity were assessed. The samples were evaluated against two bacterial strains *Staphylococcus aureus and Pseudomonas aurigenosa* for their MIC- Minimum inhibitory concentration by broth dilution method and ZOI- Zone of Inhibition by agar well diffusion. A concentration of 5 mg of each sample showed a good antimicrobial activity against both organisms. Green synthesis of C-Dots has shown good antimicrobial activity, thus paving way for studies towards formulation and its clinical use.

KEYWORDS: Antimicrobial activity, Antimicrobial, C-Dots, Green synthesis, Green precursors, Nanomedicine

*Corresponding author: kora.ramya@gmail.com

DOI: 10.1201/9781003664277-37

1. Introduction

Antimicrobial resistant pathogens or infections are becoming a major threat that affects public health.1) One of the leading causes for this is the extensive and disproportionate use of antibiotics. To combat this infection high-potency antibiotics and combination of antibiotics are preffered which further have undesired effect (Ghirardello et al.,2021; Yeh et al.,2020).

Nanocarbon science is gaining new ground owing to its substantial scope in biomedicine. From the first generation nanocarbon to the third generation, various researchers have paved way to some breakthrough applications. Among these various nano-structured carbon forms, a class of discrete quasi-spherical carbon nanoparticles with size less than 10 nm called Carbon dots (C-Dots).3,4 These C-Dots have generated considerable attention as they lend themselves to green synthesis methods. The abundant availability of raw 'green' precursors makes them environmentally benign, inexpensive and ultimately nanomaterials of the current decade (Kottam.,2021). C-Dots possess chemical and photoelectric properties that make them excellent candidates for antibacterial applications. The also have great chemical stability, high water solubility, low toxicity and excellent biocompatibility owing to the green synthesis.The ability to target pathogens at the sight of infection with C-Dots has the advantage of not using antibiotic at all. Thus, ruling out the undesired effects of systemic antibiotics.

Green synthesis of C-dots is an emerging entity which has not been explored as an antimicrobial agent for oral infections, making this a unique study with promising results. The rationale behind this study is to evaluate the antimicrobial property of naturally obtained C-dots which would pave way to targeted use as antimicrobial agents in local drug delivery to avoid the inappropriate usage of antibiotics. Hence the aim of this study to examine the efficacy of Natural Carbon nanodots as antimicrobial agents. To identify the source of natural carbon nanodots with the highest antimicrobial property.

2. Methodology

2.1 Source of Sample

Neem and Moringa leaves were locally sourced and visual inspection done to check for any damage of infestation

2.2 Preparation of the Sample

The Leaves were then washed using distilled water and made into a paste in a blender along with double distilled water. These samples were then subjected to hydrothermal treatment. This thermal-mediated approach requires pressurized autoclave vessels, reaction temperatures of 180^0 C for 8 - 12 hours. After the hydrothermal treatment was done, the sample was subjected to filtration and the C-dots were extracted from the sample. The obtained C-dots were further kept in a vacuum oven at 60 ^{0}C at 400 atm pressure to achieve a concentration of 5mg/ml.

2.3 Characterization Tests

The sample were then subjected to characterization techniques such as UV-Visible Absorption Spectroscopy, Photoluminescence spectroscopy, Fourier's Transform Infra-red spectroscopy and Transmission Electron Microscopy.

2.4 Antimicrobial Activity

The antimicrobial activity of the obtained C-dots was tested via MIC – Minimum Inhibitory concentration by broth dilution method and ZOI- Zone of Inhibition by agar well diffusion for the organism.

Minimum inhibitory concentration - The samples were evaluated for their MIC against two bacterial strains *Staphylococcus aureus and Pseudomonas aurigenosa* by broth dilution method. The test samples were taken in different concentrations 5mg, 1mg, 500ug 500 µg/mL, 250ug and 125ug/ml UV spectrometer with 600 nm wavelength was used to study the antimicrobial activity of the sample along with Ampicillin (125ug/ml) as positive control and DMSO served as negative control. The OD was then measured at 600 nm at 0 h, incubated overnight at room temperature, and measured again after 24 h. For control, 500µL of water was used in place of sample and taken as 100% growth. Then the percentage inhibition in microbial growth was calculated in comparison with control using the formula:

Where OD_i is difference between initial and final OD of PR, OD_f is difference between OD of control.

$$Percentage\ inhibition = 100 - \left[\frac{(ODi\ x100)}{ODf} \right]$$

2.5 Zone of Inhibition

The agar well diffusion method was employed to determine the antibacterial activity of samples against the selected bacterial strain *Staphylococcus aureus and Pseudomonas aurigenosa*. A subculture of bacterial strains at a volume of 200 µL, equivalent to 10^6 CFU/mL, was uniformly spread onto the surface of a petri dish containing 20ml of nutrient agar, using a sterile cotton swab and wells were punched using a sterile gel borer. On the agar, five wells with a diameter of 8mm each were created for the bacterial strains.

The first well was designated as the negative control and was loaded with 100 μL of DMSO, using which sample was dissolved in the concentrations of 5mg/mL, 1 mg/mL and 0.5 mg/mL. While second well served as the positive control and contained 100 μL of Ampicillin (an antibiotic). Rest of the wells contained 100 μL of test drug in above mentioned concentrations. Subsequently, the plates were incubated at 37°C for a duration of 24 hours for bacterial growth, following which the measurement of the zone of inhibition surrounding the wells was performed after the incubation period. The concentration of positive control used was 0.1 mg/mL.

3. Results

3.1 Characterization

UV-Visible Absorption Spectroscopy: C-dots exhibited a maximum optical absorption peak at 216 nm in Neem and 310 nm in moringa (Fig. 37.1(a) and Fig. 37.1(b)) which is in good agreement with the previous reports (5) indicating the occurrence of π-π* transition of -C=C- and n-π* transition of -C=O-.

Photoluminescence Spectroscopy- To investigate the optical properties of green synthesized C-dots, photoluminescence spectra were obtained at an excitation wavelength of 320 nm to get an emission peak centred at 400 nm. The C-dots derived from both the sources exhibited same pattern of PL spectra with the signature 'excitation-dependent emission' behaviour of C-dots as reported in the literature.

3.2 Fourier's Transform Infra-Red Spectroscopy

Vibrational spectra obtained for C- dots exhibits a strong band at 3500–3100 cm−1 due to stretching vibration of the functional group –OH/-NH. Another peak is noticed at 2400- 1500 cm−1 due to the functional group -C= (Fig. 37.3(a), Fig. 37.3(b)). This finding is in good agreement with previous studies (Sailaja et al.,2020).

3.3 Transmission Electron Microscopy

TEM images were obtained for carbon nanodots synthesized via hydrothermal technique.

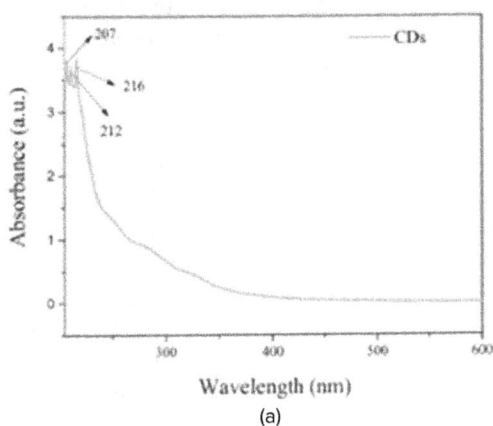

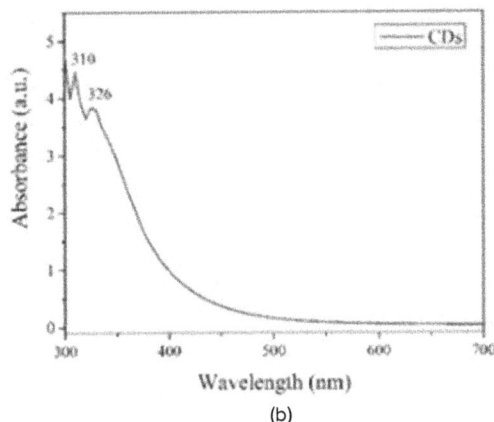

Fig. 37.1 (a) UV-Vis absorption spectrum of C-dots derived from Neem, (b) UV-Vis absorption spectrum of C-dots derived from Moringa

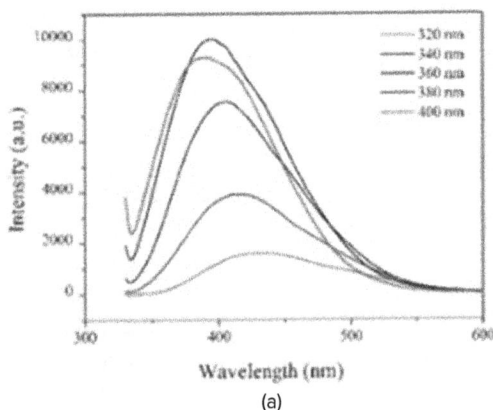

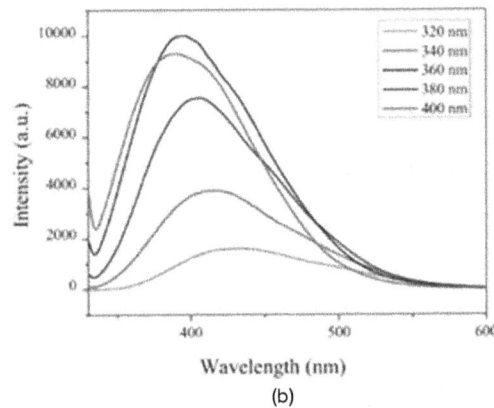

Fig. 37.2 (a) PL spectra of C-dots derived from Neem, (b) PL spectra of C-dots derived from Moringa

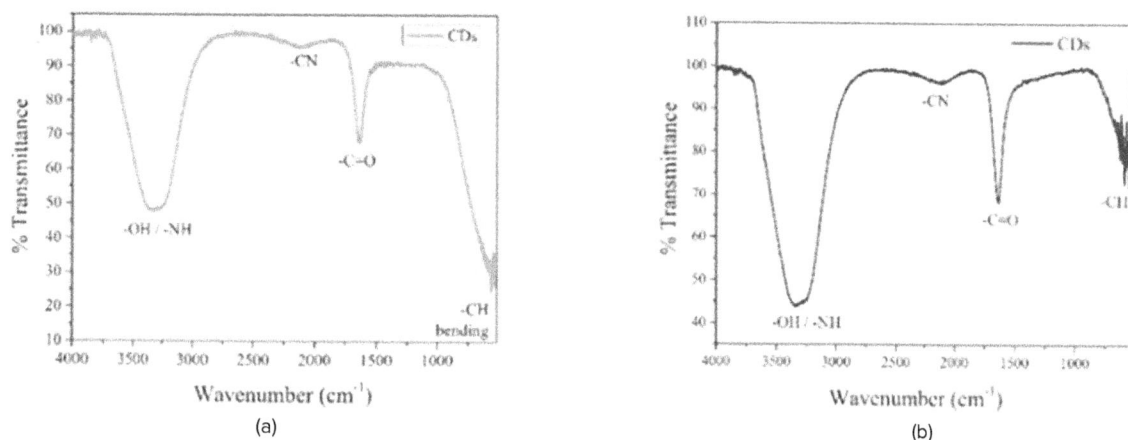

Fig. 37.3 (a) FTIR spectrum of C-dots derived from Neem, (b) FTIR spectrum of C-dots derived from Moringa

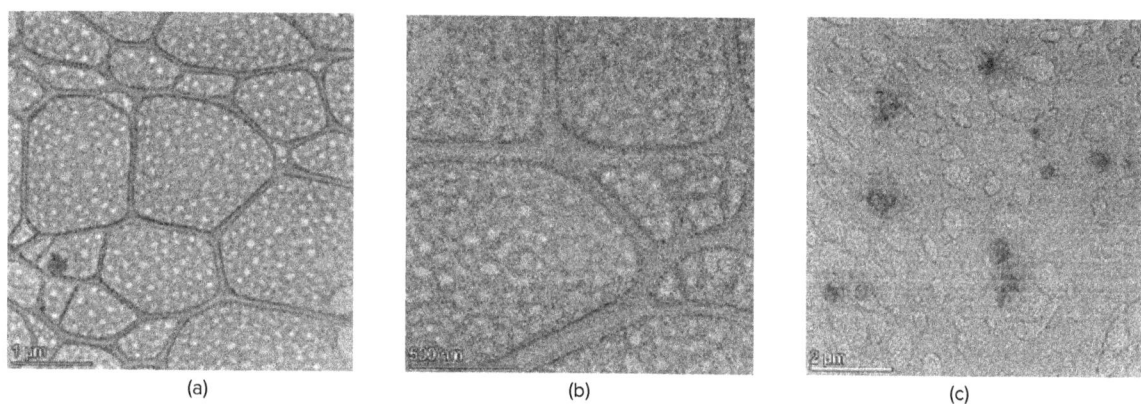

Fig. 37.4 (a), (b) and (c)). TEM images of C-dots at different resolutions. The images confirm the formation of spherical nanodots with slight agglomeration.

Table 37.1 OD value at 600nm and percentage of inhibition (N=1)

Concentration	Staphylococcus aureus		Pseudomonas aurigenosa	
	OD_i @ 600 nm	% inhibition	OD_i @ 600 nm	% inhibition
Amp 125ug/ml	0.103	90.40%	0.108	89.99%
N-5mg	0.192	82.09%	0.197	81.73%
N-1mg	0.394	63.25%	0.398	63.08%
N-500ug	0.496	53.74%	0.506	53.07%
N-250ug	0.598	44.22%	0.618	42.88%
N-125ug	0.675	37.04%	0.741	31.27%
Control	1.072	-	1.078	-
Amp 125ug/ml	0.098	90.84%	0.112	89.64%
M-5mg	0.212	80.17%	0.201	81.41%
M-1mg	0.354	66.89%	0.403	62.72%
M-500ug	0.452	57.72%	0.510	52.83%
M-250ug	0.581	45.66%	0.625	42.19%
M-125ug	0.652	39.01%	0.775	28.31%

Table 37.2 OD value at 600nm and percentage of inhibition (N=2)

Concentration	Staphylococcus aureus		Pseudomonas aurigenosa	
	OD_i @ 600 nm	% inhibition	OD_i @ 600 nm	% inhibition
Amp 125ug/ml	0.102	90.48%	0.109	89.88%
N-5mg	0.193	81.98%	0.193	82.08%
N-1mg	0.395	63.12%	0.392	63.61%
N-500ug	0.494	53.88%	0.508	52.84%
N-250ug	0.592	44.73%	0.620	42.44%
N-125ug	0.671	37.35%	0.738	31.48%
Control	1.071	-	1.077	-
Amp 125ug/ml	0.099	90.74%	0.113	89.54%
M-5mg	0.210	80.34%	0.198	81.67%
M-1mg	0.355	66.77%	0.414	61.67%
M-500ug	0.451	57.78%	0.516	52.23%
M-250ug	0.586	45.14%	0.620	42.60%
M-125ug	0.654	38.77%	0.776	28.15%
Control	1.068	-	1.080	-

3.4 Antimicrobial Activity

The antimicrobial activity of samples Neem(N) and Moringa(M) against Staphylococcus aureus and Pseudomonas aeruginosa was evaluated using the MIC method. The samples showed anti-microbial activity. MIC obtained was 1 mg for both Staphylococcus aureus and Pseudomonas aeruginosa with samples N and M.

Table 37.3 Minimum Inhibitory Concentration (MIC) of Neem (N) and Moringa (M) extracts against Staphylococcus aureus and Pseudomonas aeruginosa

Sl. No.	Extract	Zone of inhibition measured in mm				Efficiency
		Staphylococcus aureus		Pseudomonas aurigenosa		
		N=1	N=2	N=1	N=2	
1	N-5mg	16 mm	17 mm	14 mm	13.5 mm	++
2	N-1mg	11 mm	12 mm	12 mm	12.5 mm	++
3	N-0.5mg	1 mm	1.5 mm	1 mm	1.5 mm	+
4	Ampicillin (N)	26 mm	27 mm	24 mm	25 mm	+++
5	M-5mg	15 mm	14 mm	15 mm	14.5 mm	++
6	M-1mg	11 mm	10.5 mm	13 mm	12.5 mm	++
7	M-0.5mg	1 mm	1.5 mm	2 mm	1.5 mm	+
8	Ampicillin (M)	24 mm	24.5 mm	22 mm	23 mm	+++

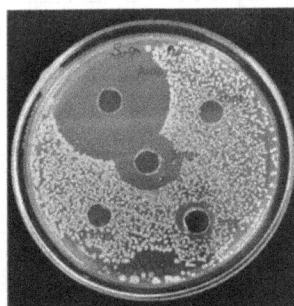
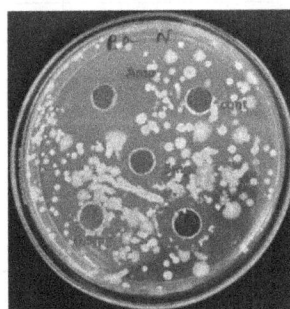

(a)   (b)

Fig. 37.5 (a) Showing inhibition zones of Neem against Staphylococcus Aureus, (b) Showing inhibition zones of Neem against Pseudomonas aurigenosa

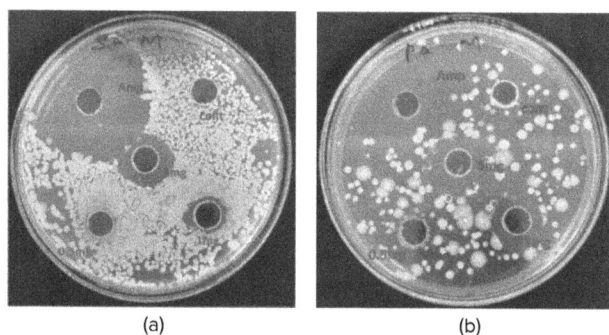

Fig. 37.6 (a) Showing inhibition zones of Moringa against *Staphylococcus aureus*, (b) Showing inhibition zones of Moringa against *Pseudomonas aurigenosa*

The antimicrobial activity of samples (N and M) against Staphylococcus aureus and Pseudomonas aeruginosa was evaluated using the zone of inhibition (ZOI) method. A concentration of 5 mg of each sample showed a good inhibition zone against both organisms, followed by 1 mg. However, 0.5 mg was not as effective.

The plethora of applications anticipated by the international research community is driving an exponential growth in the manufacture of nano-architected carbon. Nanoscale carbon is prized for having a porous structure, inexpensive, widely available, and flexible synthetic engineering. Over the years constant evolution of these carbon nanoparticles has laid a path to various sized nanoparticles. Among these various nano-structured carbon forms, a class of discrete quasi-spherical carbon nano-particles with size less than 10 nm called Carbon dots (C-Dots) (Kottam.,2021; Chahal et al., 2021). As mentioned earlier these particles are generating a lot of attention due to their green synthesis methods.

What makes a synthesis green? According to the Principles of Green Chemistry by Anastas and Warner, (Anastas and Warner.,2000) using non-toxic renewable precursors and solvents in a CD synthesis that is safe to perform, should also be chemically stable and non-toxic. Lower energy synthesis methods should be prioritized although CD synthesis is usually an energy intensive process. Most importantly disposal of the product and also the intermediates during the synthesis should be safe (Chahal et al.,2021).

In the initial stages, researchers concentrated only on carbona materials with mainly hydroxyl, carbonyl and amino group as a precursor to synthesize these luminescent C-dots. This resulted in the end product of C-dots with limited applications due to less aqueous solubility and low quantum yield. Later, green synthesis of C-dots came into picture, In the year 2010 Zao et al reported synthesis of nitrogen doped C-dots from chitosan by hydrothermal carbonization(Zhao et al 2010). In another study (Liu et al.,2012)synthesized C-dots from grass via hydrothermal treatment and synthesized C- dots from pyrolysis of coffee grounds(Hsu et al .,2012). These studies might be seen as the forerunners in the environmentally friendly synthesis of C-dots and thus they encouraged the search for more green sources of the material. The abundance of available green sources and the ease of preparation of C-Dots via green synthesis makes it not only comparatively easy but also efficient. Among the wide range of green sources fruits and vegetables contain all the essential qualities of a precursor that C-dots need and hence a lot of research groups still focus on the vast resources that nature has supplied to create a simpler synthetic pathway. Synthesis of these C-dots are basically in two ways, namely top-down and bottom-up approaches. In top-down approaches, to produce nanoparticles, large-sized carbon materials like carbon nanotubes and graphite are subjected to a variety of treatments, including oxidative cleavage, hydrothermal, solvochemical, microwave, and ultrasonic aided procedures. Furthermore, complex systems like industrial waste, plants, fungi, and bacterial derivatives that lack large polyaromatic structures, can decompose thermally through a series of carbonisation and dehydration events that eventually lead to the formation of the CD core. Conversely, bottom-up strategies use polymers and tiny molecules as carbon precursors to build CDs. The thermal breakdown of the starting components can be accomplished by a variety of techniques, such as chemical oxidation, hydrothermal treatment, reflux in basic or acidic conditions, and ultrasonic or microwave-assisted syntheses. With bottom-up techniques, virtually any organic material susceptible to thermal decomposition can be employed under thermal conditions for the manufacture of CDs, in contrast to top-down approaches that require pre-existing aromatic structures. (Wang et al., 2014)

The selection of precursors and synthetic techniques used in the manufacturing process have a direct impact on the structure of CDs. Therefore, slight variations in the kinds of precursors, solvents, and synthesis methods result in the generation of structurally distinct nanoparticles.

Because of this, it is challenging to forecast any novel CD's possible antibacterial efficacy and specificity without conducting thorough structural characterisation studies (1) In a study (Bing et al.,2016) stated that Carbon dots's (CD) surface charges are essential to their first interactions with bacterial species and, consequently, to their fluorescence labelling. The Positively charged C-Dots react with the negative charged microorganism via electrostatic interactions thus promoting internalization and bacterial apoptosis. There are multiple ways in which CDs can have a bacteriostatic or bactericidal effect on bacteria, which include both physical or mechanical damage to the bacterial membrane, bacterial cell wall breakdown leading to cytoplasmic material leakage, and fragmentation of DNA and proteins (Yang et al., 2016; Xu et al.,2020; Zhang et al.,2018). Moreover, N-doped CDs can be employed as photosensitisers to produce ROS in response to UV or visible light irradiation. This produces hydroxyl radicals ($\cdot$OH), H_2O_2, superoxide anion ($\cdot O_2-$), and singlet oxygen ($1O_2$) in response to reactions with water and dissolved O_2. This causes bacterial oxidative stress leading to apoptosis (Zang et al.,2018; Verma et al., 2019). For this reason, it has been demonstrated that CDs doped with transition metals in their core structure are useful tactics (Sun et al.,2021).

In the present study, we have chosen Neem leaves and Moringa leaves for the synthesis of C-Dots because of their innate antimicrobial property and as also mentioned above green leaves have all the precursors required for an ideal C-dots with structural and chemical properties to exhibit the desired properties. In the initial days a study conducted by Sahu et al was the use of orange juice as a carbon source via hydrothermal method the source was subjected to a heat of 120 °C for 120 minutes. They reported that 400 mg of C-dots can be produced from 400 ml of pulp free orange juice thus paving way for the possibility of large-scale production (Sahu et al.,2012). Another study by Mehta et al.,2014 where they used Sugarcane juice to synthesis C-Dots. Various other raw green resources are used to synthesize C-Dots, such as orange juice, eggshells, neem leaves and many more.

The idea behind using C-Dots as a local and targeted antimicrobial agent was to prevent systemic effects of antibiotics which are used day in and day out to combat infections. Antibiotics is one of the most inappropriately used drugs, thus leading to the rise of drug-resistant infections. To overcome this an alternative approach by using local antimicrobials was sought out. In the present study organism Stappylococus aureus and Pseudomonas aurigenosa was used, these organisms are the most common organism that cause oral infection. This results alingned with another study conducted by M Shahshahanipour et al in which C-Dots synthesized from henna leaves (Lawsonia inermis plant) showed antimicrobial activity againts Gram-positive Staphylococcus aureus and Gram-negative Escherichia coli. (Shahshahanipour et al.,2019) As mentioned in the results both the samples of C-Dots extracted from Neem and Moringa showed good antimicrobial activity against the organisms. A control group with ampicillin, the most used antibiotic revealed that although the drug has a more potent antimicrobial activity

these C-Dots nonetheless showed good antimicrobial activity. Another advantage of the C-Dots in comparison to the systemically used antibiotics is that this is locally acting and thus has a specific targeted site.

Green Synthesis of these C-Dots helps in the complete biotransformation of the drug in the system leaving behind no residue or byproducts which need to be eliminated by the body. This again has an undue effect on the various systems of the body leading to side effects. Not to forget the cost effectiveness and the ease of preparation of these C-Dots which simply gives them an added advantage over traditional antimicrobial therapy. The need for an alternate and efficient antimicrobial agent is more important now than ever before. With the ease of availability of over-the-counter antimicrobial medicine and the availability of unchanneled medical information all over the internet, self-medication and misuse of these drugs have led to rise of drug-resistant bacteria. To Combat these infection higher end antimicrobial therapies are sought out leading to systemic complications which again need medical attention.

4. Conclusion

Globally, the rise of antibiotic resistance poses a serious threat to public health and the economy. The creation of new antimicrobial medications is necessary due to the sluggish pace of antibacterial discoveries. Not only are these expensive and time consuming but also present a threat to the systemic health of the patient. The emergence of the C-Dots via green synthesis has been a game changer in the field of biomedicine. Nanomedicine is proving to be a boon in biomedicine including early cancer detection and as biosensors etc. Despite considerable advances in the green synthesis of C-Dots literature, several challenges need to be addressed to improve their range of application as antimicrobial agent. We conclude that the C-dots can be successfully synthesized via green synthesis and have good antimicrobial property. This warrants for further research and exploration into its potential application as a medicament in clinical setup.

References

1. Anastas, P.T. & Warner, J.C. (2000) Green chemistry: theory and practice [Online]. Oxford: Oxford University Press. Available at: https://academic.oup.com/book/53104 [Accessed 25 July 2024].
2. Bing, W., Sun, H., Yan, Z., Ren, J. & Qu, X. (2016) Programmed Bacteria Death Induced by Carbon Dots with Different Surface Charge, Small, 12(34), pp. 4713–4718.
3. Chahal, S., Macairan, J.R., Yousefi, N., Tufenkji, N. & Naccache, R. (2021) Green synthesis of carbon dots and

their applications, RSC Advances, 11(41), pp. 25354–25363.

4. Ghirardello, M., Ramos-Soriano, J. & Galan, M.C. (2021) 'Carbon Dots as an Emergent Class of Antimicrobial Agents', Nanomaterials, 11(8), p. 1877.

5. Hsu, P.C., Shih, Z.Y., Lee, C.H. & Chang, H.T. (2012) Synthesis and analytical applications of photoluminescent carbon nanodots, Green Chemistry, 14(4), pp. 917–920.

6. Kottam, N. & S, P.S. (2021) Luminescent carbon nanodots: Current prospects on synthesis, properties and sensing applications, Methods and Applications in Fluorescence, 9(1), p. 012001.

7. Liu, S., Tian, J., Wang, L., Zhang, Y., Qin, X. & Luo, Y. et al. (2012) Hydrothermal treatment of grass: a low-cost, green route to nitrogen-doped, carbon-rich, photoluminescent polymer nanodots as an effective fluorescent sensing platform for label-free detection of Cu^{2+} ions, Advanced Materials, 24(15), p. 2037.

8. Mehta, V.N., Jha, S. & Kailasa, S.K. (2014) One-pot green synthesis of carbon dots by using Saccharum officinarum juice for fluorescent imaging of bacteria (Escherichia coli) and yeast (Saccharomyces cerevisiae) cells, Materials Science and Engineering C, 38, pp. 20–27.

9. Sahu, S., Behera, B., Maiti, T.K. & Mohapatra, S. (2012) Simple one-step synthesis of highly luminescent carbon dots from orange juice: application as excellent bio-imaging agents, Chemical Communications, 48(70), pp. 8835–8837.

10. Sailaja Prasannakumaran Nair, S., Kottam, N. & S, G.P.K. (2020) Green Synthesized Luminescent Carbon Nanodots for the Sensing Application of Fe^{3+} Ions, Journal of Fluorescence, 30(2), pp. 357–363.

11. Shahshahanipour, M., Rezaei, B., Ensafi, A.A. & Etemadifar, Z. (2019) An ancient plant for the synthesis of a novel carbon dot and its applications as an antibacterial agent and probe for sensing of an anti-cancer drug, Materials Science and Engineering C, 98, pp. 826–833.

12. Sun, R., Chen, H., Sutrisno, L., Kawazoe, N. & Chen, G. (2021) Nanomaterials and their composite scaffolds for photothermal therapy and tissue engineering applications, Science and Technology of Advanced Materials, 22(1), pp. 404–428.

13. Verma, A., Arshad, F., Ahmad, K., Goswami, U., Samanta, S.K. & Sahoo, A.K. et al. (2019) Role of surface charge in enhancing antibacterial activity of fluorescent carbon dots, Nanotechnology, 31(9), p. 095101.

14. Wang, Y. & Hu, A. (2014) Carbon quantum dots: synthesis, properties and applications, Journal of Materials Chemistry C, 2(34), pp. 6921–6939.

15. Xu, N., Du, J., Yao, Q., Ge, H., Li, H. & Xu, F. et al. (2020) Precise photodynamic therapy: Penetrating the nuclear envelope with photosensitive carbon dots, Carbon, 159, pp. 74–82.

16. Yang, J., Zhang, X., Ma, Y.H., Gao, G., Chen, X. & Jia, H.R. et al. (2016) Carbon Dot-Based Platform for Simultaneous Bacterial Distinguishment and Antibacterial Applications, ACS Applied Materials & Interfaces, 8(47), pp. 32170–32181.

17. Yeh, Y.C., Huang, T.H., Yang, S.C., Chen, C.C. & Fang, J.Y. (2020) Nano-based drug delivery or targeting to eradicate bacteria for infection mitigation: a review of recent advances, Frontiers in Chemistry, 8, p. 286.

18. Zhang, J., Lu, X., Tang, D., Wu, S., Hou, X. & Liu, J. et al. (2018) Phosphorescent Carbon Dots for Highly Efficient Oxygen Photosensitization and as Photo-oxidative Nanozymes, ACS Applied Materials & Interfaces, 10(47), pp. 40808–40814.

19. Zhao, L., Baccile, N., Gross, S., Zhang, Y., Wei, W. & Sun, Y. et al. (2010) Sustainable nitrogen-doped carbonaceous materials from biomass derivatives, Carbon, 48(13), pp. 3778–3787.

Note: All the figures and tables in this chapter were made by the author.

Advances in Materials Science and Technology – Dr. Srikari Srinivasan et al. (eds)
© 2025 Taylor & Francis Group, London, ISBN 978-1-041-12342-2

38

Release Mechanisms of Montelukast Sodium from Transdermal Films Based on Sodium Alginate and Lignosulphonic Acid

Aashli Mary,
S. Giridhar Reddy*, B. Sivakumar
Department of Physical Sciences,
Amrita School of Engineering, Amrita Vishwa Vidyapeetham,
Bengaluru, India

Sanga Kugabalasooriar
Department of Chemistry, Northeastern University,
Boston, MA 02115, United States

ABSTRACT: Montelukast sodium, frequently prescribed for managing allergy and asthma, presents challenges in ensuring patient compliance and treatment effectiveness, primarily due to its oral administration, particularly problematic in susceptible demographics like children and the elderly. To overcome these hurdles, we previously developed transdermal films employing biodegradable polymers, notably sodium alginate and lignosulphonic acid, achieving a controlled release of montelukast sodium lasting up to 36 hours. This research aims to delve deeper into the mechanisms governing drug release and polymer behavior within these transdermal films. By employing diverse kinetic models such as Zero-order, First-order, Korsmeyer-Peppas, Kopcha, Higuchi and Sahlin-Peppas, our objective is to illuminate the underlying processes dictating drug diffusion and polymer dynamics. The insights obtained from this investigation hold significant potential for refining transdermal film formulations, thereby improving therapeutic efficacy and patient compliance in the treatment of allergies and asthma.

KEYWORDS: Asthma, Montelukast sodium, Polymeric blends, Release kinetic studies, Transdermal drug delivery

1. Introduction

Adults and children with seasonal allergic rhinitis and chronic asthma are prescribed the drug montelukast sodium (MLS). It is associated with certain disadvantages, such as hepatic first-pass metabolism, which results in a short half-life of 2.5–5.5 hours and lower bioavailability when using the traditional formulation. In this regard, improving patient compliance and addressing the disadvantage of the traditional formulation will require a sustained release

formulation for MLS. A controlled drug delivery strategy, which releases the drug at a predetermined rate to maintain therapeutic levels and lower dosage frequency, can solve issues like low bioavailability, high dosage frequency, and side effect risk (Hadi & Rao, 2012).

Developing a polymer blend that can encapsulate the drug within its matrix is one method of achieving controlled drug distribution. Natural polymers can be utilized to achieve controlled release, a technique currently being explored

*Corresponding author: s_giri@blr.amrita.edu

DOI: 10.1201/9781003664277-38

in agriculture to deliver pesticides and insecticides more effectively (Pavithran *et al.*, 2024). This technique can also be used to treat diseases like cancer (Hafsa Bahaar *et al.*, 2023), osteoporosis (Mary *et al.*, 2024), Parkinson's disease (Athira K *et al.*, 2024) and more successfully while reducing the side effects of traditional drug formulations. Transdermal delivery, a type of controlled drug delivery system, entails absorbing the drug through the skin to produce both local and systemic effects. Diffusion, degradation/erosion, and swelling are a few fundamental processes that regulate the release of the drug molecules through the polymeric formulation (Laracuente *et al.*, 2020).

In order to predict how a drug will move through the body, maximize its therapeutic efficacy, and create dosage forms that will improve patient compliance, release kinetics is crucial. In our previous research, we developed a transdermal patch made from sodium alginate and lignosulphonic acid. This patch successfully achieved controlled release of montelukast sodium for up to 36 hours (Aashli *et al.*, 2023). Therefore, to clarify the mechanisms of drug release and polymer dynamics within such transdermal films, different kinetic models were used in our current study.

2. Materials and Methods

2.1 Materials Used

Montelukast sodium (MLS), Sodium alginate (SDA), and Lignosulphonic acid (LGA) were sourced from Sigma Aldrich. Distilled water, Barium chloride ($BaCl_2$), Calcium chloride ($CaCl_2$), Hydrochloric acid (HCl) and Acetate Buffer (pH 4.6) were utilized in the experiment.

2.2 Methods

Formulation of SDA/LGA Films Loaded with MLS

A solution of 2g of SDA and LGA was prepared in 50 mL by weighing them in an 80:20 ratio, respectively (S. G. Reddy, 2021). The SDA/LGA were dissolved in MLS solution. The SDA/LGA mixture was stirred for thirty minutes using a magnetic stirrer to ensure complete dissolution. After the polymers were fully dissolved in the solution, the mixture was poured into Petri plates and dried in a hot air oven at 60 °C.

Crosslinking of the Transdermal Films

The formulated polymeric films were put in a cross-linking agent solution for ten minutes. The cross-linking agents used were Barium chloride and Calcium chloride.

2.3 *In-Vitro* Release Study

Using a Shimadzu 2600 UV-Visible Spectrophotometer, MLS solutions ranging from 0.5 to 5 mg/ml were prepared and scanned at a maximum wavelength of 285nm. Concurrently, to evaluate the kinetics of drug release, an egg membrane-covered centrifuge tube submerged in pH 4.6 buffer was checked hourly for absorbance changes using the same spectrophotometer (Ansari *et al.*, 2006).

2.4 Kinetics *of In-Vitro* Drug Release Data

To investigate the kinetics, release data was applied to zero-order, first-order Higuchi, Kopcha and Korsmeyer-Peppas kinetic models. Zero-order drug release ($X= R_o t$, R_o is the constant release for zero order and **t** is the time in hours) is the process of releasing a drug over time at a fixed rate, independent of the amount of drug still in the dosage form (Laracuente *et al.*, 2020). First-order release kinetics (Log $X=$ Log X_o -Rt/2.303, where R is the rate constant for first order, **X** is the concentration, X_o is the initial concentration and **t** is time) explains a drug release rate that is directly proportional to the drug remaining in the formulation. As the concentration of the drug decreases, the release rate also decreases (Ghavami-Lahiji *et al.*, 2021). The Higuchi model ($X_t = R_H \cdot t^{1/2}$ where X_t is released in time t, R_H is the Higuchi kinetic constant and **t** is the time hours), which is mainly applicable to systems where the release rate of the drug is proportional to the square root of time, characterizes drug release as a diffusion process based on Fick's law. Drug release from polymeric systems with the ability to swell is described by the Kopcha model ($X_t = At^{1/2} + Bt$ where, X_t is the amount of drug released in time t, R is the kinetic constant and **t** is the time hours.), which takes into consideration both diffusion and polymer relaxation mechanisms (Behafarid Ghalandari *et al.*, 2015). The Korsmeyer-Peppas model ($Vt/ V_\infty = R.t^d$, Vt is the drug released in time t, V is the overall drug amount, and **d** is the drug release mechanism related to the geometrical shape of the delivery system) is typically employed to examine the release mechanism where multiple release phenomena are present. An indicator of the release mechanism is the value of the release exponent "d". The Sahlin-Peppas model ($V_t/V_o = R_d t^f + R_r t^{2f}$ R_d is the diffusion constant, R_r is the relaxation constant, and **f** is the fickian diffusion exponent) provides a more thorough explanation of drug release by extending the Korsmeyer-Peppas model to include both case-II transport (polymer relaxation) and Fickian diffusion (Dash, 2010; Giridhar Reddy S, 2023).

3. Results and Discussion

3.1 *In-Vitro* Release Study

The study investigated the release of MLS from $CaCl_2$ crosslinked SDA/LGA films (SC), SDA/LGA non-crosslinked films (SN) and $BaCl_2$ cross-linked SDA/LGA films (SB). The MLS release for non-cross-linked SN film was observed to be more rapid and showed 100% within 3 hours. Cross-linked films, on the other hand, displayed controlled release for up to 36 hours. More specifically, SB released 95% of the drug respectively in the same period as SC released 58% of the drug respectively in 36 hours as shown in Table 38.1.

Table 38.1 Percentage of MLS release data obtained from *in-vitro* release study

Time (h)	% MLS release		
	SC	SN	SB
0	0	0	0
1	11.84	40.93	10.32
2	15.24	64.99	18.26
3	19.02	95	25.44
4	23.05	-	31.61
5	24.68	-	38.29
6	26.45	-	43.96
24	53.53	-	90.43
36	58.06	-	95.88

3.2 Kinetics of Release Data

The release kinetic parameters such as C^2 that is the correlation coefficient and kinetic constant (R), were calculated by applying the *in vitro* release data of MLS to various kinetic models. The kinetics data for zero order, first order and Higuchi model, has been shown in Table 38.2. A zero-order equation plot of the data revealed C^2 values ranging from 0.90 to 0.99 in the formulations. However, when the first order equation was used to plot the data, the formulations displayed lower correlation coefficient values than the zero-order plots i.e., 0.77 to 0.89. Therefore, the findings showed that zero-order kinetics govern the MLS's release from all formulations. The data obtained after plotting the *in-vitro* release data of MLS in Higuchi equation resulted in C^2 ranging from 0.96 to 0.99 indicating that diffusion is the predominant release mechanism in these formulations. The R value in many drug release models gives crucial information about

Table 38.2 Kinetic values obtained from zero-order, first-order and Higuchi kinetic model plots for the formulations

Formulation Code	Zero-Order		First Order		Higuchi	
	R	C^2	R	C^2	R	C^2
SC	6.04	0.90	1.35	0.77	10.89	0.99
SN	30.90	0.98	3.09	0.88	52.36	0.96
SB	8.43	0.99	1.62	0.89	14.23	0.96

the rate at which a drug is released from a dosage form. Higher R value indicates faster release rate and vice versa. Based on the R values obtained from zero-order, first-order and Higuchi model of kinetics the SC formulation has the lowest R value when compared to the other formulations, suggesting that the release from this formulation is the slowest. The data obtained after plotting the *in-vitro* release data of MLS in Kopca, Korsmeyer-Peppas and Sahlin-Peppas model resulted in C^2 value as 0.99, 0.99 to 1 and 0.99 respectively as shown in Table 38.3. The values indicate a good fit to these models. In the Kopcha kinetic model, all the formulations show a higher A value indicating diffusion as the governing mechanism of release.

In the Korsmeyer-Peppas kinetic model, Fickian diffusion is represented by the diffusion exponent (d) = 0.50 [12]. The d value of SC corresponds to quasi-Fickian diffusion (0.45>d) implying that the drug release is caused by diffusion majorly, but the release is influenced by certain other factors. The d values of SN, and SB correspond to anomalous diffusion or non-Fickian diffusion (0.45<d<0.89), suggesting that both diffusion and material changes, like swelling, are responsible for the drug release. According to Sahlin-Peppas model, for films f = 0.5 is equivalent to Fickian diffusion, 0.5<f<1 is equivalent to Anomalous transport and f =1 is equivalent to Case II transport. Based on the f value of SN and SB shows Fickian diffusion, and SC indicates the drug release is slower than what is predicted by simple Fickian diffusion. The R_r value for formulation SB, is higher compared to its R_d value. A degree of polymer relaxation and swelling in a polymer matrix is indicated when the values of R_r are greater than R_d, which supports

Table 38.3 Kinetic values obtained from Kopca, Korsmeyer-Peppas and Sahlin-Peppas kinetic model plots for the formulations

Formulation Code	Kopcha			Korsmeyer-Peppas				Sahlin-Peppas		
	A	B	C^2	R	d	C^2	R_d	R_r	f	C^2
SC	9.60	0.1	0.99	11.60	0.44	0.99	6.44	5.17	0.29	0.99
SN	19.16	1.05	0.99	39.10	0.79	0.99	20.42	19.00	0.51	0.99
SB	4.48	1.32	0.99	10.34	0.82	1	4.92	5.48	0.51	0.99

the drug's tendency to be released via non-Fickian kinetics (Baggi & Kilaru, 2016). The low RE/FI values in Fig. 38.1. shows that the Fickian diffusion governs the initial MLS release from the films. Increasing RE/FI ratio values with time indicate the increase in the relaxational contribution. Among all the formulations, SC shows the slowest release. This may be because, compared to the films crosslinked with barium chloride, those crosslinked with calcium chloride produced less swell ability at first (Agrahari & Singh, 2016).

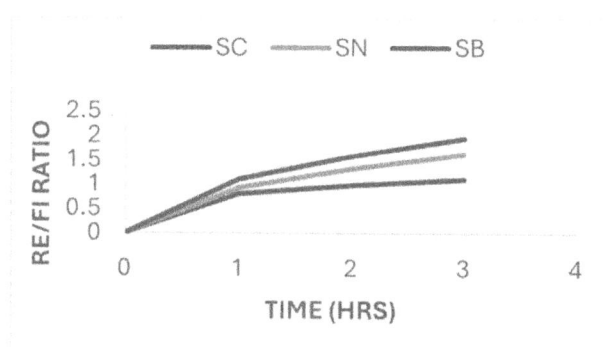

Fig. 38.1 Relaxation contribution (RE)/Fickian contribution (FI) ratio with respect to time

4. Conclusion

Drug release kinetics from formulations were examined in the study using a variety of models, including zero-order, first-order, Higuchi, Kopcha, Korsmeyer-Peppas, and Sahlin-Peppas. Zero-order kinetics were generally followed by formulations, suggesting constant MLS drug release rates. The role of crosslinking in controlled release formulations was highlighted by the faster release rates of non-crosslinked films as compared to crosslinked ones. In particular, the slowest release rate was seen in a transdermal film containing SDA/LGA, crosslinked with $CaCl_2$. This was explained by quasi-Fickian diffusion (Korsmeyer-Peppas exponent). The release mechanism was further clarified by Sahlin-Peppas modelling, which highlighted polymer relaxation in addition to diffusion. In conclusion, the results support the potential use of MLS loaded SDA/LGA crosslinked transdermal films in order to prevent the symptoms of Asthma in elderly and paediatric patients.

References

1. Aashli, S. Giridhar Reddy, Siva Kumar Belliraj, K. Prashanthi, & Murthy A. (2023). Fabricating transdermal film formulations of montelukast sodium with improved chemical stability and extended drug release. *Heliyon*, **9** (3), e14469-e14469.

2. Agrahari P., & Singh D. (2016). Effect of various electrolytes on control release of bio-molluscicides loaded in alginate as crosslinked matrices. ~ *1007* ~ *Journal of Entomology and Zoology Studies*, **4** (5), 1007-1012.

3. Ansari M., Kazemipour M., & Monireh Aklamli (2006). The study of drug permeation through natural membranes. *International journal of pharmaceutics*, **327** (1-2), 6-11.

4. Athira K, Kumar B. S., Reddy S. G., K. Prashanthi, Sanga Kugabalasooriar, & Posa J. K. (2024). A Novel Nature-Inspired Ligno-Alginate Hydrogel Coated with Fe3O4/GO for the Efficient-Sustained Release of Levodopa. *Heliyon*, e40547-e40547.

5. Baggi R. B., & Kilaru N. B. (2016). Calculation of predominant drug release mechanism using Peppas-Sahlin model, Part-I (substitution method): A linear regression approach. *Asian Journal of Pharmacy and Technology*, **6** (4), 223.

6. Behafarid Ghalandari, Adeleh Divsalar, Komeili A., Mahbube Eslami-Moghadam, Ali Akbar Saboury, & Kazem Parivar (2015). Mathematical Analysis of Drug Release for Gastrointestinal Targeted Delivery Using β-Lactoglobulin Nanoparticle. *Biomacromolecular Journal*, **1** (2), 204-211.

7. DASH S. (2010). Kinetic modeling on drug release from controlled drug delivery systems. *Acta Pol Pharm*, 67(3):217-23.

8. Ghavami-Lahiji M., Shafiei F., Kashi T. J., & Najafi F. (2021). Drug release kinetics and biological properties of a novel local drug carrier system. *Dental Research Journal*, **18** (1), 94.

9. Giridhar Reddy S (2023). Kinetic Studies for the Release of Hydroxychloroquine Sulphate Drug (HCQ) In-vitro in Simulated Gastric and Intestinal Medium from Sodium Alginate and Lignosulphonic Acid Blends. *Trends in Sciences*, **20** (5), 5318-5318.

10. Hadi M., & Rao S. (2012). Formulation and evaluation of sustained release matrix tablets of montelukast sodium Article in International Journal of Pharmacy.

11. Hafsa Bahaar, S. Giridhar Reddy, B. Siva kumar, K Prashanthi, & Murthy A. (2023). Modified Layered Double Hydroxide – PEG Magneto-Sensitive Hydrogels with Suitable Ligno-Alginate Green Polymer Composite for Prolonged Drug Delivery Applications. *Engineered science*,.

12. Laracuente M.-L., Yu M. H., & McHugh K. J. (2020). Zero-order drug delivery: State of the art and future prospects. *Journal of Controlled Release*, **327** , 834-856.

13. Mary A., Reddy S. G., Kumar B. S., & Sanga Kugabalasooriar (2024). Novel Approaches to Alendronate Delivery Beyond Oral Administration- A Review. *Engineered Science*,.

14. Pavithran R. K., Reddy S. G., Kumar B. S., & Sanga Kugabalasooriar (2024). Enhancing Sustainability in Agriculture: Natural Polymer-Based Controlled Release Systems for Effective Pest Management and Environmental Protection. *ES Food & Agroforestry*,.

15. Reddy S. G. (2021). CONTROLLED RELEASE STUDIES OF HYDROXYCHLOROQUINE SULPHATE (HCQ) DRUG-USING BIODEGRADABLE POLYMERIC SODIUM ALGINATE AND LIGNOSULPHONIC ACID BLENDS. *Rasayan Journal of Chemistry*, **14** (04).

Note: All the tables and figure in this chapter were made by the author.

Advances in Materials Science and Technology – Dr. Srikari Srinivasan et al. (eds)
© 2025 Taylor & Francis Group, London, ISBN 978-1-041-12342-2

39

Magnetically Targeted Drug Delivery of Levodopa in Parkinson's Disease using Magnetite Loaded Ligno-Alginate

Athira K, B. Siva Kumar*, S. Giridhar Reddy

Department of Physical Sciences,
Amrita School of Engineering, Amrita Vishwa Vidyapeetham,
Bengaluru, India

K. Prashanthi

Department of Biotechnology,
Ramaiah University of Applied Sciences,
Bengaluru, India

ABSTRACT: Parkinson's disease is a neurodegenerative disorder that shows the progressive loss of nerve cells. Levodopa (LD) represents a precursor to the deficient neurotransmitter dopamine. The classic formulations of LD taken orally are related to low bioavailability and non-targeted delivery. Controlled drug delivery systems consequently ensure regulated and prolonged release and thus enhanced in-vivo therapeutic efficacy and reduced toxicity based on polymer conjugates. Stimuli-responsive mechanisms, such as pH or magnetic fields trigger targeted drug release at specific sites. Alginate and Lignosulfonates in the presence of Fe_3O_4 nanoparticles form the ideal magnetic hydrogel for drug delivery. In the present investigation, formulation, characterization by UV-Vis and FTIR spectroscopy, LD loading efficiency, and *in-vitro* release kinetics evaluation studies of such hydrogels have been carried out. The results are significant and greatly show a controlled release under magnetic influence that would thus find applications in the biomedical treatment of Parkinson's disease.

KEYWORDS: Green polymers, Levodopa, Magnetic hydrogel

1. Introduction

Parkinson's disease (PD) is a neurodegenerative disorder characterized by the progressive loss of nerve cells in the central nervous system. Levodopa, also known as L-3,4-dihydroxyphenylalanine, serves as a precursor to dopamine, an essential neurotransmitter deficient in PD(Simon *et al.*, 2020). However, conventional oral formulations of levodopa suffer from limitations such as low bioavailability, potential toxicity, and non-targeted delivery (Nutt, 2008). Controlled drug delivery systems give a regulated and

hence sustained release that improves bioavailability while reducing toxicity. Normally, the system consists of conjugating the drug with polymers that allow controlled release over time (Gao *et al.*, 2023). Stimuli-responsive mechanisms such as pH, heat, or magnetic fields will trigger the targeted release of drugs from the polymer at specified sites in the body (Rahim *et al.*, 2021). Sodium Alginate is a polymeric carbohydrate extracted from brown seaweed, like Laminaria hyperborea, and comprises alpha-L-guluronate (G) and beta-D mannuronate monomers joined by 1→4 linkages. Alginate readily forms gels in the

*Corresponding author: b_sivakumar@blr.amrita.edu

DOI: 10.1201/9781003664277-39

presence of divalent—Ca^{2+}, Ba^{2+} or trivalent ions—Al^{2+} and Fe^{2+}. This is because these ions are attracted by the ionic linkage with its COO- groups (Giridhar & Pandit, 2013). Alginate displays biocompatibility and non-toxicity, is water-insoluble, but turns to water-soluble when treated with basic solutions like sodium hydroxide, which confers an advantage on this polymer for drug delivery systems (Frent et al., 2022). Lignosulfonates (LSA), derived from lignin during pulp and paper production using sulfite or bisulfite processes, are valuable due to their amphiphilic nature and water solubility facilitated by anionic sulfonate and carboxylic groups (Chen et al., 2015)(Bahaar et al., 2024). LSA's ability to adjust swelling capacity based on lignin type highlights its suitability for various industrial uses, including superplasticizers in construction materials and as components in drug delivery systems for controlled release applications (Ruwoldt, 2020)'(Reddy & Pandit, 2014). Magnetic particles, such as Fe_3O_4 nanoparticles, are employed for targeted drug delivery due to their magnetic responsiveness. By functionalizing these particles with drugs and applying external magnetic fields, they can be directed to specific sites in the body (Kondaveeti et al., 2018)(K et al., 2024). This approach reduces systemic side effects and enhances drug concentration at the target Bahaar et al. prepared a drug delivery system using magnetic nanoparticles coated with SAl, LSA, and polyethylene glycol biopolymers, then further coated with magnesium-aluminum hydroxides; the outcome showed sorafenib had a high percentage release of 99.2 % in 120 hours (Bahaar et al., 2023). A hydrogel combining SAlg, LSA, and Fe_3O_4 nanoparticles provides efficient and sustained levodopa release by exerting its biocompatibility, hydrogel-forming properties, and enhanced drug delivery characteristics.

2. Materials and Methods

2.1 Materials

SAlg and LSA were purchased from Sigma Aldrich. Ferric chloride hexahydrate [$FeCl_3.6H_2O$] and ferrous sulfate anhydrous [$FeSO_4$] were purchased from Loba Chemicals. Barium chloride was purchased from S.D. Fine Chem. Ltd. Ammonium hydroxide (NH_4OH) was purchased from Thermofisher Scientific Ltd. Levodopa, which was received as a pharmaceutical sample.

2.2 Preparation of Magnetic Hydrogel

A 2% film of SAlg and LSA blend was prepared in a ratio of 80:20. The polymers were dissolved in distilled water with the aid of a magnetic stirrer for 30 minutes to ensure proper mixing of the polymers. The contents were poured onto a glass Petri plate and dried in a hot air oven at 45°C

for 16-24 hours until completely dry. The polymeric films were then dried and, thereafter, cross-linked in a solution of 2 % barium chloride for 15 minutes, followed by drying at room temperature. $FeCl_3.6H_2O$ (1 g) and $FeSO_4$ (0.4 g) were dissolved in 50 ml of distilled water and stirred at 60°C for 15 minutes. Dried SA-LS polymeric films were immersed in this solution. The solution was nuanced to pH 9 using an ammonium solution that was added dropwise, which resulted in the precipitation of iron oxide nanoparticles on the surface of the films. After stirring at 45°C for 30 minutes, the films were washed three times with 80% methanol to remove unbound nanoparticles. The iron oxide-coated films were then air-dried. These films were further added to a 50 ml solution of levodopa (1 mg/mL) and incubated for 12-16 hours at 4°C in the dark.

3. Characterization

Characterization of the magnetic hydrogel was done using a Shimadzu 2600 UV-visible and Fourier Transform Infrared (FT-IR) spectrophotometer.

3.1 UV-VIS Spectrophotometer

The release of levodopa from the films was determined at 280 nm by measuring absorbance in 50 ml of PBS, pH 7.2.

3.2 FT-IR Spectrophotometer

The FT-IR spectra was recorded to determine drug encapsulation and identification of functional groups. Instrument calibration was done with pellets of KBr with background scans; then, each sample was scanned upon pellet preparation.

4. Results and Discussion

4.1 Analysis of Functional Groups Interaction by FT-IR

The FTIR peaks of the magnetic hydrogel are depicted in Fig. 39.1. Infrared absorption peaks at 1033 cms^{-1} indicate the elongation of C-O and sharp peaks at 1419 cm^{-1} indicate symmetric stretching of COO, which becomes characteristic features of SAlg-LSA (Aashli et al., 2023), (Liu et al., 2018). Absorbance bands at approximately 663 cm^{-1} bands indicate stretching of the Fe–O bond in the Fe_3O_4 crystalline lattice (Chaki et al., 2015). The peak at 2978 cm^{-1} in the FTIR spectrum of levodopa corresponds to the carboxylic acid O-H stretching. This absorption band indicates the hydroxyl group, OH, in the carboxylic acid moiety that exists in the molecules of levodopa (Bukhary et al., 2020). The peaks prove the presence of polymers and drugs in the magnetic hydrogel.

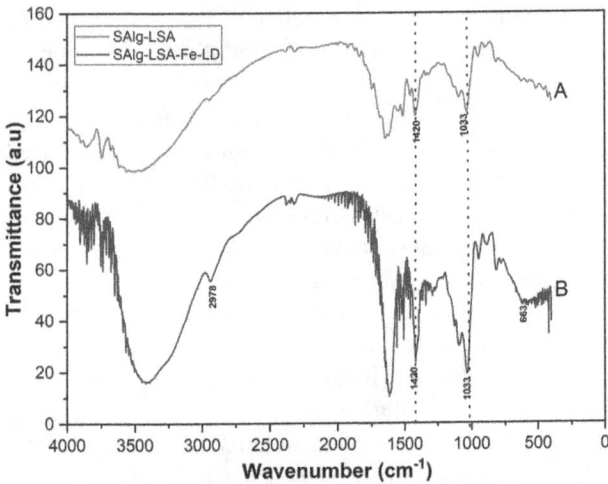

Fig. 39.1 FTIR spectra of (A) SAlg-LSA (B) SAlg-LSA-Fe-LD

4.2 Loading Efficiency of LD

The loading efficiency of LD into the polymer-nano formulation is determined by the fact of its immersion in a known concentration LD solution for 16 hours. The remaining solution was removed by removing the film and reading it at 280 nm for Ct, the amount that remained free in the supernatant.

$$Loading\ efficiency\ \% = \frac{C_0 - C_t}{C_0} \times 100 \qquad (1)$$

Co is the initial concentration of LD in the drug solution before immersion of the polymer film, and Ct is the concentration of LD in the supernatant after 16 hours. This method provides a quantitative measure of how effectively the polymer-nano formulations encapsulate LD, which is very important in assessing their suitability for drug delivery applications. The loading efficiency of LD was found to be 63.07 % in the magnetic hydrogel at pH 7.4.

4.3 *In-vitro* Release of Levodopa

Levodopa (LD) release was investigated in vitro using a UV-Vis spectrophotometer. The study involved immersing the film in a 50 mL buffer solution at pH 7.4. At two-hour intervals, 2 mL of the buffer solution was withdrawn and analyzed using spectrophotometry at 280 nm as shown in Fig. 39.2.

This method allowed for continuous monitoring of LD release kinetics from the polymer film under controlled conditions. In the presence of an external mag net placed under the beaker containing the film and PBS, a significant amount of levodopa (LD) was released. The first 10 minutes resulted in a rapid release of 15 % of the initial levodopa content. A controlled release under the effect of the magnetic field, amounting to 43 % of the levodopa content, was accomplished over a 48-hour period. This is because Fe_3O_4 delivering a slow and sustained release of levodopa under a magnetic field demonstrates this controlled delivery behavior. Fe_3O_4 incorporation promotes the biological activity of the drug because of the magnetically assisted delivery of the drug. The formulation with magnetic nanoparticles facilitates targeted delivery of the drug; hence, this newly developed hydrogel turns out to be quite an effective tool for applications related to controlled and sustained release.

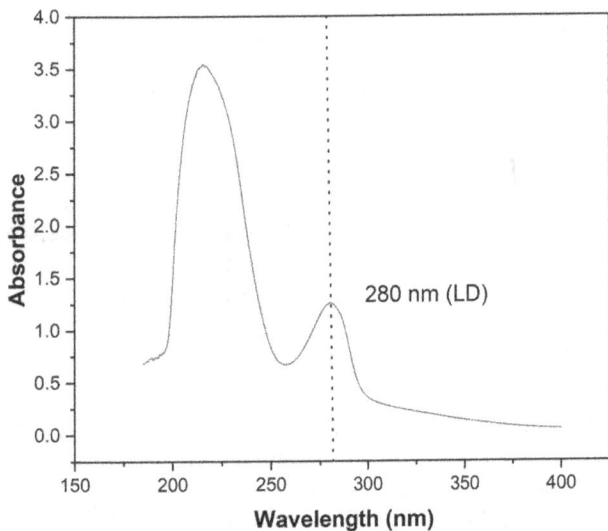

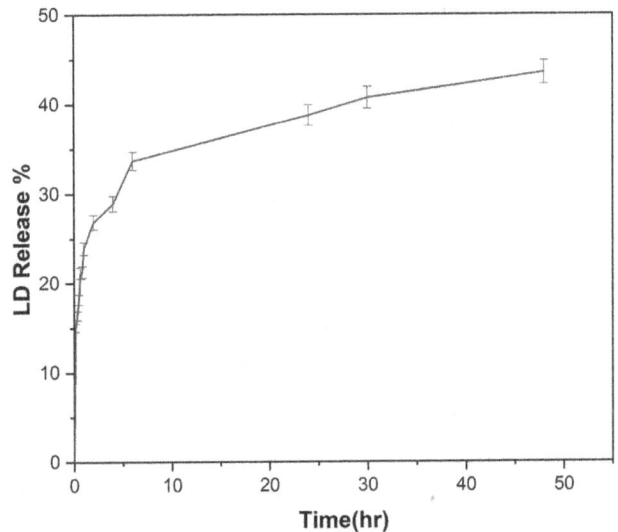

Fig. 39.2 UV-Vis spectroscopy of LD (a) UV peak of Levodopa@280nm (b) Release profile of LD

5. Conclusion

Magnetic hydrogels thus become a valuable tool in the delivery of drugs because they are environmentally friendly with the polymers SAlg and LSA. This work develops biocompatible magnetic materials with tailored LD releases to improve therapeutic efficiency and reduce side effects. The incorporation of lignin into SAlg has improved biocompatibility and pH sensitivity, while Fe_3O_4 improves the properties of the polymer synergistically, with tremendous potential as a novel LD carrier in biomedical applications. Magnetic particles allow the provision of targeted delivery, which can be further studied and validated.

References

1. Aashli, Reddy S. G., Siva Kumar B., Prashanthi K., & Murthy H. C. A. (2023). Fabricating transdermal film formulations of montelukast sodium with improved chemical stability and extended drug release. Heliyon, 9 (3), e14469.

2. Bahaar H., Kumar B. S., Reddy S. G., Guo Z., Pereira A., & Liu T. X. (2024). From Concept to Creation: Micro/Nanobot Technology from Fabrication to Biomedical Applications. Engineered Science.

3. Bahaar H., Reddy S. G., kumar B. S., K P., & H.C. A. M. (2023). Modified Layered Double Hydroxide – PEG Magneto-Sensitive Hydrogels with Suitable Ligno-Alginate Green Polymer Composite for Prolonged Drug Delivery Applications. Engineered Science.

4. Bukhary H., Williams G. R., & Orlu M. (2020). Fabrication of Electrospun Levodopa-Carbidopa Fixed-Dose Combinations. Advanced Fiber Materials, 2 (4), 194–203.

5. Chaki S. H., Malek T. J., Chaudhary M. D., Tailor J. P., & Deshpande M. P. (2015). Magnetite Fe 3 O 4 nanoparticles synthesis by wet chemical reduction and their characterization. Advances in Natural Sciences: Nanoscience and Nanotechnology, 6 (3), 035009.

6. Chen W., Peng X., Zhong L., Li Y., & Sun R. (2015). Lignosulfonic Acid: A Renewable and Effective Biomass-Based Catalyst for Multicomponent Reactions. ACS Sustainable Chemistry & Engineering, 3 (7), 1366–1373.

7. Frent O., Vicas L., Duteanu N., Morgovan C., Jurca T., Pallag A., Muresan M., Filip S., Lucaciu R.-L., & Marian E. (2022). Sodium Alginate—Natural Microencapsulation Material of Polymeric Microparticles. International Journal of Molecular Sciences, 23 (20), 12108.

8. Gao J., Karp J. M., Langer R., & Joshi N. (2023). The Future of Drug Delivery. Chemistry of Materials, 35 (2), 359–363.

9. Giridhar R. S., & Pandit A. S. (2013). Effect of Curing Agent on Sodium Alginate Blends Using Barium Chloride as Crosslinking Agent and Study of Swelling, Thermal, and Morphological Properties. International Journal of Polymeric Materials, 62 (14), 743–748.

10. K A., Kumar B. S., Reddy S. G., Prashanthi K., Kugabalasooriar S., & Posa J. K. (2024). A novel nature-inspired ligno-alginate hydrogel coated with Fe3O4/GO for the efficient-sustained release of levodopa. Heliyon, 10 (23), e40547.

11. Kondaveeti S., Semeano A. T. S., Cornejo D. R., Ulrich H., & Petri D. F. S. (2018). Magnetic hydrogels for levodopa release and cell stimulation triggered by external magnetic field. Colloids and Surfaces B: Biointerfaces, 167, 415–424.

12. Liu Q., Li Q., Xu S., Zheng Q., & Cao X. (2018). Preparation and Properties of 3D Printed Alginate–Chitosan Polyion Complex Hydrogels for Tissue Engineering. Polymers, 10 (6), 664.

13. Nutt J. G. (2008). Pharmacokinetics and pharmacodynamics of levodopa. Movement Disorders, 23 (S3), S580–S584.

14. Rahim M. A., Jan N., Khan S., Shah H., Madni A., Khan A., Jabar A., Khan S., Elhissi A., Hussain Z., Aziz H. C., Sohail M., Khan M., & Thu H. E. (2021). Recent Advancements in Stimuli Responsive Drug Delivery Platforms for Active and Passive Cancer Targeting. Cancers, 13 (4), 670.

15. Reddy S. G., & Pandit A. S. (2014). Controlled drug delivery studies of biological macromolecules: Sodium alginate and lignosulphonic acid films. Journal of Applied Polymer Science, 131 (13).

16. Ruwoldt J. (2020). A Critical Review of the Physicochemical Properties of Lignosulfonates: Chemical Structure and Behavior in Aqueous Solution, at Surfaces and Interfaces. Surfaces, 3 (4), 622–648.

17. Simon D. K., Tanner C. M., & Brundin P. (2020). Parkinson Disease Epidemiology, Pathology, Genetics, and Pathophysiology. Clinics in Geriatric Medicine, 36 (1), 1–12.

Note: All the figures in this chapter were made by the author.

Advances in Materials Science and Technology – Dr. Srikari Srinivasan et al. (eds)
© *2025 Taylor & Francis Group, London, ISBN 978-1-041-12342-2*

40

Kraft Lignin-Based Nanocarrier for Improved Sorafenib Release in Hepatocellular Carcinoma Treatment

Hafsa Bahaar,
B. Siva Kumar*, S. Giridhar Reddy
Department of Physical Sciences,
Amrita School of Engineering, Amrita Vishwa Vidyapeetham,
Bengaluru, Karnataka, India

ABSTRACT: Sorafenib is an oral multikinase inhibitor that blocks tumor growth and promotes tumor cell death. In our previous work, we created a biodegradable polymer-based nanocarrier for Sorafenib (SF) delivery to reduce cancer treatment side effects through targeted and controlled release. This nanocarrier, containing iron oxide nanoparticles (IONP), sodium alginate, lignosulfonic acid, polyethylene glycol, SF drug, and a MgAl layered double hydroxide coating, achieved 99.2% SF release in 120 hours under acidic conditions. Building on this study, our next objective was to develop a nanocarrier using Kraft lignin instead of lignosulfonic acid, as Kraft lignin improves water stability, reinforcement, and glass transition temperature (Tg). High Tg materials are glassy at room temperature, ensuring better dissolution and consistent drug release. We aimed to create a more stable drug delivery system with biocompatible polymers like sodium alginate, Kraft lignin, and polyethylene glycol, starting with a core-shell composite of Kraft lignin and iron oxide nanoparticles, followed by additional coatings of sodium alginate, polyethylene glycol, and Mg/Al LDH. The nanocomposite showed encapsulation efficiency of 97% and drug loading capacity of 59.5%. Release studies over 96 hours at pH 4.6 and pH 7.4 demonstrated sustained and controlled drug release, wherein 9.1% and 1.3% of the drug was released respectively. These promising results suggest the composite's potential for controlled Sorafenib delivery.

KEYWORDS: Controlled drug delivery, Kraft lignin, Sorafenib, Targeted drug delivery

1. Introduction

According to the 2020 GLOBOCAN report, liver cancer accounted for 830,180 deaths and 4.7% of all cases of cancer worldwide (Rumgay *et al.*, 2022), ranking as the sixth commonest cancer. Sorafenib, approved for the treatment of HCC in 2007, remains one of the systemic agents approved for use for this indication. Sorafenib is an orally active multikinase inhibitor that acts on cell surface tyrosine kinases and downstream intracellular serine/threonine kinases. All these protein kinases take part in tumor cell signaling, proliferation, angiogenesis, and apoptosis (Keating, 2017).

The downsides of the drug are many, with low bioavailability and non-specific targeting of normal cells being the primary ones. This is common for most of the drugs used in chemotherapy, which cannot differentiate between cancerous and normal cells, hence leading to side effects, some even life threatening (Alessandri *et al.*, 2019). A controlled drug delivery (CDD) approach can address these issues by releasing medication at a regulated rate,

---

*Corresponding author: b_sivakumar@blr.amrita.edu

DOI: 10.1201/9781003664277-40

maintaining therapeutic concentrations, and reducing the need for frequent dosing (Adepu & Ramakrishna, 2021; Reddy & Pandit, 2014).

Biopolymers from natural sources are increasingly preferred over synthetic ones for their biodegradability, biocompatibility, low immunogenicity, etc. (Jacob *et al.*, 2018). Lignin, together with cellulose and hemicelluloses, is a major constituent of plant biomass. The second most abundant renewable resource on Earth, next to cellulose, it accounts for 25-30% of non-fossil organic matter in nature (Thakur *et al.*, 2014). Kraft lignin, extracted from the black liquor in the pulping process, accounts for the greatest part of industrial lignin. At present, KL is mainly utilized for providing heat in the alkali recovery process, and the utilization rate remains fairly low. Other than that, KL could also be converted for biochemical and biofuel production through either chemical degradation or pyrolysis, applied in polymer materials by blending, as an adsorbent for heavy metals, or probably developed into hydrogel and various industrial dispersants through chemical modification. Whereas the use is still majorly situated in traditional industrial fields, there is huge potential for high value of KL application in nanoscience and technology (Li *et al.*, 2016). Lignin is a biopolymer, with a number of interesting advantages due to its biodegradability and biocompatibility. Having been in the spotlight recently, modifications on this polymer have been focused on finding innovative uses for carbon fibers, biofuels, bioplastics, and controlled release carriers. The presence of functional groups in lignin, for example, phenolic, hydroxyl, and carboxyl, shows potential for chemical modification to develop drug carriers and regulate the resultant drug release (Pishnamazi *et al.*, 2019). Kraft lignin also improves the stability, reinforces attributes, and increases the Tg. The Tg of a polymer is very important, more so in controlled drug delivery applications where the material in its glassy state shows slow diffusion (I Brodin, 2009 ; Siepmann & Peppas, 2012).

Sodium Alginate is a linear anionic polysaccharide which exhibits properties such as excellent solubility, viscosity, cross-linking, sol-gel transformation, biocompatibility, biodegradability, and bioadhesion. It has been widely used in pharmaceutical formulations in various dosage forms like tablets, capsules, gels, and nanoparticles (Hasnain *et al.*, 2020 ; Reddy & Pandit, 2013). Poly (ethylene glycol) (PEG) is a non-toxic, water-soluble polymer that evades detection by the immune system (Aashli *et al.*, 2023).

Nanotechnology has evolved and is being developed for use in a variety of applications, one of which is targeted drug delivery. Magnetic nanoparticles, such as iron oxide nanoparticles, find an application in the composites that facilitate the targeted drug delivery under the influence or action of an external magnetic field. The intrinsic properties of iron oxide nanoparticles (IONs), such as their magnetism and biocompatibility, along with versatile fabrication techniques, position them as excellent candidates for nanomedicine applications, including effective drug targeting and theragnostic functionalities (Laurent *et al.*, 2014).

Magnesium-aluminium layered double hydroxide, derived from magnesium hydroxide, has a layered structure with some divalent cations replaced by trivalent ones, creating a positively charged material ideal for delivering negatively charged drugs. Its hydrophilic environment also enhances the solubility of poorly water-soluble anticancer drugs (Bahaar *et al.*, 2023).

In this study, we evaluated the controlled release of sorafenib with a polymer nanocomposite as the delivery vehicle. The composite is made up of iron oxide nanoparticles coated with kraft lignin, followed by a layer of sodium alginate and polyethylene glycol. Sorafenib, an anticancer medication, was included into the composite, which was subsequently coated with Mg/Al LDH to add a positive charge. The synthesised core-shell nanoparticles were tested for drug delivery capabilities in simulated fluid conditions, demonstrating a potential approach for producing LDH-coated core-shell nanoparticles for increased therapeutic efficacy.

2. Materials and Methods

2.1 Materials

Lignin Alkaline (Lot No. E2LJE-DG, TCI), Ferric chloride (Product No. 23220, Molychem), Ferrous sulphate (Product No. 23755), 25% Ammonia Solution (37140URL05, SD Fine chemicals), Sodium Alginate (Lot No. MKBZ9710V, Sigma Aldrich), Polyethylene glycol (Product No. 39571 L05, SDFCL), DMSO (Product No. 38216L05, SDFCL), Magnesium Nitrate (Product No. 04470, Loba Chemie), Aluminium Nitrate (Product No. 00927, Loba Chemie), Sorafenib was received as a pharmaceutical sample.

2.2 Synthesis of KL-ION/SA/PEG/SF/LDH

200 mg of Kraft lignin was dissolved in 0.2M NaOH with constant stirring for 30 minutes. Iron oxide nanoparticles (IONs) were synthesized using the co-precipitation method. Separate aqueous solutions of ferric chloride and ferrous sulfate were prepared and then mixed together, continuously stirred at 60°C for 15 minutes. Ammonia solution was added dropwise while stirring at 40°C until the pH reached 9-10. The resulting black precipitate was collected by centrifugation. This precipitate was then combined with the Kraft lignin suspension.

Next, 0.8g of sodium alginate and 0.88ml of polyethylene glycol were mixed using a magnetic stirrer. This mixture was added to the KL-IONs suspension and stirred thoroughly. For the drug solution, 1g of Sorafenib was dissolved in 50 ml of DMSO and then added to the Polymer-ION suspension, which was then coated with layered double hydroxide (LDH).

LDH was synthesized by dissolving magnesium nitrate and aluminum nitrate in 50 ml of distilled water, with a few drops of NaOH added while stirring until the pH reached 9-10. This LDH solution was then added to the Polymer-ION mixture and stirred thoroughly for 30 minutes. The final suspension was centrifuged at 5000 rpm for 5 minutes and dried in a hot air oven overnight.

2.3 *In Vitro* Drug Release Studies

A calibration curve for sorafenib was plotted using standard concentrations (0.5 µg/ml, 1 µg/ml, 1.5 µg/ml, 2 µg/ml, and 2.5 µg/ml) and measuring their UV absorbance at 265 nm. Controlled release studies were conducted at two pH levels, pH 4.6 (acetate buffer) and pH 7.4 (PBS), to mimic the tumor microenvironment and normal cell conditions, respectively. The final sample was placed in a beaker containing 50 ml of the buffer solutions. The release of sorafenib (SF) from the nanocarrier was monitored using a UV-Visible Spectrophotometer (Shimadzu 2600) over a range of 190-400 nm, with the characteristic peak of SF observed at 265 nm. Calibration curves and standard graphs were plotted for each condition to determine the percentage of sorafenib released.

3. Results and Discussion

3.1 *In vitro* Drug Release of Sorafenib

The encapsulation efficiency and the Drug loading capacity of the nanocomposite were found to be 97% and 59.5% respectively. In the present study, the release kinetics of SF from the developed nanocarriers were studied in two

solutions at different pH for 96h. To understand the behavior of the nanocarriers in the tumor microenvironment and in normal cell conditions, the studies for drug release were performed at pH 4.6, acetate buffer, and at pH 7.4 in PBS. As a result, it was observed that 9.1% and 1.3% of the drug was released for over a period of 96h at pH 4.6 and pH 7.4 respectively (Fig. 40.1).

A significant difference in the release profiles at two different pH was observed. Overall, the drug was released at a higher concentration at pH 4.6, where 9.1% of the drug was released at 96h, when compared to pH 7.4, where only 1.3% of the drug was released. In this context, the pH of the surrounding environment significantly effects the drug's dissolution from the nanocomposite. This occurs because the polymers used in synthesizing the nanocomposites are sensitive to pH changes. Compared with the traditional polymeric micelle, these pH-sensitive nanosystems will change physically or chemically in an acidic environment through swelling, dissociation, or degradation and efficiently release the encapsulated drugs (Mu *et al.*, 2021). On the other hand, it was observed that, the concentration of the drug released from the nanocomposite was decreased after 48h at pH 7.4, probably due to precipitation or degradation of the drug as it is unstable in aqueous solutions.

Such pH-dependent release behavior of the drug from the nanocomposite is advantageous, since this was one of our goals: realization of targeted delivery into cancer cells. This was achieved by adding iron oxide magnetic nanoparticles that aid in directing delivery into cancer cells. Also, the pH-sensitive behavior assures that an adequate amount of the drug will be released only at the tumor site.

4. Conclusion

It is a fact that drug-encapsulated lignin-based nanoparticles have all the major properties expected for biomedical applications, including biocompatibility, biodegradability, effective hydrophobic drug loading, and favorable release

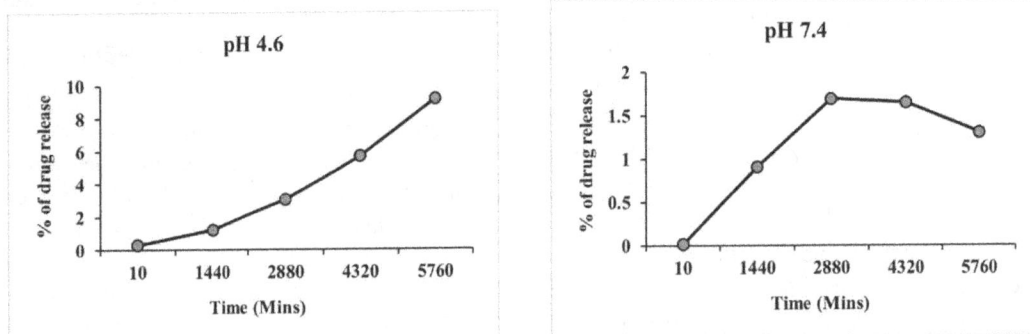

Fig. 40.1 Drug release profile of KL-ION/SA/PEG/SF/LDH at pH 4.6 and pH 7.4

profiles that ensure reduced side effects. In addition to this, IONs further impart the application of targeted drug delivery. In this regard, this study demonstrates that the synthesis of drug-encapsulated Kraft lignin nanoparticles may provide an effective means for hydrophobic drug delivery in a controlled manner, offering benefits associated with sustainability.

References

1. Aashli, Reddy S. G., Siva Kumar B., Prashanthi K., & Murthy H. C. A. (2023). Fabricating transdermal film formulations of montelukast sodium with improved chemical stability and extended drug release. *Heliyon*, **9** (3), e14469.
2. Adepu S., & Ramakrishna S. (2021). Controlled Drug Delivery Systems: Current Status and Future Directions. *Molecules*, **26** (19), 5905.
3. Alessandri G., Coccè V., Pastorino F., Paroni R., Dei Cas M., Restelli F., Pollo B., Gatti L., Tremolada C., Berenzi A., Parati E., Brini A. T., Bondiolotti G., Ponzoni M., & Pessina A. (2019). Microfragmented human fat tissue is a natural scaffold for drug delivery: Potential application in cancer chemotherapy. *Journal of Controlled Release*, **302** , 2–18.
4. Bahaar H., Reddy S. G., kumar B. S., K P., & H.C. A. M. (2023). Modified Layered Double Hydroxide – PEG Magneto-Sensitive Hydrogels with Suitable Ligno-Alginate Green Polymer Composite for Prolonged Drug Delivery Applications. *Engineered Science*,.
5. Hasnain M. S., Ahmed S. A., Alkahtani S., Milivojevic M., Kandar C. C., Dhara A. K., & Nayak A. K. (2020). Biopolymers for Drug Delivery. . p. 1–29.
6. I Brodin (2009). Chemical properties and thermal behaviour of kraft lignins.
7. Jacob J., Haponiuk J. T., Thomas S., & Gopi S. (2018). Biopolymer based nanomaterials in drug delivery systems: A review. *Materials Today Chemistry*, **9** , 43–55.
8. Keating G. M. (2017). Sorafenib: A Review in Hepatocellular Carcinoma. *Targeted Oncology*, **12** (2), 243–253.
9. Laurent S., Saei A. A., Behzadi S., Panahifar A., & Mahmoudi M. (2014). Superparamagnetic iron oxide nanoparticles for delivery of therapeutic agents: opportunities and challenges. *Expert Opinion on Drug Delivery*, **11** (9), 1449–1470.
10. Li H., Deng Y., Liu B., Ren Y., Liang J., Qian Y., Qiu X., Li C., & Zheng D. (2016). Preparation of Nanocapsules via the Self-Assembly of Kraft Lignin: A Totally Green Process with Renewable Resources. *ACS Sustainable Chemistry & Engineering*, **4** (4), 1946–1953.
11. Mu Y., Gong L., Peng T., Yao J., & Lin Z. (2021). Advances in pH-responsive drug delivery systems. *OpenNano*, **5** , 100031.
12. Pishnamazi M., Hafizi H., Shirazian S., Culebras M., Walker G., & Collins M. (2019). Design of Controlled Release System for Paracetamol Based on Modified Lignin. *Polymers*, **11** (6), 1059.
13. Reddy S. G., & Pandit A. S. (2013). Biodegradable sodium alginate and lignosulphonic acid blends: characterization and swelling studies. *Polímeros*, **23** (1), 13–18.
14. Reddy S. G., & Pandit A. S. (2014). Controlled drug delivery studies of biological macromolecules: Sodium alginate and lignosulphonic acid films. *Journal of Applied Polymer Science*, **131** (13).
15. Rumgay H., Arnold M., Ferlay J., Lesi O., Cabasag C. J., Vignat J., Laversanne M., McGlynn K. A., & Soerjomataram I. (2022). Global burden of primary liver cancer in 2020 and predictions to 2040. *Journal of Hepatology*, **77** (6), 1598–1606.
16. Siepmann J., & Peppas N. A. (2012). Modeling of drug release from delivery systems based on hydroxypropyl methylcellulose (HPMC). *Advanced Drug Delivery Reviews*, **64**, 163–174.
17. Thakur V. K., Thakur M. K., Raghavan P., & Kessler M. R. (2014). Progress in Green Polymer Composites from Lignin for Multifunctional Applications: A Review. *ACS Sustainable Chemistry & Engineering*, **2** (5), 1072–1092.

Advances in Materials Science and Technology – Dr. Srikari Srinivasan et al. (eds)
© 2025 Taylor & Francis Group, London, ISBN 978-1-041-12342-2

41

Microwave Absorption Characterization of NCQD-RGO Composites using Rectangular Waveguide for Stealth Applications

Veena Venugopal

Department of Physical Sciences,
Amrita School of Engineering, Amrita Vishwa Vidyapeetham,
India

Hanima Kannan C. H.

Department of Electronics and Communication,
Amrita School of Engineering, Amrita Vishwa Vidyapeetham,
India

Siva Kumar B.*

Department of Physical Sciences,
Amrita School of Engineering, Amrita Vishwa Vidyapeetham,
India

Parul Mathur, Dhanesh G Kurup

Department of Electronics and Communication,
Amrita School of Engineering, Amrita Vishwa Vidyapeetham,
India

ABSTRACT: Microwave-absorbing materials with stealth capabilities are crucial for aerospace platforms to minimize radar signal reflection. A common approach involves using functional material coatings to absorb electromagnetic waves, with nanofiller concentration playing a key role in their effectiveness. This study uses computational simulations to evaluate the electromagnetic absorption properties of a composite polymer matrix with Nitrogen-doped Carbon Quantum Dots and Reduced Graphene Oxide. Finite Element Method simulations were conducted using a rectangular waveguide-based setup to analyze the reflection parameters of the composite at different nanofiller concentrations. The study examined three loading percentages (3%, 5%, and 10%) inspired by the experimental work of Jungfeng et al. Results indicate that a metal-backed sample with 5% loading and 4 mm thickness achieves the best stealth performance, with a maximum return loss of 10 dB.

KEYWORDS: HFSS, Microwave absorbing materials (MAM), Nitrogen dopped carbon quantum dots (NCQDs), Rectangular waveguides, Reduced graphene oxide (RGO), Single-port system

1. Introduction

Stealth technology is vital for national security, relying on microwave-absorbing materials (MAMs) with properties like high absorption, broadband operation, light weight, thinness, and stability (Ahmad *et al.*, 2019). Carbon-based materials, including carbon fibers, nanotubes, porous carbon, graphene, reduced graphene oxide (RGO), and

*Corresponding author: b_sivakumar@blr.amrita.edu

DOI: 10.1201/9781003664277-41

carbon quantum dots (CQDs), are particularly suited for such applications (B. Wang $et\,al.$, 2021). Three-dimensional RGO enhances both conductive and dielectric losses while remaining lightweight and flexible. CQDs have also gained attention as promising components in multi-material MAMs (Liu & Huang, 2014).

Rectangular waveguides, preferred over coaxial ones for characterizing microwave absorption, provide superior contact and response for solid samples. With TE10 as the dominant propagation mode, rectangular waveguides are ideal for high-power, low-loss applications, though some attenuation may occur at higher frequencies or over long distances (X. Wang $et\,al.$, 2020).

This study uses a single-port rectangular waveguide configuration in ANSYS HFSS to analyze the electromagnetic absorption of NCQD-RGO composites with 3%, 5%, and 10% nanofiller loadings, labeled S1, S2, and S3 (Qiu $et\,al.$, 2022). Reflection losses are evaluated by incorporating dielectric parameters into the simulation. The single-port setup, designed for metal-backed applications like aircraft, allows precise reflection parameter analysis. The paper is organized into four sections: microwave absorption theory, waveguide design, simulation analysis, and conclusions with future directions.

2. Microwave Absorption Theory

The electromagnetic waves interacting with a material undergo three phenomena: reflection, transmission, and absorption. The sum of effective absorption (SE_A), multiple internal reflections (SE_M), and reflection (SE_R) gives the total effective EM shielding (SE_T) given by the equation (Jelmy $et\,al.$, 2020) :

$$SE_T = SE_A + SE_M + SE_R \tag{1}$$

For a material to become a MAM it should satisfy the following criteria: the characteristic impedance of the chosen material and the vacuum must be equal, and there should be rapid attenuation of EM wave within the material governed by the capacity to be magnetically and dielectrically lossy (Adebayo $et\,al.$, 2020). Permittivity of a material describes its ability to store energy by polarization

of the atomic charges. Therefore, impedance matching and resonance frequency of MAM can be adjusted by permittivity. The following equation gives the permittivity of the material:

$$\epsilon = \epsilon' - j\epsilon'' \tag{2}$$

where ϵ' is the real part of permittivity corresponding to energy storage of a material, where ϵ'' is the complex part of permittivity which corresponds to the energy dissipation or losses. Therefore, impedance matching and resonance frequency of MAM can be adjusted by permittivity. The amount of energy dissipated can be evaluated using loss tangent tan δ_ϵ given by the following equation

$$\tan \delta_\epsilon = \epsilon''/\epsilon' \tag{3}$$

The EM absorption coefficient also known as reflection loss is given by:

$$RL = -20 \log (\Gamma) \tag{4}$$

Where Γ is the reflection coefficient which describes the amount of EM waves being reflected by a material owing to discontinuity in impedance at the interface of the Microwave Absorber (MA).

3. Design of a Single Port Rectangular Waveguide

Rectangular waveguide techniques are non-invasive, non-destructive methods used in studying microwave absorption of materials (Venugopal et al., 2024), (Hanima Kannan et al., 2024). Dimensions used for modeling the waveguides are based on the commercially available WR340.

The electromagnetic waves inside the waveguide are characterized by reflection from the conducting walls where the medium is air. A Sub Miniature version SMA coaxial connector with 50-ohm impedance is capped on to the waveguide for signal transmission. In this paper, one-port rectangular waveguide is used to study microwave absorption characteristics of nanocomposite material. For the one-port rectangular waveguide, the sample holder is coated by Perfect Electric Conductor (PEC) on all 5 sides and is placed on the open end of the rectangular waveguide, as shown in Fig. 41.1. This set up has a metal

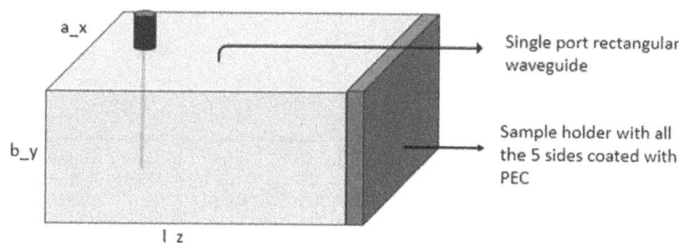

Fig. 41.1 A single port rectangular waveguide with Sample holder

backing for the material understudy and is similar to the real time application of stealth coating in an aircraft. It is clear from Fig. 41.1. that the one-port system functions as a cavity resonator.

4. Results and Discussion

Three samples with different nanofiller concentrations are studied from 2GHz to 7GHz. N-CQD acts as the nanofiller and RGO is the matrix to which different weight percentage of NCQD was added. The dielectric parameters are then fed into the HFSS and the microwave absorption properties of different NCQD-RGO samples are studied using simulations.

The graph Fig. 41.2. shows the variation in return loss (or reflection loss) for the three samples of different nanofiller concentration across various thickness. Samples S1, S2, and S3 have exhibited maximum return loss of -4dB at 4GHz, -10 dB at 5GHz and -8dB at 4GHz respectively. In comparison to samples S1 and S3, sample S2 exhibits the highest absorbance due to the variation in dielectric values resulting from the addition of the optimal quantity of nanofillers (NCQD). Beyond a certain threshold, the concentration of nanofillers increases leading to a rise in electron density and consequently causes the material to reflect more signals. S1 has a minimum concentration of nanofillers compared to S2 and S3 and therefore has minimum conduction loss. Also, S3 has increased dielectric

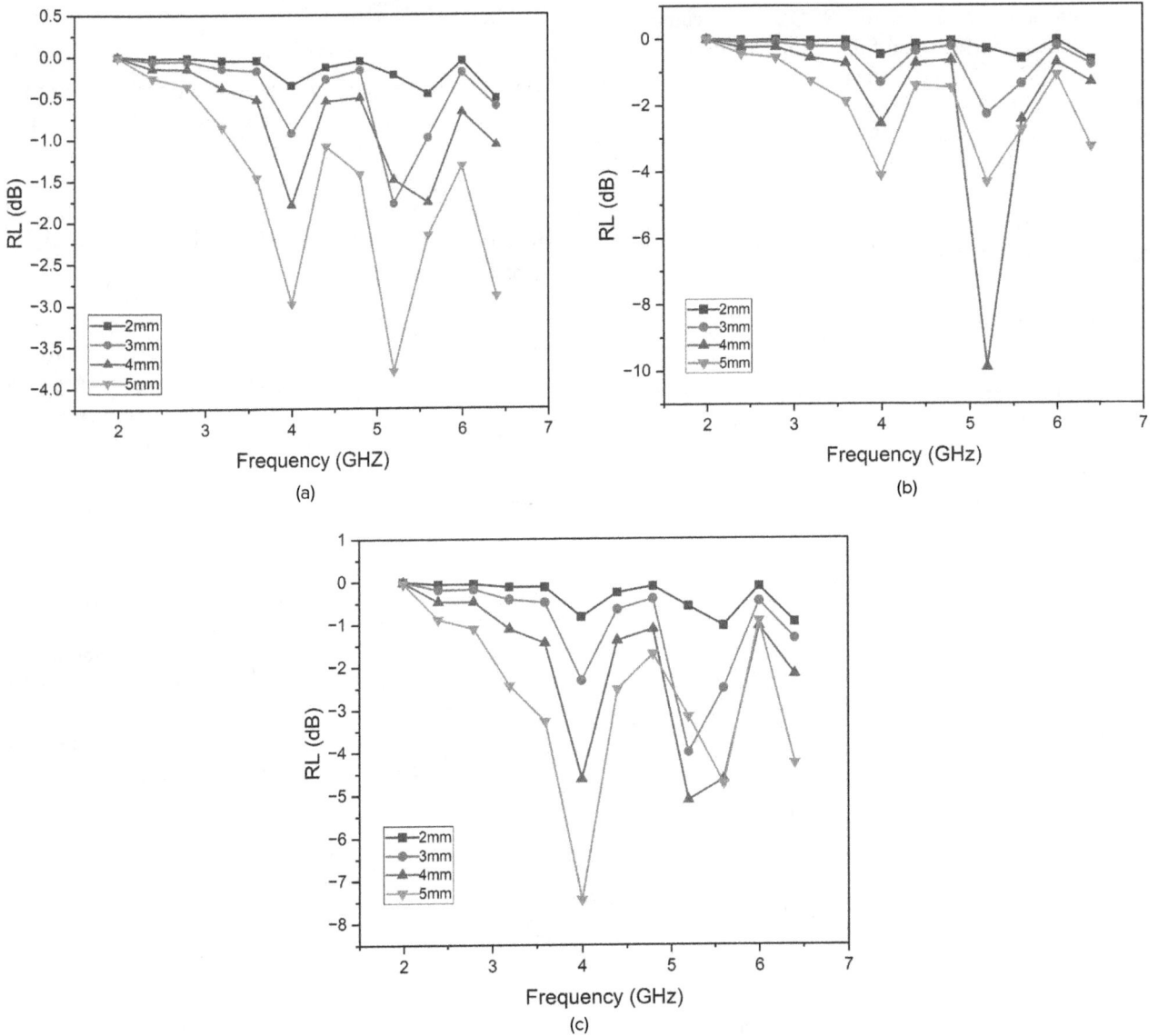

Fig. 41.2   Return loss for different thickness of a) Sample 1-S1 b) Sample 2-S2 c) Sample 3-S3

loss due to a higher concentration of nanofiller, which causes an impedance mismatch. A mismatch in impedance makes the surface reflective by preventing incoming electromagnetic waves from penetrating the substance.

The effect of nanofiller concentration over a constant thickness is represented in from Fig. 41.3. It can be seen that sample S2 has the maximum absorptive nature from analyzing only the reflection parameters. Also, as there is an increase in nanofiller loading percentage there is a shift in frequency for return loss. An illustration showing the 3-dimensional arrangement of NCQD-RGO nanocomposite is shown in Fig. 41.4, where multilayers of the RGO traps the signals thus resulting in signal

attenuation. N-CQDs present creates conductive network which increases conductivity of the material and thus results in reflective losses.

Microwave absorption properties of a material can be tweaked by introducing nanofillers which can enhance the dielectric and conduction losses of the material, thus making it a better MAM. The material's conductivity improves when CQDs are present, increasing the material's conductive losses. Additionally, the interface is increased by these NPs, leading to interfacial polarization. The presence of functionalized groups leads to dielectric polarization. Functionalization and NPs also create defects thus resulting in defect polarization. The combined effect of interfacial polarization, dielectric polarization, and defect polarization leads to dielectric losses. A diagrammatic representation of the above-mentioned defects is shown in Fig. 41.5. The microwave absorption of a material is less when ϵ'' value is low, thus making $\tan \delta_\epsilon$ decrease.

5. Conclusion

This work characterized the microwave absorption for NCQD-RGO nanocomposites with 3%, 5%, and 10% weight percent loadings using single-port and two-port rectangular waveguide setups for stealth applications at S-band frequencies (2 to 7 GHz). From the simulations, maximum absorption was shown by sample S2 with a 4 mm thickness and metal backing and thus, this may be considered an ideal MAM for metallic surfaces. Further studies can be done using a similar setup for different frequency bands like X band and Ku bands which are used in radar systems. This method can be used to reduce the amount of effort and time needed for experiments by

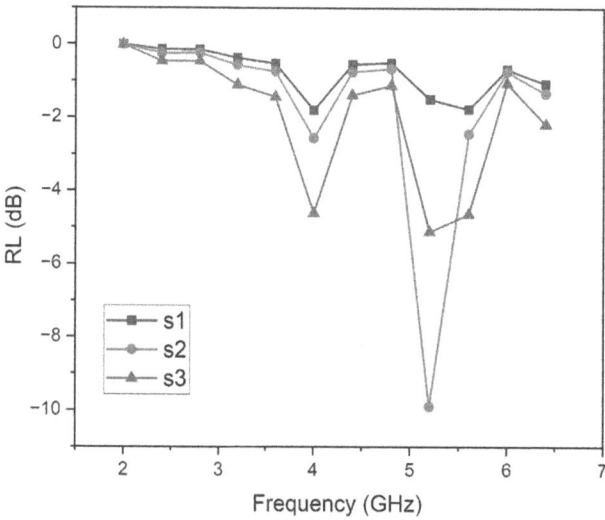

Fig. 41.3 Variation of nanofiller concentration at 4mm thickness for all three samples in one port system

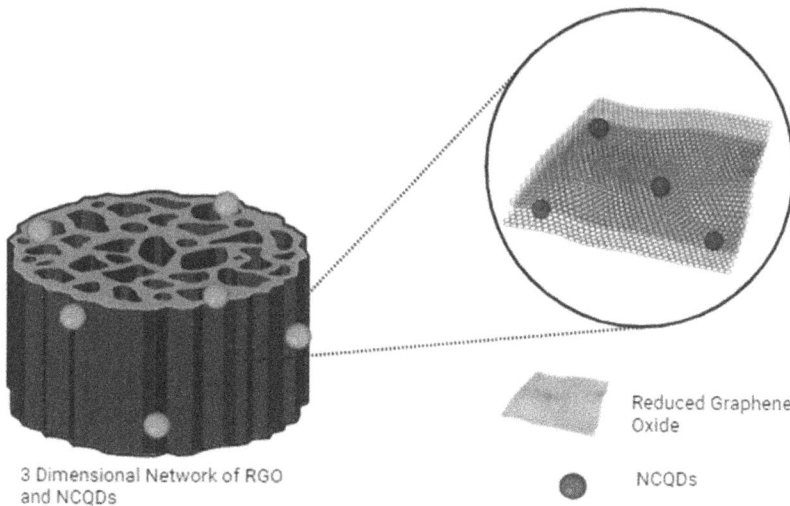

Fig. 41.4 Illustration of 3 Dimensional RGO-NCQD aerogel

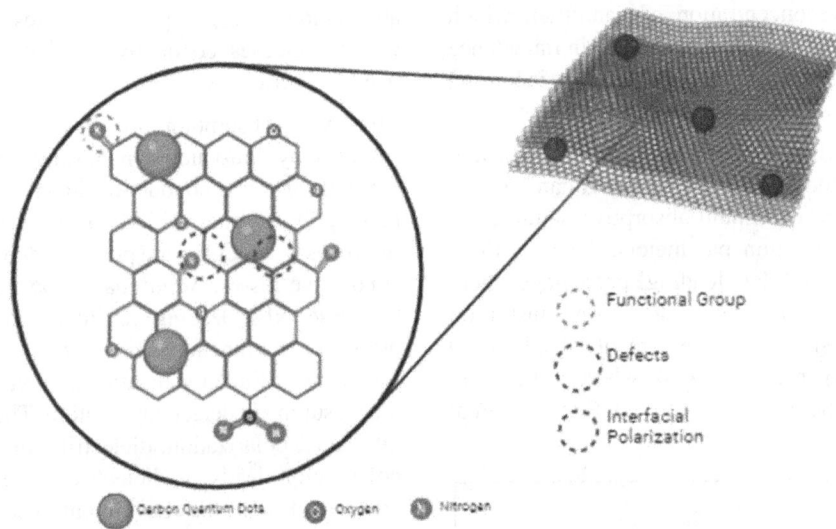

Fig. 41.5   Microwave absorption mechanism of RGO-NCQD

utilizing simulation-based studies to determine which material and at what thickness a material can be used as a stealth coating.

References

1. Adebayo L. L., Soleimani H., Yahya N., Abbas Z., Wahaab F. A., Ayinla R. T., & Ali H. (2020). Recent advances in the development OF Fe3O4-BASED microwave absorbing materials. *Ceramics International*, **46** (2), 1249–1268.

2. Ahmad H., Tariq A., Shehzad A., Faheem M. S., Shafiq M., Rashid I. A., Afzal A., Munir A., Riaz M. T., Haider H. T., Afzal A., Qadir M. B., & Khaliq Z. (2019). Stealth technology: Methods and composite materials—A review. *Polymer Composites*, **40** (12), 4457–4472.

3. Hanima Kannan C. H., Venugopal V., Mathur P., & Kurup D. G. (2024). Accurate Modeling and Performance Analysis of a Two-Port Coaxial Transmission Line for Material Characterization. In: 2024 IEEE 6th PhD Colloquium on Emerging Domain Innovation and Technology for Society (PhD EDITS). p. 1–2. IEEE.

4. Jelmy E. J., Lakshmanan M., & Kothurkar N. K. (2020). Microwave absorbing behavior of glass fiber reinforced MWCNT-PANi/epoxy composite laminates. *Materials Today: Proceedings*, **26**, 36–43.

5. Liu P., & Huang Y. (2014). Decoration of reduced graphene oxide with polyaniline film and their enhanced microwave absorption properties. *Journal of Polymer Research*, **21** (5), 430.

6. Qiu J., Liao J., Wang G., Du R., Tsidaeva N., & Wang W. (2022). Implanting N-doped CQDs into rGO aerogels with diversified applications in microwave absorption and wastewater treatment. *Chemical Engineering Journal*, **443**, 136475.

7. Venugopal V., Hanima Kannan C. H., Siva Kumar B., & Mathur P. (2024). EMI Shielding of NCQD-RGO Nanocomposites: HFSS Analysis with Two-Port Rectangular Waveguides. In: 2024 IEEE 6th PhD Colloquium on Emerging Domain Innovation and Technology for Society (PhD EDITS). p. 1–2. IEEE.

8. Wang B., Wu Q., Fu Y., & Liu T. (2021). A review on carbon/magnetic metal composites for microwave absorption. *Journal of Materials Science & Technology*, **86** , 91–109.

9. Wang X., Pan F., Xiang Z., Zeng Q., Pei K., Che R., & Lu W. (2020). Magnetic vortex core-shell Fe3O4@C nanorings with enhanced microwave absorption performance. *Carbon*, **157**, 130–139.

Note: All the figures in this chapter were made by the author.

Advances in Materials Science and Technology – Dr. Srikari Srinivasan et al. (eds)
© 2025 Taylor & Francis Group, London, ISBN 978-1-041-12342-2

42

Enhancing Pest Control Sustainability: Ligno-Alginate Blend for Controlled Pyrethrin Release

Raga K. Pavithran,
S. Giridhar Reddy*, B. Siva Kumar
Department of Physical Sciences,
Amrita School of Engineering, Amrita Vishwa Vidyapeetham,
Bengaluru, India

Sanga Kugabalasooriar
Department of Chemistry, Northeastern University,
Boston, MA 02115, United States

ABSTRACT: Pesticides play a crucial role in modern agriculture, industry, and public health, ensuring global food security. Nevertheless, the extensive utilization of pesticides leads to environmental and health concerns because of post-pesticide application losses and pesticide residues. Promising ways to address these concerns include alternative strategies such as biopesticides and controlled release systems. The objective of this work is to develop polymeric beads by combining sodium alginate and lignosulphonic acid to encapsulate synthetic pyrethrin for controlled release. Various crosslinking durations were used to maximize the rate at which the substance is released. The release studies conducted in a laboratory setting showed that the sodium alginate/lignosulphonic acid beads released pyrethrin in a controlled manner in acetate buffer with a pH of 5.5. This controlled release improved the effectiveness and long-term viability of pesticides. This approach signifies a notable advancement towards safer and more efficient pesticide applications in agriculture.

KEYWORDS: Biopolymer, Controlled pesticide delivery, Controlled release system, Polymeric beads, Pyrethrin, Sustainable agriculture

1. Introduction

Pesticides are vital to agriculture, industry, and public health, and they are also vital to maintaining food security worldwide (Khan & Rahman, 2017). However, post-application losses such as degradation, photolysis, evaporation, leaching, and surface runoff frequently reduce their efficacy and these pesticide residues can enter the environment, posing risks to both ecosystems and human health (Sabarwal et al., 2018).

The widespread use of pesticides not only affects agricultural productivity but also has detrimental effects on both the environment and human health, particularly through pesticide residues in food. Most commonly used pesticides, such as organophosphates and organochlorines, have gained popularity due to their stability over the years. However, their extensive usage resulted in considerable concerns regarding environmental and human health. As a result, there is an increasing trend towards using alternative approaches such biopesticides. Biopesticides,

*Corresponding author: s_giri@blr.amrita.edu

DOI: 10.1201/9781003664277-42

such as pyrethrins extracted from *Chrysanthemum cinerariaefolium*, have been widely acknowledged for their effectiveness in pest management (Osimitz *et al.*, 2009). However, due to stability concerns with natural compounds like pyrethrins, synthetic derivatives such as pyrethroids have emerged as highly adaptable alternatives. Pyrethroids have become one of the most popular types of pesticides due to this alteration. Consequently, pyrethrins and pyrethroids have been extensively embraced as safer substitutes over a considerable period of time (Ensley, 2007).

Another widely accepted method for sustained pesticide application is the use of controlled release systems. Incorporating controlled release technology in pesticide delivery has become increasingly popular. Controlled release technology, originally designed for drug delivery applications, has been successfully adapted and refined to optimize the effective delivery of pesticides. This evolution utilizes our expertise in formulation processes and release mechanisms, aiming to maximize efficacy while minimizing environmental impact (Aashli *et al.*, 2023; Hafsa Bahaar *et al.*, 2023; Mary *et al.*, 2024; Pavithran *et al.*, 2024; Reddy S, 2022; S. G. Reddy, 2021). Our current study aims to develop polymeric beads using sodium alginate (SA) and lignosulphonic acid (LS), encapsulating synthetic pyrethrin (PY) for controlled release. The encapsulation of synthetic pyrethrin within SA/LS/PY polymeric beads aims to achieve controlled release properties, enhancing the efficacy and sustainability of pesticide applications.

2. Materials and Methods

2.1 Materials Required

Sodium alginate (SA) (MW = 300000 g/mol) and Lignosulphonic acid (LS) (MW = 50000 g/mol) obtained from Sigma Aldrich (India), Synthetic pyrethrin (PY), 2% Calcium chloride ($CaCl_2$), Acetate buffer (pH 5.5), and Distilled water (18.2 M Ω cm^{-1}) was used in all experiments.

3. Methods

3.1 Preparation of SA/LS Polymeric Beads Loaded with Synthetic Pyrethrin

4% (w/v) of SA and LS were weighed in the ratio of 80:20, respectively. The weighed polymers were dissolved in the 10% Pyrethrin pesticide solution. The SA/LS mixture was stirred for 30 min to ensure complete dissolution using a magnetic stirrer. After the complete dissolution of the polymer in the pesticide solution, the polymeric solution was made into beads using 2% Calcium chloride as the crosslinking solution with crosslinking duration of 10 minutes, 20 minutes, and 30 minutes. It was kept in the hot air oven for drying at 50 °C SA/LS/PY beads are shown in Fig. 42.1.

3.2 In-vitro Release Studies

Various concentrations of pyrethrin were scanned using Shimadzu UV-visible spectrophotometry over a wavelength range of 200–400 nm against a blank solution. The maximum wavelength value for pyrethrin was observed to be around 222 nm which can be validated using the existing literature (PubChem, s. d., Guillory, 2007). The beads that had been crosslinked for 10 minutes, 20 minutes, and 30 minutes were kept in separate acetate buffers (pH 5.5). Every 30 minutes, the supernatant solution was examined to determine its absorbance value.

4. Results and Discussion

4.1 SA/LS/PY Beads

The beads obtained after crosslinking and drying are shown in Fig. 42.1. and Fig. 42.2 respectively.

4.2 *In-vitro* Release Studies

The release of pyrethrin from the SA/LS/PY $CaCl_2$ crosslinked beads was studied using Shimadzu UV-

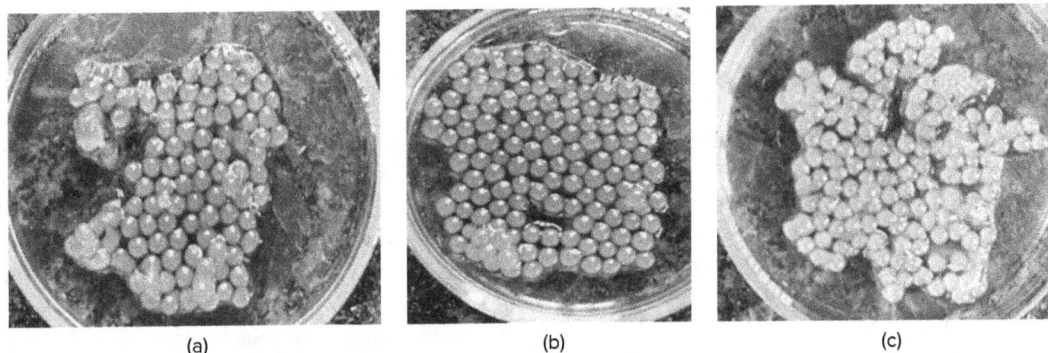

(a) (b) (c)

Fig. 42.1 The SA/LS/PY beads obtained after crosslinking are illustrated in (a) 10-minute, (b) 20-minute, and (c) 30-minute crosslinking durations

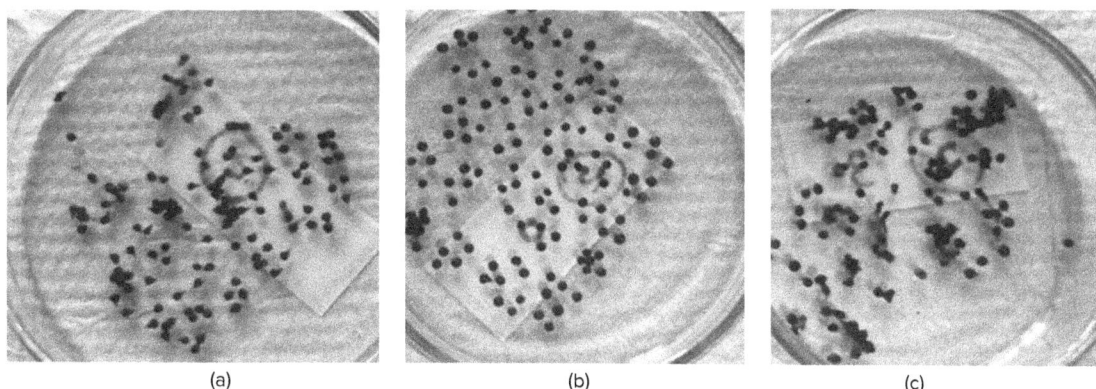

Fig. 42.2   The SA/LS/PY beads obtained after drying in a hot air oven at 50 degrees Celsius are illustrated in (a) 10-minute, (b) 20-minute, and (c) 30-minute crosslinking durations

spectrophotometer. The *in-vitro* release studies were performed in acetate buffer of pH 5.5 to replicate the pH of the soil. The release profile was studied for a period of 2-4 days. Based on the literature, pyrethrin shows a peak at 222 nm (Guillory, 2007). Figure 42.3. and Fig. 42.5. demonstrates the release of pesticide from SA/LS/PY beads, crosslinked for 10 minutes and 30 minutes respectively and the graph shows that the pesticide released in a sustained manner. Figure 42.4 demonstrates the release of pesticide from SA/LS beads crosslinked for 20 minutes respectively. An initial burst can be observed in all these beads. The initial burst may be caused due to the surface adsorbed pesticides or free pesticides that have potentially escaped from the beads. While the beads crosslinked for 10 and 30 minutes exhibited sustained pesticide release, the beads crosslinked for 20 minutes displayed a dip in the release profile after 24 hours. This anomaly could be attributed to experimental or data interpretation errors. A plausible explanation for this occurrence could be the photodegradation of the pyrethrin pesticide (Massaro *et al.*, 2022; Y. Zhang *et al.*, 2019). Consequently, this work can be expanded by further minimizing the initial burst effect

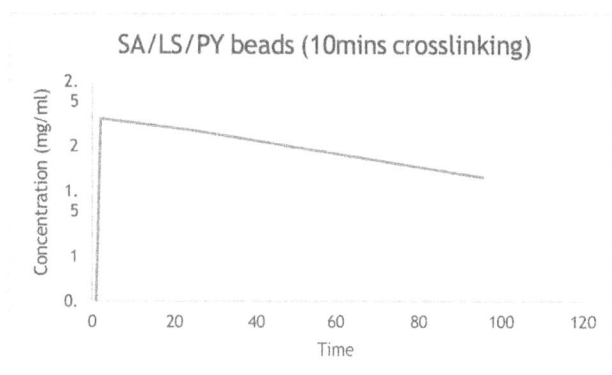

Fig. 42.4   The graph shows the release profile obtained from 20 minutes crosslinked SA/LS/PY beads

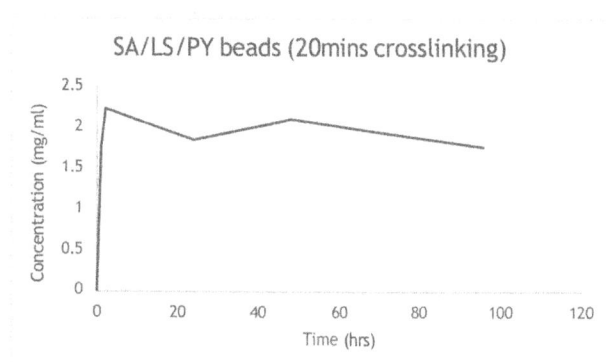

Fig. 42.5   The graph shows the release profile obtained from 30 minutes crosslinked SA/LS/PY beads

and enhancing the photostability of the pyrethrin through optimization of formulation properties.

5. Conclusion

In this study, SA/LS/PY calcium crosslinked beads with crosslinking times of 10, 20, and 30 minutes were used to study the controlled release of pyrethrin. Both the 10-

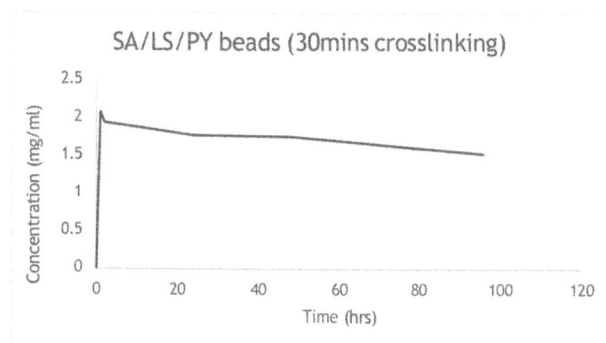

Fig. 42.3   The graph shows the release profile from 10 minutes crosslinked SA/LS/PY beads

and 30-minute crosslinked beads demonstrated sustained release over a period of two to four days, according to the results, which suggests that they can be used for controlled pesticide delivery. All the beads exhibit an initial burst release, most likely as a result of free or surface-adsorbed pyrethrin escaping from the beads. The 20-minute crosslinked beads showed an unexpected dip in release after 24 hours, which could have been caused by pyrethrin photodegradation. This result implies that additional optimization is required to decrease the initial burst effect and enhance formulation stability. Future research directions could include exploring advanced encapsulation techniques to achieve more controlled release profiles and investigating novel additives or coatings to protect the pyrethrin from photodegradation. These efforts aim to improve the overall efficacy and environmental sustainability of pesticide applications.

References

1. Aashli, S. Giridhar Reddy, Siva Kumar Belliraj, K. Prashanthi, & Murthy A. (2023). Fabricating transdermal film formulations of montelukast sodium with improved chemical stability and extended drug release. *Heliyon*, **9** (3), e14469–e14469.
2. Ensley S. (2007). Pyrethrins and pyrethroids. *Elsevier eBooks*, 494–498.
3. Guillory J. K. (2007). The Merck Index: An Encyclopedia of Chemicals,Drugs, and BiologicalsEdited by Maryadele J. O'Neil, Patricia E. Heckelman, Cherie B. Koch, and Kristin J. Roman. Merck, John Wiley & Sons, Inc., Hoboken, NJ. 2006. xiv + 2564 pp. 18 × 26 cm. ISBN-13 978-0-911910-001. $125.00. *Journal of Medicinal Chemistry*, **50** (3), 590–590.
4. Hafsa Bahaar, S. Giridhar Reddy, B. Siva kumar, K Prashanthi, & Murthy A. (2023). Modified Layered Double Hydroxide – PEG Magneto-Sensitive Hydrogels with Suitable Ligno-Alginate Green Polymer Composite for Prolonged Drug Delivery Applications. *Engineered science*.
5. Khan M. S., & Rahman M. S. (Éd.) (2017). Pesticide Residue in Foods. Springer International Publishing, Cham.
6. Mary A., Reddy S. G., Kumar B. S., & Sanga Kugabalasooriar (2024). Novel Approaches to Alendronate Delivery Beyond Oral Administration- A Review. *Engineered Science*.
7. Massaro M., Pieraccini S., Guernelli S., Dindo M. L., Francati S., Liotta L. F., Colletti G. C., Masiero S., & Riela S. (2022). Photostability assessment of natural pyrethrins using halloysite nanotube carrier system. *Applied Clay Science*, **230**, 106719.
8. Osimitz T. G., Sommers N., & Kingston R. (2009). Human exposure to insecticide products containing pyrethrins and piperonyl butoxide (2001–2003). *Food and Chemical Toxicology*, **47** (7), 1406–1415.
9. Pavithran R. K., Reddy S. G., Kumar B. S., & Sanga Kugabalasooriar (2024). Enhancing Sustainability in Agriculture: Natural Polymer-Based Controlled Release Systems for Effective Pest Management and Environmental Protection. *ES Food & Agroforestry*.
10. PubChem (s. d.) Pyrethrin I. Consulté à l'adresse https://pubchem.ncbi.nlm.nih.gov/compound/Pyrethrin-I
11. Reddy S G. (2022). Effect of Crosslinking on Control Drug Release of Hydroxychloroquine sulphate drug-using Alginate beads.
12. Reddy S. G. (2021). Controlled release studies of hydroxychloroquine sulphate (hcq) drug-using biodegradable polymeric sodium alginate and lignosulphonic acid blends. *Rasayan Journal of Chemistry*, **14** (04).
13. Sabarwal A., Kumar K., & Singh R. P. (2018). Hazardous effects of chemical pesticides on human health-Cancer and other associated disorders. *Environmental toxicology and pharmacology*, **63** (63), 103–114.
14. Zhang Y., Chen W., Jing M., Liu S., Feng J., Wu H., Zhou Y., Zhang X., & Ma Z. (2019). Self-assembled mixed micelle loaded with natural pyrethrins as an intelligent nano-insecticide with a novel temperature-responsive release mode. *Chemical Engineering Journal*, **361**, 1381–139

Note: All the figures in this chapter were made by the author.

Advances in Materials Science and Technology – Dr. Srikari Srinivasan et al. (eds)
© 2025 Taylor & Francis Group, London, ISBN 978-1-041-12342-2

43

An Static and Fatigue Study of Glass Fiber Reinforced Polymer using Carbon Nano Tubes

Shakthi Prasad M.*, L. Chikmath

KLS Gogte Institute of Technolog,
Belagavi, Karnatake

ABSTRACT: Fatigue is a common occurrence in aircraft structures. Due to the repeated cycles and frequent use, the elements of planes become weakened over time, and they will eventually require attention and repair. This weakness manifests in cracks, which are microscopic at first. The effects of stress level, stress concentration and frequency on the fatigue life of glass fiber reinforced polymer (GFRP) composites have been investigated under tension-tension fatigue at a stress ratio of 0.1. The behavior of GFRP is linear and generally failed in brittle manner, which is fundamentally different from metals where failure initiates from a single crack and propagates until failure. In this paper carbon nanotube is added to the GFRP where symmetrical orientation is used. The cnt 0.5%,1% and 1.5% will be added to the glass fiber reinforced particle. The composite laminates are prepared by hand lay up process applying pressure of 50 bar. The tensile test is carried out, then fatigue test is carried out for GFRP with and without CNT.

KEYWORDS: GFRP, CNT, Hand lay up process, Tensile test, Fatigue test

1. Introduction

Composites have been widely utilized in the construction and metallurgical sectors. In ancient Mesopotamia, straw was employed as a binding agent to strengthen mud bricks. The ancient Egyptians began utilizing plywood once they realized that the reconstructed, structured composition of the wood displayed enhanced hardness, significant thermal expansion, and increased swelling characteristics upon exposure to moisture. The opulence of the era was evident in the choice of materials and the craftsmanship employed in the production of medieval swords and armor. Plywood was invented by the Egyptians nearly a millennium prior to the construction of reinforced concrete by the Romans.

The utilization of plant fibers for clay binding predates the Iron Age, which is logical given the characteristics of composites. During approximately the same era, Assyrian archers started employing a composite material consisting of sinew, wood, bark, bone, horn, wire, and glues (1800 BC). Their composite arrows were highly efficient, contributing to their triumph against the Egyptian, Babylonian, and other empires. Polymers revolutionized the composites industry, enabling widespread industrial utilization for the first time. Historically, the sole options for adhesives consisted of plant and animal resins. During the early 20th century, various forms of plastic were developed, such as vinyl, phenolic, polystyrene, and polyester. Scientists endeavored to substitute polymers due to their subpar performance in structural applications. They observed that the reinforcements provided them with the additional vigor they required. Owens Corning introduced glass fiber, often known as fiber glass, in 1935.

*Corresponding author: prasad.shakthi@gmail.com

DOI: 10.1201/9781003664277-43

2. Objective and Methodology of Work

Based on the importance of the present research, a new set of objectives was developed. The following are the objectives that have been specifically designed.

i) To examine the mechanical properties of nano composites- polyester composites such as tensile strength, load-bearing capacity of the composite is improving.

ii) To investigate the fatigue properties of Hybrid composite reinforced polyester composites.

2.1 Selection of the Fillers Particle

Carbon Nano Tubes

Carbon nanotubes (CNTs) are carbon structures that have a diameter measured in nanometers and a length measured in micrometers, with a length to diameter ratio greater than 1000. A carbon nanotube (CNT) consists of cylindrical graphitic sheets, known as graphene, that are tightly rolled into a seamless cylinder with a diameter in the nanometer range.

Carbon nanotubes (CNTs) possess excellent electrical conductivity, thermal conductivity, and mechanical strength. Electrically conductive continuous length webs can be formed by drawing parallel arrays of multi-walled carbon nanotubes

Carbon nanotubes possess exceptional strength and elasticity, making them the most robust and rigid substances discovered to date, as measured by tensile strength and elastic modulus.

Fig. 43.1 Carbon nano tubes

Composite Preparation

The hand lay-up process was used to prepare the polyester composites. A typical hand lay-up method is depicted in Fig. 43.2.

Here, the resin was measured out into a beaker and combined with the necessary volume of biogenic silica

Fig. 43.2 Hand layup process

particle and curing hardener. After the rubber mold was secured with wax, the resin-particle admix was poured in. After the necessary number of fibers had been spread out, they were pressed together securely to create a composite material with a consistent thickness throughout. After being post-cured for 48 hours at 120 °C, the composites had been pre-cured for 24 hours at room temperature. Table provides a complete rundown of all the different composites discovered here. After removing the polyester resin-based composites from the mould, surface cleaning was performed so that they could be tested. The composite used in this study is also depicted in photographic form in Fig. 43.3.

Fig. 43.3 Composite material

2.2 Mechanical Testing of Composites

Tensile Strength Test

Tensile parameters such as tensile strength, modulus, and elongation were examined in accordance with ASTM D 3039 for Glass fiber reinforced polyesterresin composites. The testing was done with a capacity of 40 tonnes and a cross head speed of 2.5mm/sec using a universal testing machine (FTM 40F, INDIA). The ASTM D3039 standard is defined as a tensile strength testing technique for both reinforced and non-reinforced polymers. All composite samples were kept at a thickness of 3 mm. Fig. 43.4. shows the tensile test sample and universal testing machine used in this present study.

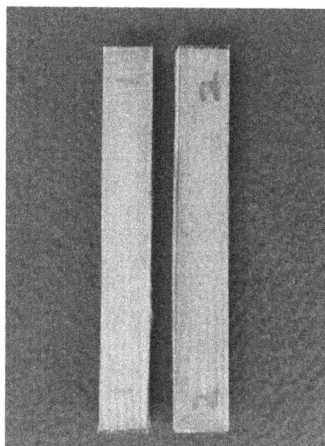

Fig. 43.4 Tensile specimens as per ASTM standards

Fatigue Test of Composites

The fatigue behavior of E-glass fibre-reinforced carbon nano tubes particle dispersed polyester resin hybrid composites were investigated by a tension-tension fatigue machine (MTS Landmark 370 load frame, USA) with hydraulic power actuated mechanical grippers. Dumbbell shaped specimens of five identical test samples were tested followed by ASTM standard ASTM D 3479 to compute average fatigue life cycles. A loading frequency of 5 Hz, stress ratio of R=0.1, maximum load of 1.28KN (50% in maximum tensile load), elastic modulus of 6.00 GPa, and working ambience of 24°C were set as process parameters. Figure 43.5 shows the ASTM sample used in this current study and Fig. 43.5 shows the fatigue testing machine.

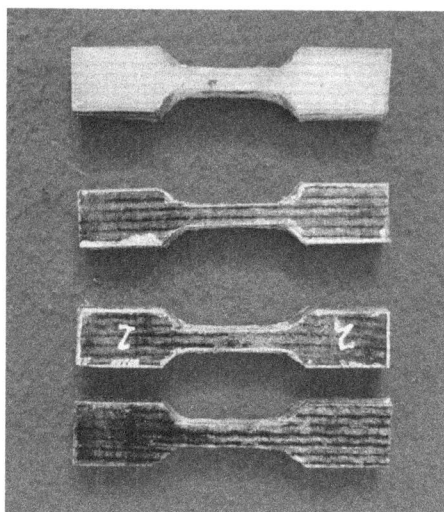

Fig. 43.5 Tensile specimens as per ASTM standards

3. Results and Discussion

3.1 Mechanical Properties

Tensile Test

As can be seen in Table the tensile strength of polyester composite reinforced with polyester glass fiberwith fillers are quite high. The tensile strength of tensile were improved when carbon nano tubes was mixed with polyester resin. In comparison, the tensile strength of the composite with the designation "PG-1" was 108.27 N/mm^2. Glass fiber higher load bearing capability in polyester composites accounts for this improvement. Fine fillers elements evenly distribute the load from the matrix when the composite is subjected to a tensile load, lowering the stress intensity factor. There is a possibility that the composite's tensile strength and modulus will improve as a result of the decreased stress intensity. The tensile strength of composites made with polyester resin and glass fiber has been found to rise dramatically(Xian et al 2022). Glass fibre reinforced polyester resin has improved tensile strength and modulus due to a reaction between the fiber and the resin bonding.

Table also reveals that the ultimate tensile strength increases. Table 43.1 also reveals that the ultimate tensile strength increases by 1.5vol.%of carbon nano tubes. when the proportion is added. There was an increase in ultimate tensile strength from 108.27 N/mm^2 to 152.29 N/mm^2. The 112.19 N/mm^2 was obtain in the designation PG-2 and the maximum tensile strength was obtained in the designation. Figure depicts the tensile strength graph of designation PG-2.

Table 43.1 Tensile properties of Glass fiber composite

S.No	Sample ID	Tensile Strength (N/mm^2)
1	PG-1	108.27
2	PG-2	112.19
3	GCN1-1	129.30
4	GCN1-2	116.48
5	GCN2-1	141.69
6	GCN2-2	132.07
7	GCN3-1	147.00
8	GCN3-2	152.29
NOTE:		
PG-PURE GLASS FIBER		
GCN1-GLASS FIBER + CARBON NANO TUBE 0.5%		
GCN2-GLASS FIBER + CARBON NANO TUBE 1.0%		
GCN3-GLASS FIBER + CARBON NANO TUBE 1.5%		

Fig. 43.7 depicts the best tensile strength graph of designation GCN1-1, Fig. 43.8 depicts the best tensile strength of designation GCN2-1 and Fig. 43.9 depicts the Tensile graph of designation GCN3-2.

PG-2
Graph Type : Stress Vs. Strain

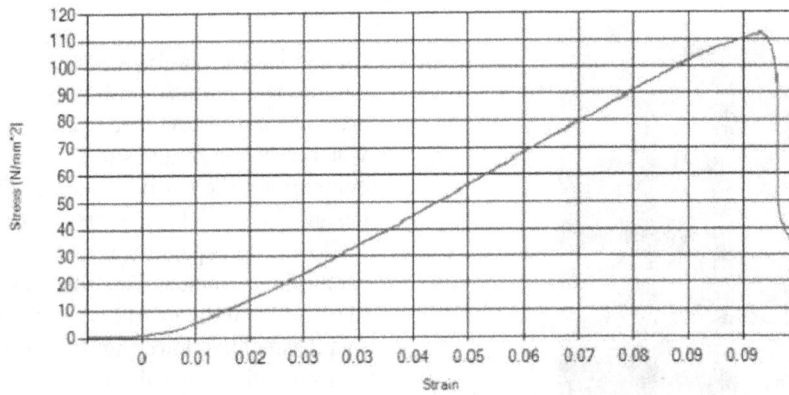

Fig. 43.6 Tensile graph of sample designation PG-2

GCN1-1
Graph Type : Stress Vs. Strain

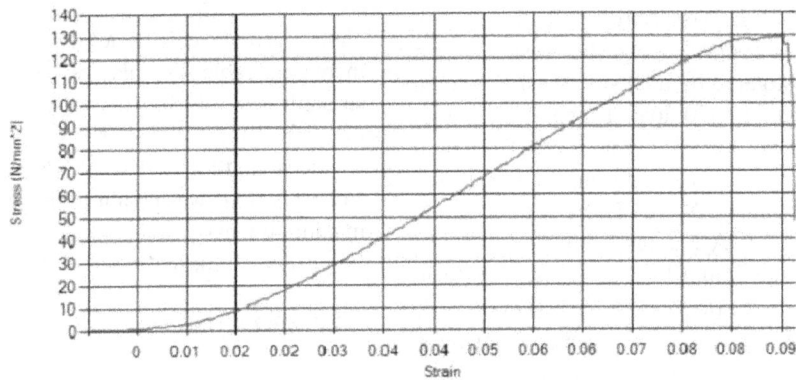

Fig. 43.7 Tensile graph of sample designation GCN1-1

GCN2-1
Graph Type : Stress Vs. Strain

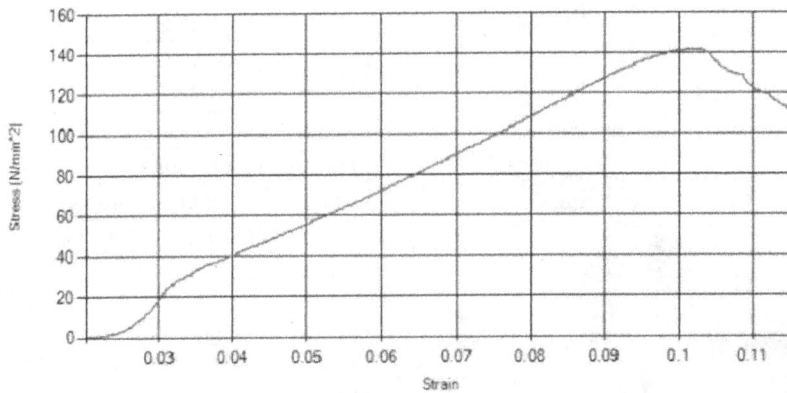

Fig. 43.8 Tensile graph of designation GCN2-1

GCN3-2
Graph Type : Stress Vs. Strain

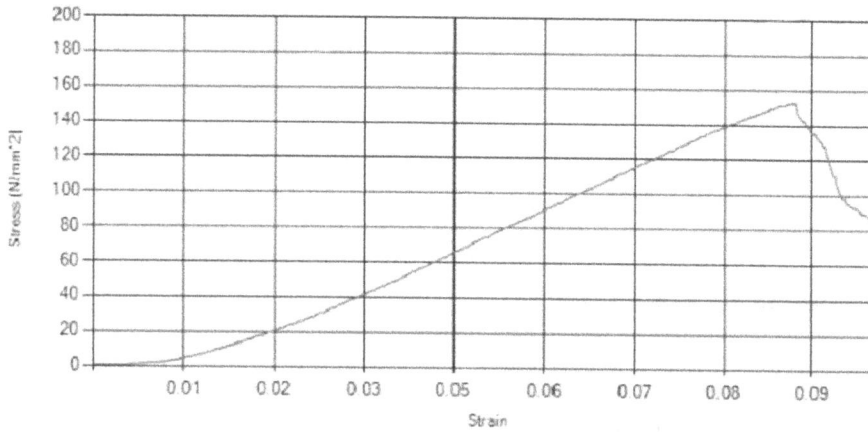

Fig. 43.9 Tensile graph of designation GCN3-2

The ultimate tensile strength was obtained in designation GCN3-2, The 1.5 vol% of carbon nano tubes distribute evenly so that bonding was improved and Strength was increased. Figure 43.10 shows the depicts of tensile testing specimen. Due to the tension apply on the material the material will bonding will be break. Figure shows the specimen after.

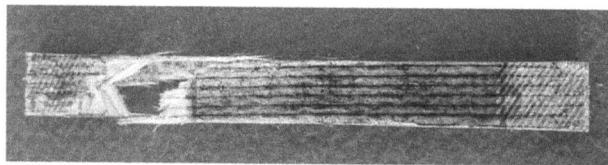

Fig. 43.10 Tensile test specimen after test

3.2 Fatigue Behaviour

This chapter examined the fatigue characteristics of different polyester composites that were manufactured.

Incorporating E-glass fibre into polyester resin resulted in an enhancement of the fatigue strength compared to the original naked polyester resin. The fatigue strength of the E-glass fibre-reinforced polyester composite was enhanced by including additional carbon nanotube particles into the polyester resin.

Figure 43.11 illustrates the fatigue life counts of polyester and its composites. It is observed that polyester glass fiber, designated as PG-1, has a fatigue life count of 47561. In designations GCN1, the fatigue life count is 48791 at 30% UTS. In designation GCN2, the fatigue life counts were increased by adding 1.0 vol % of carbon nanotubes. The lower value is attributed to the high brittleness of the molecular structure of the polyester resin, which cannot withstand repeated loading conditions. Additionally, it is noted that the addition of a significant volume % of fiber greatly improves the fatigue life counts, regardless of the addition of CNTs. Designation GCN3 shows an improved

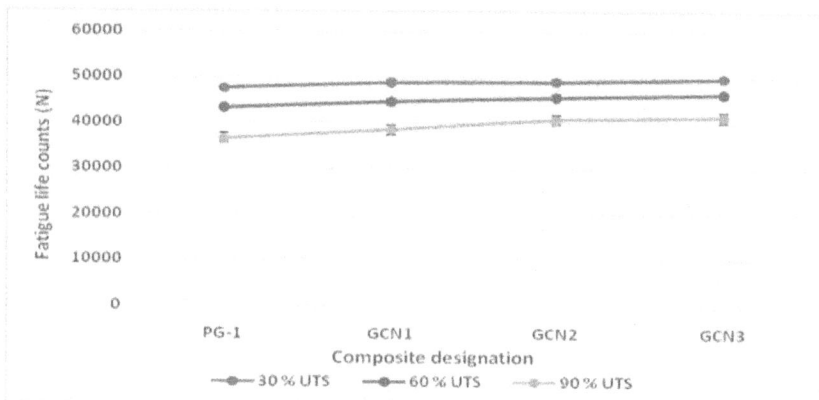

Fig. 43.11 Fatigue properties of glass fiber hybrid composites

fatigue life count of 49851. This higher count is due to the effective load-bearing ability of the composite, which contains high modulus glass fiber.

The inclusion of carbon nanotubes (CNT) at concentrations of 0.5, 1, and 1.5vol.% in the polyester composite resulted in further alterations to the fatigue life counts. The carbon nanotubes in the composite reportedly exhibited superior performance compared to the simple polyester resin composite. The composite designation 1.5vol % has the highest fatigue life counts of 46351, accounting for 60% of the ultimate tensile strength (UTS).

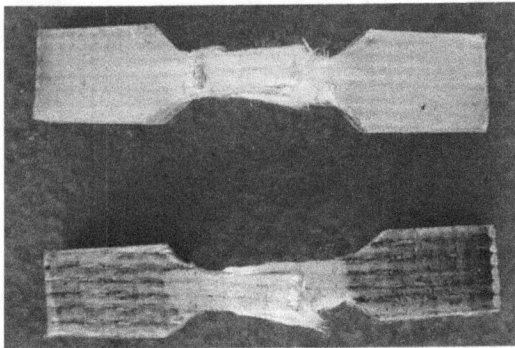

Fig. 43.12 Fatigue specimens after test of designation PG & GCN1

The incorporation of carbon nanotubes at a concentration of 1.5 volume percent resulted in an enhancement of the maximum life cycle count to 41356 at 90% UTS conditions. Additionally, this inclusion guarantees a homogeneous distribution of the particles inside the matrix. These enhanced adhesion reinforcements have the capability to effectively transfer the tension load and lower the stress concentration factor, hence improving fatigue strength.

4. Conclusion

- The objective of this study was to develop a unique hybrid composite material, utilizing a glass fiber polyester composite, specifically designed for automotive purposes. The comparative studies are conducted on polyester glass fiber and hybrid composite materials. This section provides an in-depth analysis of the research results.
- The highest tensile value was obtained in designation GCN3-2 values as 152.29 N/mm^2 tensile strength.
- The highest Fatigue strength value in obtained in 30 % UTS is designation GCN3 values as 49851 counts.
- The highest Fatigue strength value in obtained in 60 % UTS is designation GCN3 values as 46351 counts.

- The highest Fatigue strength value in obtained in 90 % UTS is designation GCN3 values as 41356 counts.
- Our hybrid composite can replace the commercial plastic material used in battery cases. This versatile hybrid composite material can be utilized in various applications such as boiler connectors, battery cases, automotive interiors, aircraft turbines, induction stove cases, and for domestic purposes.

References

1. Jalali M, Dauterstedt, Michaud & Wuthrich 2011, 'Electromagnetic shielding behaviour of polymer matrix composite with metallic nano particles', Composites Part: B Engineering, Vol. 42, no.6, pp. 1420–1426.
2. Khoramishad, H., H. Alikhani, and S. Dariushi. "An experimental study on the effect of adding multi-walled carbon nanotubes on high-velocity impact behavior of fiber metal laminates." Composite Structures 201 (2018): 561–569.
3. Naheed Saba, Paridah Md Tahir & Mohammad Jawaid 2014, 'A review on potentiality of nano filler/natural fibrefilled polymerhybridcomposites', Polymers, Vol.6, pp.2247–2273.
4. Navaneetha Krishnan, G, Selvam, V & Saravanan, C 2015 'Effect of CNTs-Fe$_2$O$_3$ hybrids on mechanical studies of glass fibre/epoxy nanocomposites', Journal of Chemical and Pharmaceutical Sciences, Vol.6, pp.196–201.
5. Nevin Gamze Karsli, Sertan Yesil, Ayse Aytac, 2014, 'Effect of hybrid carbon nanotube/short glass fibre-reinforcement on the properties of polypropylene composites', Composites: Part B Engineering, Vol.63, pp. 154–160.
6. Rajesh, S, Vijaya Ramnath, B, Elanchezhian, C, Aravind, N & Vijai Rahul, V 2014, 'Analysis of mechanical behavior of glass fibre/ Al$_2$O$_3$- SiC reinforced polymer composites', Procedia Engineering, Vol. 97, pp. 598–606.
7. Uzay, Çağrı. "Investigation of physical, mechanical, and thermal properties of glass fiber reinforced polymer composites strengthened with KH550 and KH570 silane-coated silicon dioxide nanoparticles." Journal of Composite Materials 56, no. 19 (2022): 2995–3011.
8. Wei Wu, Quanguo He & Chang Zhong Jiang 2008, 'Magnetic iron oxide nanoparticles: Synthesis and surface functionalization strategies', Nano Scale Research Letters, Vol.3, pp. 397–415.
9. Yao, Zhiqiang, Chengguo Wang, Jianjie Qin, Shunsheng Su, Yanxiang Wang, Qifen Wang, Meijie Yu, and Huazhen Wei. "Interfacial improvement of carbon fiber/epoxy composites using one-step method for grafting carbon nanotubes on the fibers at ultra-low temperatures." Carbon 164 (2020): 133–142.
10. Zhang, Xi, Xingru Yan, Jiang Guo, Zhen Liu, Dawei Jiang, Qingliang He, Huige Wei et al. "Polypyrrole doped epoxy resin nanocomposites with enhanced mechanical properties and reduced flammability." Journal of Materials Chemistry C 3, no. 1 (2015): 162–176.

Note: All the figures and table in this chapter were made by the author.

For Product Safety Concerns and Information please contact our EU
representative GPSR@taylorandfrancis.com
Taylor & Francis Verlag GmbH, Kaufingerstraße 24, 80331 München, Germany

www.ingramcontent.com/pod-product-compliance
Lightning Source LLC
Chambersburg PA
CBHW061346210326
41598CB00035B/5892

* 9 7 8 1 0 4 1 1 2 3 4 2 2 *